PARALLELOGRAM

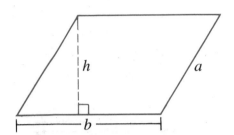

Perimeter: $P = 2a + 2b$

Area: $A = bh$

CIRCLE

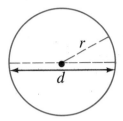

Circumference: $C = \pi d$

$C = 2\pi r$

Area: $A = \pi r^2$

RECTANGULAR SOLID

Volume: $V = LWH$

Surface Area: $A = 2HW + 2LW + 2LH$

CUBE

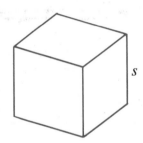

Volume: $V = s^3$

Surface Area: $A = 6s^2$

CONE

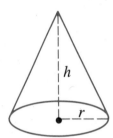

Volume: $V = \frac{1}{3}\pi r^2 h$

Surface Area: $A = \pi r \sqrt{r^2 + h^2}$

RIGHT CIRCULAR CYLINDER

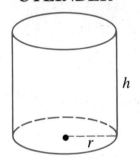

Volume: $V = \pi r^2 h$

Surface Area:

$A = 2\pi rh + 2\pi r^2$

SPHERE

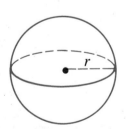

Volume: $V = \frac{4}{3}\pi r^3$

Surface Area: $A = 4\pi r^2$

OTHER FORMULAS

Distance: $d = rt$ (r = rate, t = time)

Percent: $p = br$ (p = percentage, b = base, r = rate)

Temperature: $F = \frac{9}{5}C + 32$ $C = \frac{5}{9}(F - 32)$

Simple Interest: $I = Prt$
(P = principal, r = rate, t = time in years)

Beginning Algebra

K. Elayn Martin-Gay

University of New Orleans

Prentice Hall, Englewood Cliffs, NJ 07632

Library of Congress Cataloging-in-Publication Data

Martin-Gay, K. Elayn, 1955–
    Beginning algebra / K. Elayn Martin-Gay.—Student's ed.
        p.   cm.
    Includes index.
    ISBN 0-13-073784-4 (student's)
    1. Algebra.   I. Title.
[QA152.2.M367   1993]
512'.9—dc20

92-29962
CIP

Executive Editor: Priscilla McGeehon
Editor-in-Chief: Tim Bozik
Development Editor: Steve Deitmer
Production Editor: Tom Aloisi
Marketing Manager: Paul Banks
Copy Editor: Bill Thomas
Page Layout: Jayne Conte
Designer: Judith A. Matz-Coniglio
Design Director: Florence Dara Silverman

Cover Designer: A Good Thing, Inc.
Cover Artist: Mendola, Ltd./Cliff Spohn
Prepress Buyer: Paula Massenaro
Manufacturing Buyer: Lori Bulwin
Supplements Editor: Mary Hornby
Editorial Assistant: Marisol L. Torres
Photo Editor: Lori Morris-Nantz
Photo Research: Anita Dickhuth

Photograph Credits: William Waterfall/The Stock Market, p. xvi. Dan Burns/Monkmeyer Press, p. 52. Luis Castaneda/The Image Bank, p. 112. Lawrence Migdale/The Photo Researchers, p. 156. Tom Raymond/Medichrome Div./The Stock Shop, Inc., p. 200. Benn Mitchell/The Image Bank, p. 256. Jeff Hunter/The Image Bank, p. 310. Art Stein/Photo Researchers, Inc., p. 346. Steve Krongard/The Image Bank, p. 388.

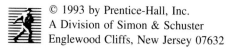

© 1993 by Prentice-Hall, Inc.
A Division of Simon & Schuster
Englewood Cliffs, New Jersey 07632

Printed in the United States of America
10   9   8   7   6   5   4   3   2   1

ISBN 0-13-073784-4

Prentice-Hall International (UK) Limited, *London*
Prentice-Hall of Australia Pty. Limited, *Sydney*
Prentice-Hall Canada Inc., *Toronto*
Prentice-Hall Hispanoamericana, S.A., *Mexico*
Prentice-Hall of India Private Limited, *New Delhi*
Prentice-Hall of Japan, Inc., *Tokyo*
Simon & Schuster Asia Pte. Ltd., *Singapore*
Editora Prentice-Hall do Brasil, Ltda., *Rio de Janeiro*

To my mother, Barbara M. Miller,
and to the memory of my father, Robert J. Martin

Contents

Preface *xi*

CHAPTER 1 **Real Numbers and Their Properties** *xvi*

1.1 Symbols *1*
1.2 Sets of Numbers *5*
1.3 Fractions *10*
1.4 Exponents and Order of Operations *16*
1.5 Introduction to Variables *21*
1.6 Adding Real Numbers *24*
1.7 Subtracting Real Numbers *30*
1.8 Multiplying and Dividing Real Numbers *34*
1.9 Properties of Real Numbers *41*
  Chapter 1 Summary *48*
  Chapter 1 Review *49*
  Chapter 1 Test *51*

CHAPTER 2 **Solving Linear Equations and Inequalities** *52*

2.1 Simplifying Algebraic Expressions *53*
2.2 The Addition Property of Equality *59*
2.3 The Multiplication Property of Equality *66*
2.4 Solving Linear Equations *73*
2.5 Formulas *81*
2.6 Applications of Linear Equations *89*
2.7 Solving Linear Inequalities *97*
  Chapter 2 Summary *106*
  Chapter 2 Review *107*
  Chapter 2 Test *109*
  Chapter 2 Cumulative Review *110*

CHAPTER 3 Exponents and Polynomials   *112*

3.1 Exponents   *113*
3.2 Negative Exponents and Scientific Notation   *122*
3.3 Adding and Subtracting Polynomials   *130*
3.4 Multiplying Polynomials   *136*
3.5 Special Products   *140*
3.6 Division of Polynomials   *145*
   Chapter 3 Summary   *151*
   Chapter 3 Review   *152*
   Chapter 3 Test   *154*
   Chapter 3 Cumulative Review   *155*

CHAPTER 4 Factoring Polynomials   *156*

4.1 The Greatest Common Factor and Factoring by Grouping   *157*
4.2 Factoring Trinomials of the Form $x^2 + bx + c$ *163*
4.3 Factoring Trinomials of the Form $ax^2 + bx + c$ *168*
4.4 Factoring Binomials   *176*
4.5 Factoring Polynomials Completely   *180*
4.6 Solving Quadratic Equations by Factoring   *183*
4.7 Applications of Quadratic Equations   *189*
   Chapter 4 Summary   *196*
   Chapter 4 Review   *197*
   Chapter 4 Test   *198*
   Chapter 4 Cumulative Review   *199*

CHAPTER 5 Rational Expressions   *200*

5.1 Simplifying Rational Expressions   *201*
5.2 Multiplying and Dividing Rational Expressions   *207*
5.3 Adding and Subtracting Rational Expressions with Common Denominators and Least Common Denominator   *213*
5.4 Adding and Subtracting Rational Expressions with Unlike Denominators   *219*
5.5 Complex Fractions   *225*
5.6 Solving Equations Containing Rational Expressions   *231*
5.7 Ratio and Proportion   *237*
5.8 Applications of Equations Containing Rational Equations   *242*
   Chapter 5 Summary   *250*
   Chapter 5 Review   *250*
   Chapter 5 Test   *253*
   Chapter 5 Cumulative Review   *254*

CHAPTER 6 Graphing Linear Equations and Inequalities 256

6.1 The Cartesian Coordinate System *257*
6.2 Graphing Linear Equations *265*
6.3 Slope *274*
6.4 Equations of Lines *283*
6.5 Graphing Linear Inequalities *289*
6.6 Functions *297*
   Chapter 6 Summary *303*
   Chapter 6 Review *304*
   Chapter 6 Test *307*
   Chapter 6 Cumulative Review *309*

CHAPTER 7 Solving Systems of Linear Equations 310

7.1 Solving Systems of Linear Equations by Graphing *311*
7.2 Solving Systems of Linear Equations by Substitution *319*
7.3 Solving Systems of Linear Equations by Addition *324*
7.4 Applications of Systems of Linear Equations *329*
7.5 Systems of Linear Inequalities *335*
   Chapter 7 Summary *340*
   Chapter 7 Review *341*
   Chapter 7 Test *344*
   Chapter 7 Cumulative Review *344*

CHAPTER 8 Roots and Radicals 346

8.1 Introduction to Radicals *347*
8.2 Simplifying Radicals *352*
8.3 Adding and Subtracting Radicals *357*
8.4 Multiplying and Dividing Radicals *361*
8.5 Solving Equations Containing Radicals *367*
8.6 Applications of Equations Containing Radicals *372*
8.7 Rational Exponents *378*
   Chapter 8 Summary *383*
   Chapter 8 Review *384*
   Chapter 8 Test *386*
   Chapter 8 Cumulative Review *386*

CHAPTER 9 **Solving Quadratic Equations**   *388*

9.1 Solving Quadratic Equations by the Square Root Method   *389*
9.2 Solving Quadratic Equations by Completing the Square   *392*
9.3 Solving Quadratic Equations by the Quadratic Formula   *397*
9.4 Summary of Methods for Solving Quadratic Equations   *402*
9.5 Complex Solutions to Quadratic Equations   *405*
9.6 Graphing Quadratic Equations   *410*
    Chapter 9 Summary   *420*
    Chapter 9 Review   *420*
    Chapter 9 Test   *422*
    Chapter 9 Cumulative Review   *423*

Appendices   *424*

A. Operations on Decimals   *424*

B. Review of Angles, Lines, and Special Triangles   *427*

C. Review of Geometric Figures   *435*

D. Squares and Square Roots   *438*

E. Selected Answers   *440*

    Index   *476*

Preface

Why This Book Was Written

This book was written to provide a solid foundation in algebra for students who might have had no previous experience in algebra. Specific care has been taken to prepare students to go on to their next course in algebra. I have tried to achieve this by writing a user-friendly text keyed to objectives containing plenty of worked-out examples. Functions are introduced in this text, and applications and geometric concepts are emphasized throughout the book.

How This Book Was Written

Throughout the writing and developing of this book, I had the help of many people. Seven instructors, who teach courses similar to this one, were involved in the actual writing of the text, contributing their ideas for helpful examples, interesting applications, and useful exercises.

Once the first draft was complete, Prentice Hall held a focus group with four reviewers, the author, and editors from Prentice Hall. We spent many hours going over the manuscript with a fine-toothed comb, refining the project's focus and enhancing its pedagogical value.

Finally, a full-time development editor worked with me to make the writing style as clear as possible while still retaining the mathematical integrity of the content.

Key Content Features

In addition to the traditional topics taught in a beginning algebra course, I have integrated as much coverage of geometry as possible. Since many students have not taken geometry in high school, this may be their first (and only) exposure to geometry. I have tried to cover those geometry concepts which are most important to a student's understanding of algebra, and to include geometric applications and exercises as much as possible. Geometric figures and a review of angles, lines and special triangles are covered in the appendix material.

Applications are emphasized by devoting single sections to them (like sections 2.5 and 2.6 on formulas and applications of linear equations) as well as application exercises throughout the book.

Key Pedagogical Features

Exercise sets.   Each exercise set is divided into two parts. Both parts contain graded problems. The first part is carefully keyed to worked examples in the text. Once a student has gained confidence in a skill, the second part contains exercises not keyed to examples. There are ample exercises throughout this book, including end-of-chapter reviews, tests, and cumulative reviews. In addition, each exercise set contains one or more of the following features:

Mental Mathematics.   These problems are found at the beginning of an exercise set. They are mental warmups that reinforce concepts found in the accompanying section and increase students' confidence before they tackle an exercise set. By relying on their own mental skills, students learn not only confidence in themselves, but also number sense and estimation ability.

Skill Review.   At the end of each section after Chapter 1, these problems are keyed to earlier sections and review concepts learned earlier in the text.

Writing in Mathematics.   These writing exercises can be used to check a student's comprehension of an algebraic concept. They are located at the end of many exercise sets, where appropriate. Guidelines recommended by the National Council of Teachers of Mathematics and other professional groups recommend incorporating writing in mathematics courses to reinforce concepts.

Applications.   This book contains a wealth of practical applications found throughout the book in worked-out examples and exercise sets.

A Look Ahead.   These are examples and problems similar to those found in college algebra books. "A Look Ahead" is presented as a natural extension of the material and contains an example followed by advanced exercises. I strongly suggest that any student who plans to take another algebra course work these problems.

Calculator Boxes.   Calculator Boxes are placed appropriately throughout the text to instruct students on proper use of the calculator. These boxes, entirely optional, contain examples and exercises to reinforce the material introduced.

Critical Thinking.   Each chapter opens with a critical thinking problem. The student does not need to work this problem immediately. Rather, the skills needed to solve each critical thinking problems are developed throughout the chapter. At the close of the chapter, the student is asked to apply the skills he or she has learned to answer the problem. The critical thinking problems, based on real-life situations relevant to students, require thinking "beyond the numbers." On occasion, a critical thinking problem may have more than one answer. These problems are excellent for cooperative learning situations. You can assign the problem to a groups of students who can then present their solution to the rest of the class for discussion.

Helpful Hint Boxes.   These boxes contain practical advice on problem-solving. Helpful Hints appear in the context of material in the chapter, and give students extra help in understanding and working problems. They are set off in a box for easy referral.

Chapter Glossary and Summary.   Found at the end of each chapter, the chapter glossary contains a list of definitions of new terms introduced in the chapter, and the summary contains a list of important rules, properties, or steps introduced in the chapter.

Chapter Review and Test. The end of each chapter contains a review of topics introduced in the chapter. These review problems are keyed to sections. The chapter test is not keyed to sections.

Cumulative Review. Each chapter after the first contains a cumulative review. Each problem contained in the cumulative review is actually an earlier worked example in the text which is referenced in the back of the book along with the answer. Students who need to see a complete worked-out solution with explanation, can do so by turning to the appropriate example in the text.

Supplements

The following supplements are available to qualified adopters of *Beginning Algebra*:

For the Instructor

Annotated Instructor's Edition has answers to all exercises displayed on the same page with the exercises.

Instructor's Manual with Tests, Syllabus and Instructor's Disk contains 9 tests per chapters (5 are free-response, 4 are multiple choice), suggested syllabi and homework assignments, and an ASCII disk which allows customization of syllabi.

Instructor Solutions Manual provides even-numbered solutions.

IPS Testing (IBM) generates test questions and drill worksheets from algorithms keyed to the learning objectives in the book. Available free upon adoption in 3.5″ and 5.25″ IBM formats.

PHTestmanager testing (IBM). A bank of test items designed specifically for the book, in both free-response and multiple choice form; fully editable for flexible use. A Macintosh version will also be available.

Test Item File contains a hard copy of test questions on PHTestmanager.

For the Student

Student Solutions Manual contains odd-numbered solutions and solutions to all chapter tests and cumulative tests.

Math Master Tutor software (IBM and Mac) provides text-specific tutorial, exercises graduated in difficulty which are generated new each time, fully worked-out examples, and a timed quiz.

Videotapes with class lectures by the author are closely keyed to the book itself. (One set available with an adoption of 100 copies or more.)

Acknowledgements

Writing this book has been a humbling experience, an effort requiring the help of many more people than I originally imagined. I will attempt to thank them here.

First, I would like to thank my husband, Clayton. Without his constant encouragement, this project would not have become a reality. I would also like to thank my children, Eric and Bryan, for eating my burnt bacon. Writing a book while raising two small children is an experience that requires an infinite amount of patience and a good sense of humor.

I would like to thank my extended family for their invaluable help. Their contributions are too numerous to list. They are Peter, Karen, Michael, Christopher, Matthew and Jessica Callac; Stuart, Earline, Melissa, and Mandy Martin; Mark Martin; Barbara and Leo Miller; and Jewett Gay.

I would like to thank the following excellent writers for their work. Creating the first draft manuscript would not have been possible without them. —Ned Schillow, Cynthia Miller, Lea Campbell, Jan Vandever, Cathy Pace, Myrna Mitchell and Curtis McKnight. Each provided invaluable help.

I would like to thank the following reviewers for their suggestions:

Carol Achs, *Mesa Community College*
Gabrielle Andries, *University of Wisconsin–Milwaukee*
Jan Archibald, *Ventura College*
Carol Atnip, *University of Louisville*
Sandra Beken, *Horry-Georgetown Technical College*
Nancy J. Bray, *San Diego Mesa College*
Helen Burrier, *Kirkwood Community College*
Dee Ann Christianson, *The University of the Pacific*
John Coburn, *St. Louis Community College*
Iris DeLoach-Johnson, *Miami University*
Catherine Folio, *Brookdale Community College*
Robert W. Gesell, *Cleary College*
Dauhrice Gibson, *Gulf Coast Community College*
Marian Glasby, *Anne Arundel Community College*
Margaret (Peg) Greene, *Florida Community College @ Jacksonville*
Frank Gunnip, *Oakland Community College*
Mike Mears, *Manatee Community College*
James W. Newsom, *Tidewater Community College*
Mary Kay Schippers, *Fort Hays State University*
Mary Lee Seitz, *Erie Community College–City Campus*
Ken Seydel, *Skyline College*
Edith Silver, *Mercer County Community College*
Debbie Singleton, *Lexington Community College*
Ronald Smith, *Edison Community College*
Richard Spangler, *Tacoma Community College*
Diane Trimble, *Collin County Community College*
John C. Wenger, *City College of Chicago–Harold Washington College*
Jerry Wilkerson, *Missouri Western State College.*

Richard Semmler did an excellent job of working with the galley proofs and page proofs, insuring that they be as error-free as possible. Assisting him were Cheryl Roberts, Fran Hopf, and William Radulovich. Rick Ponticelli, Howard Sorkin and Karen Schwitters wrote the answers and the solutions manual, contributing to the accuracy as well.

Finally, I would like to thank Laurie Golson, Steve Deitmer, and Christine Peckaitis for their invaluable contributions; production editor Tom Aloisi; and executive editor Priscilla McGeehon, who is always there when I need her. Paul Banks' efforts as marketing manager, even before the book was published, are much appreciated.

About the Author

Elayn Martin-Gay has taught mathematics at the University of New Orleans for 14 years. She now coordinates the developmental mathematics program. In 1985, she won the local University Alumni Association's Award for Excellence in Teaching.

Over the years, Elayn has developed videotaped lecture series to help her students understand algebra material better. This highly successful video material is the basis for the three-book series, *Prealgebra, Beginning Algebra*, and *Intermediate Algebra*. Her updated video lecture series, which fits specifically with the books, is available to adopters through Prentice Hall.

1.1 Symbols

1.2 Sets of Numbers

1.3 Fractions

1.4 Exponents and Order of Operations

1.5 Introduction to Variables

1.6 Adding Real Numbers

1.7 Subtracting Real Numbers

1.8 Multiplying and Dividing Real Numbers

1.9 Properties of Real Numbers

CHAPTER **1**

Real Numbers and Their Properties

Shirley is debating whether a remote Bahamian island is the ideal place for a convention she is planning. Given the limitations of local transportation, she's having second thoughts. (See Critical Thinking, page 47.)

INTRODUCTION

In arithmetic, everyday situations are described using numbers. Algebra differs from arithmetic in that letters are used to represent unknown numbers. An important part of learning algebra is learning the symbols and words—the language—of algebra. Much of this language is familiar to you already as the language of arithmetic. We begin our study of algebra with a review of arithmetic: its symbols, words, and patterns. This review is essential in forming the tools needed to learn the language of algebra.

1.1
Symbols

OBJECTIVES

Tape 1

1 Identify the symbols used for natural and whole numbers, and picture them on a number line.

2 Define the meaning of the symbols $=$, $\neq$, $<$, $>$, $\leq$, and $\geq$.

3 Translate sentences into mathematical statements.

4 Define the meaning of the symbols used for addition, subtraction, multiplication, and division.

1 We begin with a review of natural numbers and whole numbers and how we use symbols to compare these numbers.

The **natural numbers** are 1, 2, 3, 4, 5, 6, 7, 8, 9, 10, 11, 12, and so on.
The **whole numbers** are the natural numbers together with zero.

The whole numbers are 0, 1, 2, 3, 4, 5, 6, 7, 8, 9, 10, 11, 12, and so on. Whole numbers can be pictured with a **number line.** We will use the number line often to help us visualize objects and relationships. Visualizing mathematical concepts is an important skill and tool, and later we will develop and explore other visualizing tools.

To draw a number line, first draw a line. Choose a point on the line and label it 0. To the right of 0, label any other point 1. Being careful to use the same distance as from 0 to 1, mark off equally spaced distances. Label these points 2, 3, 4, 5, and so on. Since the whole numbers continue indefinitely, it is not possible to show every whole number on the number line. The arrow at the right end of the line indicates that the pattern continues indefinitely.

```
◄──┼──┼──┼──┼──┼──►
   0  1  2  3  4  5
```

2 Picturing whole numbers on a number line helps us to see the order of the numbers. Symbols can be used to concisely describe what we see.

The **equal symbol,** $=$, states that one value "is equal to" another.

The **not equal symbol,** ≠, states that one value "is not equal to" another. For example,

$$2 = 2 \quad \text{states that "two is equal to two"}$$

$$2 \neq 6 \quad \text{states the "two is not equal to six"}$$

Using these symbols forms a **mathematical statement.** The statement might be true or it might be false. The above two statements are both true.

If two numbers are not equal, then one number is larger than the other. The **greater than symbol,** >, states that one value "is greater than" another. For example,

$$2 > 0 \quad \text{states that "two is greater than zero"}$$

The **less than symbol,** <, states that one value "is less than" another. For example,

$$3 < 5 \quad \text{states that "three is less than five"}$$

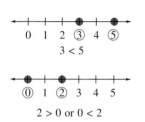

$3 < 5$

$2 > 0$ or $0 < 2$

On the number line, we see that a number **to the right of** another number is **larger.** Similarly, a number **to the left of** another number is smaller. For example, 3 is to the left of 5 on the number line, which means that 3 is less than 5, or $3 < 5$. Similarly, 2 is to the right of 0 on the number line, which means 2 is greater than 0, or $2 > 0$. Since 0 is to the left of 2, we can also say that 0 is less than 2, or $0 < 2$.

HELPFUL HINT

Notice that $2 > 0$ has exactly the same meaning as $0 < 2$. Switching the order of the numbers and reversing the "direction of the inequality symbol" does not change the meaning. For example,

$$5 > 3 \quad \text{has the same meaning as} \quad 3 < 5.$$

Also notice that, when the statement is true, the inequality arrow "points" to the smaller number.

EXAMPLE 1   Insert <, >, or = in the space between the paired numbers to make each statement true.

a. 2   3   **b.** 7   4   **c.** 72   27

Solution: **a.** $2 < 3$ since 2 is to the left of 3 on the number line.

b. $7 > 4$ since 7 is to the right of 4 on the number line.

c. $72 > 27$ since 72 is to the right of 27 on the number line.   ∎

Two other symbols are used to compare numbers. The **less than or equal to** symbol, ≤, states that one value "is less than or equal to" another value. The **greater than or equal to** symbol, ≥, states that one value "is greater than or equal to" another value. For example,

$$7 \leq 10 \quad \text{states that "seven is less than or equal to ten"}$$

This statement is true since $7 < 10$. If either $7 < 10$ or $7 = 10$ is true, then $7 \leq 10$ is true.

$$3 \geq 3 \quad \text{states that "three is greater than or equal to three"}$$

This statement is true since $3 = 3$. If either $3 > 3$ or $3 = 3$ is true, then $3 \geq 3$ is true.

The statement $6 \geq 10$ is false since neither $6 > 10$ nor $6 = 10$.

The symbols <, >, ≤, and ≥ are called **inequality symbols.**

EXAMPLE 2   Tell whether the statement is true or false.

a. $8 \geq 8$ **b.** $8 \leq 8$ **c.** $23 \leq 0$ **d.** $23 \geq 0$

Solution: **a.** True, since $8 = 8$.   **b.** True, since $8 = 8$.
c. False, since neither $23 < 0$ nor $23 = 0$.   **d.** True, since $23 > 0$.   ∎

3   Now, let's use the symbols discussed above to translate sentences into mathematical statements.

EXAMPLE 3   Translate each sentence into a mathematical statement.
a. 9 is less than or equal to 11.
b. 8 is greater than 1.
c. 3 is not equal to 4.

Solution:

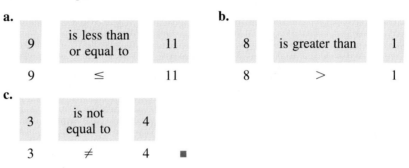

a.

| 9 | is less than or equal to | 11 |

$$9 \qquad \leq \qquad 11$$

b.

| 8 | is greater than | 1 |

$$8 \qquad > \qquad 1$$

c.

| 3 | is not equal to | 4 |

$$3 \qquad \neq \qquad 4 \qquad ∎$$

4   Symbols are also used to represent the sum, difference, product, and quotient of numbers, corresponding to the operations of addition, subtraction, multiplication, and division. Before continuing further, you should feel completely comfortable performing these operations on whole numbers. The following table summarizes the symbols and meanings of these basic operations.

Operation Symbols		
	Symbols	**Meanings**
Addition	$a + b$	The sum of a and b or a plus b
Subtraction	$a - b$	The difference of a and b or a minus b
Multiplication	$a \times b,\ a \cdot b,\ ab,\ a(b),\ (a)b,\ (a)(b)$	The product of a and b or a times b
Division	$\dfrac{a}{b},\ a/b,\ a \div b$	The quotient of a and b or a divided by b

EXAMPLE 4   Translate each sentence into a mathematical statement.
a. The product of 2 and 3 is 6.
b. The difference of 8 and 4 is less than or equal to 4.
c. The quotient of 10 and 2 is not equal to 3.

Solution: **a.**

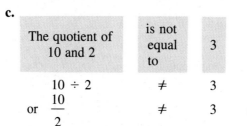

| The product of 2 and 3 | is | 6 |

$$2 \times 3 = 6$$
$$\text{or} \quad 2 \cdot 3 = 6$$
$$\text{or} \quad 2(3) = 6$$

b.

| The difference of 8 and 4 | is less than or equal to | 4 |

$$8 - 4 \quad \leq \quad 4$$

c.

| The quotient of 10 and 2 | is not equal to | 3 |

$$10 \div 2 \quad \neq \quad 3$$
$$\text{or} \quad \frac{10}{2} \quad \neq \quad 3 \quad \blacksquare$$

The following is a summary of symbols used to compare numbers.

Symbols Used to Compare Numbers			
$=$	is equal to	$\neq$	is not equal to
$<$	is less than	$\leq$	is less than or equal to
$>$	is greater than	$\geq$	is greater than or equal to

EXERCISE SET 1.1

Insert $<$, $>$, or $=$ in the space between the paired numbers to make each statement true. See Example 1.

1. 4   10   **2.** 8   5   **3.** 7   3   **4.** 9   15

5. 6.26   6.26   **6.** 2.13   1.13   **7.** 0   7   **8.** 20   0

Are the following statements true or false? See Example 2.

9. $11 \leq 11$ **10.** $4 \geq 7$ **11.** $10 > 11$ **12.** $17 > 16$

13. $3 + 8 \geq 3(8)$ **14.** $8 \cdot 8 \leq 8 \cdot 7$ **15.** $7 > 0$ **16.** $4 < 7$

Write each sentence as a mathematical statement. See Examples 3 and 4.

17. 8 is less than 12.

18. 5 is less than 15.

19. 5 is greater than or equal to 4.

20. 10 is greater than or equal to 7.

21. The sum of 2 and 3 is less than 6.

22. The difference of 16 and 4 is greater than 10.

23. The quotient of 10 and 2 is 5.

24. The quotient of 18 and 6 is 3.

25. 4 is less than or equal to the product of 3 and 5.

26. 5 is less than or equal to the product of 4 and 9.

Insert <, >, or = in the appropriate space to make each statement true.

27. 5   7

28. 8   2

29. 7   5

30. 2   8

31. 3 · 8   25

32. 4 · 8   30

33. 15 − 9   6

34. 19 − 8   11

35. 15 − 9   5

36. 19 − 8   12

Are the following statements true or false?

37. 5 − 0 = 5

38. 8(1) = 9

39. 5(0) = 5

40. 12 ≠ 12

41. 0 ≤ 7

42. 7 ≥ 4

Rewrite the following inequalities so that the inequality symbol points in the opposite direction and the resulting statement is equivalent to the given one.

43. 25 ≥ 25

44. 13 ≤ 13

45. 0 < 6

46. 5 > 3

47. $a \leq b$

48. $c \geq d$

49. $x > y$

50. $m < n$

Write each sentence as a mathematical statement.

51. 4 is greater than 2.5.

52. 5 is greater than 3.2.

53. 8 is less than or equal to 12.

54. 5 is less than or equal to 15.

55. 5 is less than or equal to 6.

56. 8 is less than or equal to 10.

57. The sum of 5 and 6 is greater than 10.

58. The difference of 15 and 7 is less than 9.

59. The product of 3 and 5 is greater than 12.

60. The product of 4 and 7 is less than 30.

61. The quotient of 12 and 6 is greater than 1.

62. The quotient of 25 and 5 is less than 6.

63. 3 is greater than 2.

64. 4 is equal to 3 plus 1.

65. 4 is equal to 4 plus 0.

66. 25 times 1 equals 25.

67. a is less than or equal to 5.

68. b is greater than or equal to 3.

69. c is less than d.

70. d is greater than the product of 3 and c.

1.2
Sets of Numbers

OBJECTIVES

Tape 1

1   Identify natural numbers, whole numbers, integers, rational numbers, irrational numbers, and real numbers.

2   Find the absolute value of real numbers.

1   We now define and explore other **sets** of numbers commonly used in algebra. A set is a collection of objects called **members** or **elements.** A pair of brace symbols { } encloses the list of elements and is translated as "the set of" or "the set containing."

> **Natural Numbers**
>
> The set of **natural numbers** is {1, 2, 3, 4, 5, 6, . . . }.

The three dots (an ellipsis) at the end of the list of elements of a set means that the list continues in the same manner indefinitely.

> **Whole Numbers**
>
> The set of **whole numbers** is $\{0, 1, 2, 3, 4, \ldots\}$.

Whole numbers are not sufficient to describe many situations in the real world. For example, quantities smaller than zero must sometimes be represented, such as temperatures less than 0 degrees.

We can picture numbers smaller than zero on the number line as follows:

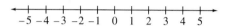

Numbers less than 0 are to the left of 0 and are labeled -1, -2, -3, and so on. A $-$ sign such as the one in -1 tells us that the number is to the left of 0 on the number line. In words, -1 is read "negative one." A $+$ sign or no sign tells us that a number lies to the right of 0 on the number line. For example, 3 and $+3$ both mean positive three.

The numbers we have pictured are called the set of **integers.** Integers to the left of 0 are called **negative integers;** integers to the right of 0 are called **positive integers.** The integer 0 is neither positive nor negative.

> **Integers**
>
> The set of integers is $\{\ldots -3, -2, -1, 0, 1, 2, 3, \ldots\}$.

Notice the three dots to the left and to the right of the list for the integers. This indicates that the positive integers and the negative integers continue indefinitely.

A problem with integers in real-life settings arises when quantities are smaller than some integer, but greater than the next smallest integer. On the number line, these quantities are points between integers. Some of these quantities between integers can be represented as a quotient of integers. For example,

The point representing $\dfrac{1}{2}$ lies halfway between 0 and 1.

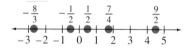

The point representing $-\dfrac{1}{2}$ lies halfway between 0 and -1. Other quotients of integers and their graphs are shown.

The set of numbers, each of which can be represented as a quotient of integers, is called the set of **rational numbers.** Notice that every integer is also a rational number since each integer can be expressed as a quotient of integers. For example, the integer 5 is also a rational number since $5 = \dfrac{5}{1}$.

> **Rational Numbers**
>
> The set of **rational numbers** is the set of all numbers that can be expressed as a quotient $\dfrac{a}{b}$, where a and b are integers and $b \neq 0$.

The number line also contains points that cannot be represented by rational numbers. These numbers are called **irrational numbers** because they cannot be expressed as quotients of integers. For example, $\sqrt{2}$ and π are irrational numbers.

Irrational Numbers

The set of **irrational numbers** is the set of all numbers that correspond to a point on the number line but that are not rational numbers. That is, an irrational number is a number that cannot be expressed as a quotient of integers.

The set of numbers, each of which corresponds to a point on the number line, is called the set of **real numbers.** One and only one point on the number line corresponds to each real number.

Real Numbers

The set of **real numbers** is the set of all numbers each of which corresponds to a point on the number line.

On the following number line, we see that real numbers can be positive, negative, or 0. Numbers to the left of 0 are called **negative numbers;** numbers to the right of 0 are called **positive numbers.** Positive and negative numbers are also called **signed numbers.**

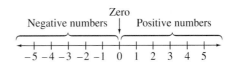

Several different sets of numbers have been discussed in this section. The following diagram shows the relationships between these sets of real numbers.

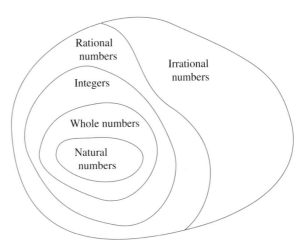

Real numbers

EXAMPLE 1   Given the set $\left\{-2, 0, \frac{1}{4}, 112, -3, 11, \sqrt{2}\right\}$, list the numbers in this set that belong to the set of:

a. Natural numbers   b. Whole numbers   c. Integers

d. Rational numbers   e. Irrational numbers   f. Real numbers

Solution: a. The natural numbers are 11 and 112.

b. The whole numbers are 0, 11, and 112.

c. The integers are -3, -2, 0, 11, and 112.

d. Recall that integers are rational numbers also. The rational numbers are -3, -2, 0, $\frac{1}{4}$, 11, and 112.

e. The irrational number is $\sqrt{2}$.

f. The real numbers are all numbers in the given set.   ∎

We can now extend the meaning and use of inequality symbols such as $<$ and $>$ to apply to all real numbers.

> **Order Property for Real Numbers**
>
> Given any two real numbers a and b, $a < b$ if a is to the left of b on the number line. Similarly, $a > b$ if a is to the right of b on the number line.

EXAMPLE 2   Insert $<$, $>$, or $=$ in the space between the paired numbers to make each statement true.

a. -1 ___ 0   b. 7 ___ $\frac{14}{2}$ c. -5 ___ -6

Solution: a. $-1 < 0$ since -1 is to the left of 0 on the number line.

b. $7 = \frac{14}{2}$ since $\frac{14}{2}$ simplifies to 7.

c. $-5 > -6$ since -5 is to the right of -6 on the number line.   ∎

2   The number line not only gives us a picture of the real numbers, but it also helps us visualize the distance between numbers. The distance between a real number a and 0 is given a special name called the **absolute value** of a. "The absolute value of a" is written in symbols as $|a|$.

> **Absolute Value**
>
> The absolute value of a real number a, denoted by $|a|$, is the distance between a and 0.

For example, $|3| = 3$ and $|-3| = 3$ since both 3 and -3 are a distance of 3 units from 0 on the number line.

$$|-3| = 3 \qquad |3| = 3$$

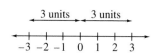

> **HELPFUL HINT**
>
> Since $|a|$ is a distance, $|a|$ will always be either positive or 0, never negative. That is, **for any real number a, $|a| \geq 0$.**

EXAMPLE 3   Find the absolute value of each number.

   a. $|4|$ **b.** $|-5|$ **c.** $|0|$

   Solution: **a.** $|4| = 4$ since 4 is 4 units from 0 on the number line.

   b. $|-5| = 5$ since -5 is 5 units from 0 on the number line.

   c. $|0| = 0$ since 0 is 0 units from 0 on the number line.   ■

EXAMPLE 4   Insert $<$, $>$, or $=$ in the appropriate space to make the statement true.

   a. $|0|$ 2   **b.** $|-5|$ 5   **c.** $|-3|$ $|-2|$ **d.** $|5|$ $|6|$

   e. $|-7|$ $|6|$

   Solution: **a.** $|0| < 2$ since $|0| = 0$ and $0 < 2$.

   b. $|-5| = 5$.

   c. $|-3| > |-2|$ since $3 > 2$.

   d. $|5| < |6|$ since $5 < 6$.

   e. $|-7| > |6|$ since $7 > 6$.   ■

EXERCISE SET 1.2

Tell which set or sets each number belongs to. Choose among the sets of natural numbers, whole numbers, integers, rational numbers, irrational numbers, and real numbers. See Example 1.

1. 0   **2.** $\dfrac{1}{4}$ **3.** -2 **4.** $-\dfrac{1}{2}$

5. 6   **6.** 5   **7.** $\dfrac{2}{3}$ **8.** $\sqrt{3}$

Insert $<$, $>$, or $=$ in the appropriate space to make a true statement. See Example 2.

9. 10   20   **10.** 100   10   **11.** 0 -2 **12.** 0   2

13. 0.01   0.01   **14.** 0.05   0   **15.** -1.5 2.8   **16.** $\dfrac{14}{2}$ 7

Insert $<$, $>$, or $=$ in the appropriate space to make a true statement. See Examples 3 and 4.

17. $|20|$ 20   **18.** $|-52|$ $|52|$ **19.** -7 -12

20. $|-2|$ $|-3|$ **21.** -500 $|-50|$ **22.** $|0|$ $|-8|$

Tell which set or sets each number belongs to. Choose among the sets of natural numbers, whole numbers, integers, rational numbers, irrational numbers, and real numbers.

23. -9 **24.** $|-8|$ **25.** π

26. $\dfrac{3}{8}$ **27.** 2   **28.** $|2|$

Tell whether each statement is true or false.

29. $5 < 6$

30. $7 > 8$

31. $-5 < -6$

32. $-7 > -8$

33. $|-5| < |-6|$

34. $|-7| > |-8|$

35. $|-5| \geq |5|$

36. $|-3| < |0|$

37. $-3 > 2$

38. $-5 < 5$

39. $|8| = |-8|$

40. $|9| = |-9|$

41. $|0| > |-4|$

42. $|0| \leq |0|$

Insert $<$, $>$, or $=$ in the appropriate space to make a true statement.

43. $-10 \quad\quad -100$

44. $-200 \quad\quad -20$

45. $32 \quad\quad 5.2$

46. $7 \quad\quad -7$

47. $\dfrac{18}{3} \quad\quad \dfrac{24}{3}$

48. $\dfrac{8}{2} \quad\quad \dfrac{12}{3}$

49. $-51 \quad\quad -50$

50. $|-20| \quad\quad -200$

51. $|-5| \quad\quad -4$

52. $0 \quad\quad |0|$

53. $|-1| \quad\quad |1|$

54. $\left|\dfrac{2}{5}\right| \quad\quad \left|-\dfrac{2}{5}\right|$

Tell whether each statement is true or false.

55. Every rational number is also an integer.

56. Every natural number is positive.

57. 0 is a real number.

58. Every whole number is an integer.

59. Every negative number is also a rational number.

60. Every rational number is also a real number.

61. Every real number is also a rational number.

62. $\dfrac{1}{2}$ is an integer.

Writing in Mathematics

63. In your own words, explain what absolute value is.

64. Give an example of a real-life situation that can be described with integers but not with whole numbers.

1.3
Fractions

OBJECTIVES

Tape 1

1 Write fractions in simplest form.

2 Multiply and divide fractions.

3 Add and subtract fractions.

1 A quotient of two integers such as $\dfrac{2}{9}$ is called a **fraction.** In the fraction $\dfrac{2}{9}$, the top number, 2, is called the **numerator** and the bottom number, 9, is called the **denominator.**

In the statement $3 \cdot 5 = 15$, 3 and 5 are called **factors** and 15 is the **product.**

$$3 \quad \cdot \quad 5 \quad = \quad 15$$
$$\uparrow \qquad \uparrow \qquad\quad \uparrow$$
$$\text{factor} \quad \text{factor} \quad\ \text{product}$$

To **factor** 15 means to write it as a product. The number 15 can be factored as $3 \cdot 5$ or as $1 \cdot 15$.

A fraction is said to be **simplified** or in **lowest terms** when the numerator and the denominator have no factors in common other than 1. For example, the fraction $\dfrac{5}{11}$ is in lowest terms since 5 and 11 have no common factors other than 1.

To help us simplify fractions, we will write the numerator and the denominator as a product of **prime numbers.** A prime number is a whole number, other than 1, whose only factors are 1 and itself. The first few prime numbers are

$$2, 3, 5, 7, 11, 13, 17, 19, 23, 29, \text{ and so on}$$

EXAMPLE 1   Write each of the following numbers as a product of primes.
a. 40   **b.** 63

Solution: **a.** Write 40 as the product of any two whole numbers.

$$40 = 4 \cdot 10$$

Next, factor each of these numbers. Continue this process until all of the factors are prime numbers.

$$40 = \underset{2 \cdot 2}{4} \cdot \underset{2 \cdot 5}{10}$$
$$= 2 \cdot 2 \cdot 2 \cdot 5$$

All the factors are now prime numbers. Then 40 written as a product of primes is

$$40 = 2 \cdot 2 \cdot 2 \cdot 5$$

b. $63 = \underset{3 \cdot 3}{9} \cdot 7$
$$= 3 \cdot 3 \cdot 7 \quad \blacksquare$$

To use prime factors to write a fraction in lowest terms, follow these steps:

Writing a Fraction in Lowest Terms

Step 1   Write the numerator and the denominator as a product of primes.
Step 2   Divide the numerator and the denominator by their common factors.

EXAMPLE 2   Write each fraction in lowest terms.
a. $\dfrac{42}{49}$ **b.** $\dfrac{11}{27}$ **c.** $\dfrac{88}{20}$

Solution: **a.** Write the numerator and the denominator as products of primes; then divide both by the common factor 7.

$$\frac{42}{49} = \frac{2 \cdot 3 \cdot \boxed{7}}{7 \cdot \boxed{7}} = \frac{6}{7}$$

b. $\dfrac{11}{27} = \dfrac{11}{3 \cdot 3 \cdot 3}$

There are no common factors other than 1, so $\dfrac{11}{27}$ is already in lowest terms.

c. $\dfrac{88}{20} = \dfrac{\boxed{2} \cdot \boxed{2} \cdot 2 \cdot 11}{\boxed{2} \cdot \boxed{2} \cdot 5} = \dfrac{22}{5} \quad \blacksquare$

2   To multiply two fractions, multiply numerator times numerator to obtain the numerator of the product and denominator times denominator to obtain the denominator of the product.

Multiplying Fractions

$$\frac{a}{b} \cdot \frac{c}{d} = \frac{a \cdot c}{b \cdot d}, \qquad \text{if } b \neq 0 \text{ and } d \neq 0$$

EXAMPLE 3   Find the product of $\dfrac{2}{15}$ and $\dfrac{5}{13}$. Write the answer in lowest terms.

Solution:
$$\frac{2}{15} \cdot \frac{5}{13} = \frac{2 \cdot 5}{15 \cdot 13} \qquad \begin{array}{l}\text{Multiply numerators.}\\ \text{Multiply denominators.}\end{array}$$

Next, simplify the product by dividing the numerator and the denominator by any common factors.

$$= \frac{2 \cdot 5}{3 \cdot 5 \cdot 13}$$

$$= \frac{2}{39} \qquad \blacksquare$$

Before dividing fractions, we first define **reciprocals.** Two fractions are reciprocals of one another if their product is 1. For example $\dfrac{2}{3}$ and $\dfrac{3}{2}$ are reciprocals since $\dfrac{2}{3} \cdot \dfrac{3}{2} = 1$. Also, the reciprocal of 5 is $\dfrac{1}{5}$ since $5 \cdot \dfrac{1}{5} = \dfrac{5}{1} \cdot \dfrac{1}{5} = 1$.

To divide fractions, multiply the first fraction by the reciprocal of the second fraction.

Dividing Fractions

$$\frac{a}{b} \div \frac{c}{d} = \frac{a}{b} \cdot \frac{d}{c}, \qquad \text{if } b \neq 0, d \neq 0, \text{ and } c \neq 0$$

EXAMPLE 4   Find each quotient. Write all answers in lowest terms.

a. $\dfrac{4}{5} \div \dfrac{5}{16}$ **b.** $\dfrac{7}{10} \div 14$ **c.** $\dfrac{3}{8} \div \dfrac{3}{10}$

Solution: **a.** $\dfrac{4}{5} \div \dfrac{5}{16} = \dfrac{4}{5} \cdot \dfrac{16}{5} = \dfrac{4 \cdot 16}{5 \cdot 5} = \dfrac{64}{25}$

b. $\dfrac{7}{10} \div 14 = \dfrac{7}{10} \div \dfrac{14}{1} = \dfrac{7}{10} \cdot \dfrac{1}{14} = \dfrac{7 \cdot 1}{2 \cdot 5 \cdot 2 \cdot 7} = \dfrac{1}{20}$

c. $\dfrac{3}{8} \div \dfrac{3}{10} = \dfrac{3}{8} \cdot \dfrac{10}{3} = \dfrac{3 \cdot 2 \cdot 5}{2 \cdot 2 \cdot 2 \cdot 3} = \dfrac{5}{4}$ $\blacksquare$

3   To add or subtract fractions with the same denominator, combine numerators and place the sum or difference over the common denominator.

> **Adding and Subtracting Fractions with the Same Denominator**
>
> $$\frac{a}{b} + \frac{c}{b} = \frac{a + c}{b}, \qquad \text{if } b \neq 0$$
>
> $$\frac{a}{b} - \frac{c}{b} = \frac{a - c}{b}, \qquad \text{if } b \neq 0$$

EXAMPLE 5   Add or subtract as indicated. Write each answer in lowest terms.

a. $\dfrac{2}{7} + \dfrac{4}{7}$ **b.** $\dfrac{3}{10} + \dfrac{2}{10}$ **c.** $\dfrac{9}{7} - \dfrac{2}{7}$ **d.** $\dfrac{5}{3} - \dfrac{1}{3}$

Solution: **a.** $\dfrac{2}{7} + \dfrac{4}{7} = \dfrac{2 + 4}{7} = \dfrac{6}{7}$

b. $\dfrac{3}{10} + \dfrac{2}{10} = \dfrac{3 + 2}{10} = \dfrac{5}{10} = \dfrac{5}{2 \cdot 5} = \dfrac{1}{2}$

c. $\dfrac{9}{7} - \dfrac{2}{7} = \dfrac{9 - 2}{7} = \dfrac{7}{7} = 1$

d. $\dfrac{5}{3} - \dfrac{1}{3} = \dfrac{5 - 1}{3} = \dfrac{4}{3}$ ∎

To add or subtract fractions without the same denominator, first write the fractions as **equivalent fractions** with a common denominator. Equivalent fractions are fractions that represent the same quantity. For example, $\dfrac{3}{4}$ and $\dfrac{12}{16}$ are equivalent fractions since they represent the same portion of a whole, as the diagram shows. Count the larger squares and the shaded portion is $\dfrac{3}{4}$. Count the smaller squares and the shaded portion is $\dfrac{12}{16}$. Thus $\dfrac{3}{4} = \dfrac{12}{16}$.

We can write equivalent fractions by multiplying a given fraction by 1, as shown in the next example. Multiplying a number by 1 does not change the value of the number.

Whole

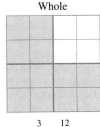

$\dfrac{3}{4} = \dfrac{12}{16}$

EXAMPLE 6   Write $\dfrac{2}{5}$ as an equivalent fraction with a denominator of 20.

Solution: Since $5 \cdot 4 = 20$, multiply the fraction by $\dfrac{4}{4}$. Multiplying by $\dfrac{4}{4} = 1$ does not change the value of the fraction.

$$\frac{2}{5} = \frac{2}{5} \cdot \frac{4}{4} = \frac{2 \cdot 4}{5 \cdot 4} = \frac{8}{20}$$ ∎

EXAMPLE 7   Add or subtract as indicated. Write each answer in lowest terms.

$$\textbf{a. } \frac{2}{5} + \frac{1}{4} \quad \textbf{b. } \frac{1}{2} + \frac{17}{22} - \frac{2}{11} \quad \textbf{c. } 3\frac{1}{6} - 1\frac{11}{12}$$

Solution: **a.** Fractions must have a common denominator before they can be added or subtracted. Since 20 is the smallest number that both 5 and 4 divide into evenly, 20 is the **least common denominator.** Write both fractions as equivalent fractions with denominators of 20. Since

$$\frac{2}{5} \cdot \frac{4}{4} = \frac{2 \cdot 4}{5 \cdot 4} = \frac{8}{20} \quad \text{and} \quad \frac{1}{4} \cdot \frac{5}{5} = \frac{1 \cdot 5}{4 \cdot 5} = \frac{5}{20}$$

then

$$\frac{2}{5} + \frac{1}{4} = \frac{8}{20} + \frac{5}{20} = \frac{13}{20}$$

b. The least common denominator for denominators 2, 22, and 11 is 22. First, write each fraction as an equivalent fraction with a denominator of 22.

$$\frac{1}{2} = \frac{1}{2} \cdot \frac{11}{11} = \frac{11}{22}, \quad \frac{17}{22} = \frac{17}{22}, \quad \text{and} \quad \frac{2}{11} = \frac{2}{11} \cdot \frac{2}{2} = \frac{4}{22}$$

Then

$$\frac{1}{2} + \frac{17}{22} - \frac{2}{11} = \frac{11}{22} + \frac{17}{22} - \frac{4}{22} = \frac{24}{22} = \frac{12}{11}$$

c. To find $3\frac{1}{6} - 1\frac{11}{12}$, first rewrite each number as follows:

$$3\frac{1}{6} = 3 + \frac{1}{6} = \frac{18}{6} + \frac{1}{6} = \frac{19}{6}$$

$$1\frac{11}{12} = 1 + \frac{11}{12} = \frac{12}{12} + \frac{11}{12} = \frac{23}{12}$$

Then

$$3\frac{1}{6} - 1\frac{11}{12} = \frac{19}{6} - \frac{23}{12} = \frac{38}{12} - \frac{23}{12} = \frac{15}{12} = \frac{5}{4} \quad \blacksquare$$

EXERCISE SET 1.3

Write each of the following numbers as a product of primes. See Example 1.

1. 20   **2.** 56   **3.** 75
4. 32   **5.** 45   **6.** 24

Write the following fractions in lowest terms. See Example 2.

7. $\frac{2}{4}$ **8.** $\frac{3}{6}$ **9.** $\frac{10}{15}$ **10.** $\frac{15}{20}$

11. $\frac{3}{7}$ **12.** $\frac{5}{9}$ **13.** $\frac{18}{30}$ **14.** $\frac{42}{45}$

Multiply or divide as indicated. Write the answer in lowest terms. See Examples 3 and 4.

15. $\dfrac{1}{2} \cdot \dfrac{3}{4}$

16. $\dfrac{10}{6} \cdot \dfrac{3}{5}$

17. $\dfrac{2}{3} \cdot \dfrac{3}{4}$

18. $\dfrac{7}{8} \cdot \dfrac{3}{21}$

19. $\dfrac{1}{2} \div \dfrac{7}{12}$

20. $\dfrac{7}{12} \div \dfrac{1}{2}$

21. $\dfrac{3}{4} \div \dfrac{1}{20}$

22. $\dfrac{3}{5} \div \dfrac{9}{10}$

23. $\dfrac{7}{10} \cdot \dfrac{5}{21}$

24. $\dfrac{3}{35} \cdot \dfrac{10}{63}$

25. $2\dfrac{7}{9} \cdot \dfrac{1}{3}$

26. $\dfrac{1}{4} \cdot 5\dfrac{5}{6}$

Add or subtract as indicated. Write the answer in lowest terms. See Example 5.

27. $\dfrac{4}{5} - \dfrac{1}{5}$

28. $\dfrac{6}{7} - \dfrac{1}{7}$

29. $\dfrac{4}{5} + \dfrac{1}{5}$

30. $\dfrac{6}{7} + \dfrac{1}{7}$

31. $\dfrac{17}{21} - \dfrac{10}{21}$

32. $\dfrac{18}{35} - \dfrac{11}{35}$

33. $\dfrac{23}{105} + \dfrac{4}{105}$

34. $\dfrac{13}{132} + \dfrac{35}{132}$

Write each of the following fractions as an equivalent fraction with the given denominator. See Example 6.

35. $\dfrac{7}{10}$ with a denominator of 30

36. $\dfrac{2}{3}$ with a denominator of 9

37. $\dfrac{2}{9}$ with a denominator of 18

38. $\dfrac{8}{7}$ with a denominator of 56

39. $\dfrac{4}{5}$ with a denominator of 20

40. $\dfrac{4}{5}$ with a denominator of 25

Add or subtract as indicated. Write the answer in lowest terms. See Example 7.

41. $\dfrac{2}{3} + \dfrac{3}{7}$

42. $\dfrac{3}{4} + \dfrac{1}{6}$

43. $\dfrac{13}{15} - \dfrac{1}{5}$

44. $\dfrac{2}{9} - \dfrac{1}{6}$

45. $\dfrac{5}{22} - \dfrac{5}{33}$

46. $\dfrac{7}{10} - \dfrac{8}{15}$

47. $\dfrac{12}{5} - 1$

48. $2 - \dfrac{3}{8}$

Perform the following operations. Write answers in lowest terms.

49. $\dfrac{10}{21} + \dfrac{5}{21}$

50. $\dfrac{11}{35} + \dfrac{3}{35}$

51. $\dfrac{10}{3} - \dfrac{5}{21}$

52. $\dfrac{11}{7} - \dfrac{3}{35}$

53. $\dfrac{2}{3} \cdot \dfrac{3}{5}$

54. $\dfrac{2}{3} \div \dfrac{3}{4}$

55. $\dfrac{3}{4} \div \dfrac{7}{12}$

56. $\dfrac{3}{5} + \dfrac{2}{3}$

57. $\dfrac{5}{12} + \dfrac{4}{12}$

58. $\dfrac{2}{7} + \dfrac{4}{7}$

59. $5 + \dfrac{2}{3}$

60. $7 + \dfrac{1}{10}$

61. $\dfrac{7}{8} \div 3\dfrac{1}{4}$

62. $3 \div \dfrac{3}{4}$

63. $\dfrac{7}{18} \div \dfrac{14}{36}$

64. $4\dfrac{3}{7} \div \dfrac{31}{7}$

65. $\dfrac{23}{105} - \dfrac{2}{105}$

66. $\dfrac{57}{132} - \dfrac{13}{132}$

67. $1\dfrac{1}{2} + 3\dfrac{2}{3}$

68. $2\dfrac{3}{5} + 4\dfrac{7}{10}$

69. $\dfrac{2}{3} - \dfrac{5}{9} + \dfrac{5}{6}$

70. $\dfrac{8}{11} - \dfrac{1}{4} + \dfrac{1}{2}$

Applications

71. John began the Slimfast diet to lose 30 pounds. He has lost $17\dfrac{1}{4}$ pounds so far. How many more pounds must he lose to reach his goal?

72. One of the New Orleans Saints guards was told to lose 20 pounds during the summer. So far, the guard has lost $11\dfrac{1}{2}$ pounds. How many more pounds must he lose?

73. A carpenter finds that a board to be used for a window sill is $\dfrac{1}{3}$ too long. If the board is $\dfrac{11}{12}$ of a yard, how much must be cut off?

74. In an English class at Cartez College, Stuart was told to read a 340-page book. So far, he has read $\dfrac{2}{5}$ of it. How many pages does he *have left to read*?

1.4
Exponents and Order of Operations

Tape 2

OBJECTIVES

1	Use and simplify exponential expressions.
2	Define and use the order of operations.

1   A notation frequently used in algebra is the **exponent.** An exponent is a shorthand notation for repeated multiplication of the same factor. For instance in $2 \cdot 2 \cdot 2 = 8$ the factor 2 appears three times. Using exponents, $2 \cdot 2 \cdot 2$ can be written as 2^3. The 2 in 2^3 is called the **base;** it is the repeated factor. The 3 in 2^3 is called the **exponent** and is the number of times the base is used as a factor. The expression 2^3 is called an **exponential expression.**

$$2^3 = 2 \cdot 2 \cdot 2 = 8$$

exponent

base (2 is a factor 3 times)

EXAMPLE 1   Evaluate the following:

a. 3^2 [read as "3 squared" or "3 to the second power"]

b. 5^3 [read as "5 cubed" or as "5 to the third power"]

c. 2^4 [read as "2 to the fourth power"]

d. 7^1 **e.** $\left(\dfrac{3}{7}\right)^2$

Solution: **a.** $3^2 = 3 \cdot 3 = 9$,   "3 squared equals 9"

b. $5^3 = 5 \cdot 5 \cdot 5 = 125$ **c.** $2^4 = 2 \cdot 2 \cdot 2 \cdot 2 = 16$

d. $7^1 = 7$ **e.** $\left(\dfrac{3}{7}\right)^2 = \left(\dfrac{3}{7}\right)\left(\dfrac{3}{7}\right) = \dfrac{9}{49}$ ■

HELPFUL HINT

$2^3 \neq 2 \cdot 3$ since 2^3 is a shorthand notation for repeated **multiplication.**

$$2^3 = 2 \cdot 2 \cdot 2 = 8, \text{ whereas } 2 \cdot 3 = 6.$$

2   Using symbols for mathematical operations is a great convenience. The more operation symbols presented in an expression, the more careful we must be when simplifying. For example, in the expression $2 + 3 \cdot 7$, do we add first or multiply first. To eliminate confusion, **grouping symbols** may be used. Examples of grouping symbols are parentheses ( ), brackets [ ], braces { }, and the fraction bar. If we wish $2 + 3 \cdot 7$ to be simplified by adding first, enclose $2 + 3$ with parentheses.

$$(2 + 3) \cdot 7 = 5 \cdot 7 = 35$$

If we wish to multiply first, $3 \cdot 7$ may be enclosed with parentheses.

$$2 + (3 \cdot 7) = 2 + 21 = 23$$

To eliminate confusion when no grouping symbols are present, use the following agreed upon order of operations.

Order of Operations

Simplify expressions using the following order. If grouping symbols such as parentheses are present, simplify expressions within those first, starting with the innermost set. If fraction bars are present, simplify the numerator and the denominator separately.

1. Simplify exponential expressions.
2. Perform multiplications or divisions in order from left to right.
3. Perform additions or subtractions in order from left to right.

Now simplify $2 + 3 \cdot 7$. There are no grouping symbols and no exponents, so we multiply and then add.

$$2 + \boxed{3 \cdot 7} = 2 + \boxed{21} \quad \text{Multiply.}$$
$$= 23 \quad \text{Add.}$$

EXAMPLE 2   Simplify each expression.

a. $6 \div 3 + 5^2$ **b.** $\dfrac{2(12 + 3)}{|-15|}$ **c.** $3 \cdot 10 - 7 \div 7$ **d.** $3 \cdot 4^2$ **e.** $\dfrac{3}{2} \cdot \dfrac{1}{2} - \dfrac{1}{2}$

Solution: **a.** Evaluate 5^2 first.

$$6 \div 3 + \boxed{5^2} = 6 \div 3 + \boxed{25}$$

Next divide, then add.

$$\boxed{6 \div 3} + 25 = \boxed{2} + 25 \quad \text{Divide.}$$
$$= 27 \quad \text{Add.}$$

b. First, simplify the numerator and the denominator separately.

$$\frac{2(\boxed{12 + 3})}{|-15|} = \frac{2(\boxed{15})}{15} \quad \text{Simplify numerator and denominator separately.}$$

$$= \frac{30}{15}$$

$$= 2 \quad \text{Simplify.}$$

c. Multiply and divide from left to right. Then subtract.

$$\boxed{3 \cdot 10} - \boxed{7 \div 7} = \boxed{30} - \boxed{1}$$
$$= 29 \quad \text{Subtract.}$$

d. In this example, only the 4 is raised to the second power. The factor of 3 is not part of the base because no grouping symbols include it as part of the base.

$$3 \cdot \boxed{4^2} = 3 \cdot \boxed{16} \quad \text{Evaluate the exponential expression.}$$
$$= 48 \quad \text{Multiply.}$$

e. The order of operations applies to fractions in exactly the same way as it applies to whole numbers.

$$\frac{3}{2} \cdot \frac{1}{2} - \frac{1}{2} = \frac{3}{4} - \frac{1}{2} \qquad \text{Multiply.}$$

$$= \frac{3}{4} - \frac{2}{4} \qquad \text{The least common denominator is 4.}$$

$$= \frac{1}{4} \qquad \text{Subtract.} \quad \blacksquare$$

HELPFUL HINT

Be careful when evaluating an exponential expression. In $3 \cdot 4^2$, the exponent 2 applies only to the base of 4. In $(3 \cdot 4)^2$, we multiply first because of parentheses, so the exponent of 2 applies to the product $3 \cdot 4$.

$$3 \cdot 4^2 = 3 \cdot 16 = 48 \qquad (3 \cdot 4)^2 = (12)^2 = 144$$

Expressions that include many grouping symbols can be confusing. When simplifying these expressions, keep in mind that grouping symbols separate the expression into distinct parts. Each is then simplified separately.

EXAMPLE 3 Simplify $\dfrac{3 + |4 - 3| + 2^2}{6 - 3}$.

Solution: The fraction bar serves as a grouping symbol and separates the numerator and denominator. Simplify each separately. Also, the absolute value bars here serve as a grouping symbol. We begin in the numerator by simplifying within the absolute value bars.

$$\frac{3 + |4 - 3| + 2^2}{6 - 3} = \frac{3 + |1| + 2^2}{6 - 3} \qquad \begin{array}{l}\text{Simplify the expression inside} \\ \text{absolute value bars.}\end{array}$$

$$= \frac{3 + 1 + 2^2}{3} \qquad \begin{array}{l}\text{Find the absolute value and simplify} \\ \text{the denominator.}\end{array}$$

$$= \frac{3 + 1 + 4}{3} \qquad \text{Evaluate the exponential expression.}$$

$$= \frac{8}{3} \qquad \text{Simplify the numerator.} \quad \blacksquare$$

EXAMPLE 4 Simplify $3[4(5 + 2) - 10]$.

Solution: Notice that both parentheses and brackets are used as grouping symbols. Start with the innermost set of grouping symbols.

$$3[4(5 + 2) - 10] = 3[4(7) - 10] \qquad \text{Evaluate the expression in parentheses.}$$

$$= 3[28 - 10] \qquad \text{Multiply.}$$

$$= 3[18] \qquad \text{Subtract inside the brackets.}$$

$$= 54 \qquad \text{Multiply.} \quad \blacksquare$$

EXAMPLE 5   Simplify $\dfrac{8 + 2 \cdot 3}{2^2 - 1}$.

Solution: $\dfrac{8 + \boxed{2 \cdot 3}}{\boxed{2^2} - 1} = \dfrac{8 + \boxed{6}}{\boxed{4} - 1} = \dfrac{14}{3}$ ■

CALCULATOR BOX

Exponents

To evaluate exponential expressions on a calculator, find the key marked $\boxed{y^x}$.   To evaluate, for example, $\boxed{3^5}$, press the following keys: $\boxed{3}$ $\boxed{y^x}$ $\boxed{5}$ $\boxed{=}$. The display should read $\boxed{243}$.

Order of Operations

Some calculators follow the order of operations, and others do not. To see whether or not your calculator has the order of operations built in, use your calculator to find $2 + 3 \cdot 4$. To do this, press the following sequence of keys:

$\boxed{2}$ $\boxed{+}$ $\boxed{3}$ $\boxed{\times}$ $\boxed{4}$ $\boxed{=}$.

The correct answer is 14 because the order of operations tells us to multiply before we add. If the calculator displays $\boxed{14}$, then it has order of operations built in.

Even if the order of operations is built in, parentheses must sometimes be inserted. For example, to simplify $\dfrac{5}{12 - 7}$, press the keys

$\boxed{5}$ $\boxed{\div}$ $\boxed{(}$ $\boxed{1}$ $\boxed{2}$ $\boxed{-}$ $\boxed{7}$ $\boxed{)}$ $\boxed{=}$.

The display should read $\boxed{1}$.

Use a calculator to simplify each expression.

1. 5^3	**2.** 7^4
3. 9^5	**4.** 8^6
5. $2(20 - 5)$	**6.** $3(14 - 7) + 21$
7. $24(862 - 455) + 89$	**8.** $99 + (401 + 962)$
9. $\dfrac{4623 + 129}{36 - 34}$	**10.** $\dfrac{956 - 452}{89 - 86}$

EXERCISE SET 1.4

Evaluate. See Example 1.

1. 3^5	**2.** 5^3	**3.** 3^3	**4.** 4^4
5. 1^5	**6.** 1^8	**7.** 5^1	**8.** 8^1
9. $\left(\dfrac{1}{5}\right)^3$	**10.** $\left(\dfrac{6}{11}\right)^2$	**11.** $\left(\dfrac{2}{3}\right)^4$	**12.** $\left(\dfrac{1}{2}\right)^5$

Simplify each expression. See Example 2.

13. $5 + 6 \cdot 2$

14. $8 + 5 \cdot 3$

15. $4 \cdot 8 - 6 \cdot 2$

16. $12 \cdot 5 - 3 \cdot 6$

17. $2(8 - 3)$

18. $5(6 - 2)$

19. $2 + (5 - 2) + 4^2$

20. $5 \cdot 3^2$

21. $2 \cdot 5^2$

22. $\dfrac{1}{4} \cdot \dfrac{2}{3} - \dfrac{1}{6}$

23. $\dfrac{3}{4} \cdot \dfrac{1}{2} + \dfrac{2}{3}$

Evaluate each expression. See Examples 3 through 5.

24. $\dfrac{19 - 3 \cdot 5}{6 - 4}$

25. $\dfrac{|6 - 2| + 3}{8 + 2 \cdot 5}$

26. $\dfrac{15 - |3 - 1|}{12 - 3 \cdot 2}$

27. $\dfrac{3(2 + 5)}{6 + 2}$

28. $\dfrac{4(8 - 3)}{6 + 3}$

29. $\dfrac{4 \cdot 3 + 2}{4 + 3 \cdot 2}$

30. $5[3(2 + 1) + 4]$

31. $4[5(2 + 4) - 8]$

32. $6[5(2 + 6) - 9]$

Evaluate.

33. 7^2

34. 9^2

35. 2^5

36. 2^6

37. 0^3

38. 0^2

39. 4^2

40. 2^4

41. $(1.2)^2$

42. $(0.07)^2$

Insert $<$, $>$, or $=$ in the appropriate space to make the following true.

43. $8^1 \qquad 8(1)$

44. $9^1 \qquad 9(1)$

45. $2^3 \qquad 2 \cdot 3$

46. $2^4 \qquad 2 \cdot 4$

47. $\left(\dfrac{2}{3}\right)^3 \qquad \dfrac{1}{9} + \dfrac{1}{27}$

48. $\left(\dfrac{3}{10}\right)^2 \qquad \dfrac{1}{10} \cdot \dfrac{2}{10}$

49. $(0.6)^2 \qquad 36$

50. $(0.3)^3 \qquad 27$

Simplify each expression.

51. $6 + 3(7 - 2)$

52. $48 - 3(6 + 5)$

53. $3 + (7 - 3) + 3^2$

54. $\dfrac{6 - 4}{9 - 2}$

55. $\dfrac{8 - 5}{24 - 20}$

56. $\dfrac{7 + 4 \cdot 2}{4 - 1}$

57. $2[5 + 2(8 - 3)]$

58. $3[4 + 3(6 - 4)]$

59. $(5 \cdot 3)^2$

60. $(2 \cdot 5)^2$

61. $8 + 24 \div 3 \cdot 2 - 6$

62. $\dfrac{6 + 2 \cdot 3}{(6 + 2) \cdot 3}$

63. $\dfrac{3 + 2 \cdot 3^2}{3 + 2^2}$

64. $\dfrac{4 + 3 \cdot 2^2}{4 + 3^2}$

65. $9 + 16 \div 8 \cdot 2 - 4$

66. $\dfrac{1}{4} \cdot \dfrac{4}{3} - \dfrac{1}{6}$

67. $\dfrac{3}{4} \cdot \dfrac{1}{2} + \dfrac{1}{3}$

68. $\dfrac{3 + 3(5 + 3)}{3^2 + 1}$

69. $\dfrac{3 + 6(8 - 5)}{4^2 + 2}$

70. $6[2(3 + 1) + 2]$

71. $6(12 \div 4 + 2) - 3(10 \div 2 + 3)$

72. $5(18 \div 6 + 3) - 2(9 \div 3 + 6)$

73. $\dfrac{6 + |8 - 2| + 3^2}{18 - 3}$

74. $\dfrac{16 + |13 - 5| + 4^2}{17 - 5}$

75. $\dfrac{3 \cdot 6 + 3^2}{12 \cdot 5 - 5}$

76. $\dfrac{5.1 + 2^2}{9 - 2}$

77. $2.6[(1.3(4 + 7) - 8]$

78. $5.2[1.8(2 + 7) - 6]$

79. $\dfrac{6 + 2(1.3)^2}{0.55 + 3(0.65)}$

80. $\dfrac{4 + 2(2.3)^2}{1.8 + 1.6(0.75)}$

Writing in Mathematics

81. Why are the rules for the order of operations needed?

82. Are parentheses necessary in the expression $2 + (3 \cdot 5)$? Explain your answer.

83. Are parentheses necessary in the expression $(2 + 3) \cdot 5$? Explain your answer.

1.5
Introduction to Variables

OBJECTIVES

Tape 2

1 Simplify algebraic expressions, given replacement values for variables.

2 Translate word phrases into algebraic expressions.

3 Translate word statements into equations.

---

1 In earlier sections we used letters to represent numbers. A letter that is used to represent a number is called a **variable.** An **algebraic expression** is a collection of numbers, variables, operation symbols, and grouping symbols. For example,

$$2x, \quad -3, \quad 2x - 10 \quad 5(p^2 + 1), \quad \text{and} \quad \frac{3y^2 - 6y + 1}{5}$$

are algebraic expressions. The expression $2x$ means $2 \cdot x$. Also $5(p^2 + 1)$ means $5 \cdot (p^2 + 1)$ and $3y^2$ means $3 \cdot y^2$. If we give a specific value to a variable, we can **evaluate an algebraic expression.** To evaluate an algebraic expression means to find its numerical value. Make sure the order of operations is followed when evaluating an expression.

EXAMPLE 1   Find the value of each expression if $x = 3$ and $y = 2$.

a. $2x - y$ **b.** $\dfrac{3x}{2y}$ **c.** $\dfrac{x}{y} + \dfrac{y}{2}$ **d.** $x^2 - y^2$

Solution: **a.** Replace x with 3 and y with 2.

$$2x - y = \boxed{2(3)} - 2 \qquad \text{Let } x = 3 \text{ and } y = 2.$$
$$= \boxed{6} - 2 \qquad \text{Multiply.}$$
$$= 4 \qquad\qquad \text{Subtract.}$$

b. $\dfrac{3x}{2y} = \dfrac{3 \cdot 3}{2 \cdot 2} = \dfrac{9}{4} \qquad$ Let $x = 3$ and $y = 2$.

c. Replace x with 3 and y with 2. Then simplify.

$$\frac{x}{y} + \frac{y}{2} = \frac{3}{2} + \frac{2}{2} = \frac{5}{2}$$

d. Replace x with 3 and y with 2.

$$x^2 - y^2 = 3^2 - 2^2 = 9 - 4 = 5 \qquad \blacksquare$$

2 Now that we can represent an unknown number by a variable, let's practice translating phrases containing unknown numbers into algebraic expressions.

EXAMPLE 2 Write an algebraic expression that represents each of the following phrases. Let x represent the unknown number.
   a. The sum of a number and 3
   b. The product of a 3 and a number
   c. Twice a number
   d. 10 decreased by a number
   e. 7 more than 5 times a number

Solution: **a.** $x + 3$ since "sum" means to add.
   b. $3 \cdot x$ or $3x$ are both ways to denote the product of x and 3.
   c. $2 \cdot x$ or $2x$
   d. $10 - x$ because "decreased by" means to subtract.
   e. $\underbrace{5x}_{\substack{\text{5 times} \\ \text{a number}}} + 7$ ∎

3 An **equation** is a mathematical statement that two expressions are equal. For example, $3 + 2 = 5$, $7x = 35$, $\dfrac{2(x - 1)}{3} = 0$, and $I = PRT$ are all equations. We now practice translating sentences into equations.

EXAMPLE 3 Write each sentence as an equation. Let x represent the unknown number.
   a. The quotient of 15 and a number is 4.
   b. Three subtracted from 12 is a number.
   c. Four times a number added to 17 is 21.

Solution: **a.**

The quotient of 15 and a number	is	4
$\dfrac{15}{x}$	$=$	4

   b.

Three subtracted **from** 12	is	a number
$12 - 3$	$=$	x

Care must be taken when the operation is subtraction. The expression $3 - 12$ would be incorrect. Notice that $3 - 12 \neq 12 - 3$.

   c.

4 times a number	added to	17	is	21
$4x$	$+$	17	$=$	21

∎

EXERCISE SET 1.5

Write each of the following phrases as algebraic expressions. Let x represent the unknown number. See Example 1.

1. Fifteen more than a number.

2. One-half times a number.

3. Five subtracted from a number.

4. The quotient of a number and 9.

5. Three times a number increased by 22.

6. The product of 8 and a number.

Evaluate each expression if x = 1, y = 3, and z = 5. See Example 2.

7. $3x - 2$

8. $6y - 8$

9. $|2x + 3y|$

10. $5z - 2y$

11. $xy + z$

12. $yz - x$

13. $x^2 + y^2$

14. $|y^2 + z^2|$

15. $|3z - 10y|$

Write each of the following sentences as equations. Use x to represent the unknown number. See Example 3.

16. The sum of 5 and a number is 20.

17. Twice a number is 17.

18. Thirteen minus three times a number is 13.

19. Seven subtracted from a number is 0.

20. The quotient of 12 and a number is $\frac{1}{2}$.

21. The sum of 8 and twice a number is 42.

Evaluate each expression if x = 2, y = 6, and z = 3.

22. $3x + 8y$

23. $|4z - 2y|$

24. $\dfrac{4x}{3y}$

25. $\dfrac{6z}{5x}$

26. $\dfrac{y}{x} + \dfrac{y}{x}$

27. $\dfrac{9}{z} + \dfrac{4z}{y}$

28. $3x^2 + y$

29. $y + 4z^2$

30. $x + yz$

31. $z + xy$

Find the value of each expression if x = 12, y = 8, and z = 4.

32. $\dfrac{x}{z} + 3y$

33. $\dfrac{y}{z} + 8x$

34. $3z + z^2$

35. $7y + y^2$

36. $x^2 - 3y + x$

37. $y^2 - 3x + y$

38. $\dfrac{2x + z}{3y - z}$

39. $\dfrac{3x - y}{4z + x}$

40. $\dfrac{x^2 + z}{y^2 + 2z}$

41. $\dfrac{y^2 + x}{x^2 + 3y}$

Write each of the following as algebraic expressions or equations.

42. A number divided by 13.

43. Seven times the sum of a number and 19.

44. Ten subtracted from twice a number is 18.

45. A number subtracted from four times that number is 75.6.

46. Twenty less the product of 30 and a number.

47. Twelve decreased by a number.

48. A number times 0.02 equals 1.76.

49. The product of $\frac{3}{4}$ and the sum of a number and 1 equals 9.

50. A number subtracted from 19 is three times that number.

51. Three times a number added to $1\frac{11}{12}$ equals the number increased by 2.

Solve the following.

52. The perimeter of a figure is the distance around the figure. The equation $P = 2l + 2w$ can be used to find the perimeter, P, of a rectangle, where l is its length and w is its width. Find the perimeter of the following rectangle.

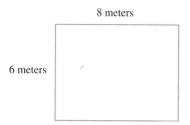

8 meters

6 meters

53. The equation $P = a + b + c$ can be used to find the perimeter, P, of a triangle, where a, b, and c, are the lengths of its sides. Find the perimeter of the following triangle.

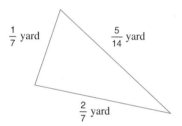

$\frac{1}{7}$ yard

$\frac{5}{14}$ yard

$\frac{2}{7}$ yard

54. The **area** of a figure is the total enclosed surface of the figure. Area is measured in square units. The equation $A = lw$ is used to find the area of a rectangle where l is its length and w is its width. Find the area of the following rectangular-shaped lot.

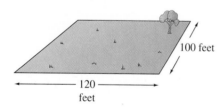

100 feet

120 feet

55. A trapezoid is a four-sized figure with exactly one pair of parallel sides. The equation $A = \dfrac{(B + b)h}{2}$ can be used to find its area, where B and b are the lengths of the two parallel sides and h is the height between these sides. Find the area if $B = 15$ inches, $b = 7$ inches, and $h = 5$ inches.

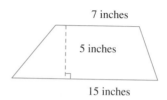

7 inches

5 inches

15 inches

56. The equation $R = \dfrac{I}{PT}$ can be used to find the rate of interest being charged if a loan of P dollars for T years required I dollars in interest to be paid. Find the interest rate if a \$650 loan for 3 years to buy a used IBM personal computer requires \$126.75 in interest to be paid.

57. The equation $r = \dfrac{d}{t}$ is used to find the average speed r in miles per hour if a distance of d miles is traveled in t hours. Find the rate to the nearest whole number if the distance between Dallas, Texas, and Kaw City, Oklahoma, is 432 miles, and it takes Barbara Goss 8.5 hours to drive the distance.

58. Peter Callac earns a base salary plus a commission on all sales he makes at St. Joe Brick Company. The equation $I = B + RS$ calculates his gross income, where B is the base income, R is the commission rate, and S is the amount sold. Find Peter's income if $B = \$300$, $R = 8\%$, $S = \$500$.

1.6
Adding Real Numbers

Tape 3

OBJECTIVES

1 Add real numbers with the same sign.

2 Add real numbers with unlike signs.

3 Find the opposite of a number.

1 Real numbers can be added, subtracted, multiplied, divided, and raised to powers, just as whole numbers can. We will use the number line to help picture the addition of real numbers.

On a number line, a positive number can be represented anywhere by an arrow of appropriate length pointing to the right and a negative number by an arrow pointing to the left. For example, to add $3 + 2$ on a number line, start at 0 on the number line and draw an arrow representing 3. This arrow will be three units long pointing to the right, since 3 is positive. From the tip of this arrow, draw another arrow representing 2. The tip of the second arrow ends at their sum, 5.

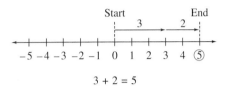

$$3 + 2 = 5$$

To add $-1 + (-2)$, start at 0 and draw an arrow representing -1. This arrow is one unit long pointing to the left. At the tip of this arrow, draw an arrow representing -2. The tip of the second arrow ends at their sum, -3.

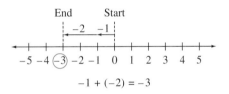

$$-1 + (-2) = -3$$

Using a number line each time we add two numbers can be time consuming. Instead, we can notice patterns in the previous examples and write rules for adding signed numbers. When adding two numbers with the same sign, notice that the sign of the sum is the same as the sign of the addends.

To Add Two Numbers with the Same Sign

Step 1 Find the sum of their absolute values.
Step 2 Use their common sign as the sign of the sum.

Add $(-7) + (-6)$. First, find the sum of their absolute values.

$$|-7| = 7, \quad |-6| = 6, \quad \text{and} \quad 7 + 6 = 13$$

Next, use their common sign as the sign of the sum. This means that

$$(-7) + (-6) = -13$$
$$\text{└── common sign}$$

Thinking of signed numbers as money earned or lost might help you understand this rule. If \$1 is earned and later another \$3 is earned, the total amount earned is \$4. Earnings can be thought of as positive numbers: $1 + 3 = 4$.

On the other hand, if \$1 is lost and later another \$3 is lost, a total of \$4 is lost. Losses can be thought of as negative numbers: $(-1) + (-3) = -4$.

EXAMPLE 1   Find each sum.

a. $-3 + (-7)$ b. $5 + (+12)$ c. $(-1) + (-20)$ d. $-2 + (-10)$

Solution: a. $-3 + (-7) = -10$ b. $5 + (+12) = 17$ c. $(-1) + (-20) = -21$
d. $-2 + (-10) = -12$ ■

2   Adding numbers whose signs are not the same can be pictured on the number line, also. To find the sum of $-4 + 6$, begin at 0 and draw an arrow representing -4. This arrow is 4 units long and pointing to the left. At the tip of this arrow, draw an arrow representing 6. The tip of the second arrow ends at their sum, 2.

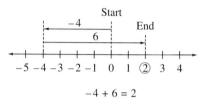

$$-4 + 6 = 2$$

Using temperature as an example, if the thermometer registers 4 degrees below zero and then rises 6 degrees, the new temperature is 2 degrees above zero. Thus, it is reasonable that $-4 + 6 = 2$.

Find the sum of $3 + (-4)$ on the number line. From the number line, the answer is -1.

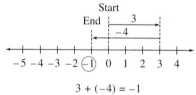

$$3 + (-4) = -1$$

To Add Two Numbers with Unlike Signs

Step 1   Find the difference of the larger absolute value and the smaller absolute value.

Step 2   Use the sign of the addend whose absolute value is larger as the sign of the answer.

To add $4 + (-9)$, two numbers with unlike signs, first recall that $|4| = 4$ and $|-9| = 9$. Their difference is $9 - 4 = 5$. The sign of the sum $4 + (-9)$ will be negative since -9 has the larger absolute value. Therefore, $4 + (-9) = -5$.

EXAMPLE 2   Find each sum.

a. $(+3) + (-7)$ b. $(-2) + (10)$ c. $2 + (-5)$

Solution: a. Since $|-7|$ is larger than $|3|$, the sum will have the same sign as -7. That is, the sum is negative. Then $|-7| = 7$, $|3| = 3$, and $7 - 3 = 4$. The answer is -4.

$$(+3) + (-7) = -4$$

b. Since $|10|$ is larger than $|-2|$, the sign of the sum is the same as the sign of 10, positive. Then $|10| = 10$, $|-2| = 2$, and $10 - 2 = 8$. The answer is $+8$ or 8.

$$10 + (-2) = 8$$

c. $2 + (-5) = -3$ ■

EXAMPLE 3   Simplify the following:
   a. $3 + (-7) + (-8)$ **b.** $[7 + (-10)] + [-2 + (-4)]$

Solution: **a.** Perform the additions from left to right.

$$3 + (-7) \; + (-8) = \; -4 \; + (-8) \qquad \text{Adding unlike signs.}$$
$$= -12 \qquad \text{Adding like signs.}$$

b. Simplify inside brackets first.

$$[\; 7 + (-10) \;] + [\; -2 + (-4) \;] = [\; -3 \;] + [\; -6 \;]$$
$$= -9 \qquad \text{Add.} \qquad ■$$

Negative numbers are used in everyday life. Stock market returns show gains and losses as positive and negative numbers. Temperatures in cold climates often dip into the negative range, commonly referred to as "below zero" temperatures. Bank statements report deposits and withdrawals as positive and negative numbers.

EXAMPLE 4   The temperature at the Winter Olympics was a frigid 14 degrees below zero in the morning, but by noon it had risen 31 degrees. What was the temperature at noon?

Solution: Use -14 to represent 14 degrees below zero. Because the temperature rose, add a positive 31, representing 31 degrees, to the morning temperature of -14.

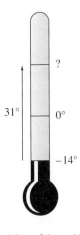

31°   ?   0° -14°

$$-14 + 31 = 17$$

The temperature at noon was 17 degrees.   ■

3 Let's look at 4 and -4 on anumber line.

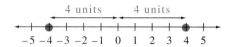

Notice that 4 and -4 lie on opposite sides of 0, and each is 4 units away from 0.

This relationship between -4 and $+4$ is an important one. Such numbers are known as **opposites,** or **additive inverses** of each other.

> ### Opposites or Additive Inverses
>
> Two numbers that are the same distance from 0 but on opposite sides of 0 are called opposites or additive inverses of each other.

The opposite of 10 is -10.
The opposite of -3 is 3.
The opposite of $\dfrac{1}{2}$ is $-\dfrac{1}{2}$.

Notice that the sum of a number and its opposite is 0.

$$10 + (-10) = 0$$

$$-3 + 3 = 0$$

$$\frac{1}{2} + \left(-\frac{1}{2}\right) = 0$$

In general, **if a is a number, the opposite or additive inverse of a is $-a$.** This means that the opposite of -3 is $-(-3)$. But we said above that the opposite of -3 is 3. This can only be true if $-(-3) = 3$.

> If a is a number, then $-(-a) = a.$

For example, $-(-10) = 10$, $-\left(-\dfrac{1}{2}\right) = \dfrac{1}{2}$, and $-(-2x) = 2x$.

EXAMPLE 5 Find the opposite or additive inverse of each number.
a. 5   **b.** 0   **c.** -6

Solution: **a.** The opposite of 5 is -5. Notice that 5 and -5 are on opposite sides of 0 when plotted on a number line and are equal distances away.
b. The opposite of 0 is 0 since $0 + 0 = 0$.
c. The opposite of -6 is 6. ∎

EXAMPLE 6 Simplify the following.
a. $-(-6)$ **b.** $-|-6|$

Solution: **a.** $-(-6) = 6$ **b.** $-\;\boxed{|-6|}\;=\;-\;6$ ∎

EXERCISE SET 1.6

Find the following sums. See Examples 1 through 3.

1. $6 + 3$

2. $9 + (-12)$

3. $-6 + (-8)$

4. $-6 + (-4)$

5. $-8 + (-7)$

6. $6 + (-4)$

7. $-52 + 36$

8. $-94 + 27$

9. $6 + (-4) + 9$

10. $-18 + |-53|$

11. $2\dfrac{3}{4} + -\dfrac{1}{8}$

12. $4\dfrac{4}{5} + -\dfrac{3}{10}$

Solve the following. See Example 4.

13. The low temperature in Lander, Wyoming, was -15 degrees last night. During the day it rose only 9 degrees. Find the high temperature for the day.

14. On January 2, 1943, the temperature was -4 degrees at 7:30 A.M. in Spearfish, South Dakota. Incredibly, it got 49 degrees warmer in the next 2 minutes. To what temperature did it rise by 7:32?

Find the additive inverse or the opposite. See Example 5.

15. 6

16. 4

17. -2

18. -8

19. 0

20. $-\dfrac{1}{4}$

21. $|-6|$

22. $|-11|$

Simplify the following. See Example 6.

23. $-|-2|$

24. $-(-3)$

25. $-|0|$

26. $|-\dfrac{2}{3}|$

27. $-|-\dfrac{2}{3}|$

28. $-(-7)$

29. $-3 + (-5)$

30. $-7 + (-4)$

31. $-9 + (-3)$

32. $+8 + (-6)$

33. $+9 + (-3)$

34. $-9 + (+4)$

35. $-7 + (+3)$

36. $-5 + (+9)$

37. $-3 + (+10)$

38. $3 + (-8)$

39. $8 + 4$

40. $12 + (+2)$

41. $-15 + 9 + (-2)$

42. $-9 + 15 + (-5)$

43. $-21 + (-16) + (-22)$

44. $-14 + |-16|$

45. $|-8| + (-16)$

46. $|-6| + (-61)$

47. $-\dfrac{7}{16} + \dfrac{1}{4}$

48. $-\dfrac{5}{9} + \dfrac{1}{3}$

49. $-\dfrac{7}{10} + -\dfrac{3}{5}$

50. $-33 + (-14)$

51. $27 + (-46)$

52. $53 + (-37)$

53. $-18 + 49$

54. $-26 + 14$

55. $126 + (-67)$

56. $-\dfrac{5}{6} + -\dfrac{2}{3}$

57. $6.3 + (-8.4)$

58. $9.2 + (-11.4)$

59. $117 + (-79)$

60. $-114 + (-88)$

61. $-214 + (-86)$

62. $18 + (-6) + 4$

63. $-23 + 16 + (-2)$

64. $-14 + (-3) + 11$

65. $-9.6 + (-3.5)$

66. $-6.7 + (-7.6)$

67. $|5 + (-10)|$

68. $[-3 + 5] + (-11)$

69. $[-2 + (-7)] + [-11 + (-4)]$

70. $8 + [-5 + 5] + (-12)$

71. $|7 + (-10)| + |-16|$

72. $\left| -\dfrac{3}{10} + \dfrac{1}{10} \right| - \left| \dfrac{2}{5} - \dfrac{1}{10} \right|$

73. $\left(-\dfrac{1}{8} + \dfrac{3}{8} \right) \cdot \dfrac{2}{7}$

Solve the following.

74. The lowest elevation on Earth is -1312 feet (that is, 1312 feet below sea level) at the Dead Sea. If we are standing 658 feet above the Dead Sea, what is our elevation?

75. The lowest point in Africa is -512 feet at Lake Assal in Djibouti. If we are standing at a point 658 feet above Lake Assal, what is our elevation?

76. In checking the stock market results, Alexis discovers our stock posted changes of $-1\frac{5}{8}$ and $-2\frac{1}{2}$ over the last two days. What is the combined change?

77. Yesterday our stock posted a change of $-1\frac{1}{4}$, but today it showed a gain of $+\frac{7}{8}$. Find the overall change for the two days.

78. In golf, scores that are under par for the entire round are shown as negative scores; positive scores are shown for scores that are over par. In two rounds in the United States Open, Arnold Palmer had overall scores of -6 and $+4$. What was his overall score?

Writing in Mathematics

79. In your own words, what is an opposite?

80. Explain why 0 is the only number that is its own opposite.

81. Explain why adding a negative number to another negative number produces a negative sum.

82. When a positive and a negative number are added, sometimes the answer is positive and sometimes it is negative. Explain why this happens.

1.7
Subtracting Real Numbers

OBJECTIVES

Tape 3

1 Subtract real numbers.

2 Add and subtract real numbers.

3 Evaluate algebraic expressions using real numbers.

1 Now that addition of signed numbers has been discussed we explore subtraction. We know that $9 - 7 = 2$. Notice that $9 + (-7) = 2$, also. This means that

$$9 - 7 = 9 + (-7)$$

In general, any subtraction problem can be written as an equivalent addition problem. To do so, we use an important relationship discussed in the last section, that of opposites.

> **Subtracting Two Real Numbers**
>
> If a and b are real numbers, then $a - b = a + (-b)$.

In other words, to find the difference of two numbers, add the first number to the opposite of the second number.

EXAMPLE 1 Find each difference.

a. $-13 - (+4)$ **b.** $5 - (-6)$ **c.** $3 - 6$ **d.** $-1 - (-7)$

Solution:

add

a. $-13 \; - \; (\; +4 \;) = -13 \; + \; (\; -4 \;)$ Add the opposite of $+4$, which is -4.

$= -17$ opposite

add

b. $5 \; - \; (\; -6 \;) = 5 \; + \; (\; 6 \;)$ Add the opposite of -6, which is 6.

$= 11$ opposite

c. $3 - 6 = 3 + (-6)$ Add 3 to the opposite of 6, which is -6.

$= -3$

d. $-1 - (-7) = -1 + (7) = 6$ ■

EXAMPLE 2 Subtract 8 from -4.

Solution: Be careful when interpreting. The order of numbers in subtraction is important. 8 is to be subtracted **from** -4.

$$-4 - 8 = -4 + (-8) = -12 \quad ■$$

2 Expressions containing both sums and differences make good use of grouping symbols. In simplifying these expressions, remember to rewrite differences as sums and follow the standard order of operations.

EXAMPLE 3 Simplify each expression.

a. $-3 + [(-2 - 5) - 2]$ **b.** $2^3 - |10| + [-6 - (-5)]$

Solution: **a.** Start with the innermost sets of parentheses. Rewrite $-2 - 5$ as a sum.

$-3 + [(\; -2 - 5 \;) - 2] = -3 + [(\; -2 + (-5) \;) - 2]$

$= -3 + [(\; -7 \;) - 2]$ Add $-2 + (-5)$.

$= -3 + [-7 + (-2)]$ Write $-7 - 2$ as a sum.

$= -3 + [-9]$ Add.

$= -12$ Add.

b. Start simplifying the expression inside the brackets by writing $-6 - (-5)$ as a sum.

$2^3 - |10| + [\; -6 - (-5) \;] = 2^3 - |10| + [\; -6 + 5 \;]$

$= 2^3 - 10 + [\; -1 \;]$ Add. Write $|10|$ as 10.

$= 8 - 10 + (-1)$ Evaluate 2^3.

$= \; 8 + (-10) \; + (-1)$ Write $8 - 10$ as a sum.

$= \; -2 \; + (-1)$ Add.

$= -3$ Add. ■

3

EXAMPLE 4   If $x = 2$ and $y = -5$, find the value of the following expressions.

a. $\dfrac{x - y}{12 + x}$ b. $x^2 - y$

Solution: **a.** Replace x with 2 and y with -5. Be sure to put parentheses around -5 to separate signs. Then simplify the resulting expression.

$$\frac{x - y}{12 + x} = \frac{2 - (-5)}{12 + 2} = \frac{2 + 5}{14} = \frac{7}{14} = \frac{1}{2}$$

b. Replace the x with 2 and y with -5 and simplify.

$$x^2 - y = 2^2 - (-5) = 4 - (-5) = 4 + 5 = 9 \quad \blacksquare$$

EXAMPLE 5   The lowest point in North America is in Death Valley, at an elevation of 282 feet below sea level. Nearby, Mount Whitney reaches 14,494 feet, the highest point in the United States outside of Alaska. How much of a variation in elevation is there between these two extremes?

Solution: To find the variation in elevation between the two heights, find the difference of the high point and the low point.

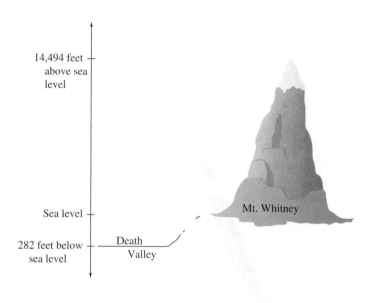

high point	minus	low point	
14,494	$-$	(-282)	$= 14,494 + 282$
			$= 14,776$ feet

Thus, the variation in elevation is 14,776 feet. $\blacksquare$

EXAMPLE 6   At 6:00 P.M., the temperature at the Winter Olympics was 14 degrees; by morning, the temperature dropped to -23 degrees. Find the overall change in temperature.

Solution: To find the overall change, find the difference of the morning temperature and the 6:00 P.M. temperature.

morning temp.	minus	6:00 P.M. temp.

$$-23° \quad - \quad 14° \quad = -23° + (-14°) = -37°$$

Thus, the overall change is $-37°$. That is, the temperature dropped by 37 degrees.   ∎

EXERCISE SET 1.7

Find each difference. See Example 1.

1. $-6 - 4$

2. $-12 - 8$

3. $4 - 9$

4. $8 - 11$

5. $16 - (-3)$

6. $12 - (-5)$

7. $18 - 33$

8. $14 - 27$

9. $-16 - 18$

Find the following. See Example 2.

10. Subtract -5 from 8.

11. Subtract 3 from -2.

12. Subtract -1 from -6.

Simplify each expression. (Remember the order of operations.) See Example 3.

13. $-6 - (2 - 11)$

14. $-9 - (3 - 8)$

15. $3^3 - 8 \cdot 9$

16. $2^3 - 6 \cdot 3$

17. $2 - 3(8 - 6)$

18. $4 - 6(7 - 3)$

19. $|-3| + 2^2 + [-4 - (-6)]$

20. $|-2| + 6^2 + (-3 - 8)$

If $x = -6$, $y = -3$, and $t = 2$, evaluate each expression. See Example 4.

21. $x - y$

22. $3t - x$

23. $x + t - 12$

24. $y - 3t - 8$

25. $\dfrac{x - (-12)}{y + 6}$

26. $\dfrac{x - (-9)}{t + 1}$

27. $x - t^2$

28. $y - t^3$

29. $\dfrac{4t - y}{3t}$

30. $\dfrac{5t - x}{5t}$

Solve the following. See Examples 5 and 6.

31. Within 24 hours in 1916, the temperature in Browning, Montana, fell from 44 degrees to -56 degrees. How large a drop in temperature was this?

32. Much of New Orleans is just barely above sea level. If George descends 12 feet from an elevation of 5 feet above sea level, what will his new elevation be?

33. In a series of plays, a football team gains 2 yards, loses 5 yards, and then loses another 20 yards. What is the total gain or loss of yardage?

34. In some card games, it is possible to have a negative score. Lavonne currently has a score of 15 points. She then loses 24 points. What is her new score?

Simplify each expression.

35. $-6 - 5$

36. $-8 - 4$

37. $7 - (-4)$

38. $3 - (-6)$

39. $-6 - (-11)$

40. $-4 - (-16)$

41. $6\dfrac{2}{5} - \dfrac{7}{10}$

42. $-\dfrac{3}{5} - \dfrac{5}{6}$

43. $-\dfrac{3}{4} - \dfrac{1}{9}$

44. $-6.1 - (-5.3)$

45. $-2.6 - (-6.7)$

46. $6 - 11 - (-8)$

47. $4 - (-6) - 9$

48. $-6 - 14 - (-3)$

49. $-11 - (-8) - 4$

50. $-13 - 27$

51. $16 - (-21)$

52. $15 - (-33)$

53. $-44 - (-27)$

54. $-36 - (-51)$

55. $4\dfrac{2}{3} - \dfrac{5}{12}$

56. $-\dfrac{1}{6} - \left(-\dfrac{3}{4}\right)$

57. $-\dfrac{1}{10} - \left(-\dfrac{7}{8}\right)$

58. $8.3 - 11.2$

59. $9.7 - 16.1$

60. $8.3 - (-0.62)$

61. $4.3 - (-0.87)$

62. $3 - (6[3 - (-2)] - 8)$

63. $4 - (3[9 - (-8)] - 11)$

64. $3 - 6 \cdot 4^2$

65. $2 - 3 \cdot 5^2$

66. $(3 - 6) \cdot 4^2$

67. $(2 - 3) \cdot 5^2$

68. $3 - (6 \cdot 4)^2$

69. $2 - (3 \cdot 5)^2$

70. Subtract 8 from 7.

71. Subtract 9 from -4.

72. Decrease -8 by 15.

73. Decrease 11 by -14.

74. $-2 + [(8 - 11) - (-2 - 9)]$

75. $-5 + [(4 - 15) - (-6) - 8]$

If $x = -5$, $y = 4$, and $t = 10$, simplify each expression.

76. $x - y$

77. $y - x$

78. $|x| + 2t - 8y$

79. $|x + t - 7y|$

80. $\dfrac{9 - x}{y + 6}$

81. $\dfrac{15 - x}{y + 2}$

82. $y^2 - x$

83. $t^2 - x$

84. $\dfrac{|x - (-10)|}{2t}$

85. $\dfrac{|5y - x|}{6t}$

Solve the following.

86. Aristotle died in the year -322 (or 322 B.C.). When was he born, if he was 62 years old when he died?

87. Augustus Caesar died in 14 A.D. in his 77th year. When was he born?

88. Your stock posted a loss of $1\dfrac{5}{8}$ yesterday. If it drops another $\dfrac{3}{4}$ points today, find its overall change for the two days.

89. A jet liner hits an air pocket and drops 250 feet. After climbing 120 feet, it drops another 178 feet. What is its overall vertical change?

Writing in Mathematics

90. In your own words, explain why $5 - 8$ is negative.

91. Explain why $6 - 11$ is the same as $6 + (-11)$.

1.8
Multiplying and Dividing Real Numbers

Tape 4

OBJECTIVES		
	1	Multiply and divide real numbers.
	2	Find the value of algebraic expressions.

1   In this section, we discover rules for multiplying and dividing signed numbers. The sign rules for multiplying and dividing these numbers will be the same since a quotient can be rewritten as a product.

For example, $\dfrac{10}{2}$ can be rewritten as $10 \cdot \dfrac{1}{2}$. In general, we have the following:

> If a and b are real numbers and $b \neq 0$, then
>
> $$\frac{a}{b} = a \cdot \frac{1}{b}$$

To discover sign rules for multiplication (as well as division), recall that multiplication is repeated addition. Thus $3 \cdot 2$ means 2 is used as an addend 3 times. That is,

$$2 + 2 + 2 = 3 \cdot 2$$

which equals 6. Let's use this fact to help us find a rule for multiplying a positive number and a negative number. To do this, apply the same reasoning to using -2 as an addend 3 times. That is,

$$(-2) + (-2) + (-2) = 3 \cdot (-2)$$

Since $(-2) + (-2) + (-2) = -6$, then $3 \cdot (-2) = -6$. This suggests that the product or quotient of a positive number and a negative number is a negative number.

What about the product of two negative numbers? To find out, consider the following pattern.

$-3 \cdot 2 = -6$ Factor decreases by 1 each time

$-3 \cdot 1 = -3$ Product increases by 3 each time.

$-3 \cdot 0 = 0$

This pattern continues as

$-3 \cdot -1 = 3$

$-3 \cdot -2 = 6$

This suggests that the product or quotient of two negative numbers gives a positive number.

The rules for multiplying and dividing signed numbers are summarized in the following box.

> **Multiplying and Dividing Real Numbers**
>
> **1.** The product or quotient of two numbers having the same sign is a positive number.
> **2.** The product or quotient of two numbers having unlike signs is a negative number.

EXAMPLE 1   Find the product.
  a. $(-6)(4)$ **b.** $2(-1)$ **c.** $(-5)(-10)$

Solution: **a.** $(-6)(4) = -24$ **b.** $2(-1) = -2$ **c.** $(-5)(-10) = 50$ ■

We know that every whole number multiplied by zero, equals zero. This remains true for signed numbers.

Multiplying by Zero

If b is a real number, then $b \cdot 0 = 0$. Also, $0 \cdot b = 0$.

EXAMPLE 2   Find each product.
  a. $(7)(0)(-6)$ **b.** $(-2)(-3)(-4)$ **c.** $(-1)(5)(-9)$ **d.** $(-2)^3$
  e. $(-4)(-11) - (5)(-2)$

Solution: **a.** By the order of operations, we multiply from left to right. Notice that, because one of the factors is 0, the product is 0.

$$(7)(0)(-6) = 0(-6) = 0$$

  b. Multiply two factors at a time, from left to right.

$$(-2)(-3)(-4) = (6)(-4) \qquad \text{Multiply } (-2)(-3).$$
$$= -24$$

  c. Multiply from left to right.

$$(-1)(5)(-9) = (-5)(-9) \qquad \text{Multiply } (-1)(5).$$
$$= 45$$

  d. $(-2)^3 = (-2)(-2)(-2)$
$$= -8 \qquad\qquad \text{Multiply.}$$

  e. Follow the rules for order of operation.

$$(-4)(-11) - (5)(-2) = 44 - (-10) \qquad \text{Find the products.}$$
$$= 44 + 10 \qquad\quad \text{Add 44 to the opposite of } -10.$$
$$= 54 \qquad\qquad \text{Add.} \quad ■$$

Multiplying signed decimals or fractions is carried out exactly the same way as multiplying by integers.

EXAMPLE 3   Find each product.
  a. $(-1.2)(0.05)$ **b.** $\dfrac{2}{3} \cdot -\dfrac{7}{10}$

Solution: **a.** The product of two numbers with opposite signs is negative.

$$(-1.2)(0.05) = -[(1.2)(0.05)]$$
$$= -0.06$$

b. $\dfrac{2}{3} \cdot -\dfrac{7}{10} = -\dfrac{2 \cdot 7}{3 \cdot 10} = -\dfrac{2 \cdot 7}{3 \cdot \boxed{2} \cdot 5} = -\dfrac{7}{15}$ ■

EXAMPLE 4   Find each quotient.
a. $-18 \div 3$ **b.** $-14 \div -2$ **c.** $20 \div -4$

Solution: **a.** $\dfrac{-18}{3} = -6$ **b.** $\dfrac{-14}{-2} = 7$ **c.** $\dfrac{20}{-4} = -5$ ■

The definition of division does not allow for division by 0. How do we interpret $\dfrac{3}{0}$? If $\dfrac{3}{0} = a$, then $a \cdot 0$ must equal 3. But recall that, for any real number a, $a \cdot 0 = 0$ and can never equal 3. Thus, we say that division by 0 is not allowed or not defined since $\dfrac{3}{0}$ does not represent a real number. The denominator of a fraction can never be 0.

Quotients With Zero

1. Division of any nonzero real number by 0 is undefined. In symbols, if $a \neq 0$, $\dfrac{a}{0}$ is **undefined.**

2. Division of 0 by any real number except 0 is 0. In symbols, if $a \neq 0$, $\dfrac{0}{a} = 0$.

EXAMPLE 5   Simplify the following if possible.
a. $\dfrac{1}{0}$ **b.** $\dfrac{0}{-3}$ **c.** $\dfrac{0(-8)}{2}$

Solution: **a.** $\dfrac{1}{0}$ is undefined   **b.** $\dfrac{0}{-3} = 0$ **c.** $\dfrac{0(-8)}{2} = \dfrac{0}{2} = 0$ ■

Notice that $\dfrac{12}{-2} = -6$, $-\dfrac{12}{2} = -6$, and $\dfrac{-12}{2} = -6$. This means that $\dfrac{12}{-2} = -\dfrac{12}{2} = \dfrac{-12}{2}$.

In other words, a single negative sign in a fraction can be written in the denominator, in the numerator, or in front of the fraction without changing the value of the fraction. Thus,

$$\dfrac{1}{-7} = \dfrac{-1}{7} = -\dfrac{1}{7}$$

In general, if a and b are real numbers, $b \neq 0$, $\dfrac{a}{-b} = \dfrac{-a}{b} = -\dfrac{a}{b}$.

Examples combining the four basic arithmetic procedures along with the principles of order of operations will help us to review these concepts.

EXAMPLE 6 Simplify each expression.

a. $\dfrac{(-12)(-3) + 4}{-7 - (-2)}$

b. $\dfrac{2(-3)^2 - 20}{-5 + 4}$

Solution: a. First, simplify the numerator and denominator separately.

$$\frac{(-12)(-3) + 4}{-7 - (-2)} = \frac{36 + 4}{-7 + 2}$$

$$= \frac{40}{-5}$$

$$= -8 \qquad \text{Divide.}$$

b. Simplify the numerator and denominator separately; then divide.

$$\frac{2(-3)^2 - 20}{-5 + 4} = \frac{2 \cdot 9 - 20}{-5 + 4} = \frac{18 - 20}{-5 + 4} = \frac{-2}{-1} = 2 \quad \blacksquare$$

2

EXAMPLE 7 If $x = -2$ and $y = -4$, find the value of each expression.

a. $5x - y$

b. $x^3 - y^2$

c. $\dfrac{3x}{2y}$

Solution: a. Replace x with -2 and y with -4 and simplify.

$$5x - y = 5(-2) - (-4) = -10 - (-4) = -10 + 4 = -6$$

b. Replace x with -2 and y with -4.

$$x^3 - y^2 = (-2)^3 - (-4)^2 \qquad \text{Substitute the given values for the variables.}$$

$$= -8 - (16) \qquad \text{Evaluate exponential expressions.}$$

$$= -8 + (-16) \qquad \text{Write as an equivalent addition problem.}$$

$$= -24 \qquad \text{Add.}$$

c. Replace x with -2 and y with -4 and simplify.

$$\frac{3x}{2y} = \frac{3(-2)}{2(-4)} = \frac{-6}{-8} = \frac{3}{4} \quad \blacksquare$$

CALCULATOR BOX

Entering Negative Numbers

To enter a negative number on a calculator, find a key marked $\boxed{+/-}$.(On some calculators, this key is marked $\boxed{\text{CHS}}$ for "change sign.") To enter -8, for example, press the keys $\boxed{8}$ $\boxed{+/-}$. The display will read $\boxed{-8}$.

Operations with Real Numbers

To evaluate $-2(7 - 9) - 20$ on a calculator, press the keys

$\boxed{2}$ $\boxed{+/-}$ $\boxed{\times}$ $\boxed{(}$ $\boxed{7}$ $\boxed{-}$ $\boxed{9}$ $\boxed{)}$ $\boxed{-}$ $\boxed{2}$ $\boxed{0}$ $\boxed{=}$.

The display will read $\boxed{-16}$.

Use a calculator to simplify each expression.

1. $-8(26 - 27)$

2. $-9(-8) + 17$

3. $134 + 25(68 - 91)$

4. $45(32) - 8(218)$

5. $\dfrac{3 - 13}{17 - 22}$

6. $\dfrac{-14 + 36}{77 - 88}$

7. $9^3 - 455$

8. $5^4 - 625$

9. $(-3)^2$ Be careful.

10. -3^2 Be careful.

EXERCISE SET 1.8

Find the following products. See Examples 1 through 3.

1. $(-3)(+4)$

2. $(+8)(-2)$

3. $-6(-7)$

4. $(-3)(-8)$

5. $(-2)(-5)(0)$

6. $(7)(0)(-3)$

7. $2(-9)$

8. $(-5)(3)$

9. $\left(-\dfrac{3}{4}\right)\left(\dfrac{8}{9}\right)$

10. $\left(\dfrac{5}{6}\right)\left(-\dfrac{3}{10}\right)$

11. $\left(-1\dfrac{1}{5}\right)\left(-1\dfrac{2}{3}\right)$

12. $\left(-\dfrac{5}{6}\right)\left(-\dfrac{3}{10}\right)$

13. $(-1)(2)(-3)(-5)$

14. $(-2)(-3)(-4)(-2)$

15. $(2)(-1)(-3)(5)(3)$

16. $(3)(-5)(-2)(-1)(-2)$

17. $(-4)^2$

18. $(-3)^3$

Find the following quotients. See Examples 4 and 5.

19. $\dfrac{18}{-2}$

20. $-\dfrac{14}{7}$

21. $\dfrac{-12}{-4}$

22. $-\dfrac{20}{5}$

23. $\dfrac{-45}{-9}$

24. $\dfrac{30}{-2}$

25. $\dfrac{0}{-3}$

26. $-\dfrac{4}{0}$

27. $-\dfrac{3}{0}$

28. $\dfrac{0}{-4}$

Simplify the following. See Example 6.

29. $\dfrac{-6^2 + 4}{-2}$

30. $\dfrac{3^2 + 4}{5}$

31. $\dfrac{8 + (-4)^2}{4 - 12}$

32. $\dfrac{6 + (-2)^2}{4 - 9}$

33. $\dfrac{22 + (3)(-2)}{-5 - 2}$

34. $\dfrac{-20 + (-4)(3)}{1 - 5}$

If $x = -5$ and $y = -3$, evaluate each expression. See Example 7.

35. $3x + 2y$

36. $4x + 5y$

37. $2x^2 - y^2$

38. $x^2 - 2y^2$

39. $x^3 + 3y$

40. $y^3 + 3x$

41. $\dfrac{2x - 5}{y - 2}$

42. $\dfrac{2y - 12}{x - 4}$

43. $\dfrac{6 - y}{x - 4}$

44. $\dfrac{4 - 2x}{y + 3}$

Evaluate the following.

45. $(-6)(-2)$

46. $5(-3)$

47. $(-7)(2)$

48. $(-3)(-9)$

49. $\dfrac{18}{-3}$

50. $\dfrac{-16}{-4}$

51. $-\dfrac{6}{0}$

52. $-\dfrac{16}{2}$

53. $-\dfrac{15}{-3}$

54. $\dfrac{48}{-12}$

55. $\dfrac{0}{-7}$

56. $-\dfrac{48}{-8}$

57. $(-6)(3)(-2)(-1)$

58. $(-3)(-2)(-1)(-2)$

59. $(-5)^3$

60. $(-2)^5$

61. $(-4)^2$

62. $(-6)^2$

63. -4^2

64. -6^2

65. $\dfrac{-3 - 5^2}{2(-7)}$

66. $\dfrac{-2 - 4^2}{3(-6)}$

67. $\dfrac{6 - 2(-3)}{4 - 3(-2)}$

68. $\dfrac{8 - 3(-2)}{2 - 5(-4)}$

69. $\dfrac{-3 - 2(-9)}{-15 - 3(-4)}$

70. $\dfrac{-4 - 8(-2)}{-9 - 2(-3)}$

71. $-3(2 - 8)$

72. $-4(3 - 9)$

73. $6(3 - 8)$

74. $4(8 - 11)$

75. $-3[(2 - 8) - (-6 - 8)]$

76. $-2[(3 - 5) - (2 - 9)]$

77. $\left(\dfrac{2}{5}\right)\left(-1\dfrac{1}{4}\right)$

78. $\left(-4\dfrac{2}{3}\right)\left(-\dfrac{8}{21}\right)$

79. $-4\dfrac{1}{6} + \left(-\dfrac{1}{30}\right)$

80. $-3\dfrac{1}{5} + 5$

81. $(1.82)(-4.6)$

82. $(-3.6)(-0.61)$

83. $-22.4 \div (-1.6)$

84. $15.3 - (-2.4)$

If $x = 2$ and $y = -6$, find the value of each expression.

85. $2x - 3y$

86. $-3x + 2y$

87. $2x^2 + y$

88. $2y^2 + x$

89. $\dfrac{xy}{-4}$

90. $\dfrac{-48}{2x - 4}$

91. $\dfrac{2x + 8}{y + 2}$

92. $\dfrac{3y - 6}{x - 4}$

93. $\dfrac{x - y}{3x + 6}$

94. $\dfrac{8 - y}{2x - 3}$

Applications

95. Jean decided to sell 280 shares of stock, which decreased in value by $1.50 per share yesterday. How much money did she lose?

Writing in Mathematics

96. Explain why $\dfrac{6}{0}$ is undefined.

97. Why must the product of an even number of negative values be positive?

1.9
Properties of Real Numbers

OBJECTIVES

Tape 4

1   Identify the commutative property.

2   Identify the associative property.

3   Identify the distributive property.

4   Identify the additive and multiplicative identities.

5   Identify the additive and multiplicative inverse properties.

Specialized terms occur in every area of study. A biologist or ecologist will often be concerned with "symbiotic relationships." "Torque" is a major concern to the physicist. An economist or business analyst might be interested in "marginal revenue." A nurse needs to be familiar with "hemostasis." Mathematics, too, has its own specialized terms and principles.

This section introduces the basic properties of the real number system. Throughout this section, a, b, and c represent real numbers.

When we add, subtract, multiply, or divide two real numbers (except for division by zero), the result is a real number. This is guaranteed by the closure properties.

> **Closure Properties**
>
> If x and y are real numbers then $x + y$, $x - y$, and xy are real numbers. Also, $\dfrac{x}{y}$, $y \neq 0$, is a real number.

1   Next, we look at the commutative properties. These properties state that the order in which any two real numbers are added or multiplied does not change the value of the sum or product.

Commutative Properties

Addition: $\qquad\qquad a + b = b + a$

Multiplication: $\qquad a \cdot b = b \cdot a$

For example, if we let $a = 3$ and $b = 5$, then the commutative properties guarantee that

$$3 + 5 = 5 + 3 \quad \text{and} \quad 3 \cdot 5 = 5 \cdot 3$$

HELPFUL HINT

Is subtraction also commutative? Try an example. Is $3 - 2 = 2 - 3$? **No!** The left side of this statement equals 1; the right side equals -1. There is no commutative property of subtraction. Similarly, there is no commutative property for division. For example, $\dfrac{10}{2}$ does not equal $\dfrac{2}{10}$.

EXAMPLE 1   If $a = -2$, and $b = 5$, show that:
a. $a + b = b + a$ **b.** $a \cdot b = b \cdot a$

Solution: **a.** Replace a with -2 and b with 5. Then

$$a + b = b + a$$

becomes

$$-2 + 5 = 5 + (-2)$$

or

$$3 = 3$$

Since both sides represent the same number, $-2 + 5 = 5 + (-2)$ is a true statement.

b. Replace a with -2 and b with 5. Then

$$a \cdot b = b \cdot a$$

becomes

$$-2 \cdot 5 = 5 \cdot (-2)$$

or

$$-10 = -10$$

Since both sides represent the same number, the statement is true. ■

2   When adding or multiplying three numbers, does it matter how we group the numbers? This question is answered by the associative properties. These properties state that when adding or multiplying three numbers, any two adjacent numbers may be grouped together without changing the answer.

> **Associative Properties**
>
> **Addition:** $(a + b) + c = a + (b + c)$
>
> **Multiplication:** $(a \cdot b) \cdot c = a \cdot (b \cdot c)$

Illustrate these properties by working the following example.

EXAMPLE 2   If $a = -3$, $b = 2$, and $c = 4$, show that:

a. $(a + b) + c = a + (b + c)$ **b.** $(a \cdot b) \cdot c = a \cdot (b \cdot c)$

Solution: Replace a with -3, b with 2, and c with 4.

a. $(a + b) + c = a + (b + c)$

$(-3 + 2) + 4 = -3 + (2 + 4)$ Replace a with -3, b with 2, and c with 4.

$-1 + 4 = -3 + 6$ Simplify inside parentheses.

$3 = 3$ Add.

b. $(a \cdot b) \cdot c = a \cdot (b \cdot c)$

$(-3 \cdot 2) \cdot 4 = -3 \cdot (2 \cdot 4)$ Replace a with -3, b with 2, and c with 4.

$-6 \cdot 4 = -3 \cdot 8$ Simplify inside parentheses.

$-24 = -24$ Multiply.   ■

3   The **distributive property of multiplication over addition** is used repeatedly through algebra. It is useful because it allows us to write a product as a sum or a sum as a product.

> **Distributive Property of Multiplication over Addition**
>
> $$a(b + c) = ab + ac$$

Since multiplication is commutative, the distributive property can also be written

$$(b + c)a = ba + ca$$

The truth of this property can be illustrated by letting $a = 3$, $b = 2$, and $c = 5$.

$$a(b + c) = ab + ac$$

becomes

$$3(2 + 5) \stackrel{?}{=} 3 \cdot 2 + 3 \cdot 5$$

$$3 \cdot 7 \stackrel{?}{=} 6 + 15$$

$$21 = 21$$

Notice in this example that 3 is "being distributed to" each addend inside the parentheses. That is, 3 is multiplied by each addend.

The distributive property can be extended so that factors can be distributed to more than two addends in parentheses. For example,

$$3(x + y + 2) = 3(x) + 3(y) + 3(2)$$
$$= 3x + 3y + 6$$

EXAMPLE 3   Use the distributive property to write each expression without parentheses.
a. $2(x + y)$ **b.** $-5(-3 + z)$ **c.** $5(x + y - z)$ **d.** $-1(2 - y)$
e. $-(3 + x - w)$

Solution: **a.** $2(x + y) = 2 \cdot x + 2 \cdot y$
$$= 2x + 2y$$

b. $-5(-3 + z) = -5(-3) + (-5) \cdot z$
$$= 15 - 5z$$

c. $5(x + y - z) = 5 \cdot x + 5 \cdot y + 5(-z)$
$$= 5x + 5y - 5z$$

d. $-1(2 - y) = (-1)(2) + (-1)(-y)$
$$= -2 + y$$

e. $-(3 + x - w) = -1(3 + x - w)$
$$= (-1)(3) + (-1)(x) + (-1)(-w)$$
$$= -3 - x + w \quad \blacksquare$$

Notice in the last example that $-(3 + x - w)$ is rewritten as $-1(3 + x - w)$.

4   Next, we look at the **identity properties.** These properties guarantee that two special numbers exist. These numbers are called the **identity element for addition** and the **identity element for multiplication.**

Identities for Addition and Multiplication

0 is the identity element for addition.
$$a + 0 = 0 + a = a$$

1 is the identity element for multiplication.
$$a \cdot 1 = 1 \cdot a = a$$

Notice that 0 is the only number that can be added to any real number with the result that the sum is the same real number. Also, 1 is the only number that can be multiplied by any other real number with the result that the product is the same real number.

5   We were introduced to **additive inverses** or **opposites** in Section 1.6. Two numbers are called additive inverses or opposites if their sum is 0. The additive inverse or opposite of 6 is -6 because $6 + (-6) = 0$. The additive inverse or opposite of -5 is 5 because $-5 + 5 = 0$.

Reciprocals or **multiplicative inverses** were introduced in Section 1.3. Two nonzero numbers are called reciprocals or multiplicative inverses if their product is 1. The reciprocal or multiplicative inverse of $\frac{2}{3}$ is $\frac{3}{2}$ because $\frac{2}{3} \cdot \frac{3}{2} = 1$. Likewise, the reciprocal of -5 is $-\frac{1}{5}$ because $-5\left(-\frac{1}{5}\right) = 1$.

Additive and Multiplicative Inverses

The numbers a and $-a$ are additive inverses or opposites of each other because their sum is 0; that is,

$$a + (-a) = 0$$

The numbers b and $\frac{1}{b}$ (for $b \neq 0$) are called reciprocals or multiplicative inverses of each other because their product is 1; that is,

$$b \cdot \frac{1}{b} = 1$$

EXAMPLE 4   Find the additive inverse or opposite of each number.
  a. -3 **b.** 5   **c.** 0

Solution: **a.** The additive inverse of -3 is 3 because $-3 + 3 = 0$.
  b. The additive inverse of 5 is -5 because $5 + (-5) = 0$.
  c. The additive inverse of 0 is 0 because $0 + 0 = 0$.   ∎

EXAMPLE 5   Find the multiplicative inverse or reciprocal of each number.
  a. 7   **b.** $\frac{-1}{9}$

Solution: **a.** The multiplicative inverse of 7 is $\frac{1}{7}$ because $7 \cdot \frac{1}{7} = 1$.

  b. The multiplicative inverse of $\frac{-1}{9}$ is $\frac{9}{-1}$, or -9, because $\left(\frac{-1}{9}\right)(-9) = 1$.   ∎

EXAMPLE 6   Name the property illustrated.
  a. $2 \cdot 3 = 3 \cdot 2$ **b.** $3(x + 5) = 3x + 15$ **c.** $2 + (4 + 8) = (2 + 4) + 8$

Solution: **a.** The commutative property of multiplication
  b. The distributive property
  c. The associative property of addition   ∎

EXAMPLE 7   Use the indicated property to write each expression in an equivalent form.
  a. $2 + 9$; the commutative property of addition
  b. $(5 \cdot 8) \cdot 9$; the associative property of multiplication
  c. $x + 0$; the additive identity property

Solution: **a.** $2 + 9 = 9 + 2$ **b.** $(5 \cdot 8) \cdot 9 = 5 \cdot (8 \cdot 9)$ **c.** $x + 0 = x$ ∎

EXERCISE SET 1.9

Name the properties illustrated by each of the following. See Examples 1, 2, and 6.

1. $3 \cdot 5 = 5 \cdot 3$

2. $4(3 + 8) = 4 \cdot 3 + 4 \cdot 8$

3. $2 + (8 + 5) = (2 + 8) + 5$

4. $4 + 9 = 9 + 4$

5. $9(3 + 7) = 9 \cdot 3 + 9 \cdot 7$

6. $1 \cdot 9 = 9$

Use the distributive property to write each expression without parentheses. See Example 3.

7. $3(6 + x)$

8. $2(x - 5)$

9. $-2(y - z)$

10. $-3(z - y)$

11. $-7(3y - 5)$

12. $-5(2r + 11)$

Find the additive inverse or opposite of each of the following numbers. See Example 4.

13. 16

14. 14

15. -8

16. -3

17. $|-9|$

18. $|11|$

Find the multiplicative inverse or reciprocal of each of the following numbers. See Example 5.

19. $\dfrac{2}{3}$

20. $\dfrac{3}{4}$

21. $-\dfrac{5}{6}$

22. $-\dfrac{7}{8}$

23. 6

24. 3

25. -2

26. -5

Use the indicated property to write each expression in an equivalent form. See Example 7.

27. $\dfrac{2}{3} \cdot \dfrac{3}{2}$; multiplicative inverse property

30. $-4 + 4$; additive inverse property

31. $3 + (8 + 9)$; associative property of addition

28. $8 + 16$; commutative property of addition

32. $(4 \cdot 3) \cdot 9$; associative property of multiplication

29. $(-4)(-3)$; commutative property of multiplication

Name the properties illustrated by each of the following.

33. $(4 \cdot 8) \cdot 9 = 4 \cdot (8 \cdot 9)$

34. $6 \cdot \dfrac{1}{6} = 1$

35. $0 + 6 = 6$

36. $(4 + 9) + 6 = 4 + (9 + 6)$

37. $-4(3 + 7) = -4 \cdot 3 + (-4) \cdot 7$

38. $11 + 6 = 6 + 11$

39. $-4 \cdot (8 \cdot 3) = (-4 \cdot 8) \cdot 3$

40. $10 + 0 = 10$

Use the distributive property to write each of the following without parentheses.

41. $5(x + 4m + 2)$

42. $8(3y + z - 6)$

43. $-4(1 - 2m + n)$

44. $-4(4 + 2p + 5)$

45. $-(5x + 2)$

46. $-(9r + 5)$

47. $-(r - 3 - 7p)$

48. $-(-q - 2 + 6r)$

Find the additive inverse or opposite of each of the following numbers.

49. 0

50. 7

51. $|2|$

52. $|-5|$

53. $|8|$

54. $|6|$

55. $-(-3)$

56. $-(-4)$

57. $-|-2|$

58. $-|-9|$

Find the multiplicative inverse or reciprocal of each of the following numbers.

59. $\dfrac{1}{5}$ **60.** $\dfrac{1}{8}$ **61.** $\dfrac{3}{9}$ **62.** $\dfrac{2}{8}$

63. -1 **64.** 1 **65.** 0 **66.** 100

67. $-\left|-\dfrac{3}{5}\right|$ **68.** $-\left|-\dfrac{2}{5}\right|$ **69.** $3\dfrac{5}{6}$ **70.** $2\dfrac{3}{5}$

Use the indicated property to write each expression in an equivalent form.

71. $y + 0$; additive identity property

72. $1 \cdot x$; multiplicative identity property

73. $x(a + b)$; distributive property

74. $(m + n)y$; distributive property

75. $a(b + c)$; commutative property of multiplication

76. $x + 2$; commutative property of addition

Writing in Mathematics

77. Use an example to show that division is not commutative.

78. Explain why 0 does not have a multiplicative inverse.

79. Define the identity for addition.

CRITICAL THINKING

Shirley has taken on the job of organizing this year's 3-day state convention of orthopedic surgeons. The location she has chosen for the convention is a remote Bahamian island. Shirley is convinced the island is nearly perfect for the 600-member group, but admits one problem: the island has a single airstrip allowing only one plane holding 26 passengers to land each half-hour.

Investigating other methods of transportation, Shirley finds three local fishermen who could transport 15 people at a time in each of their boats. The one-way passage could take as long as three-quarters of an hour from the main island where conventioneers would wait their turns for a boat or a plane.

Should Shirley look for another convention location?

On what basis should Shirley make her decision? What other information might you need to help Shirley decide?

CHAPTER 1 GLOSSARY

The **absolute value** of a real number a, $|a|$, is the distance between a and 0.

The **integers** are $\ldots, -3, -2, -1, 0, 1, 2, 3, \ldots$.

Irrational numbers are numbers represented by points on the number line that are not rational numbers. That is, irrational numbers cannot be expressed as a ratio of integers.

The **natural numbers** are $1, 2, 3, 4, 5, 6, \ldots$.

If the sum of two numbers is zero, the two numbers are said to be **opposites** or **additive inverses** of one another.

Rational numbers are numbers that can be expressed as a quotient $\dfrac{a}{b}$, where a and b are integers and $b \neq 0$.

Real numbers are rational numbers along with irrational numbers.

Two numbers are **reciprocals** or **multiplicative inverses** of each other if their product is 1.

A letter that is used to represent a number is called a **variable.**

The **whole numbers** are $0, 1, 2, 3, 4, 5, \ldots$.

CHAPTER 1 SUMMARY

SYMBOLS USED TO COMPARE NUMBERS   (1.1)

$=$	is equal to	$\neq$	is not equal to
$<$	is less than	$\leq$	is less than or equal to
$>$	is greater than	$\geq$	is greater than or equal to

MULTIPLYING AND DIVIDING FRACTIONS   (1.3)

$$\frac{a}{b} \cdot \frac{c}{d} = \frac{a \cdot c}{b \cdot d}, \qquad b \neq 0, \quad d \neq 0$$

$$\frac{a}{b} \div \frac{c}{d} = \frac{a}{b} \cdot \frac{d}{c} = \frac{a \cdot d}{b \cdot c}, \qquad b \neq 0, \quad d \neq 0, \quad c \neq 0$$

ADDING AND SUBTRACTING FRACTIONS

$$\frac{a}{b} + \frac{c}{b} = \frac{a + c}{b}, \qquad b \neq 0$$

$$\frac{a}{b} - \frac{c}{b} = \frac{a - c}{b}, \qquad b \neq 0$$

ORDER OF OPERATIONS   (1.4)

Simplify expressions in the order below. If grouping symbols such as parentheses are present, simplify expressions within those first; start with the innermost set. If fraction bars are present, simplify the numerator and the denominator separately.

Step 1   Simplify exponential expressions.
Step 2   Perform multiplications or divisions in order from left to right.
Step 3   Perform additions or subtractions in order from left to right.

ADDING TWO NUMBERS WITH THE SAME SIGN   (1.6)

To add two numbers with the same sign, add the absolute values of the two numbers and use their common sign as the sign of the sum.

ADDING TWO NUMBERS WITH UNLIKE SIGNS

To add two numbers with unlike signs, find the difference of the larger absolute value and the smaller absolute value. The sign of the answer will be the sign of the number with the larger absolute value.

SUBTRACTING TWO REAL NUMBERS   (1.7)

If a and b are real numbers, then $a - b = a + (-b)$.

MULTIPLYING AND DIVIDING SIGNED NUMBERS   (1.8)

The product or quotient of two numbers having the same sign is a positive number, and the product or quotient of two numbers having unlike signs is a negative number.

MULTIPLYING BY ZERO

If a is a real number, then $a \cdot 0 = 0$.

QUOTIENTS WITH ZERO

$\frac{a}{0}$ is undefined, $a \neq 0$. Also, $\frac{0}{a} = 0$, $a \neq 0$.

COMMUTATIVE PROPERTIES (1.9)

$a + b = b + a$

$a \cdot b = b \cdot a$

ASSOCIATIVE PROPERTIES

$(a + b) + c = a + (b + c)$

$(a \cdot b) \cdot c = a \cdot (b \cdot c)$

DISTRIBUTIVE PROPERTY

$a(b + c) = ab + ac$

CHAPTER 1 REVIEW

(**1.1**)   *Insert* $<$, $>$, *or* $=$ *in the appropriate space to make the following statements true.*

1. 2   6
2. 1.2   0.951
3. 10.6   10.6
4. 0.07   0.7

Translate each statement into symbols.

5. 4 is greater than or equal to 3.
6. 6 is not equal to 5.
7. The sum of 8 and 4 is less than or equal to 12.

8. 32 is greater than three squared.
9. 7 is equal to 3 plus 4.
10. 0.03 is less than 0.3.

(**1.2**)   *Given the following sets of numbers, list the numbers in each set that also belong to the set of:*

a. Natural numbers
d. Rational numbers

b. Whole numbers
e. Irrational numbers

c. Integers
f. Real numbers

11. $\left\{ -6, 0, 1, 1\frac{1}{2}, 3, \pi, 9.62 \right\}$

12. $\left\{ -3, -1.6, 2, 5, \frac{11}{2}, 15.1, \sqrt{5}, 2\pi \right\}$

Insert $<$, $>$, *or* $=$ *in the appropriate space to make the statement true.*

13. -4 -5
14. 6 -8
15. -2 2
16. $-\dfrac{3}{2}$ $-\dfrac{3}{4}$

(**1.3**)   *Write the number as a product of prime factors.*

17. 36

18. 120

Perform the indicated operations. Write answers in lowest terms.

19. $\dfrac{8}{15} \cdot \dfrac{27}{30}$

20. $\dfrac{7}{8} \div \dfrac{21}{32}$

21. $\dfrac{7}{15} + \dfrac{5}{6}$

22. $\dfrac{3}{4} - \dfrac{3}{20}$

23. $2\dfrac{3}{4} + 6\dfrac{5}{8}$

24. $7\dfrac{1}{6} - 2\dfrac{2}{3}$

(1.4) *Simplify each expression.*

25. $6 \cdot 3^2 + 2 \cdot 8$

26. $24 - 5 \cdot 2^3$

27. $3(1 + 2 \cdot 5) + 4$

28. $8 + 3(2 \cdot 6 - 1)$

29. $\dfrac{4 + |6 - 2| + 8^2}{4 + 6 \cdot 4}$

30. $5[3(2 - 5) - 5]$

(1.5) *Evaluate each expression if x = 6, y = 2, and z = 8.*

31. $2x + 3y$

32. $x(y + 2z)$

33. $\dfrac{x}{y} + \dfrac{z}{2y}$

34. $x^2 - 3y^2$

(1.6) *Find the additive inverse or the opposite.*

35. -9

36. -3

Find the following sums.

37. $-15 + 4$

38. $-6 + (-11)$

39. $16 + (-4)$

40. $-8 + |-3|$

41. $-4.6 + (-9.3)$

42. $-2.8 + 6.7$

(1.7) *Simplify the expression.*

43. $6 - 20$

44. $-3 - 8$

45. $-6 - (-11)$

46. $4 - 15$

47. $-21 - 16 + 3(8 - 2)$

48. $\dfrac{11 - (-9) + 6(8 - 2)}{2 + 3 \cdot 4}$

If x = 3, y = −6, and z = −9, evaluate each expression.

49. $2x^2 - y + z$

50. $\dfrac{y - x + 5x}{2x}$

(1.8) *Simplify each expression.*

51. $6(-8)$

52. $(-2)(-14)$

53. $\dfrac{-18}{-6}$

54. $\dfrac{42}{-3}$

55. $-3(-6)(-2)$

56. $(-4)(-3)(0)(-6)$

57. $\dfrac{4 \cdot (-3) + (-8)}{2 + (-2)}$

58. $\dfrac{3(-2)^2 - 5}{-14}$

(1.9) *Find the additive inverse or opposite.*

59. -6

60. $-|-7|$

Find the multiplicative inverse or reciprocal.

61. -6

62. $\dfrac{3}{5}$

Name the property illustrated.

63. $-6 + 5 = 5 + (-6)$

64. $6 \cdot 1 = 6$

65. $3(8 - 5) = 3 \cdot 8 + 3 \cdot -5$

66. $4 + (-4) = 0$

67. $2 + (3 + 9) = (2 + 3) + 9$

68. $2 \cdot 8 = 8 \cdot 2$

69. $6(8 + 5) = 6 \cdot 8 + 6 \cdot 5$

70. $(3 \cdot 8) \cdot 4 = 3 \cdot (8 \cdot 4)$

71. $4 \cdot \dfrac{1}{4} = 1$

72. $8 + 0 = 8$

7. $3 + 3) = 4(3 + 8)$

CHAPTER 1 TEST

Translate the statement into symbols.

1. The absolute value of negative seven is greater than five.

2. The sum of nine and five is greater than or equal to four.

Simplify the expression.

3. $-13 + 8$

4. $-13 - (-2)$

5. $6 \cdot 3 - 8 \cdot 4$

6. $(13)(-3)$

7. $(-6)(-2)$

8. $\dfrac{|-16|}{-8}$

9. $\dfrac{-8}{0}$

10. $\dfrac{-6 + 2}{5 - 6}$

11. $\dfrac{1}{2} - \dfrac{5}{6}$

12. $-1\dfrac{1}{8} + 5\dfrac{3}{4}$

13. $-\dfrac{3}{5} + \dfrac{15}{8}$

14. $3(-4)^2 - 80$

15. $6[5 + 2(3 - 8) - 3]$

16. $\dfrac{-12 + 3 \cdot 8}{4}$

17. $\dfrac{(-2)(0)(-3)}{-6}$

Insert $<$, $>$, or $=$ in the appropriate space to make each of the following statements true.

18. $-3 \qquad -7$

19. $4 \qquad -8$

20. $|-3| \qquad 2$

21. $|-2| \qquad -1 - (-3)$

22. Given $\left\{-5, -1, \dfrac{1}{4}, 0, 1, 7, 11.6, \sqrt{7}, 3\pi\right\}$, list the numbers in this set that also belong to the set of:

    a. Natural numbers
    b. Whole numbers
    c. Integers

    d. Rational numbers
    e. Irrational numbers
    f. Real numbers

If $x = 6$, $y = -2$, and $z = -3$, evaluate each expression.

23. $x^2 + y^2$

24. $x + yz$

25. $2 + 3x - y$

26. $\dfrac{y + z - 1}{x}$

Identify the property illustrated by each expression.

27. $8 + (9 + 3) = (8 + 9) + 3$

28. $6 \cdot 8 = 8 \cdot 6$

29. $-6(2 + 4) = -6 \cdot 2 + (-6) \cdot 4$

30. $\dfrac{1}{6}(6) = 1$

31. Find the opposite of -9.

32. Find the reciprocal of $-\dfrac{1}{3}$.

2.1 Simplifying Algebraic Expressions

2.2 The Addition Property of Equality

2.3 The Multiplication Property of Equality

2.4 Solving Linear Equations

2.5 Formulas

2.6 Applications of Linear Equations

2.7 Solving Linear Inequalities

CHAPTER 2

Solving Linear Equations and Inequalities

The Upper New York Community Ice Hockey Association is looking for a fair way to decide which teams to invite to the season playoffs. (See Critical Thinking, page 106.)

INTRODUCTION

Much of mathematics relates to deciding which statements are true and which are false. When a statement, such as an equation, contains variables, it is usually not possible to decide whether the equation is true or false until the variable has been replaced by a value. For example, the statement $x + 7 = 15$ is an equation stating that the sum $x + 7$ is the same quantity as 15. Is this statement true or false? It is false for some values of x and true for just one value of x, that is 8. Our purpose in this chapter is to learn ways of deciding which values make an equation or an inequality true. To begin, we spend a bit more time learning about simplifying expressions.

2.1
Simplifying Algebraic Expressions

OBJECTIVES

Tape 5

1	Identify terms, like terms, and unlike terms.
2	Combine like terms.
3	Use the distributive property to remove parentheses.
4	Write word phrases as algebraic expressions.

1 Before we practice simplifying expressions, some new language is presented. A **term** is a number or the product of a number and variables raised to powers. For example,

$$-y, \ 2x^3, \ -5, \ 3xz^2, \ \frac{2}{y}, \ 0.8z$$

are terms. The **numerical coefficient** of the term $3x$ is 3. Recall that $3x$ means $3 \cdot x$.

Terms	*Numerical Coefficients*
$3x$	3
-5	-5
$\dfrac{y^3}{5}$	$\dfrac{1}{5}$ since $\dfrac{y^3}{5}$ means $\dfrac{1}{5} \cdot y^3$
$-0.7ab^3c^5$	-0.7
$-y$	-1
z	1

HELPFUL HINT

The term $-y$ means $-1y$ and thus has a numerical coefficient of -1. The term z means $1z$ and thus has a numerical coefficient of 1.

EXAMPLE 1 Find the numerical coefficient.

a. $-3y$ **b.** $22z^4$ **c.** $\dfrac{2}{y}$ **d.** $-x$

Solution: **a.** The numerical coefficient of $-3y$ is -3.

b. The numerical coefficient of $22z^4$ is 22.

c. The numerical coefficient of $\dfrac{2}{y}$ is 2.

d. The numerical coefficient of $-x$ is -1 since $-x$ is $-1x$. ■

Terms with the same variables raised to exactly the same powers are **like terms.**

Like Terms	*Unlike Terms*	
$3x, 2x$	$5x, 5x^2$	Same variables, but different powers
$-6x^2y, 2x^2y, 4x^2y$	$7y, 3z, 8x^2$	Different variables
$2ab^2c^3, ac^3b^2$	$6abc^3, 6ab^2$	Different variables and different powers

Each variable and its exponent must match exactly in like terms, but like terms need not have the same numerical coefficients, nor do their factors need to be in the same order. For example, $2x^2y$ and $-yx^2$ are like terms.

EXAMPLE 2 Tell whether the terms are like or unlike.

a. $-x^2, 3x^3$ **b.** $4x^2y, x^2y, -2x^2y$ **c.** $-2yz, -3zy$ **d.** $-x^4, x^4$

Solution: **a.** Unlike terms, since the exponents on x are not the same.

b. Like terms, since each variable and its exponent match.

c. Like terms, since $zy = yz$ by the commutative property.

d. Like terms. ■

2   An expression can be simplified by **combining like terms.** For example, by the distributive property, we rewrite like terms $3x + 2x$ as

$$3x + 2x = (3 + 2)x = 5x$$

Also,

$$-y^2 + 5y^2 = (-1 + 5)y^2 = 4y^2$$

This suggests the following rule for combining like terms.

> To **combine like terms,** add the numerical coefficients and multiply the result by the common variables.

EXAMPLE 3 Simplify the following by combining like terms.

a. $7x - 3x$ **b.** $10y^2 + y^2$ **c.** $8x^2 + 2x - 3x$

Solution: **a.** $7x - 3x = (7 - 3)x = 4x$

b. $10y^2 + y^2 = (10 + 1)y^2 = 11y^2$

c. $8x^2 + 2x - 3x = 8x^2 + (2 - 3)x = 8x^2 - x$ ■

EXAMPLE 4 Simplify each expression by combining like terms.

　　a. $2x + 3x + 5 + 2$ **b.** $-5a - 3 + a + 2$ **c.** $4y - 3y^2$ **d.** $2.3x + 5x - 6$

Solution: Use the distributive property to combine the numerical coefficients of like terms.

　　a. $2x + 3x + 5 + 2 = (2 + 3)x + (5 + 2)$
$$= 5x + 7$$

　　b. $-5a - 3 + a + 2 = -5a + 1a + (-3 + 2)$
$$= (-5 + 1)a + (-3 + 2)$$
$$= -4a - 1$$

　　c. $4y - 3y^2$　　These two terms cannot be combined because they are unlike terms.

　　d. $2.3x + 5x - 6 = (2.3 + 5)x - 6$
$$= 7.3x - 6　■$$

3　　Simplifying expressions makes frequent use of the distributive property to remove parentheses.

EXAMPLE 5 Use the distributive property to remove parentheses.

　　a. $5(x + 2)$ **b.** $-2(y + 0.3z - 1)$ **c.** $-(x + y - 2z + 6)$

Solution: **a.** $5(x + 2) = 5(x) + 5(2)$　　　　　　　　　Apply the
　　　　　　　　　　　　　　　　　　　　　　　　distributive property.

　　　　　　　　$= 5x + 10$　　　　　　　　　　　　Multiply.

　　b. $-2(y + 0.3z - 1) = -2(y) - 2(0.3z) - 2(-1)$　　Apply the
　　　　　　　　　　　　　　　　　　　　　　　　distributive property

　　　　　　　　　　$= -2y - 0.6z + 2$　　　　　　Multiply.

　　c. $-(x + y - 2z + 6) = -1(x + y - 2z + 6)$
　　　　　　　　　　　　$= -1(x) - 1(y) - 1(-2z) - 1(6)$　　Distribute -1 over
　　　　　　　　　　　　　　　　　　　　　　　　each term.

　　　　　　　　　　　　$= -x - y + 2z - 6$　■

HELPFUL HINT

When a "−" sign precedes parentheses, the sign of each term inside the parentheses is changed when the distributive property is applied to remove parentheses.

Examples:

　　$-(2x + 1) = -2x - 1$　　　　　　　$-(x - 2y) = -x + 2y$
　　$-(-5x + y - z) = 5x - y + z$　　　　$-(-3x - 4y - 1) = 3x + 4y + 1$

EXAMPLE 6 Simplify the following expressions.

　　a. $3(2x - 5) + 1$ **b.** $8 - (7x + 2) + 3x$ **c.** $-2(4x + 7) - (3$

Solution: **a.** $3(2x - 5) + 1 = 6x - 15 + 1$ Apply the distributive property.

$$= 6x - 14$$ Combine like terms.

b. $8 - (7x + 2) + 3x = 8 - 7x - 2 + 3x$ Apply the distributive property.

$$= -7x + 3x + 8 - 2$$

$$= -4x + 6$$ Combine like terms.

c. $-2(4x + 7) - (3x - 1) = -8x - 14 - 3x + 1$ Apply the distributive property.

$$= -11x - 13$$ Combine like terms.   ∎

EXAMPLE 7   Subtract $4x - 2$ from $2x - 3$.

Solution: "Subtract $4x - 2$ **from** $2x - 3$" translates into $(2x - 3) - (4x - 2)$. Next, simplify the algebraic expression.

$$(2x - 3) - (4x - 2) = 2x - 3 - 4x + 2$$ Apply the distributive property.

$$= -2x - 1$$ Combine like terms.   ∎

4

EXAMPLE 8   Write the following phrases as algebraic expressions. Let x represent the unknown number.

a. Twice a number, added to six.

b. The difference of a number and four, divided by 7.

Solution: **a.**

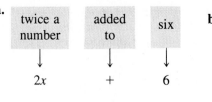

b.

EXAMPLE 9   Write each of the following as an algebraic expression.

a. The sum of two numbers is 8. If one number is 3, find the other number.

b. The sum of two numbers is 8. If one number is x, find the other number in terms of x.

Solution: **a.** The other number is the difference, $8 - 3 = 5$.

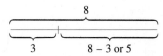

b. The other number is their difference, $8 - x$.

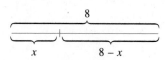

∎

EXAMPLE 10   If x is the first of three consecutive integers, express the sum of the three integers in terms of x. Simplify if possible.

Solution: An example of three consecutive integers is 7, 8, 9. The second consecutive integer is always one unit more than the first, and the third consecutive integer is two units more than the first. If x is the first of three consecutive integers, the three consecutive integers are

$$x, \quad x + 1, \quad x + 2$$

Their sum is $x + (x + 1) + (x + 2)$, which simplifies to $3x + 3$.   ∎

MENTAL MATH

Give the numerical coefficient of each term. See Example 1.

1. $-7y$

2. $3x$

3. x

4. $-y$

5. $17x^2y$

6. $1.2xyz$

Indicate whether the following lists of terms are like or unlike. See Example 2.

7. $5y, -y$

8. $-2x^2y, 6xy$

9. $2z, 3z^2$

10. $ab^2, -7ab^2$

11. $8wz, 7zw$

12. $7.4p^3q^2, 6.2p^3q^2r$

EXERCISE SET 2.1

Simplify each expression by combining any like terms. See Examples 3 and 4.

1. $7y + 8y$

2. $5x - 2x$

3. $8w - w + 6w$

4. $c - 7c + 2c$

5. $3b - 5 - 10b - 4$

6. $6g + 5 - 3g - 7$

7. $m - 4m + 2m - 6$

8. $a + 3a - 2 - 7a$

Simplify each expression. Use the distributive property to remove any parentheses. See Examples 5 and 6.

9. $5(y - 4)$

10. $7(r - 3)$

11. $7(d - 3) + 10$

12. $9(z + 7) - 15$

13. $-(3x - 2y + 1)$

14. $-(y + 5z - 7)$

15. $5(x + 2) - (3x - 4)$

16. $4(2x - 3) - 2(x + 1)$

Write each of the following as an algebraic expression. Simplify if possible. See Example 7.

17. Add $6x + 7$ to $4x - 10$.

18. Add $3y - 5$ to $y + 16$.

19. Subtract $7x + 1$ from $3x - 8$.

20. Subtract $4x - 7$ from $12 + x$.

Write each of the following phrases as algebraic expressions. Let x represent the unknown number. See Example 8.

21. Twice a number decreased by four.

22. The difference of a number and two, divided by five.

23. Three-fourths of a number increased by twelve.

24. Eight more than triple the number.

Write each algebraic expression described. See Example 9.

25. The sum of two numbers is 12. If one number is z, express the other number in terms of z.

26. A 10-foot board is cut into two pieces. If one piece is x feet long, express the other length in terms of x.

27. In a mayoral election, Eric Martin received 284 more votes than Charles Pecot. If Charles received n votes, how many votes did Eric receive?

28. A 5-foot piece of string is cut into two pieces. If one piece is x feet long, express the other length in terms of x.

Write each algebraic expression described. See Example 10.

29. If x represents the first odd integer, express the next odd integer in terms of x.

30. If x represents the first even integer, express the next even integer in terms of x.

31. If x represents the first of two consecutive even integers, express the sum of the two integers in terms of x.

32. If x represents the first of two consecutive odd integers, express the sum of the two integers in terms of x.

Simplify each expression.

33. $7x^2 + 8x^2 - 10x^2$

34. $8x + x - 11x$

35. $6x - 5x + x - 3 + 2x$

36. $8h + 13h - 6 + 7h - h$

37. $-5 + 8(x - 6)$

38. $-6 + 5(r - 10)$

39. $5g - 3 - 5 - 5g$

40. $8p + 4 - 8p - 15$

41. $6.2x - 4 + x - 1.2$

42. $7.9y - 0.7 - y + 0.2$

43. $2k - k - 6$

44. $7c - 8 - c$

45. $0.5(m + 2) + 0.4m$

46. $0.2(k + 8) - 0.1k$

47. $-4(3y - 4)$

48. $-3(2x + 5)$

49. $3(2x - 5) - 5(x - 4)$

50. $2(6x - 1) - (x - 7)$

51. $3.4m - 4 - 3.4m - 7$

52. $2.8w - 0.9 - 0.5 - 2.8w$

53. $6x + 0.5 - 4.3x - 0.4x + 3$

54. $0.4y - 6.7 + y - 0.3 - 2.6y$

55. $-2(3x - 4) + 7x - 6$

56. $8y - 2 - 3(y + 4)$

Write each of the following as an algebraic expression. Simplify if possible.

57. Subtract $5m - 6$ from $m - 9$.

58. Subtract $m - 3$ from $2m - 6$.

59. Eight times the sum of a number and six.

60. Five less than four times the number.

61. Double a number minus the sum of the number and ten.

62. Half a number minus the product of the number and eight.

63. The perimeter of a figure is the total distance around the figure. Given the following rectangle, express the perimeter as an algebraic expression in x.

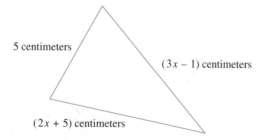

$5x$ feet

$(4x - 1)$ feet

$(4x - 1)$

$5x$ feet

64. Given the following triangle, express its perimeter as an algebraic expression in x.

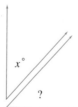

5 centimeters

$(3x - 1)$ centimeters

$(2x + 5)$ centimeters

65. Seven multiplied by the quotient of a number and six.

66. The product of a number and ten, less twenty.

67. Two angles are supplementary if their sum is $180°$. If one angle measures $x°$, express the measure of its supplement in terms of x.

$x°$?

68. Two angles are complementary if their sum is $90°$. If one angle measures $x°$, express the measure of its complement in terms of x.

$x°$

?

69. The sum of the angles of a triangle is $180°$. If one angle of a triangle measures $x°$ and a second angle measures $(2x + 7)°$, express the measure of the third angle in terms of x.

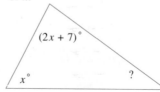

$(2x + 7)°$

$x°$?

70. A quadrilateral is a four-sided figure like the one shown next whose angle sum is 360°. If one angle measures $x°$, a second angle measures $3x°$, and a third angle measures $5x°$, express the measure of the fourth angle in terms of x.

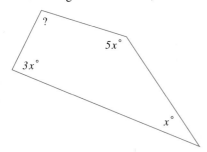

71. If x is the first of three consecutive integers, express the sum of the first integer and the third integer as an algebraic expression in terms of x.

72. If x is the first of three consecutive integers, express the sum of 20 and the second consecutive integer as an algebraic expression in terms of x.

A Look Ahead

Simplify each expression. See the following example.

> **EXAMPLE**   Simplify $-3xy + 2x^2y - (2xy - 1)$.
>
> **Solution:** $-3xy + 2x^2y - (2xy - 1) = -3xy + 2x^2y - 2xy + 1$
>
> $= -5xy + 2x^2y + 1$ ∎

73. $5b^2c^3 + 8b^3c^2 - 7b^3c^2$

74. $4m^4p^2 + m^4p^2 - 5m^2p^4$

75. $3x - (2x^2 - 6x) + 7x^2$

76. $9y^2 - (6xy^2 - 5y^2) - 8xy^2$

77. $-(2x^2y + 3z) + 3z - 5x^2y$

78. $-(7c^3d - 8c) - 5c - 4c^3d$

Writing in Mathematics

79. Explain why decimal points are lined up vertically when adding or subtracting decimals.

80. In your own words, explain how to combine like terms.

Skill Review

Evaluate the following expressions for the given values. See Section 1.8.

81. If $x = -1$ and $y = 3$, find $y - x^2$.

82. If $a = 2$ and $b = -5$, find $a - b^2$.

83. If $y = -5$ and $z = 0$, find $yz - y^2$.

84. If $g = 0$ and $h = -4$, find $gh - h^2$.

85. If $x = -3$, find $x^3 - x^2 + 4$.

86. If $x = -2$, find $x^3 - x^2 - x$.

2.2
The Addition Property of Equality

Tape 5

OBJECTIVES

1 Define the solution or root of an equation.

2 Define linear equation in one variable and equivalent equations.

3 Use the addition property of equality to solve linear equations.

4 Write sentences as equations and solve.

1   An equation has the form **expression = expression.** An equation can be labeled as

$$x + 4 = 7$$

left side | right side

is equal to

When an equation contains a variable, deciding which values of the variable make an equation true is called **solving** an equation for the variable. A **solution** or **root** of an equation is a value for the variable that makes the equation true. For example, 3 is a solution of the equation $x + 4 = 7$, because if x is replaced by 3 the statement is true.

$$x + 4 = 7$$
$$\downarrow$$
$$3 + 4 = 7 \qquad \text{Replace } x \text{ with 3.}$$
$$7 = 7 \qquad \text{True.}$$

Similarly, 1 is not a solution of the equation $x + 4 = 7$, because $1 + 4 = 7$ is **not** a true statement.

EXAMPLE 1   Decide whether 2 is a solution of $3x - 10 = -2x$.

Solution: Replace x with 2 and see if a true statement results.

$$3x - 10 = -2x \qquad \text{Original equation.}$$
$$3(2) - 10 = -2(2) \qquad \text{Replace } x \text{ with 2.}$$
$$6 - 10 = -4 \qquad \text{Simplify each side.}$$
$$-4 = -4 \qquad \text{True.}$$

Since we arrived at a true statement after replacing x with 2 and simplifying both sides of the equation, 2 is a solution of the equation.   ■

2   In this chapter, we will solve linear equations in one variable.

Linear Equation in One Variable

A linear equation in one variable can be written in the form

$$ax + b = c$$

where a, b, and c are real numbers and $a \neq 0$.

To solve a linear equation in x, we will write a series of simpler equations, all equivalent to the original equation, so that the final equation has the form

$$x = \textbf{number} \quad \textbf{or} \quad \textbf{number} = x$$

Equivalent equations have the same solution so that the "number" above will be the solution to the original equation.

3   The first property of equality that will help us write simpler equations is the **addition property of equality.**

> **Addition Property of Equality**
>
> If a, b, and c are real numbers, then
>
> $$a = b \quad \text{and} \quad a + c = b + c$$
>
> are equivalent equations.

This property guarantees that adding the same number to both sides of an equation does not change the solution of the equation. Since subtraction is defined in terms of addition, we may also **subtract the same number from both sides** without changing the solution.

A good way to picture a true equation is as a balanced scale. Since it is balanced, each side of the scale weighs the same amount.

If the same weight is added to or subtracted from each side, the scale remains balanced.

We use the addition property of equality to write equivalent equations until the variable is isolated (by itself on one side of the equation) and the equation looks like "$x = $ number" or "number $= x$."

EXAMPLE 2   Solve $x - 7 = 10$ for x.

Solution: To solve for x, isolate x on one side of the equation. To do this, add 7 to both sides of the equation.

$$x - 7 = 10$$
$$x - 7 \boxed{+ 7} = 10 \boxed{+ 7} \qquad \text{Add 7 to both sides.}$$
$$x = 17 \qquad \text{Simplify.}$$

To check, replace x with 17 in the original equation.

$$x - 7 = 10$$
$$17 - 7 = 10 \qquad \text{Replace } x \text{ with 17 in the original equation.}$$
$$10 = 10 \qquad \text{True.}$$

Since the statement is true, 17 is the solution.   ■

EXAMPLE 3   Solve $y + 0.6 = -1.0$.

Solution: To solve for y, subtract 0.6 from both sides of the equation.

$$y + 0.6 = -1.0$$
$$y + 0.6 \boxed{- 0.6} = -1.0 \boxed{- 0.6} \qquad \text{Subtract 0.6 from both sides.}$$
$$y = -1.6 \qquad \text{Combine like terms.}$$

Check:

$$y + 0.6 = -1.0$$

$$-1.6 + 0.6 = -1.0 \qquad \text{Replace } y \text{ with } -1.6 \text{ in the original equation.}$$

$$-1.0 = -1.0 \qquad \text{True.}$$

The solution is -1.6. ■

EXAMPLE 4   Solve $5t - 5 = 6t + 2$ for t.

Solution: To solve for t, we first want all terms containing a t on one side of the equation and all other terms on the other side of the equation. To do this, first subtract $5t$ from both sides of the equation.

$$5t - 5 = 6t + 2$$

$$5t - 5 \boxed{-5t} = 6t + 2 \boxed{-5t} \qquad \text{Subtract } 5t \text{ from both sides.}$$

$$-5 = t + 2 \qquad \text{Combine like terms.}$$

Next, subtract 2 from both sides and the variable t will be isolated.

$$-5 = t + 2$$

$$-5 \boxed{-2} = t + 2 \boxed{-2} \qquad \text{Subtract 2 from both sides.}$$

$$-7 = t$$

Check the solution, -7, in the original equation. ■

HELPFUL HINT

We may isolate the variable on either side of the equation. $-7 = t$ is equivalent to $t = -7$.

Many times, it is best to simplify one or both sides of an equation before applying the addition property of equality.

EXAMPLE 5   Solve $2x + 3x - 5 + 7 = 10x + 3 - 6x - 4$ for x.

Solution: First, simplify both sides of the equation.

$$2x + 3x - 5 + 7 = 10x + 3 - 6x - 4$$

$$5x + 2 = 4x - 1 \qquad \begin{array}{l}\text{Combine like terms on each} \\ \text{side of the equation.}\end{array}$$

$$5x + 2 \boxed{-4x} = 4x - 1 \boxed{-4x} \qquad \text{Subtract } 4x \text{ from both sides.}$$

$$x + 2 = -1 \qquad \text{Combine like terms.}$$

$$x + 2 \boxed{-2} = -1 \boxed{-2} \qquad \text{Subtract 2 from both sides.}$$

$$x = -3 \qquad \text{Combine like terms.}$$

Check by replacing x with -3 in the original equation.

$$2x + 3x - 5 + 7 = 10x + 3 - 6x - 4$$

$$2(-3) + 3(-3) - 5 + 7 = 10(-3) + 3 - 6(-3) - 4 \qquad \text{Replace } x \text{ with } -3.$$

$$-6 - 9 - 5 + 7 = -30 + 3 + 18 - 4 \qquad \text{Multiply.}$$
$$-13 = -13 \qquad \text{True.}$$

The solution is -3. ■

If an equation contains parentheses, use the distributive property to remove them.

EXAMPLE 6 Solve $-5(2a - 1) - (-11a + 6) = 7$ for a.

Solution: $-5(2a - 1) - (-11a + 6) = 7$

$$-10a + 5 + 11a - 6 = 7 \qquad \text{Apply the distributive property.}$$
$$a - 1 = 7 \qquad \text{Combine like terms.}$$
$$a - 1 \;\boxed{+ 1} = 7 \;\boxed{+ 1} \qquad \text{Add 1 to both sides to isolate } a.$$
$$a = 8 \qquad \text{Combine like terms.}$$

Check to see that 8 is the solution. ■

When solving equations, we may sometimes encounter an equation such as

$$-x = 5$$

This equation is not solved for x because x is not isolated. To solve this equation for x, recall that

"−" can be read as "the opposite of"

We can read the equation $-x = 5$ then as "the opposite of $x = 5$." If the opposite of x is 5, this means that x is the opposite of 5 or -5.

In summary,

$$-x = 5 \quad \text{and} \quad x = -5$$

are equivalent equations and $x = -5$ is solved for x.

4 We now apply our equation-solving skills to solving problems written in words. In Chapter 1, we provided a chart of common translations, words that are other ways of expressing mathematical symbols. Here, we provide a more extended chart of translations.

Addition	Subtraction	Multiplication	Division	Equality
sum	difference of	product	quotient	equals
plus	minus	times	divide	gives
added to	subtracted from	multiply	into	is/was
more than	less than	twice	ratio	yields
increased by	decreased by	of		amounts to
total	less			represents
				is the same as

Many word problems involve a direct translation from a sentence into an equation.

EXAMPLE 7 Twice a number added to seven is the same as three subtracted from the number.

Solution: Translate the sentence into an equation and solve.

twice a number	added to	seven	is the same as	three subtracted from the number
↓	↓	↓	↓	↓
$2x$	$+$	7	$=$	$x - 3$

To solve, begin by subtracting x on both sides to isolate the variable term.

$$2x + 7 = x - 3$$

$$2x + 7 \;\boxed{-\, x} = x - 3 \;\boxed{-\, x} \qquad \text{Subtract } x \text{ from both sides.}$$

$$x + 7 = -3 \qquad \text{Simplify.}$$

$$x + 7 \;\boxed{-\, 7} = -3 \;\boxed{-\, 7} \qquad \text{Subtract 7 from both sides.}$$

$$x = -10 \qquad \text{Simplify.}$$

Check the solution in the **original stated problem.** To do so, replace the unknown number with -10. Twice -10 added to 7 is the same as 3 subtracted from -10.

$$-13 \quad \text{is the same as} \quad -13$$

The unknown number is -10. ∎

HELPFUL HINT

When checking solutions, go back to the original stated problem, rather than to your equation in case errors have been made setting up the equation.

MENTAL MATH

Solve each equation mentally. See Example 2.

1. $x + 4 = 6$

2. $x + 7 = 10$

3. $n + 18 = 30$

4. $z + 22 = 40$

5. $b - 11 = 6$

6. $d - 16 = 5$

EXERCISE SET 2.2

Decide whether the given number is a solution to the given equation. See Example 1.

1. $3x - 6 = 9$; 5

2. $2x + 7 = 3x$; 6

3. $2x + 6 = 5x - 1$; 0

4. $-4x + 2 = x - 8$; 2

5. $x^2 + 2x + 1 = 0$; -1

6. $x^2 + 2x - 15 = 0$; -5

Solve each equation. See Examples 2 and 3.

7. $x + 11 = -2$

8. $y - 5 = -9$

9. $5y + 14 = 4y$

10. $8x - 7 = 9x$

11. $8x = 7x - 8$

12. $x = 2x + 3$

Solve each equation. See Examples 4 and 5.

13. $3x - 6 = 2x + 5$

14. $7y + 2 = 6y + 2$

15. $3t - t - 7 = t - 7$

16. $4c + 8 - c = 8 + 2c$

17. $7x + 2x = 8x - 3$

18. $3n + 2n = 7 + 4n$

Solve each equation. See Example 6.

19. $2(x - 4) = x + 3$

20. $3(y + 7) = 2y - 5$

21. $7(6 + w) = 6(2 + w)$

22. $6(5 + c) = 5(c - 4)$

23. $10 - (2x - 4) = 7 - 3x$

24. $15 - (6 - 7k) = 2 + 6k$

Write an equation for each of the following and solve. See Example 7.

25. The sum of twice a number and 7 is equal to the sum of a number and 6.

26. The difference of three times a number and 1 is the same as twice a number.

27. Three times the difference of a number and 10 is the same as 14.

28. Twice the sum of a number and 3 is equal to 8.

Solve the following.

29. $x - 2 = -4$

30. $y + 7 = 5$

31. $2y + 10 = y$

32. $4x - 4 = 3x$

33. $y + 0.8 = 9.7$

34. $w + 0.9 = 3.6$

35. $5b - 0.7 = 6b$

36. $8n + 1.5 = 9n$

37. $5x - 6 = 6x - 5$

38. $2x + 7 = x - 10$

39. $7t - 12 = 6t$

40. $9m + 14 = 8m$

41. $-5(n - 2) = 8 - 4n$

42. $-4(z - 3) = 2 - 3z$

43. $y - 5y + 0.6 = 0.8 - 5y$

44. $6z + z - 0.9 = 6z + 0.9$

45. $-3(x - 4) = -4x$

46. $-2(x - 1) = -3x$

47. $3(n - 5) - (6 - 2n) = 4n$

48. $5(3 + z) - (8z + 9) = -4z$

49. $c + \dfrac{1}{4} = \dfrac{3}{8}$

50. $\dfrac{1}{2} + f = \dfrac{3}{4}$

51. $-2(t - 1) - 3t = 8 - 4t$

52. $-4(r + 5) - 7r = 12 - 10r$

53. $4y - 6(y + 4) = 1 - y$

54. $-7k + 3(k - 1) = 6 - 5k$

55. $7(m - 2) - 6(m + 1) = -20$

56. $-4(x - 1) - 5(2 - x) = -6$

57. $0.8t + 0.2(t - 0.4) = 1.75$

58. $0.6v + 0.4(0.3 + v) = 2.34$

Write an equation for each of the following and solve. Check the solution.

59. Three times a number minus 6 is equal to two times a number plus 8.

60. The sum of 4 times a number and -2 is equal to the sum of 5 times a number and -2.

61. Twice the difference of a number and 8 is equal to three times the sum of a number and 3.

62. Five times the sum of a number and -1 is the same as 6 times a number.

Writing in Mathematics

63. Explain what we mean by the solution of an equation.

64. Make up an equation that has no solution and explain why.

Skill Review

Multiply or divide the following fractions. See Section 1.3.

65. $\dfrac{3}{5} \cdot \dfrac{10}{21}$

66. $\dfrac{7}{8} \cdot \dfrac{24}{29}$

67. $\dfrac{12}{15} \div \dfrac{2}{3}$

68. $\dfrac{9}{10} \div \dfrac{4}{3}$

69. $\dfrac{2}{9} \cdot \dfrac{12}{14} \cdot \dfrac{7}{4}$

70. $\dfrac{6}{15} \cdot \dfrac{45}{18} \cdot \dfrac{4}{5}$

2.3
The Multiplication Property of Equality

OBJECTIVES

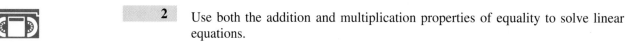

1   Use the multiplication property of equality to solve linear equations.

2   Use both the addition and multiplication properties of equality to solve linear equations.

Tape 6

3   Solve word problems.

1   As useful as the addition property of equality is, it will not help us solve every type of linear equation in one variable. For example, adding or subtracting a value to both sides of the equation does not help us solve

$$\frac{5}{2}x = 15$$

Fortunately, there is a second important property of equality, the **multiplication property of equality.**

> **Multiplication Property of Equality**
>
> If a, b, and c are real numbers and $c \neq 0$, then
>
> $$a = b \quad \text{and} \quad ac = bc$$
>
> are equivalent equations.

This property guarantees that multiplying both sides of an equation by the same nonzero number does not change the solution of the equation. Since division is defined in terms of multiplication, we may also divide both sides of the equation by the same nonzero number without changing the solution.

EXAMPLE 1   Solve for x: $\frac{5}{2}x = 15$.

Solution: To isolate x, multiply both sides of the equation by the reciprocal of $\frac{5}{2}$, which is $\frac{2}{5}$.

$$\frac{5}{2}x = 15$$

$$\frac{2}{5} \cdot \frac{5}{2}x = \frac{2}{5} \cdot 15 \qquad \text{Multiply both sides by } \frac{2}{5}.$$

$$\left(\frac{2}{5} \cdot \frac{5}{2}\right)x = \frac{2}{5} \cdot 15 \qquad \text{Apply the associative property.}$$

$$1x = 6 \qquad\qquad \text{Simplify.}$$

or

$$x = 6$$

The solution is 6. Check the solution in the original equation.   ∎

Why did we multiply both sides by the reciprocal of the coefficient of x? Multiplying the coefficient $\frac{5}{2}$ by $\frac{2}{5}$ leaves a coefficient of 1 and thus isolates the variable.

In general, multiplying by the reciprocal of the variable's coefficient is a way to isolate a variable. Multiplying by the reciprocal of a number is, of course, the same as dividing by the number.

EXAMPLE 2   Solve $-3x = 33$ for x.

Solution: Recall that $-3x$ means $-3 \cdot x$. To isolate x, divide both sides by the coefficient of x, that is, -3.

$$-3x = 33$$

$$\frac{-3x}{-3} = \frac{33}{-3} \qquad \text{Divide both sides by } -3.$$

$$1x = -11 \qquad \text{Simplify.}$$

$$x = -11$$

To check, replace x with -11 in the original equation.

$$-3x = 33$$

$$-3(-11) = 33 \qquad \text{Replace } x \text{ with } -3 \text{ in the original equation.}$$

$$33 = 33 \qquad \text{True.}$$

The solution is -11.   ■

EXAMPLE 3   Solve $\frac{y}{7} = 20$ for y.

Solution: Recall that $\frac{y}{7} = \frac{1}{7}y$. To isolate y, multiply both sides of the equation by 7, the reciprocal of $\frac{1}{7}$.

$$\frac{y}{7} = 20$$

$$7 \cdot \frac{y}{7} = 7 \cdot 20 \qquad \text{Multiply each side by 7.}$$

$$y = 140 \qquad \text{Simplify.}$$

To check, replace y with 140 in the original equation.

$$\frac{y}{7} = 20 \qquad \text{Original equation.}$$

$$\frac{140}{7} = 20 \qquad \text{Replace } y \text{ with 140.}$$

$$20 = 20 \qquad \text{True.}$$

The solution is 140.   ■

EXAMPLE 4   Solve $-\dfrac{2}{3}x = -5$ for x.

Solution: To isolate x, multiply both sides of the equation by $-\dfrac{3}{2}$, the reciprocal of the coefficient of x.

$$-\frac{2}{3}x = -5$$

$$\boxed{-\frac{3}{2}} \cdot \frac{-2}{3}x = \boxed{-\frac{3}{2}} \cdot -5 \quad \text{Multiply both sides by the reciprocal of } -\frac{2}{3}.$$

$$x = \frac{15}{2} \qquad\qquad \text{Simplify.}$$

The solution is $\dfrac{15}{2}$. Check this solution in the original equation.   ■

2   We are now ready to combine the skills learned in the last section with the skills learned from this section to solve equations by applying more than one property.

EXAMPLE 5   Solve $-z - 4 = 6$ for z.

Solution: First, add 4 to both sides of the equation.

$$-z - 4 \boxed{+ 4} = 6 \boxed{+ 4} \qquad \text{Add 4 to both sides.}$$

$$-z = 10 \qquad\qquad \text{Simplify.}$$

Next, recall that $-z$ means $-1 \cdot z$. To isolate z, either multiply or divide both sides of the equation by -1. In this example, we divide.

$$-z = 10$$

$$\frac{-z}{\boxed{-1}} = \frac{10}{\boxed{-1}} \qquad \text{Divide both sides by the coefficient } -1.$$

$$z = -10 \qquad \text{Simplify.}$$

Check:
$$-z - 4 = 6$$
$$-(-10) - 4 = 6 \qquad \text{Replace } z \text{ with } -10.$$
$$10 - 4 = 6$$
$$6 = 6 \qquad \text{True.}$$

Since this is a true statement, -10 is the solution.   ■

EXAMPLE 6   Solve $5x - 2 = 18$ for x.

Solution: First, we isolate $5x$, the term containing the variable. To do so, add 2 to both sides of the equation.

$$5x - 2 = 18$$

$$5x - 2 \boxed{+ 2} = 18 \boxed{+ 2} \qquad \text{Add 2 to both sides.}$$

$$5x = 20 \qquad\qquad \text{Simplify.}$$

Having isolated the variable term, we now use the multiplication property of equality to achieve a coefficient of 1.

$$\frac{5x}{5} = \frac{20}{5} \qquad \text{Divide both sides by 5.}$$

$$x = 4 \qquad \text{Simplify.}$$

The solution is 4. As usual, we can check this solution by replacing x with 4 in the original equation.

$$5x - 2 = 18$$
$$5(4) - 2 = 18 \qquad \text{Replace } x \text{ with 4.}$$
$$20 - 2 = 18 \qquad \text{Simplify.}$$
$$18 = 18 \qquad \text{True.} \quad \blacksquare$$

EXAMPLE 7   Solve for a: $2a + 5a - 10 + 7 = 5a - 13$.

Solution: First, simplify both sides of the equation by combining like terms.

$$2a + 5a - 10 + 7 = 5a - 13$$

$7a - 3 = 5a - 13$	Combine like terms.
$7a - 3 \;-\; 5a = 5a - 13 \;-\; 5a$	Subtract $5a$ from both sides.
$2a - 3 = -13$	Combine like terms.
$2a - 3 \;+\; 3 = -13 \;+\; 3$	Add 3 to both sides.
$2a = -10$	Simplify.
$\dfrac{2a}{2} = \dfrac{-10}{2}$	Divide both sides by 2.
$a = -5$	Simplify.

To check, replace a with -5 in the original equation. $\blacksquare$

3   The next example continues our practice in solving problems written in words.

EXAMPLE 8   When a number is multiplied by -7, the result is 21. Find the number.

Solution: Translate the sentence into an equation and solve.

a number multiplied by -7	the result is	21
↓	↓	↓
$-7x$	$=$	21

To solve, divide both sides by -7.

$$\frac{-7x}{-7} = \frac{21}{-7} \qquad \text{Divide both sides by } -7.$$

$$x = -3 \qquad \text{Simplify.}$$

The unknown number is -3. Check the solution in the original stated problem. When -3 is multiplied by -7, the result is 21. True. $\blacksquare$

EXAMPLE 9 A 10-foot piece of wire is to be cut into two pieces so that the longer piece is 4 times the shorter. If x represents the length of the shorter piece, find the length of each piece.

Solution: If the shorter length is represented by x, the longer piece is 4 times the shorter, or $4x$. Draw and label a diagram.

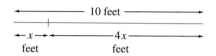

Since the board is 10 feet long, the sum of the lengths should be 10 feet, or

$$x + 4x = 10$$

Now, solve the equation.

$$x + 4x = 10$$
$$5x = 10$$
$$\frac{5x}{5} = \frac{10}{5}$$
$$x = 2$$

The length of the shorter piece is 2 feet.
The length of the longer piece is $4x = 4(2 \text{ feet}) = 8$ feet.

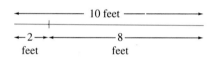

To check, notice that the longer piece is 4 times the shorter piece and that their sum is 10 feet. ■

MENTAL MATH

Solve the following equations mentally.

1. $3a = 27$ **2.** $9c = 54$ **3.** $5b = 10$
4. $7t = 14$ **5.** $6x = -30$ **6.** $8r = -64$

EXERCISE SET 2.3

Solve the following equations. See Example 2.

1. $-5x = 20$ **2.** $-7x = -49$ **3.** $3x = 0$
4. $-2x = 0$ **5.** $-x = -12$ **6.** $-y = 8$

Solve the following equations. See Examples 1, 3, and 4.

7. $\frac{2}{3}x = -8$　　　　**8.** $\frac{3}{4}n = -15$　　　　**9.** $\frac{1}{6}d = \frac{1}{2}$　　　　**10.** $\frac{1}{8}v = \frac{1}{4}$

11. $\frac{a}{-2} = 1$　　　　**12.** $\frac{d}{15} = 2$　　　　**13.** $\frac{k}{7} = 0$　　　　**14.** $\frac{f}{-5} = 0$

Solve the following equations. See Example 5.

15. $2x - 4 = 16$　　　　　　**16.** $3x - 1 = 26$　　　　　　**17.** $-5x + 2 = 22$

18. $7x + 4 = -24$　　　　　　**19.** $6x + 10 = -20$　　　　　**20.** $-10y + 15 = 5$

Solve the following equations. See Example 6.

21. $-4y + 10 = -6y - 2$　　　　　　**22.** $-3z + 1 = -2z + 4$

23. $9x - 8 = 10 + 15x$　　　　　　　**24.** $15t - 5 = 7 + 12t$

25. $2x - 7 = 6x - 27$　　　　　　　**26.** $3 + 8y = 3y - 2$

Solve the following equations. See Example 7.

27. $6 - 2x + 8 = 10$　　　　　　**28.** $-5 - 6y + 6 = 19$

29. $-3a + 6 + 5a = 7a - 8a$　　　　　**30.** $4b - 8 - b = 10b - 3b$

Write an equation for each of the following and solve. See Example 8.

31. Find a number such that half the number is negative five.

32. Negative four is half a number. Find the number.

33. Three times a number is equal to twice the number. Find the number.

34. The product of a number and six is triple the number. Find the number.

Solve the following. See Example 9.

35. A 12-foot board is to be divided into two pieces so that one piece is twice as long as the other. If x represents the length of the shorter piece, find the length of each piece.

36. A woman's $15,000 estate is to be divided so that her husband receives twice as much as her son. If x represents the amount of money that her son receives, find the amount of money that her husband receives and the amount of money that her son receives.

37. Two angles are supplementary if their sum is 180°. One angle measures three times the measure of a smaller angle. If x represents the measure of the smaller angle and

these two angles are supplementary, find the measure of each angle.

38. Two angles are complementary if their sum is 90°. Given the measures of the complementary angles shown, find the measure of each angle.

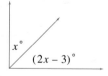

Solve the following equations.

39. $-3w = 18$　　　　　**40.** $5j = -45$　　　　　**41.** $-0.2z = -0.8$

42. $-0.1m = 3.6$　　　　**43.** $-h = -\frac{3}{4}$　　　　**44.** $-b = \frac{4}{7}$

45. $6a + 3 = 3$　　　　　**46.** $8t + 5 = 5$　　　　　**47.** $5 - 0.3k = 5$

48. $2 + 0.4p = 2$　　　　**49.** $2x + \frac{1}{2} = \frac{7}{2}$　　　　**50.** $3n - \frac{1}{3} = \frac{8}{3}$

51. $\frac{x}{3} - 2 = 5$

52. $\frac{b}{4} + 1 = 7$

53. $10 = 2x - 1$

54. $12 = 3j - 4$

55. $4 - 12x = 7$

56. $24 + 20b = 10$

57. $-\frac{2}{3}x = \frac{5}{9}$

58. $-\frac{3}{8}y = -\frac{1}{16}$

59. $10 = -6n + 16$

60. $-5 = -2m + 7$

61. $z - 5z = 7z - 9 - z$

62. $t - 6t = -13 + t - 3t$

63. $5x + 20 = 8 - x$

64. $6t - 18 = t - 3$

65. $5y - y = 2y - 14$

66. $k - 4k = 20 - 5k$

67. $6z - 8 - z + 3 = 0$

68. $4a + 1 + a - 11 = 0$

69. $10 - n - 2 = 2n + 2$

70. $4 - 10 - 5c = 3c - 12$

71. $0.4x - 0.6x - 5 = 1$

72. $0.4x - 0.9x - 6 = 19$

Solve the following.

73. One-third of a number is five sixths. Find the number.

74. Seven-eighths of a number is one-half. Find the number.

75. A 21-foot beam is to be divided so that the longer piece is 1 foot more than 3 times the shorter piece. If x represents the length of the shorter piece, find the lengths of both pieces.

76. A 40-inch board is to be cut into three pieces so that the second piece is twice as long as the first piece and the third piece is 5 times as long as the first piece. If x represents the length of the first piece, find the lengths of all three pieces.

77. The difference of a number and four is twice the number. Find the number.

78. The sum of double a number and six is four times the number. Find the number.

79. The perimeter of a geometric figure is the sum of the lengths of its sides. If the perimeter of the following pentagon (five-sided figure) is 28 centimeters, find the length of each side.

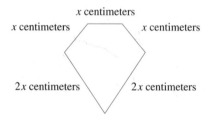

80. The perimeter of the following triangle is 35 meters. Find the length of each side.

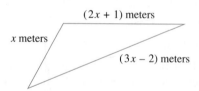

81. Five times a number subtracted from ten is triple the number. Find the number.

82. Nine is equal to ten subtracted from double a number. Find the number.

Writing in Mathematics

83. Explain why it is important to check the solution to a word problem in the original stated problem rather than in the equation that was written.

Skill Review

Insert $<$, $>$, or $=$ in the appropriate space to make the following true. See Section 1.8.

84. $(-3)^2$ _____ -3^2

85. $(-2)^4$ _____ -2^4

86. $(-2)^3$ _____ -2^3

87. $(-4)^3$ _____ -4^3

88. $-|-6|$ _____ 6

89. $-|-0.7|$ _____ -0.7

90. $|-8| - |-5|$ _____ $-|5 - 8|$

91. $-|2| - |-7|$ _____ $|2 - 7|$

2.4
Solving Linear Equations

OBJECTIVES

Tape 6

| **1** | Learn the general strategy for solving a linear equation. |
| **2** | Solve word problems. |

1   In this section, we present a general strategy for solving linear equations. One new piece of strategy is a suggestion to "clear an equation of fractions" as a first step. Since operating on integers is more convenient than operating on fractions, doing so makes the equation more manageable.

To Solve Linear Equations in One Variable

Step 1   Clear the equation of fractions by multiplying both sides of the equation by the lowest common denominator (LCD) of all denominators in the equation.

Step 2   Use the distributive property to remove any grouping symbols, such as parentheses.

Step 3   Combine like terms on each side of the equation.

Step 4   Use the addition property of equality to write the equation as an equivalent equation with variable terms on one side and numbers on the other side.

Step 5   Use the multiplication property of equality to isolate the variable.

Step 6   Check the solution by substituting it in the original equation.

EXAMPLE 1   Solve for x: $\dfrac{x}{2} - 1 = \dfrac{2}{3}x - 3$.

Solution: This equation contains fractions, so we begin by clearing fractions. To do this, multiply both sides of the equation by the LCD of 2 and 3, which is 6.

$$\frac{x}{2} - 1 = \frac{2}{3}x - 3$$

Step 1 $\quad 6\left(\dfrac{x}{2} - 1\right) = 6\left(\dfrac{2}{3}x - 3\right)\qquad$ Multiply both sides by 6.

$$6\left(\frac{x}{2}\right) - 6(1) = 6\left(\frac{2}{3}x\right) - 6(3)\qquad \text{Apply the distributive property.}$$

$$3x - 6 = 4x - 18 \qquad\qquad\qquad \text{Simplify.}$$

There are no grouping symbols and no like terms on either side of the equation, so we continue with step 4.

$$3x - 6 = 4x - 18$$

Step 4 $3x - 6 \boxed{- 3x} = 4x - 18 \boxed{- 3x}$ Subtract $3x$ from both sides.

$$-6 = x - 18$$ Simplify.

$$-6 \boxed{+ 18} = x - 18 \boxed{+ 18}$$ Add 18 to both sides.

$$12 = x$$ Simplify.

The equation is solved for x and the solution is 12. To check, replace x with 12 in the original equation.

Step 6 $\dfrac{x}{2} - 1 = \dfrac{2}{3}x - 3$ Original equation.

$$\dfrac{12}{2} - 1 = \dfrac{2}{3} \cdot 12 - 3$$ Replace x with 12.

$$6 - 1 = 8 - 3$$ Simplify.

$$5 = 5$$ True.

The solution is 12.   ■

EXAMPLE 2   Solve $4(2x - 3) + 7 = 3x + 5$.

Solution: There are no fractions to clear, so begin with step 2.

$$4(2x - 3) + 7 = 3x + 5$$

Step 2 $8x - 12 + 7 = 3x + 5$ Apply the distributive property.

Step 3 $8x - 5 = 3x + 5$ Combine like terms.

Step 4 $8x - 5 \boxed{- 3x} = 3x + 5 \boxed{- 3x}$ Subtract $3x$ from both sides.

$$5x - 5 = 5$$ Simplify.

$$5x - 5 \boxed{+ 5} = 5 \boxed{+ 5}$$ Add 5 to both sides.

$$5x = 10$$ Simplify.

Step 5 $\dfrac{5x}{5} = \dfrac{10}{5}$ Divide both sides by 5.

$$x = 2$$ Simplify.

Step 6   *Check:* $4(2x - 3) + 7 = 3x + 5$

$$4[2(2) - 3] + 7 = 3(2) + 5$$ Replace x with 2.

$$4(4 - 3) + 7 = 6 + 5$$

$$4(1) + 7 = 11$$

$$4 + 7 = 11$$

$$11 = 11$$ True.

The solution is 2.   ■

EXAMPLE 3   Solve $8(2 - t) = -5t$.

Solution: First, apply the distributive property.

$$8(2 - t) = -5t$$

Step 2 $16 - 8t = -5t$ Use the distributive property.

Step 4 $16 - 8t \boxed{+ 8t} = -5t \boxed{+ 8t}$ Add $8t$ to both sides.

$$16 = 3t$$ Combine like terms.

$$\frac{16}{\boxed{3}} = \frac{3t}{\boxed{3}}$$ Divide both sides by 3.

$$\frac{16}{3} = t$$ Simplify.

Step 6   *Check*: $8(2 - t) = -5t$

$$8\left(2 - \frac{16}{3}\right) = -5\left(\frac{16}{3}\right)$$ Replace t with $\frac{16}{3}$.

$$8\left(\frac{6}{3} - \frac{16}{3}\right) = -\frac{80}{3}$$ The LCD is 3.

$$8\left(-\frac{10}{3}\right) = -\frac{80}{3}$$ Subtract fractions.

$$-\frac{80}{3} = -\frac{80}{3}$$ True.

The solution is $\frac{16}{3}$.   ■

EXAMPLE 4   Solve $\dfrac{2(a + 3)}{3} = 6a + 2$.

Solution: We will clear the equation of fractions first.

$$\frac{2(a + 3)}{3} = 6a + 2$$

$$\boxed{3} \cdot \frac{2(a + 3)}{3} = \boxed{3}(6a + 2)$$ Clear fraction by multiplying both sides by the LCD 3.

$$2(a + 3) = 3(6a + 2)$$

Next, use the distributive property and remove parentheses.

$2a + 6 = 18a + 6$ Apply the distributive property.

$2a + 6 \boxed{- 6} = 18a + 6 \boxed{- 6}$ Subtract 6 from both sides.

$2a = 18a$

$2a \boxed{- 18a} = 18a \boxed{- 18a}$ Subtract $18a$ from both sides.

$-16a = 0$

$$\frac{-16a}{-16} = \frac{0}{-16}$$ Divide both sides by -16.

$a = 0$ Write the fraction in simplest form.

To check, replace a with 0 in the original equation.   ■

Not every linear equation in one variable has a single solution. Some equations have no solution, while others have an infinite number of solutions. For example,

$$x + 5 = x + 7$$

has no solution since, no matter what **real number** we replace x with,

$$(\textbf{real number}) + 5 \neq (\text{same } \textbf{real number}) + 7$$

On the other hand,

$$x + 6 = x + 6$$

has infinitely many solutions since x can be replaced by any real number and the equation is always true. The equation $x + 6 = x + 6$ is called an **identity.** The next few examples illustrate equations like these.

EXAMPLE 5   Solve $-2(x - 5) + 10 = -3(x + 2) + x$.

Solution:
$$-2(x - 5) + 10 = -3(x + 2) + x$$

$$-2x + 10 + 10 = -3x - 6 + x \qquad \text{Distribute on both sides.}$$

$$-2x + 20 = -2x - 6 \qquad \text{Combine like terms.}$$

$$-2x + 20 \;\boxed{+\, 2x} = -2x - 6 \;\boxed{+\, 2x} \qquad \text{Add } 2x \text{ to both sides.}$$

$$20 = -6 \qquad \text{Combine like terms.}$$

Notice that no value for x will make $20 = -6$ a true equation. We conclude that there is **no solution** to this equation.   ■

EXAMPLE 6   Solve $3(x - 4) = 3x - 12$.

Solution:
$$3(x - 4) = 3x - 12$$

$$3x - 12 = 3x - 12 \qquad \text{Apply the distributive property.}$$

The left side of the equation is now identical to the right side. Every real number may be substituted for x and a true statement will result. We arrive at the same conclusion if we continue.

$$3x - 12 = 3x - 12 \qquad \text{Add.}$$

$$3x = 3x \qquad \text{Add 12 to both sides.}$$

$$0 = 0 \qquad \text{Subtract } 3x \text{ from both sides.}$$

Again, one side of the equation is identical to the other side. Thus, $3(x - 4) = 3x - 12$ is an **identity** and every real number is a solution.   ■

2   As our ability to solve equations becomes more sophisticated, so does our ability to solve problems written in words. The next two examples give additional practice.

EXAMPLE 7   Twice the sum of a number and 4 is the same as four times the number minus 12. Find the number.

Solution: Translate the sentence into an equation and solve.

twice	sum of a number and 4	is the same as	four times the number	minus	12
↓	↓	↓	↓	↓	↓
2	$(x + 4)$	=	$4x$	−	12

Now solve the equation.

$$2(x + 4) = 4x - 12$$

$$2x + 8 = 4x - 12 \qquad \text{Apply the distributive property.}$$

$$2x + 8 - 4x = 4x - 12 - 4x \qquad \text{Subtract } 4x \text{ from both sides.}$$

$$-2x + 8 = -12 \qquad \text{Simplify.}$$

$$-2x + 8 - 8 = -12 - 8 \qquad \text{Subtract 8 from both sides.}$$

$$-2x = -20 \qquad \text{Simplify.}$$

$$\frac{-2x}{-2} = \frac{-20}{-2} \qquad \text{Divide both sides by } -2.$$

$$x = 10 \qquad \text{Simplify.}$$

Check this solution in the original stated problem. Twice the sum of   10   and 4 is 28, which is the same as 4 times   10   minus 12.   ■

The next example mentions consecutive integers.

An example of three consecutive integers is 10, 11, and 12. Notice that each integer is one more than the previous integer.

An example of three consecutive **even** integers is 14, 16, and 18. Notice that each integer is two more than the previous integer.

An example of three consecutive **odd** integers is 5, 7, and 9. Notice that each integer is again two more than the previous integer.

In general, if x is the first integer, we have the following:

Three consecutive integers: $x, x + 1, x + 2$
Three consecutive odd integers: $x, x + 2, x + 4$
Three consecutive even integers: $x, x + 2, x + 4$

EXAMPLE 8   Find three consecutive integers whose sum is 414.

Solution: Let x = first integer. Then $x + 1$ = second integer and $x + 2$ = third integer. Next, write an equation and solve.

first integer	plus	second integer	plus	third integer	is	414
x	+	$(x + 1)$	+	$(x + 2)$	=	414

$$3x + 3 = 414 \qquad \text{Combine like terms.}$$

$$3x = 411 \qquad \text{Subtract 3 from both sides.}$$

$$\frac{3x}{3} = \frac{411}{3} \qquad \text{Divide both sides by 3.}$$

$$x = 137 \qquad \text{Simplify.}$$

If $x = 137$, then $x + 1 = 138$ and $x + 2 = 139$. The integers are 137, 138, and 139. To check, see if their sum is 414. ■

CALCULATOR BOX

Checking Equations

We can use a calculator to check possible solutions of equations. To do this, replace the variable by the possible solution and evaluate both sides of the equation separately.

Equation: $3x - 4 = 2(x + 6)$ 　　　Solution: $x = 16$

$$3x - 4 = 2(x + 6) \qquad \text{Original equation.}$$

$$3(16) - 4 \stackrel{?}{=} 2(16 + 6) \qquad \text{Replace } x \text{ with 16.}$$

Now evaluate each side with your calculator.

Evaluate left side: $\boxed{3}\;\boxed{\times}\;\boxed{16}\;\boxed{-}\;\boxed{4}\;\boxed{=}$　　　Display: $\boxed{ 44}$

Evaluate right side: $\boxed{2}\;\boxed{(}\;\boxed{16}\;\boxed{+}\;\boxed{6}\;\boxed{)}\;\boxed{=}$　　　Display: $\boxed{ 44}$

Since the left side equals the right side, the equation checks.

Use a calculator to check the possible solutions to each of the following equations.

1. $2x = 48 + 6x$; $x = -12$

2. $-3x - 7 = 3x - 1$; $x = -1$

3. $5x - 2.6 = 2(x + 0.8)$; $x = 4.4$

4. $-1.6x - 3.9 = -6.9x - 25.6$; $x = 5$

5. $\dfrac{564x}{4} = 200x - 11(649)$; $x = 121$

6. $20(x - 39) = 5x - 432$; $x = 23.2$

EXERCISE SET 2.4

Solve each equation. See Example 1.

1. $\dfrac{3}{4}x - \dfrac{1}{2} = 1$

2. $\dfrac{2}{3}x + \dfrac{5}{3} = \dfrac{5}{3}$

3. $x + \dfrac{5}{4} = \dfrac{3}{4}x$

4. $\dfrac{7}{8}x + \dfrac{1}{4} = \dfrac{3}{4}x$

5. $\dfrac{x}{2} - 1 = \dfrac{x}{5} + 2$

6. $\dfrac{x}{5} - 2 = \dfrac{x}{3}$

Solve each equation. See Examples 2 and 3.

7. $-2(3x - 4) = 2x$

8. $-(5x - 1) = 9$

9. $4(2n - 1) = (6n + 4) + 1$

10. $3(4y + 2) = 2(1 + 6y) + 8$

11. $5(2x - 1) - 2(3x) = 4$

12. $3(2 - 5x) + 4(6x) = 12$

13. $6(x - 3) + 10 = -8$

14. $-4(2 + n) + 9 = 1$

Solve each equation. See Example 4.

15. $\dfrac{6(3 - z)}{5} = -z$

16. $\dfrac{4(5 - w)}{3} = -w$

17. $\dfrac{2(x + 1)}{4} = 3x - 2$

18. $\dfrac{3(y + 3)}{5} = 2y + 6$

Solve each equation. See Examples 5 and 6.

19. $5x - 5 = 2(x + 1) + 3x - 7$

20. $3(2x - 1) + 5 = 6x + 2$

21. $\dfrac{x}{4} + 1 = \dfrac{x}{4}$

22. $\dfrac{x}{3} - 2 = \dfrac{x}{3}$

23. $3x - 7 = 3(x + 1)$

24. $2(x - 5) = 2x + 10$

Write each of the following sentences as equations. Then solve. See Example 7.

25. The sum of twice a number and $\dfrac{1}{5}$ is equal to the difference between three times a number and $\dfrac{4}{5}$. Find the number.

26. The sum of four times a number and $\dfrac{2}{3}$ is equal to the difference of five times the number and $\dfrac{5}{6}$. Find the number.

27. If the sum of a number and five is tripled, the result is one less than twice the number. Find the number.

28. Twice the sum of a number and six equals three times the sum of the number and four. Find the number.

Solve the following. See Example 8.

29. Find two consecutive odd integers such that twice the larger is 15 more than three times the smaller.

30. Find three consecutive even integers whose sum is negative 114.

Solve each equation.

31. $4x + 3 = 2x + 11$

32. $6y - 8 = 3y + 7$

33. $-2y - 10 = 5y + 18$

34. $7n + 5 = 10n - 10$

35. $6x - 1 = 5x + 2$

36. $2x - 1 = 6x - 21$

37. $2y + 2 = y$

38. $7y + 4 = -3$

39. $3(5c - 1) - 2 = 13c + 3$

40. $4(3t + 4) - 20 = 3 + 5t$

41. $x + \dfrac{7}{6} = 2x - \dfrac{7}{6}$

42. $\dfrac{5}{2}x - 1 = x + \dfrac{1}{4}$

43. $2(x - 5) = 7 + 2x$

44. $-3(1 - 3x) = 9x - 3$

45. $\dfrac{2(z + 3)}{3} = 5 - z$

46. $\dfrac{3(w + 2)}{4} = 2w + 3$

47. $\dfrac{4(y - 1)}{5} = -3y$

48. $\dfrac{5(1 - x)}{6} = -4x$

49. $8 - 2(a - 1) = 7 + a$

50. $5 - 6(2 + b) = b - 14$

51. $2(x + 3) - 5 = 5x - 3(1 + x)$

52. $4(2 + x) + 1 = 7x - 3(x - 2)$

53. $\dfrac{5x - 7}{3} = x$

54. $\dfrac{7n + 3}{5} = -n$

55. $\dfrac{9 + 5v}{2} = 2v - 4$

56. $\dfrac{6 - c}{2} = 5c - 8$

57. $-3(t - 5) + 2t = 5t - 4$

58. $-(4a - 7) - 5a = 10 + a$

59. $2(6t - 3) = 5(t - 2) + 2$

60. $3(m + 7) = 2(5 - m) + 3$

61. $6 - (x + 1) = -2(2 - x)$

62. $-(5x + 4) = 4 - (x + 4)$

63. $\dfrac{3(x - 5)}{2} = \dfrac{2(x + 5)}{3}$

64. $\dfrac{5(x - 1)}{4} = \dfrac{3(x + 1)}{2}$

Solve the following.

65. Twice the sum of a number and three is the same as the difference of the number and six.

66. Six times the difference of a number and one equals five times the sum of the number and two.

67. The sum of two consecutive integers is 31. What are the numbers?

68. The sum of two consecutive odd integers is 52. What are the numbers?

69. The product of twice a number and three is the same as the difference of five times the number and $\dfrac{3}{4}$. Find the number.

70. If the difference of a number and four is doubled, the result is $\dfrac{1}{4}$ less than the number. Find the number.

71. Find three consecutive integers such that one-third the sum of the largest and the smallest is equal to the middle integer.

72. Find three consecutive even integers such that the difference of the largest and the smallest is six more than the middle integer.

73. If the quotient of a number and 4 is added to $\dfrac{1}{2}$, the result is $\dfrac{3}{4}$.

74. If $\dfrac{3}{4}$ is added to three times a number, the result is $\dfrac{1}{2}$ subtracted from twice a number.

A Look Ahead

Solve each equation. See example below.

EXAMPLE Solve $t(t + 4) = t^2 - 2t + 10$.

Solution:

$$t(t + 4) = t^2 - 2t + 10$$

$$t^2 + 4t = t^2 - 2t + 10$$

$$t^2 + 4t - t^2 = t^2 - 2t + 10 - t^2$$

$$4t = -2t + 10$$

$$4t + 2t = -2t + 10 + 2t$$

$$6t = 10$$

$$\frac{6t}{6} = \frac{10}{6}$$

$$t = \frac{5}{3} \quad \blacksquare$$

75. $x(x - 3) = x^2 + 5x + 7$
77. $t^2 - 6t = t(8 + t)$
79. $2z(z + 6) = 2z^2 + 12z - 8$

76. $y^2 - 4y + 10 = y(y - 5)$
78. $n(3 + n) = n^2 + 4n$
80. $3c^2 - 8c + 2 = c(3c - 8)$

Writing in Mathematics

81. Explain the difference between simplifying an expression and solving an equation.
82. When solving an equation, if the final equivalent equation is $0 = 5$, what can we conclude? If the final equivalent equation is $-2 = -2$, what can we conclude?

83. Make up an equation that has no solution and an equation for which every real number is a solution. Explain the solutions of each.

Skill Review

Simplify by combining like terms. See Section 2.1.

84. $2x - 3x + x$
86. $5(x - 2) + y$
88. $-(a - 5) + 2(3a + 6)$
90. $5x - 7(x - 2) + 10$

85. $5p + 6 - p$
87. $-(y + 1) - 12y$
89. $3(x + 12) - (5x - 1)$
91. $4x - 3(-2x + 5) - 5$

2.5
Formulas

Tape 7

OBJECTIVES	**1**	Use formulas to solve word problems.
	2	Solve a formula for one of its variables.

1   A **formula** is an equation that describes a known relationship among quantities, such as distance, time, volume, weight, and money. These quantities are represented by letters and are thus variables of the formula. Here are some common formulas and their meanings.

A = lw
Area of a rectangle = length · width

I = PRT
Simple Interest = **P**rincipal · **R**ate · **T**ime

P = a + b + c
Perimeter of a triangle = side **a** + side **b** + side **c**

d = rt
distance = **r**ate · **t**ime

V = lwh

Volume of a rectangular solid = **length · width · height**

F = (9/5)C + 32

degrees **Fahrenheit** = (9/5) · degrees **Celsius** + 32

Formulas are valuable tools because they allow us to calculate measurements as long as we know certain other measurements. For example, if we know we traveled a distance of 100 miles at a rate of 40 miles per hour, we can replace the variables d and r in the formula $d = rt$ and find our time, t.

$$d = rt \qquad \text{Formula.}$$

$$100 = 40t \qquad \text{Replace } d \text{ with 100 and } r \text{ with 40.}$$

This is a linear equation in one variable, t. To solve for t, divide both sides of the equation by 40.

$$\frac{100}{40} = \frac{40t}{40} \qquad \text{Divide both sides by 40.}$$

$$\frac{5}{2} = t \qquad \text{Simplify.}$$

The time traveled is $\frac{5}{2}$ hours or $2\frac{1}{2}$ hours.

EXAMPLE 1   Given is the following rectangular solid whose volume is 432 cubic feet. If the length of the solid is 9 feet and its height is 8 feet, find its width.

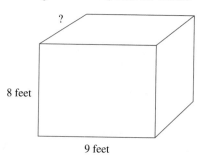

Solution: The formula that relates the volume of a rectangular solid to its height, length, and width is $V = lwh$. Replace the variables V, l, and h by their given values and solve the equation for w.

$$V = lwh$$

$$432 = 9 \cdot w \cdot 8 \qquad \text{Replace } V \text{ with 432, } l \text{ with 9, and } h \text{ with 8.}$$

$$432 = 72w \qquad \text{Multiply.}$$

$$\frac{432}{72} = \frac{72w}{72} \qquad \text{Divide both sides by 72.}$$

$$6 = w \qquad \text{Simplify.}$$

The width is 6 feet.

To check, notice that the volume, 432 cubic feet, is equal to lwh or (9 feet)(6 feet)(8 feet). ∎

EXAMPLE 2   If the current Fahrenheit temperature is 59°, find the equivalent temperature in degrees Celsius.

Solution: The formula $F = \dfrac{9}{5}C + 32$ relates degrees Fahrenheit to degrees Celsius. Replace the variable F with 59 and solve the equation for C.

$$F = \frac{9}{5}C + 32 \qquad \text{Original formula.}$$

$$59 = \frac{9}{5}C + 32 \qquad \text{Replace } F \text{ with 59.}$$

$$59 \;-\; 32 = \frac{9}{5}C + 32 \;-\; 32 \qquad \text{Subtract 32 from both sides.}$$

$$27 = \frac{9}{5}C \qquad \text{Simplify.}$$

$$\frac{5}{9} \cdot 27 = \frac{5}{9} \cdot \frac{9}{5}C \qquad \text{Multiply both sides by } \frac{5}{9}.$$

$$15 = C \qquad \text{Simplify.}$$

Then 59° Fahrenheit is equivalent to 15° Celsius.   ■

The following steps may be used when using formulas to solve word problems.

To Use Formulas to Solve Problems

Step 1   Determine the appropriate formula. See the proper appendix, if needed, for a list of common formulas.

Step 2   Replace variables with their known measurement.

Step 3   Solve the formula for the remaining variable.

Step 4   State the solution to the problem.

Step 5   Check the solution from the original stated problem.

EXAMPLE 3   The area of the plot of land pictured is 60 square meters and its width is 5 meters. Find the length of the plot of land.

5 meters

Solution: *Step 1*   The appropriate formula is for the area of a rectangle: $A = lw$

Step 2   Replace A with 60 and w with 5. $60 = l \cdot 5$

Step 3   Solve for l. $\dfrac{60}{5} = \dfrac{l \cdot 5}{5}$ Divide both sides by 5.

Step 4   The length of the rectangle is 12 meters. $12 = l$ Simplify.

Step 5   *Check:*

12 meters

5 meters

The area of this rectangle is $5(12) = 60$ square meters.   ■

> **HELPFUL HINT**
>
> Make sure that appropriate units are attached to the solution.

EXAMPLE 4   Find how long it takes for $500 in a savings account that pays 5% simple interest to earn $75.

Solution: *Step 1*   The appropriate formula is the simple interest formula:

$$I = PRT$$

Step 2   Replace P with 500, I with 75, and R with 5% or 0.05.

$$75 = 500 \cdot 0.05 \cdot T$$

Step 3   Solve for T:

$$75 = 25T \qquad \text{Multiply.}$$

$$\frac{75}{25} = \frac{25T}{25} \qquad \text{Divide both sides by 25.}$$

$$3 = T \qquad \text{Simplify.}$$

Step 4   The time is 3 years.

Step 5   *Check:* If $500 is invested for 3 years in an account paying 5% simple interest, the interest would be $500(0.05)(3) = $75. ∎

EXAMPLE 5   The distance from Pensacola, Florida, to New Orleans, Louisiana, is 290 miles. It takes a person 5 hours to drive the distance. Find the average driving speed.

Solution: *Step 1*   The appropriate formula is the distance formula:

$$d = r \cdot t$$

Step 2   Replace d with 290 miles and t with 5 hours.

$$290 = r \cdot 5$$

Step 3   Solve for r.

$$\frac{290}{5} = \frac{5r}{5} \qquad \text{Divide both sides by 5.}$$

$$58 = r \qquad \text{Simplify.}$$

Step 4   The speed is 58 miles per hour (mph).

Step 5   The speed of 58 mph is reasonable. To check, substitute 58 for r in the original formula. ∎

2   One important skill in algebra is to solve a formula for one of its variables. Remember that to solve for a variable means to isolate that variable on one side of the equation. We say the formula $d = rt$ is solved for d in terms of r and t. We can also solve $d = rt$ for t in terms of d and r. To solve for t, divide both sides of the equation by r.

$$d = rt$$

$$\frac{d}{r} = \frac{rt}{r} \qquad \text{Divide both sides by } r.$$

$$\frac{d}{r} = t \qquad \text{Simplify.} \quad \blacksquare$$

EXAMPLE 6   Solve $V = lwh$ for l.

Solution: This formula is used to find the volume of a box. To solve for l, divide both sides by wh.

$$V = lwh$$

$$\frac{V}{wh} = \frac{lwh}{wh} \qquad \text{Divide both sides by } wh.$$

$$\frac{V}{wh} = l \qquad \text{Simplify.}$$

Since we have isolated l on one side of the equation, we have solved for l in terms of V, w, and h. Remember that it does not matter on which side of the equation we isolate the variable. $\blacksquare$

EXAMPLE 7   Solve $y = mx + b$ for x.

Solution: First, isolate mx by subtracting b from both sides.

$$y = mx + b$$

$$y \; - b \; = mx + b \; - b \qquad \text{Subtract } b \text{ from both sides.}$$

$$y - b = mx \qquad \text{Simplify.}$$

Next, solve for x by dividing both sides by m.

$$\frac{y - b}{m} = \frac{mx}{m}$$

$$\frac{y - b}{m} = x \qquad \text{Simplify.} \quad \blacksquare$$

EXAMPLE 8   Solve $P = 2l + 2w$ for w.

Solution: This formula relates the perimeter of a rectangle to its length and width. To solve for w, begin by subtracting $2l$ from both sides.

$$P = 2l + 2w$$

$$P - 2l = 2l + 2w - 2l \qquad \text{Subtract } 2l \text{ from both sides.}$$

$$P - 2l = 2w$$

$$\frac{P - 2l}{2} = \frac{2w}{2} \qquad \text{Divide both sides by 2.}$$

$$\frac{P - 2l}{2} = w \qquad \text{Simplify.} \quad \blacksquare$$

The next example has an equation containing a fraction. We will first clear the equation of fractions and then solve for the specified variable.

EXAMPLE 9 Solve $F = \frac{9}{5}C + 32$ for C.

Solution:

$$F = \frac{9}{5}C + 32$$

$$5(F) = 5\left(\frac{9}{5}C + 32\right)$$ Clear the fraction by multiplying both sides by the LCD.

$$5F = 9C + 160$$ Distribute the 5.

$$5F - 160 = 9C + 160 - 160$$ Subtract 160 from both sides.

$$5F - 160 = \qquad 9C$$

$$\frac{5F - 160}{9} = \frac{9C}{9}$$ Divide both sides by 9.

$$\frac{5F - 160}{9} = C \quad \blacksquare$$

EXERCISE SET 2.5

Substitute the given values into the given formulas and solve for the unknown variable. See Examples 1 through 5.

1. $A = bh$; $A = 45$, $b = 15$

2. $D = rt$; $D = 195$, $t = 3$

3. $I = PRT$; $P = 5000$, $R = 0.08$, $T = 2$

4. $V = lwh$; $l = 14$, $w = 8$, $h = 3$,

5. $C = 2\pi r$; $C = 94.2$ (use the approximation 3.14 for π)

6. $A = \pi r^2$; $r = 4$ (use the approximation 3.14 for π)

Solve the following applications by using known formulas. See Examples 1 and 3.

7. Find the amount of fencing needed for a rectangular yard 40 feet by 30 feet.

8. Find the amount of wallpaper border needed for a square room that measures 12 feet on a side if the width of the border is 3 feet.

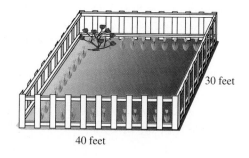

30 feet

40 feet

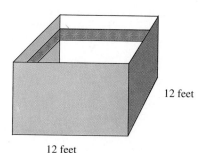

12 feet

12 feet

9. If the length of a rectangularly shaped garden is 6 meters and its width is 4.5 meters, find the amount of fencing required.

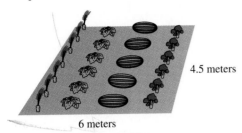

4.5 meters

6 meters

10. If the area of a right triangularly shaped sail is 20 square feet, and its base is 5 feet, find the height of the sail.

?

5 feet

See Example 2.

11. Convert Paris, France's low temperature of $-5°C$ to Fahrenheit.

12. Convert Nome, Alaska's 14°F high temperature to Celsius.

See Example 4.

13. How much interest will $3000 in a passbook savings account earn in 2 years at a bank that pays 6% simple interest?

14. Ellen wishes to win the lottery and live off the interest earned by placing it into a savings account that pays 8% annual interest. How much would she have to win in order to have $25,000 per year to live on?

See Example 5.

15. How many hours will it take a person to drive 385 miles if the cruise control is set at 55 miles per hour?

16. A 5-hour nonstop flight from Atlanta, Georgia, to Seattle, Washington, averages 470 miles per hour. Find the "Frequent Flyer" miles earned on this flight. (Assume this airline rounds up to the nearest thousand.)

Solve the following equations for the indicated variable. See Examples 6 through 9.

17. $f = 5gh$ for h

18. $C = 2\pi r$ for r

19. $V = LWH$ for W

20. $T = mnr$ for n

21. $3x + y = 7$ for y

22. $-x + y = 13$ for y

23. $A = p + PRT$ for R

24. $A = p + PRT$ for T

25. $V = \frac{1}{3}Ah$ for A

26. $D = \frac{1}{4}fk$ for k

Substitute the given values into the given formulas and solve for the unknown variable.

27. $y = mx + b$; $y = 2$, $m = 3$, and $x = -1$

28. $y - y_1 = m(x - x_1)$; $y_1 = 5$, $m = 0.5$, $x = 4$, $x_1 = 2$

29. $A = 0.5h(b_1 + b_2)$; $A = 40$, $b_1 = 11$, $b_2 = 9$

30. $A = 0.5bh$; $A = 35$, $h = 7$

31. $V = \frac{1}{3}\pi r^2 h$; $V = 565.2$, $r = 6$ (use the approximation 3.14 for π)

32. $V = \frac{4}{3}\pi r^3$; $r = 2$ (use the approximation 3.14 for π)

Solve the following by using known formulas.

33. Chattanooga, Tennessee, is about 120 miles from Nashville, Tennessee. How long would it take John to drive roundtrip if he averages 50 mph?

34. It took Maria 5 hours roundtrip to drive from her house to her aunt's house 155 miles away. What was her average speed?

35. A package of floor tiles contains 25 square-foot tiles. Find the number of packages needed to cover a rectangular floor that measures 12 feet by 14 feet.

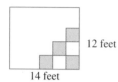

12 feet

14 feet

36. One-foot-square ceiling tiles are sold in packages of 50. How many packages are needed for a square room whose side is 15 feet?

37. Driving at an average speed of 60 mph, how long will it take to drive 150 miles?

38. If Sally entered a Florida turnpike at 11:00 A.M. and exited 180 miles later at 2:00 P.M., what was her average driving speed?

39. A gallon of latex paint covers 500 square feet. How many gallons are needed to paint one coat on each wall of a rectangular room 18 feet by 24 feet? (Assume 8-foot ceilings.)

40. A gallon of enamel paint covers 300 square feet. How many gallons are needed to paint two coats on a wall 42 feet by 8 feet?

41. How much interest will be paid on a $7000 car loan in 3 years at a bank that charges 11% simple interest?

42. A couple would like to eventually live off the interest earned on a savings account that pays 7% simple interest. How much do they need to place in the savings account in order to have $35,000 per year to live on?

43. Paolo's Pizza sells one 16-inch cheese pizza or two 10-inch cheese pizzas for $9.99. Which choice will give us more pizza?

44. A lawn is in the shape of a trapezoid with a height of 60 feet and bases of 70 feet and 130 feet. How many bags of fertilizer must be purchased to cover the lawn if each bag covers 4000 square feet?

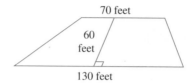

70 feet

60 feet

130 feet

45. Find the volume of a fish tank that measures 12 inches by 6 inches by 4 inches.

46. Find the surface area of a fish tank that measures 12 inches by 6 inches by 4 inches. (*Hint:* Use the formula: $S = 2lw + 2wh + 2lh$, where 12 inches is the length, 6 inches is the width, and 4 inches is the height.)

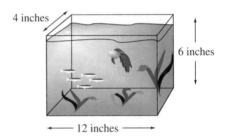

4 inches

6 inches

12 inches

47. Find the volume of a cube whose sides measure 8 inches.

48. Find the surface area of a cube with 8-inch sides.

Solve the following equations for the given variable.

49. $P = a + b + c$ for a

50. $PR = s_1 + s_2 + s_3 + s_4$ for s_3

51. $5x - 4y = 13$ for x

52. $6x + 5y = 10$ for y

53. $A = 0.5bh$ for h

54. $g = 0.3ct$ for c

55. $C = \dfrac{5}{9}(F - 32)$ for F

56. $t = \dfrac{2}{3}(n - 7)$ for n

57. $2x - 7y + 3 = 0$ for y

58. $-5x + 2y - 8 = 0$ for x

59. $\dfrac{3}{4}x + \dfrac{3}{4}y - 2 = 0$ for y

60. $\dfrac{7}{8}x - \dfrac{2}{3}y + 3 = 0$ for x

61. $T = \dfrac{r}{4} - 3(T - 1)$ for r

62. $K = \dfrac{t}{3} - 5(2 - K)$ for t

Writing in Mathematics

63. Make up one example of a formula that might be used when cooking.

64. When solving an equation, why does it not matter on which side you isolate the variable?

Skill Review

Write the following phrases as algebraic expressions. See Section 1.5.

65. Nine divided by the sum of a number and 5.

66. Three times the sum of a number and four.

67. Double the sum of ten and four times the number.

68. Triple the difference of a number and twelve.

69. Half the product of a number and five.

70. One-third of the quotient of a number and six.

71. Twice a number divided by three times the number.

72. A number minus the sum of the number and six.

2.6
Applications of Linear Equations

OBJECTIVES

Tape 7

1 Solve word problems using geometric formulas.

2 Solve word problems involving the value of coins.

3 Solve word problems involving simple interest.

4 Solve word problems involving distance.

5 Solve word problems involving solutions.

This section is devoted to solving word problems in the categories listed. These problems provide an opportunity to judge whether a formula may or may not apply. It is also an opportunity to use the equation-solving skills that we have been developing in the last few sections. We will use the following steps to solve word problems in this section.

Solving a Word Problem

Step 1 Read and then reread the problem. Choose a variable to represent one unknown quantity.

Step 2 Use this variable to represent any other unknown quantities.

Step 3 Draw a diagram if possible to visualize the known facts.

Step 4 Translate the word problem into an equation.

Step 5 Solve the equation.

Step 6 Answer the question asked and check to see if the answer is **reasonable.**

Step 7 Check the solution in the originally stated problem.

1

EXAMPLE 1 The length of a rectangular athletic field is three times its width. If 640 yards of fencing surround the field, find the dimensions of the field.

Solution: *Step 1* Let a variable represent one unknown quantity.
We are asked to find the dimensions of the rectangle.
Let x represent the width of the rectangle.

Step 2 Use this variable to represent other unknown quantities.
Since the length is 3 times the width, the length is represented by $3x$.

Step 3 Draw a diagram.

$3x$

Step 4 Write an equation from the stated problem.

Fencing is placed along the edge or perimeter of the rectangle, so we use the formula for the perimeter of a rectangle, $P = 2l + 2w$. To find the unknown dimensions, let **Perimeter** = 640, **width** = x, and **length** = $3x$, and solve for x.

$$P = 2l + 2w$$
$$640 = 2(3x) + 2(x) \qquad \text{Let } P = 640,\ w = x,\ \text{and } l = 3x.$$

Step 5 Solve the equation.

$$640 = 6x + 2x \qquad \text{Multiply.}$$
$$640 = 8x \qquad \text{Combine like terms.}$$
$$\frac{640}{8} = \frac{8x}{8} \qquad \text{Divide both sides by 8.}$$
$$80 = x \qquad \text{Simplify.}$$

Step 6 State the solution in words.
The width of the field is 80 yards and the length is $3(80) = 240$ yards.

Step 7 Check the solution in the original stated problem.
To check, see if our 80-yard by 240-yard rectangle has a perimeter of 640. It does, since $2(80) + 2(240) = 640$. Also, notice that the length is three times the width. ■

2

EXAMPLE 2 When Hiroto empties the coke machine, he finds 130 coins with a total value of \$25.75. If the machine accepts only quarters and dimes, find the number of quarters and dimes Hiroto emptied out of the machine.

Solution: *Step 1* Let a variable represent an unknown quantity.
Let x represent the number of quarters.

Step 2 Use this variable to represent other unknown quantities.
The number of dimes is 130 less x, or $130 - x$.

Step 3 Draw a diagram.
Since the value of a quarter is \$0.25, the value of x quarters is $0.25x$.
The value of a dime is \$0.10, so the value of $(130 - x)$ dimes is $0.10(130 - x)$.

	Number of Coins	Value of One Coin	Total Value
Dimes	$130 - x$	0.10	$0.10(130 - x)$
Quarters	x	0.25	$0.25x$
Total	130		25.75

Step 4   Use the stated problem to write an equation.
The value of all the coins, 25.75, is the sum of the value of the quarters and the value of the dimes.

$$\boxed{\text{money in quarters}} \quad + \quad \boxed{\text{money in dimes}} \quad = \quad \boxed{\text{total money}}$$

$$0.25x \quad + \; 0.10(130 - x) = \quad 25.75$$

Step 5   Solve the equation.

$$0.25x + 13 - 0.10x = 25.75 \qquad \text{Apply the distributive property.}$$
$$0.15x + 13 = 25.75 \qquad \text{Combine like terms.}$$
$$0.15x = 12.75 \qquad \text{Subtract 13 from both sides.}$$
$$x = 85 \qquad \text{Divide both sides by 0.15.}$$

Step 6   State the solution in words.
There are 85 quarters in the machine. The number of dimes in the machine is $130 - x$ or $130 - 85 = 45$.

Step 7   Check the solution in the original stated problem.
These solutions are reasonable since their sum is 130, the required total. Notice that

$$\text{Value of 85 quarters is } 0.25(85) = \$21.25$$
$$\text{Value of 45 dimes is } 0.10(45) = \$\ 4.50$$
$$\text{Their sum is the required sum} = \$25.75 \qquad \blacksquare$$

3

EXAMPLE 3   Rajiv invested part of his $20,000 inheritance in a mutual funds account that pays 7% simple interest yearly, and the rest in a certificate of deposit that pays 9% simple interest yearly. At the end of one year, Rajiv's investments earned $1550. Find the amount he invested at each rate.

Solution: *Step 1*   Let a variable represent the unknown quantity.
Let x represent the amount invested at 7% interest.

Step 2   Use this variable to represent other unknown quantities.
The rest of the invested money is $20,000 less x or $20,000 - x$, the amount at 9%.

Step 3 Draw a diagram.

We apply the simple interest formula $I = PRT$ and organize our information in the following chart. Since there are two different rates of interest and two different amounts invested, we apply the formula twice.

	Principal	· Rate	· Time	=	Interest
7% Fund	x	0.07	1		$x(0.07)(1)$ or $0.07x$
9% Fund	$20,000 - x$	0.09	1		$(20,000 - x)(0.09)(1)$ or $0.09(20,000 - x)$
Total	20,000				1550

Step 4 Use the stated problem to write an equation.

The total interest earned, $1550, is the sum of the interest earned at 7% and the interest earned at 9%.

$$\boxed{\begin{array}{c}\text{interest}\\\text{at 7\%}\end{array}} \quad + \quad \boxed{\begin{array}{c}\text{interest}\\\text{at 9\%}\end{array}} \quad = \quad \boxed{\text{total interest}}$$

$$0.07x \quad + \quad 0.09(20,000 - x) = \quad 1550$$

Step 5 Solve the equation.

$$0.07x + 0.09(20,000 - x) = 1550$$

$0.07x + 1800 - 0.09x = 1550$ Apply the distributive property.

$1800 - 0.02x = 1550$ Combine like terms.

$-0.02x = -250$ Subtract 1800 from both sides.

$x = 12,500$ Divide both sides by -0.02.

Step 6 State the solution in words.

The amount invested at 7% is $12,500. The rest of the money, $20,000 - x = 20,000 - 12,500 = 7500$, was invested at 9%.

Step 7 Check the solution in the original stated problem.

These solutions are reasonable since their sum is $20,000 as required. The annual interest on $12,500 at 7% is $875; the annual interest on $7500 at 9% is $675, and $875 + $675 = 1550. ∎

4

EXAMPLE 4 A jogger starts running at 6 miles per hour. One hour later a biker starts riding along the same path at 15 mph. How long will it take the biker to catch up to the jogger?

Solution: Let t represent the amount of time the biker rides before catching the jogger. Then the amount of time the jogger jogs before the biker catches up can be represented by $t + 1$ because the jogger started 1 hour earlier.

We use the formula $d = rt$ and the organize the information in the following diagram:

	r	$\cdot$	t	$=$	d
Biker	15		t		$15t$
Jogger	6		$t + 1$		$6(t + 1)$

When the biker catches up with the jogger, their distances will be equal. This leads to

$$\text{biker's distance} = \text{jogger's distance}$$

$$15t = 6(t + 1)$$

Solve for t

$15t = 6t + 6$	Apply the distributive property.
$9t = 6$	Subtract $6t$ from both sides.
$t = \dfrac{2}{3}$	Divide both sides by 9.

It takes $\frac{2}{3}$ hour or 40 minutes for the biker to catch up to the jogger. First, notice that this solution is reasonable. If the biker rides $\frac{2}{3}$ hour, his distance is $r \cdot t = 15 \cdot \frac{2}{3} = 10$ miles. Since the jogger started 1 hour earlier, the jogger's time is $1\frac{2}{3} = \frac{5}{3}$ hour, and the jogger's distance is $r \cdot t = 6 \cdot \frac{5}{3} = 10$ miles. The distances traveled are the same. ■

5

EXAMPLE 5 A graduate student in chemistry needs a 40% salt solution for an experiment. How much of a 20% salt solution should he add to 10 liters of a 50% salt solution in order to get the needed 40% salt solution.

Solution: Let x represent the number of liters of the 20% salt solution. The following diagram will help organize our information.

x liters of a 20% salt solution contains 0.2(x) liters of pure salt.
10 liters of a 50% salt solution contains 0.5(10) or 5 liters of pure salt.

Strength of Solution	Liters of Solution	Liters of Salt
20%	x	$0.2(x)$
50%	10	0.5(10) or 5
40%	$10 + x$	$0.4(10 + x)$

The sum of the liters of pure salt in the 20% solution and the 50% solution is equal to the liters of pure salt in the 40% solution.

salt in 20% solution	+	salt in 50% solution	=	salt in 40% solution
$0.2x$	+	5	=	$0.4(10 + x)$

Solve the equation.

$$0.2x + 5 = 0.4(10 + x)$$

$$0.2x + 5 = 4 + 0.4x \qquad \text{Apply the distributive property.}$$

$$0.2x + 5 - 0.2x = 4 + 0.4x - 0.2x \qquad \text{Subtract 0.2 from both sides.}$$

$$5 = 4 + 0.2x \qquad \text{Simplify.}$$

$$5 - 4 = 4 + 0.2x - 4 \qquad \text{Subtract 4 from both sides.}$$

$$1 = 0.2x \qquad \text{Simplify.}$$

$$\frac{1}{0.2} = \frac{0.2x}{0.2} \qquad \text{Divide both sides by 0.2.}$$

$$5 = x \qquad \text{Simplify.}$$

Thus, if 5 liters of a 20% salt solution are mixed with 10 liters of a 50% salt solution, the result is $10 + 5 = 15$ liters of a 40% salt solution, the needed strength. ∎

EXERCISE SET 2.6

Solve each word problem. See Example 1.

1. An architect designs a rectangular flower garden such that the width is exactly two-thirds of the length. If 260 feet of antique picket fencing are to be used, find the dimensions of the garden.

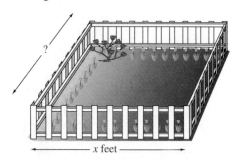

x feet

2. If the length of a rectangular parking lot is 10 meters less than twice its width, and the perimeter is 400 meters, find the length of the parking lot.

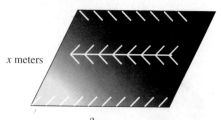

x meters

?

3. A 10-inch pipe cleaner is bent to form a triangle with two equal sides. If the two equal sides are each 3.7 inches long, how long is the third side?

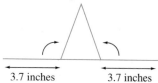

3.7 inches   3.7 inches

4. The perimeter of a yield sign in the shape of an isosceles triangle is 22 feet. If the shortest side is 2 feet less than the other two sides, find the length of the shortest side. (*Hint:* An isosceles triangle has two sides the same length.)

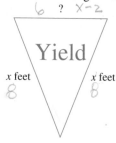

? $x-2$

Yield

x feet   *x* feet

Solve each word problem. See Example 2.

5. The bottom of Maggie's purse contains 20 coins in nickels and dimes. If the coins have a total value of $1.85, find the number of each type of coin.

6. A university drama production sold a total of 400 tickets for $1850. If student tickets sold for $4 and guest tickets sold for $6, how many tickets of each type were sold?

7. A bank teller cashed a $200 check and gave Zita 12 bills using twenty-dollar bills and ten-dollar bills. How many bills of each type were given?

8. A postal drawer contains 15-cent stamps and 25-cent stamps with a total value of $142.50. If the number of 25-cent stamps is fifty more than twice the number of 15-cent stamps, find the number of 25-cent stamps.

Solve the following. See Example 3.

9. Zoya invested part of her $25,000 advance at 8% interest and the rest at 9% interest. If her total yearly interest from both accounts was $2135, find the amount invested at each rate.

10. Michael invested part of his $10,000 bonus in a fund that paid an 11% profit and invested the rest in stock that suffered a 4% loss. Find the amount of each investment if his overall net profit was $650.

11. Shirley invested some money at 9% interest and $250 more than that amount at 10%. If her total yearly interest was $101, how much was invested at each rate?

12. Bruce invested a sum of money at 10% interest and invested twice that amount at 12% interest. If his total yearly income from both investments was $2890, how much was invested at each rate?

Solve each word problem. See Example 4.

13. A jet plane traveling at 500 mph overtakes a propeller plane traveling at 200 mph that had a 2-hour head start. How far from the starting point are the planes?

14. How long will it take a bus traveling at 60 miles per hour to overtake a car traveling at 40 mph if the car had a 1.5-hour head start?

15. The Jones family drove to Disneyland at 50 miles per hour and returned on the same route at 40 mph. Find the distance to Disneyland if the total driving time was 7.2 hours.

16. A bus traveled on a level road for 3 hours at an average speed of 20 miles per hour faster than it traveled on a winding road. The time spent on the winding road was 4 hours. Find the average speed on the level road if the entire trip was 305 miles.

Solve each word problem. See Example 5.

17. How much pure acid should be mixed with 2 gallons of a 40% acid solution in order to get a 70% acid solution?

18. How many cubic centimeters of a 25% antibiotic solution should be added to 10 cubic centimeters of a 60% antibiotic solution in order to get a 30% antibiotic solution?

19. Planter's Peanut Company wants to mix 20 pounds of peanuts worth $3 a pound with cashews worth $5 a pound in order to make an experimental mix worth $3.50 a pound. How many pounds of cashews should be added to the peanuts?

20. Community Coffee Company wishes to mix a new flavor of Cajun coffee. How many pounds of coffee worth $7 a pound should be added to 14 pounds of coffee worth $4 a pound in order to obtain a mixture worth $5 a pound?

21. A flower bed is in the shape of a triangle with one side twice the length of the shortest side, and the third side is 30 feet more than the length of the shortest side. Find the dimensions if the perimeter is 102 feet.

22. A square animal pen and a pen shaped like an equilateral triangle have equal perimeters. Find the length of the sides of each pen if the sides of the triangular pen are fifteen less than twice a side of the square pen. (*Hint:* An equilateral triangle has three sides the same length.)

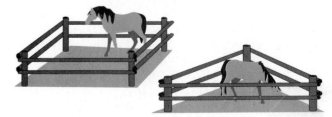

23. How can $54,000 be invested, part at 8% and the remainder at 10%, so that the interest earned by the two accounts will be equal?

24. Ms. Mills invested her $20,000 bonus in two accounts. She took a 4% loss on one investment and made a 12% profit on another investment, but ended up breaking even. How much was invested at each rate?

25. If $3000 is invested at 6%, how much should be invested at 9% so that the total income from both investments is $585?

26. A financial planner invested a certain amount of money at 9%, twice that amount at 10%, and three times that amount at 11%. Find the amount invested at each rate if her total yearly income from the investments was $2790.

27. A piggy bank contains nickels, dimes, and quarters totaling $12.90. The number of dimes is fifteen more than the number of nickels, and the number of quarters is twice the number of dimes. How many nickels were in the bank?

28. At the beginning of the day a cash drawer contained $150 in one-, five-, and ten-dollar bills. There were twice as many ones as fives. If there were 38 bills in the drawer, find the number of each type of bill in the cash drawer.

29. A mother spent $32.25 to take her daughter's birthday party to the movies. Adult tickets cost $5.75 and children tickets cost $3.00. If 8 persons were at the party, how many adult tickets were bought?

30. A vending machine contains $51.75 in nickels, dimes, and quarters. The number of quarters is five more than twice the number of dimes, and the number of nickels is thirty less than the number of dimes. Find the total number of coins in the machine.

31. Jane and Jake leave simultaneously from the same point hiking in opposite directions, Jane walking at 4 miles per hour and Jake at 5 mph. How long can they talk on their walkie-talkies if the walkie-talkies have a 20-mile radius?

32. Alan and Dave leave from the same point driving in opposite directions, Alan driving at 55 miles per hour and Dave at 65 mph. Alan has one-hour head start. How long will they be able to talk on their car phones if the phones have a 250-mile range?

33. How much of an alloy that is 20% copper should be mixed with 200 ounces of an alloy that is 50% copper in order to get an alloy that is 30% copper?

34. How much water should be added to 30 gallons of a solution that is 70% antifreeze in order to get a mixture that is 60% antifreeze?

35. Alice and Linda are 12 miles apart hiking toward each other. How long will it take them to meet if Alice walks at 3 miles per hour and Linda walks 1 mph faster?

36. Two hikers are 11 miles apart and walking toward each other. They meet in 2 hours. Find the rate of each hiker if one hiker walks 1.1 miles per hour faster than the other.

37. On a 255-mile trip, Gary traveled at an average speed of 70 miles per hour, got a speeding ticket, and then traveled at 60 mph for the remainder of the trip. If the entire trip took 4.5 hours and the speeding ticket stop took 30 minutes, how long did Gary speed before getting stopped?

38. Mark can row upstream at 5 miles per hour and downstream at 11 mph. If Mark starts rowing upstream until he gets tired, and then rows downstream to his starting point, how far did Mark row if the entire trip took 4 hours?

Writing in Mathematics

39. Give an example of how you solved a mathematical application in the last week.

Skill Review

Add or subtract the following. See Sections 1.6 and 1.7.

40. $3 + (-7)$

41. $-2 + (-8)$

42. $4 - 10$

43. $-11 + 2$

44. $-5 - (-1)$

45. $-12 - 3$

2.7
Solving Linear Inequalities

OBJECTIVES

Tape 8

1	Define linear inequality.
2	Graph intervals on a number line.
3	Solve linear inequalities.
4	Solve compound inequalities.
5	Solve inequality applications.

1   In Chapter 1, we reviewed these inequality symbols and their meanings:

$<$ means "is less than" $\leq$ means "is less than or equal to"

$>$ means "is greater than" $\geq$ means "is greater than or equal to"

A linear inequality is similar to a linear equation except that the equality symbol is replaced with an inequality symbol.

Linear Equations	*Linear Inequalities*
$x = 3$	$x < 3$
$5n - 6 = 14$	$5n - 6 \geq 14$
$12 = 7 - 3y$	$12 \leq 7 - 3y$
$\dfrac{x}{4} - 6 = 1$	$\dfrac{x}{4} - 6 > 1$

Linear Inequality in One Variable

A linear inequality in one variable is an inequality that can be written in the following form:

$$ax + b < c$$

where a, b, and c are real numbers and a is not 0.

This definition and all other definitions, properties, and steps in this section also hold true for the inequality symbols, $>$, $\geq$, and $\leq$.

2   A **solution of an inequality** is a value of the variable that makes the inequality a true statement. In the inequality $x < 3$, replacing x with any number smaller than 3 or to the left of 3 on the number line makes the resulting inequality true. This means that any number smaller than 3 is a solution of the inequality $x < 3$. Since there are infinitely many such numbers, we cannot list all the solutions of the inequality. We can

picture them, though, on a number line. To do so, shade the portion of the number line corresponding to numbers smaller than 3.

Recall that all the numbers less than 3 lie to the left of 3 on the number line. An open circle about the point representing 3 indicates that 3 is not a solution of the inequality: 3 **is not** less than 3. The shaded arrow indicates that the solutions of $x <$ 3 continue indefinitely to the left of 3.

Picturing the solutions of an inequality on a number line is called **graphing** the solutions or graphing the inequality, and the picture is called the **graph** of the inequality.

To graph $x \le 3$, shade the numbers to the left of 3 and place a closed circle on the point representing 3.

The closed circle indicates that 3 is a solution of the inequality $x \le 3$ and is part of the graph.

EXAMPLE 1   Graph $x \ge -1$.

Solution: We place a closed circle at -1 since the inequality symbol is $\ge$ and -1 is greater than or equal to -1. Then shade to the right of -1.

■

Inequalities containing one inequality symbol are called **simple inequalities,** while inequalities containing two inequality symbols are called **compound inequalities.** A compound inequality is really two simple inequalities in one. The compound inequality

$$3 < x < 5 \quad \text{means} \quad 3 < x \text{ and } x < 5$$

This can be read "x is greater than 3 and less than 5." To graph $3 < x < 5$, place open circles at both 3 and 5 and shade between.

EXAMPLE 2   Graph $2 < x \le 4$.

Solution: Graph all numbers greater than 2 and less than or equal to 4. Place an open circle at 2, a closed circle at 4, and shade between.

■

3   When solutions to a linear inequality are not immediately obvious, they are found through a process similar to the one used to solve a linear equation. Our goal is to isolate the variable, and we use properties of inequality similar to properties of equality.

Addition Property of Inequality

If a, b, and c are real numbers, then

$$a < b \quad \text{and} \quad a + c < b + c$$

are equivalent inequalities.

This property also holds true for subtracting values since subtraction is defined in terms of addition. In other words, adding or subtracting the same quantity from both sides of an inequality does not change the solutions of the inequality.

EXAMPLE 3   Solve $x + 4 \leq -6$ for x. Graph the solution.

Solution: To solve for x, subtract 4 from both sides of the inequality.

$$x + 4 \leq -6 \qquad \text{Original inequality.}$$

$$x + 4 \boxed{-4} \leq -6 \boxed{-4} \qquad \text{Subtract 4 from both sides.}$$

$$x \leq -10 \qquad \text{Simplify.}$$

The solution $x \leq -10$ is graphed as shown.   ■

An important difference between linear equations and linear inequalities is shown when we multiply or divide both sides of an inequality by a nonzero real number. Let us start with the true statement $6 < 8$ and multiply both sides by 2.

$$6 < 8 \qquad \text{True.}$$

$$2(6) < 2(8) \qquad \text{Multiply both sides by 2.}$$

$$12 < 16 \qquad \text{True.}$$

The inequality remains true.

But if we start with the same true statement $6 < 8$ and multiply both sides by -2, the resulting inequality is no longer a true statement.

$$6 < 8 \qquad \text{True.}$$

$$-2(6) < -2(8) \qquad \text{Multiply both sides by } -2.$$

$$-12 < -16 \qquad \textbf{False.}$$

Notice, however, that if we reverse the direction of the inequality symbol, the resulting inequality is true.

$$-12 < -16 \qquad \textbf{False.}$$

$$-12 > -16 \qquad \textbf{True.}$$

This demonstrates the multiplication property of inequality.

> **Multiplication Property of Inequality**
>
> **1.** If a, b, and c are real numbers, and c is **positive,** then
>
> $$a < b \quad \text{and} \quad ac < bc$$
>
> are equivalent inequalities.
>
> **2.** If a, b, and c are real numbers, and c is **negative,** then
>
> $$a < b \quad \text{and} \quad ac > bc$$
>
> are equivalent inequalities.

Because division is defined in terms of multiplication, this property also holds true when dividing both sides of an inequality by a nonzero number: If we multiply or divide both sides of an inequality by a negative number, **the direction of the inequality sign must be reversed for the inequalities to remain equivalent.**

EXAMPLE 4   Solve $-2x \le -4$, and graph the solution.

Solution: Remember to reverse the direction of the inequality symbol when dividing by a negative number.

$$-2x \le -4$$

$$\frac{-2x}{-2} \ge \frac{-4}{-2} \qquad \text{Divide both sides by } -2 \text{ and reverse the inequality sign.}$$

$$x \ge 2 \qquad \text{Simplify.}$$

The solution $x \ge 2$ is graphed as shown.   ∎

EXAMPLE 5   Solve $2x < -4$, and graph the solution.

Solution: $2x < -4$

$$\frac{2x}{2} < \frac{-4}{2} \qquad \text{Divide both sides by 2. Do not reverse the inequality sign.}$$

$$x < -2 \qquad \text{Simplify.}$$

The graph of $x < -2$ is shown.   ∎

Follow these steps to solve linear inequalities.

To Solve Linear Inequalities in One Variable

Step 1   Clear the inequality of fractions by multiplying both sides of the inequality by the lowest common denominator (LCD) of all fractions in the inequality.

Step 2   Remove grouping symbols.

Step 3   Combine like terms on each side of the inequality.

Step 4   Use the addition property of inequality to write the inequality as an equivalent inequality with variable terms on one side and numbers on the other side.

Step 5   Use the multiplication property of inequality to isolate the variable.

Don't forget that if both sides of an inequality are multiplied or divided by a negative number, the direction of the inequality sign must be reversed.

EXAMPLE 6   Solve $-4x + 7 \ge -9$, and graph the solution.

Solution: $-4x + 7 \ge -9$

$$-4x + 7 - 7 \ge -9 - 7 \qquad \text{Subtract 7 from both sides.}$$

$$-4x \ge -16 \qquad \text{Simplify.}$$

$$\frac{-4x}{-4} \le \frac{-16}{-4}$$ Divide both sides by -4 and reverse the direction of the inequality sign.

$$x \le 4$$ Simplify.

The graph of $x \le 4$ is shown.   ■

EXAMPLE 7   Solve $2x + 7 \le x - 11$, and graph the solution.

Solution:

$$2x + 7 \le x - 11$$

$$2x + 7 - x \le x - 11 - x$$ Subtract x from both sides.

$$x + 7 \le -11$$ Simplify.

$$x + 7 - 7 \le -11 - 7$$ Subtract 7 from both sides.

$$x \le -18$$ Simplify.

The solution $x \le -18$ is graphed.   ■

EXAMPLE 8   Solve $-5x + 7 < 2(x - 3)$, and graph the solution.

Solution:

$$-5x + 7 < 2(x - 3)$$

$$-5x + 7 < 2x - 6$$ Use the distributive property.

$$-5x + 7 - 2x < 2x - 6 - 2x$$ Subtract $2x$ from both sides.

$$-7x + 7 < -6$$

$$-7x + 7 - 7 < -6 - 7$$ Subtract 7 from both sides.

$$-7x < -13$$

$$\frac{-7x}{-7} > \frac{-13}{-7}$$ Divide both sides by -7 and reverse the direction of the inequality sign.

$$x > \frac{13}{7}$$ Simplify.

The solution $x > \dfrac{13}{7}$ is graphed.   ■

EXAMPLE 9   Solve $2(x - 3) - 5 \le 3(x + 2) - 18$, and graph the solution.

Solution:

$$2(x - 3) - 5 \le 3(x + 2) - 18$$

$$2x - 6 - 5 \le 3x + 6 - 18$$ Apply the distributive property.

$$2x - 11 \le 3x - 12$$ Simplify.

$$-x - 11 \le -12$$ Subtract $3x$ from both sides.

$$-x \le -1$$ Add 11 to both sides.

$$\frac{-x}{-1} \ge \frac{-1}{-1}$$ Divide both sides by -1 and reverse the direction of the inequality sign.

$$x \ge 1$$ Simplify.

The graph of the solution $x \ge 1$ is as shown.   ■

4 When we solve a simple inequality, we isolate the variable on one side of the inequality. When we solve a compound inequality, we isolate the variable in the middle part of the inequality. Also, when solving a compound inequality, we must perform the same operation to all **three** parts of the inequality: left, middle, and right.

EXAMPLE 10 Solve $-1 \le 2x - 3 < 5$, and graph the solution.

Solution:

$$-1 \le 2x - 3 < 5$$

$$-1 + 3 \le 2x - 3 + 3 < 5 + 3 \qquad \text{Add 3 to all three parts.}$$

$$2 \le 2x < 8 \qquad \text{Simplify.}$$

$$\frac{2}{2} \le \frac{2x}{2} < \frac{8}{2} \qquad \text{Divide all three parts by 2.}$$

$$1 \le x < 4 \qquad \text{Simplify.}$$

The solution $1 \le x < 4$ is graphed. ■

EXAMPLE 11 Solve $3 \le \dfrac{3x}{2} + 4 \le 5$, and graph the solution.

Solution:

$$3 \le \frac{3x}{2} + 4 \le 5$$

$$2(3) \le 2\left(\frac{3x}{2} + 4\right) \le 2(5) \qquad \text{Multiply all three parts by 2.}$$

$$6 \le 3x + 8 \le 10 \qquad \text{Distribute.}$$

$$-2 \le 3x \le 2 \qquad \text{Subtract 8 from all three parts.}$$

$$\frac{-2}{3} \le \frac{3x}{3} \le \frac{2}{3} \qquad \text{Divide all three parts by 3.}$$

$$\frac{-2}{3} \le x \le \frac{2}{3} \qquad \text{Simplify.}$$

The graph of $-\dfrac{2}{3} \le x \le \dfrac{2}{3}$ is as shown. ■

5 Now that we can solve linear inequalities, this skill can be used to solve word problems that involve inequalities. These word problems will be solved using a procedure similar to the one used to solve word problems using linear equations, except that an inequality will be formed instead of an equation.

EXAMPLE 12 If three times a number is added to twice the sum of the number and 3, the result is less than or equal to 0. Find all such numbers.

Solution: Let $x =$ the number(s). Next, write an inequality and solve.

three times a number	added to	twice	the sum of the number and 3	$\le$	0
↓	↓	↓	↓	↓	↓
$3x$	$+$	2	$(x + 3)$	$\le$	0

$$3x + 2x + 6 \leq 0 \qquad \text{Use the distributive property.}$$
$$5x + 6 \leq 0 \qquad \text{Combine like terms.}$$
$$5x \leq -6 \qquad \text{Subtract 6 from both sides.}$$
$$x \leq \frac{-6}{5} \qquad \text{Divide both sides by 5.}$$

Any number less than or equal to $-\dfrac{6}{5}$ is a solution, as shown in the graph.   ∎

MENTAL MATH

Mentally solve each of the following inequalities.

1. $5x > 10$ **2.** $4x < 20$ **3.** $2x \geq 16$ **4.** $9x \leq 63$

EXERCISE SET 2.7

Graph each inequality on a number line. See Examples 1 and 2.

1. $x \leq -1$ **2.** $y < 0$ **3.** $x > \dfrac{1}{2}$

4. $z \geq -\dfrac{2}{3}$ **5.** $-1 < x < 3$ **6.** $2 \leq y \leq 3$

7. $0 \leq y < 2$ **8.** $-1 \leq x \leq 4$

Solve each inequality and graph the solution. See Examples 3 through 5.

9. $2x < -6$ **10.** $3x > -9$ **11.** $x - 2 \geq -7$

12. $x + 4 \leq 1$ **13.** $-8x \leq 16$ **14.** $-5x < 20$

Solve each inequality and graph the solution. See Examples 6 and 7.

15. $3x - 5 > 2x - 8$ **16.** $3 - 7x \geq 10 - 8x$ **17.** $4x - 1 \leq 5x - 2x$

18. $7x + 3 < 9x - 3x$ **19.** $x - 7 < 3(x + 1)$ **20.** $3x + 9 \geq 5(x - 1)$

Solve each inequality and graph the solution. See Examples 8 and 9.

21. $-6x + 2 \geq 2(5 - x)$ **22.** $-7x + 4 > 3(4 - x)$

23. $4(3x - 1) \leq 5(2x - 4)$ **24.** $3(5x - 4) \leq 4(3x - 2)$

25. $3(x + 2) - 6 > -2(x - 3) + 14$

26. $7(x - 2) + x \leq -4(5 - x) - 12$

Solve each inequality, and then graph the solution. See Examples 10 and 11.

27. $-3 < 3x < 6$

28. $-5 < 2x < -2$

29. $2 \leq 3x - 10 \leq 5$

30. $4 \leq 5x - 6 \leq 19$

31. $-4 < 2(x - 3) < 4$

32. $0 < 4(x + 5) < 8$

Solve each application. Write the solution as an inequality. See Example 11.

33. Six more than twice a number is greater than negative fourteen. Find all numbers that make this statement true.

34. Five times a number increased by one is less than or equal to ten. Find all such numbers.

35. The perimeter of a rectangle is to be no greater than 100 centimeters and the width must be 15 centimeters. Find the maximum length of the rectangle.

36. One side of a triangle is four times as long as another side, and the third side is 12 inches long. If the perimeter can be no longer than 87 inches, find the maximum lengths of the other two sides.

Solve the following inequalities. Graph each solution.

37. $-2x \leq -40$

38. $-7x > 21$

39. $-9 + x > 7$

40. $y - 4 \leq 1$

41. $3x - 7 < 6x + 2$

42. $2x - 1 \geq 4x - 5$

43. $5x - 7x \leq x + 2$

44. $4 - x < 8x + 2x$

45. $\frac{3}{4}x > 2$

46. $\frac{5}{6}x \geq -8$

47. $3(x - 5) < 2(2x - 1)$

48. $5(x + 4) < 4(2x + 3)$

49. $4(2x + 1) > 4$

50. $6(2 - x) \geq 12$

51. $-5x + 4 \leq -4(x - 1)$

52. $-6x + 2 < -3(x + 4)$

53. $-2 < 3x - 5 < 7$

54. $1 < 4 + 2x \leq 7$

55. $-2(x - 4) - 3x < -(4x + 1) + 2x$

56. $-5(1 - x) + x \leq -(6 - 2x) + 6$

57. $-3x + 6 \geq 2x + 6$

58. $-(x - 4) < 4$

59. $-6 < 3(x - 2) < 8$

60. $-5 \leq 2(x + 4) < 8$

Solve the following.

61. Ben bowled 146 and 201 in his first two games. What must he bowl in his third game to have an average of at least 180?

62. On an NBA team the two forwards measure 6'8" and 6'6" and the two guards measure 6'0" and 5'9" tall. How tall a center should they hire if they wish to have a starting team average height of at least 6'5"?

63. Twice a number increased by one is between negative five and seven. Find all such numbers.

64. Half a number decreased by four is between two and three. Find all such numbers.

65. A financial planner has a client with $15,000 to invest. If he invests $10,000 in a certificate of deposit paying 11%

interest, at what rate does the remainder of the money need to be invested so that the two investments together yield at least $1600 in yearly interest?

66. Alex earns $600 per month plus a 4% commission on all sales over $1000. Find the minimum sales that will allow Alex to earn at least $3000 per month.

A Look Ahead

Solve each inequality and then graph the solution.

EXAMPLE Solve $x(x - 6) > x^2 - 5x + 6$, and then graph the solution.

Solution:

$$x(x - 6) > x^2 - 5x + 6$$
$$x^2 - 6x > x^2 - 5x + 6$$
$$x^2 - 6x - x^2 > x^2 - 5x + 6 - x^2$$
$$-6x > -5x + 6$$
$$-x > 6$$
$$\frac{-x}{-1} < \frac{6}{-1}$$
$$x < -6$$

The solution $x < -6$ is graphed as shown. ■

67. $x(x + 4) > x^2 - 2x + 6$

68. $x(x - 3) \geq x^2 - 5x - 8$

69. $x^2 + 6x - 10 < x(x - 10)$

70. $x^2 - 4x + 8 < x(x + 8)$

71. $x(2x - 3) \leq 2x^2 - 5x$

72. $x(4x + 1) < 4x^2 - 3x$

Writing in Mathematics

73. Explain how solving a linear inequality is similar to solving a linear equation.

74. Explain how solving a linear inequality is different from solving a linear equation.

Skill Review

Evaluate the following. See Section 1.4.

75. $(2)^3$

76. $(3)^3$

77. $(1)^{12}$

78. 0^5

79. $\left(\frac{4}{7}\right)^2$

80. $\left(\frac{2}{3}\right)^3$

CRITICAL THINKING The Upper New York Community Ice Hockey Association recently decided to use a point system to determine which teams to invite to the season playoffs. For each game it wins during the season, a team is awarded two playoff points; for each tie, one playoff point. At the end of the season, the team in each league with the most playoff points is invited to the playoffs.

Additionally, the association wants to invite to the playoffs any team that accumulates at least as many playoff points as some fixed, predetermined number. During the season, each team plays 24 games, and there are six teams in each of five leagues.

What fixed number of playoff points might the Association use as the cutoff for inviting teams to the playoffs?

What criticisms might a team manager make of the association's playoff point system? Does it make any difference if teams play during the regular season only within their leagues?

If a team manager claims the cutoff number is not fair, what defense can the association offer?

If a team wins w games and ties t games, what inequality expresses the role of the cutoff number? If a team has six games left to play and is eight points shy of the cutoff number, what combinations of wins and ties will allow the team to exceed the cutoff number?

CHAPTER 2 GLOSSARY

A **compound inequality** is of the form $a < x < b$ and means $x > a$ and $x < b$.

An **equation** states that two algebraic expressions are equal.

A **formula** is a known equation that describes a relation among quantities.

Terms with the same variable raised to the exact same powers are called **like terms.**

A **linear equation** in one variable can be written in the form $ax + b = c$, where a, b, and c are real numbers and $a \neq 0$.

A **linear inequality** in one variable is similar to a linear equation in one variable except that the equality symbol is replaced by an inequality symbol.

A **solution** or root of an equation is a value for the variable that makes the equation true.

A **term** is a number or the product of a number and variables raised to powers.

CHAPTER 2 SUMMARY

ADDITION PROPERTY OF EQUALITY (2.2)

If a, b, and c are real numbers, then $a = b$ and $a + c = b + c$ are equivalent equations.

MULTIPLICATION PROPERTY OF EQUALITY (2.3)

If a, b, and c are real numbers with $c \neq 0$, then $a = b$ and $ac = bc$ are equivalent equations.

TO SOLVE LINEAR EQUATIONS   (2.4)

Step 1   Clear the equation of fractions by multiplying each side of the equation by the lowest common denominator (LCD) of all denominators in the equation.

Step 2   Remove any grouping symbols.

Step 3   Combine any like terms on each side of the equation.

Step 4   Use the addition property of equality to write the equation as an equivalent equation with variable terms on one side and numbers on the other side.

Step 5   Use the multiplication property of equality to isolate the variable.

Step 6   Check the answer by substituting it into the original equation.

SOLVING A WORD PROBLEM OR APPLICATION   (2.6)

Step 1   Read and then reread the problem. Choose a variable to represent one unknown quantity.

Step 2   Use this variable to represent any other unknown quantities.

Step 3   Draw a diagram.

Step 4   Translate the word problem into an equation.

Step 5   Solve the equation.

Step 6   Answer the question asked in the application and then check to see if your answer is a reasonable one.

Step 7   Check the solution in the original stated problem.

ADDITION PROPERTY OF INEQUALITY   (2.7)

If a, b, and c are real numbers, and if $a < b$, then $a + c < b + c$.

MULTIPLICATION PROPERTY OF INEQUALITY   (2.7)

If a, b, and c are real numbers and $a < b$ and c is **positive,** then $ac < bc$.
If a, b, and c are real numbers and $a < b$ and c is **negative,** then $ac > bc$.

CHAPTER 2 REVIEW

(2.1)   *Simplify the following expressions.*

1. $5x - x + 2x$

2. $z - 4x - 7z$

3. $\frac{1}{2}x + 3 + \frac{7}{2}x - 5$

4. $\frac{4}{5}y + 1 + \frac{6}{5}y + 2$

5. $2(n - 4) + n - 10$

6. $3(w + 2) - 12 - w$

7. Subtract $7x - 2$ from $x + 5$

8. Subtract $1.4y - 3$ from $y - 0.7$

Write each of the following as algebraic expressions.

9. Three times a number decreased by 7.

10. Twice the sum of a number and 2.8.

11. The sum of two numbers is 10. If one number is x, express the other number in terms of x.

12. Mandy is 5 inches taller than Melissa. If x inches represents the height of Mandy, express Melissa's height in terms of x.

(2.2)   *Solve the following.*

13. $8x + 4 = 9x$

14. $5y - 3 = 6y$

15. $3x - 5 = 4x + 1$

16. $2x - 6 = x - 6$

17. $4(x + 3) = 3(1 + x)$

18. $6(3 + n) = 5(n - 1)$

19. Find a number such that twice the number decreased by 5 is equal to 6 subtracted from three times the number.

20. If twice the sum of a number and 4 is the same as the difference of the number and 8, find the number.

(2.3)   *Solve each equation.*

21. $\dfrac{3}{4}x = -9$

22. $\dfrac{x}{6} = \dfrac{2}{3}$

23. $-3x + 1 = 19$

24. $5x + 25 = 20$

25. $5x - 6 + x = 9 + 4x - 1$

26. $8 - y + 4y = 7 - y - 3$

27. Eighteen subtracted from five times a number is the opposite of three. Find the number.

28. Twice a number increased by four is the opposite of ten. Find the number.

29. Double the sum of a number and six is the opposite of the number. Find the number.

(2.4)   *Solve the following.*

30. $\dfrac{2}{7}x - \dfrac{5}{7} = 1$

31. $\dfrac{5}{3}x + 4 = \dfrac{2}{3}x$

32. $-(5x + 1) = -7x + 3$

33. $-4(2x + 1) = -5x + 5$

34. $-6(2x - 5) = -3(9 + 4x)$

35. $3(8y - 1) = 6(5 + 4y)$

36. $\dfrac{3(2 - z)}{5} = z$

37. $\dfrac{4(n + 2)}{5} = -n$

38. $5(2n - 3) - 1 = 4(6 + 2n)$

39. $-2(4y - 3) + 4 = 3(5 - y)$

40. $9z - z + 1 = 6(z - 1) + 7$

41. $5t - 3 - t = 3(t + 4) - 15$

42. $-n + 10 = 2(3n - 5)$

43. $-9 - 5a = 3(6a - 1)$

44. $\dfrac{5(c + 1)}{6} = 2c - 3$

45. $\dfrac{2(8 - a)}{3} = 4 - 4a$

46. $0.2x - x = 2.4$

47. $y - 0.4y = 0.42$

48. Half the difference of a number and three is the opposite of the number. Find the number.

49. If the sum of a number and 5 is divided by 4, the result is twice the number.

50. The quotient of a number and 3 is the same as the difference of the number and two. Find the number.

51. Find three consecutive odd integers such that three times the smallest integer is the sum of the other two integers decreased by 1.

(2.5)   *Solve each of the following for the indicated variable.*

52. $y = mx + b$ for m

53. $r = vst - 5$ for s

54. $2y - 5x = 7$ for x

55. $3x - 6y = -2$ for y

56. $C = \pi D$ for π

57. $C = 2\pi r$ for π

58. A swimming pool holds 900 cubic meters of water. If its length is 20 meters and its height is 3 meters, find its width.

59. The high temperature in Slidell, Louisiana, one day was 90° Fahrenheit. Convert this temperature to degrees Celsius.

60. How long will it take to run/walk a 10K race (10 kilometers or 10,000 meters) if your average pace is 125 **meters** per minute?

(2.6) *Solve each of the following.*

61. A $50,000 retirement pension is to be invested into two accounts: a money market fund that pays 8.5% and a certificate of deposit that pays 10.5%. How much should be invested at each rate in order to provide a yearly interest income of $4550?

62. A pay phone is holding its maximum number of 500 coins consisting of nickels, dimes, and quarters. The number of quarters is twice the number of dimes. If the value of all the coins is $88.00, how many nickels were in the pay phone?

63. If the length of a rectangular swimming pool is twice the width, and the perimeter is 90 meters, find the length of the pool.

64. How long will it take an Amtrak passenger train to catch up to a freight train if their speeds are 60 and 45 miles per hour and the freight train had an hour and a half head start?

65. Jerry rides a bicycle up a mountain trail at 8 miles per hour and down the same trail at 12 mph. Find the round-trip distance traveled if the total travel time was 5 hours.

(2.7) *Solve and graph the solution of each of the following inequalities.*

66. $x \leq -2$

67. $x > 0$

68. $-1 < x < 1$

69. $0.5 \leq y < 1.5$

70. $-2x \geq -20$

71. $-3x > 12$

72. $5x - 7 > 8x + 5$

73. $x + 4 \geq 6x - 16$

74. $2 \leq 3x - 4 < 6$

75. $-3 < 4x - 1 < 2$

76. $-2(x - 5) > 2(3x - 2)$

77. $4(2x - 5) \leq 5x - 1$

78. Tina earns $175 per week plus a 5% commission on all her sales. Find the minimum amount of sales to ensure that she earns at least $300 per week.

79. Ellen shot rounds of 76, 82, and 79 golfing. What must she shoot on her next round so that her average will be below 80?

CHAPTER 2 TEST

Simplify each of the following expressions.

1. $2y - 6 - y - 4$

2. $x + 7 + x - 16$

3. $4(x - 2) - 3(2x - 6)$

4. $-5(y + 1) + 2(3 - 5y)$

Solve each of the following equations.

5. $-\dfrac{4}{5}x = 4$

6. $4(n - 5) = -(4 - 2n)$

7. $5y - 7 + y = -(y + 3y)$

8. $4z + 1 - z = 1 + z$

9. $\dfrac{2(x + 6)}{3} = x - 5$

10. $\dfrac{4(y - 1)}{5} = 2y + 3$

11. $\dfrac{1}{2} - x + \dfrac{3}{2} = x - 4$

12. $\dfrac{2}{3} + n + \dfrac{1}{3} = n + 2$

13. $\dfrac{1}{3}(y + 3) = 4y$

14. $\dfrac{2}{3}(1 - c) = 5c$

15. $-3(x - 4) + x = 5(3 - x)$

16. $-4(a + 1) - 3a = -7(2a - 3)$

Solve each of the following applications.

17. A number increased by two-thirds of the number is 35. Find the number.

18. A gallon of water seal covers 200 square feet. How many gallons are needed to paint two coats of water seal on a deck that measures 20 feet by 35 feet?

19. How should a $90,000 retirement pension be invested in two accounts so that the yearly interest from the two accounts is equal? One account pays 8% interest and the other pays 10% interest.

20. Tim cashes a $350 payroll check and gets 23 bills in tens and twenties. How many ten-dollar bills were given?

21. Two trains leave Los Angeles simultaneously traveling on the same track in opposite directions at speeds of 50 and 64 miles per hour. How long will it take before they are 287.5 miles apart?

Solve each of the following literal equations for the indicated variable.

22. $V = \pi r^2 h$ for h

23. $W = 6bt$ for t

24. $5g - 2h = p$ for h

25. $3x - 4y = 10$ for y

Solve and graph each of the following inequalities.

26. $3x - 5 > 7x + 3$

27. $x + 6 > 4x - 6$

28. $-2 < 3x + 1 < 8$

29. $0 < 4x - 7 < 9$

30. $\dfrac{2(5x + 1)}{3} > 2$

CHAPTER 2 CUMULATIVE REVIEW

1. Find the absolute value of each number:
   a. $|4|$ **b.** $|-5|$ **c.** $|0|$

2. Write each fraction in lowest terms:
   a. $\dfrac{42}{49}$ **b.** $\dfrac{11}{27}$ **c.** $\dfrac{88}{20}$

3. Add or subtract as indicated. Write the answer in lowest terms:
   a. $\dfrac{2}{7} + \dfrac{4}{7}$ **b.** $\dfrac{3}{10} + \dfrac{2}{10}$ **c.** $\dfrac{9}{7} - \dfrac{2}{7}$
   d. $\dfrac{5}{3} - \dfrac{1}{3}$

4. Simplify $\dfrac{3 + |4 - 3| + 2^2}{6 - 3}$.

5. Find the value of each expression if $x = 3$ and $y = 2$:
   a. $2x - y$ **b.** $\dfrac{3x}{2y}$ **c.** $\dfrac{x}{y} + \dfrac{y}{2}$
   d. $x^2 - y^2$

6. Write each sentence as an equation. Let x represent the unknown number:
   a. The quotient of 15 and a number is 4.
   b. Three subtracted from 12 is a number.
   c. Four times a number added to 17 is 21.

7. Find each sum:

   a. $-3 + (-7)$ **b.** $5 + (+12)$

   c. $(-1) + (-20)$ **d.** $-2 + (-10)$

8. Simplify the following:

   a. $-(-6)$ **b.** $-|-6|$

9. Simplify each expression:

   a. $-3 + [(-2 - 5) - (6 - 4)]$

   b. $2^3 - |10| + [-6 - (-5)]$

10. Find each product:

   a. $(-6)(4)$ **b.** $(2)(-1)$ **c.** $(-5)(-10)$

11. If $x = -2$ and $y = -4$, find the value of each expression:

   a. $5x - y$ **b.** $x^3 - y^2$

   c. $\dfrac{3x}{2y}$

12. Simplify the following by combining like terms:

   a. $7x - 3x$ **b.** $10y^2 + y^2$

   c. $8x^2 + 2x - 3x$

13. Use the distributive property to remove parentheses:

   a. $5(x + 2)$

   b. $-2(y + 0.3z - 1)$

   c. $-(x + y - 2x + 6)$

14. Decide whether 2 is a solution of $3x - 10 = -2x$.

15. Solve $5t - 5 = 6t + 2$ for t.

16. Solve $\dfrac{y}{7} = 20$ for y.

17. Solve for a: $2a + 5a - 10 + 7 = 5a - 13$.

18. Solve for x: $\dfrac{x}{2} - 1 = \dfrac{2}{3}x - 3$.

19. Solve $8(2 - t) = -5t$.

20. Twice the sum of a number and 4 is the same as four times the number minus 12. Find the number.

21. Given the following rectangular solid whose volume is 432 cubic feet, if the length of the solid is 9 feet and its height is 8 feet, find its width.

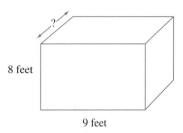

8 feet

9 feet

22. Solve $y = mx + b$ for x.

23. Solve $2x - 7 \le x - 11$, and graph the solution.

3.1 Exponents
3.2 Negative Exponents and Scientific Notation
3.3 Adding and Subtracting Polynomials
3.4 Multiplying Polynomials
3.5 Special Products
3.6 Division of Polynomials

CHAPTER 3

Exponents and Polynomials

Many people enjoy chain letters, but certain chain letters violate laws of the U.S. Postal Service. Subtle reasoning involving exponential growth justifies the law (See Critical Thinking, page 150.)

INTRODUCTION

Recall from Chapter 1 that an exponent is a shorthand notation for repeated factors. This chapter explores additional concepts about exponents and exponential expressions. An especially useful type of exponential expression is a polynomial. Polynomials model many real-world phenomena. Our goal in this chapter is to become proficient with operations on polynomials.

3.1
Exponents

OBJECTIVES

Tape 9

1 Evaluate a number raised to a power.

2 Use the product rule for exponents.

3 Use the power rule for exponents.

4 Use power rules for products and quotients.

5 Use the quotient rule for exponents, and define a number raised to the 0 power.

1 As we reviewed in Section 1.4, an exponent is a shorthand notation for repeated factors. For example, $2 \cdot 2 \cdot 2 \cdot 2 \cdot 2$ can be written as 2^5.

$$\underbrace{5 \cdot 5 \cdot 5 \cdot 5 \cdot 5 \cdot 5}_{6 \text{ factors of } 5} = 5^6 \quad \text{and} \quad \underbrace{(-3) \cdot (-3) \cdot (-3) \cdot (-3)}_{4 \text{ factors of } -3} = (-3)^4$$

The **base** of an exponential expression is the repeated factor. The **exponent** is the number of times that the base is a factor.

$$5^6 \quad \begin{matrix} \text{exponent} \\ \text{base} \end{matrix} \qquad (-3)^4 \quad \begin{matrix} \text{exponent} \\ \text{base} \end{matrix}$$

Definition of a^n

If a is a real number and n is a positive integer, then **a raised to the nth power,** written a^n, is the product of n factors of a.

$$a^n = \underbrace{a \cdot a \cdot a \cdot a \cdot a \cdot \ldots a}_{n \text{ factors of } a}$$

EXAMPLE 1   Evaluate each expression.

     a. 2^3 **b.** 3^1 **c.** $(-4)^2$ **d.** -4^2

Solution: **a.** $2^3 = 2 \cdot 2 \cdot 2 = 8$

     b. To raise 3 to the first power means that 3 will be used as a factor only once. Therefore, $3^1 = 3$. Also, when no exponent is shown, the exponent is assumed to be 1.

     c. $(-4)^2 = (-4)(-4) = 16$

     d. $-4^2 = -(4 \cdot 4) = -16$ ∎

     Notice how similar -4^2 is to $(-4)^2$. The difference between the two is the parentheses. In $(-4)^2$ the parentheses tells us that the base, or repeated factor, is -4. In -4^2 only 4 is the base.

HELPFUL HINT

Be careful when identifying the base of an exponent.

$(-3)^2$	-3^2	$2 \cdot 3^2$
Base is -3	Base is 3	Base is 3
$(-3)^2 = (-3)(-3) = 9$	$-3^2 = -(3 \cdot 3) = -9$	$2 \cdot 3^2 = 2 \cdot 3 \cdot 3 = 18$

---

2   Exponents used with variables have the same meaning as they do with numbers. If x is a real number and n is a positive integer, then x^n is the product of n factors of x.

$$x^n = \underbrace{x \cdot x \cdot x \cdot x \cdot x \ldots x}_{n \text{ factors of } x}$$

     Exponential expressions can be multiplied, divided, added, subtracted, and themselves raised to powers. Suppose we multiply $5^4 \cdot 5^3$.

$$5^4 \cdot 5^3 = \underbrace{(5 \cdot 5 \cdot 5 \cdot 5)}_{4 \text{ factors of } 5} \cdot \underbrace{(5 \cdot 5 \cdot 5)}_{3 \text{ factors of } 5}$$

$$= \underbrace{5 \cdot 5 \cdot 5 \cdot 5 \cdot 5 \cdot 5 \cdot 5}_{7 \text{ factors of } 5}$$

$$= 5^7$$

Or we might multiply x^2 by x^3:

$$x^2 \cdot x^3 = (x \cdot x) \cdot (x \cdot x \cdot x)$$

$$= x \cdot x \cdot x \cdot x \cdot x$$

$$= x^5$$

In both cases, notice that the result is exactly the same if the exponents are added.

$$5^4 \cdot 5^3 = 5^{4+3} = 5^7 \quad \text{and} \quad x^2 \cdot x^3 = x^{2+3} = x^5$$

The following property states this result.

Product Rule for Exponents

If m and n are positive integers, and a is a real number, then

$$a^m \cdot a^n = a^{m+n}$$

This rule says that to multiply two exponential expressions with a **common base,** keep the base and add the exponents.

EXAMPLE 2   Simplify each product.

a. $4^2 \cdot 4^5$ **b.** $x^2 \cdot x^5$ **c.** $y^3 \cdot y$ **d.** $y^3 \cdot y^2 \cdot y^7$ **e.** $(-5)^7 \cdot (-5)^8$

Solution: **a.** $4^2 \cdot 4^5 = 4^{2+5} = 4^7$

b. $x^2 \cdot x^5 = x^{2+5} = x^7$

c. $y^3 \cdot y = y^3 \cdot y^1$ Recall that if no exponent is

$\quad\quad = y^{3+1}$ written, it is assumed to be 1.

$\quad\quad = y^4$

d. $y^3 \cdot y^2 \cdot y^7 = y^{3+2+7} = y^{12}$

e. $(-5)^7 \cdot (-5)^8 = (-5)^{7+8} = (-5)^{15}$. Although we do not want to evaluate $(-5)^{15}$, we can simplify this expression. Because $(-5)^{15}$ is the product of an odd number of negative numbers, the product will be negative so that

odd number

$(-5)^{15}$ can also be written as -5^{15}.

Both expressions have the same value.   ∎

EXAMPLE 3   Simplify $(2x^2)(-3x^5)$.

Solution: Recall that $2x^2$ means $2 \cdot x^2$ and $-3x^5$ means $-3 \cdot x^5$.

$(2x^2)(-3x^5) = 2 \cdot x^2 \cdot -3 \cdot x^5$ Remove parentheses.

$\quad\quad = 2 \cdot -3 \cdot x^2 \cdot x^5$ Group factors with like bases.

$\quad\quad = -6x^7$ Multiply.   ∎

3   Exponential expressions can themselves be raised to powers. Let's try to discover a rule that simplifies an expression like $(7^2)^3$. By the definition of a^n,

$$(7^2)^3 = \underbrace{(7^2)(7^2)(7^2)}_{3 \text{ factors of } 7^2}$$

which can be simplified by the product rule by adding exponents. Then

$$(7^2)^3 = (7^2)(7^2)(7^2) = 7^{2+2+2} = 7^6$$

Notice that the result is exactly the same if we multiply the exponents.

$$(7^2)^3 = 7^{2 \cdot 3} = 7^6$$

The following property states this result.

Power Rule for Exponents

If m and n are positive integers and a is a real number, then

$$(a^m)^n = a^{mn}$$

This rule says that to raise an exponential expression to a power, keep the base and multiply the exponents.

EXAMPLE 4 Simplify each of the following expressions.

a. $(x^2)^5$ **b.** $(y^8)^2$ **c.** $[(-5)^3]^4$

Solution: **a.** $(x^2)^5 = x^{2 \cdot 5} = x^{10}$

b. $(y^8)^2 = y^{8 \cdot 2} = y^{16}$

c. $[(-5)^3]^4 = (-5)^{12}$. Because $(-5)^{12}$ is the product of an even number of negative numbers, their product is a positive number so that

$$\overset{\text{even number}}{(-5)^{12}} \quad \text{can be written as} \quad 5^{12}$$

Both expressions have the same value. ■

4 When a base contains more than one factor, the definition of a^n still applies. To simplify $(x^2y)^3$, for example,

$$(x^2y)^3 = \underbrace{(x^2y)(x^2y)(x^2y)}_{3 \text{ factors of } x^2y}$$

$$= (x^2 \cdot x^2 \cdot x^2)(y^1 \cdot y^1 \cdot y^1) \qquad \text{Group common bases.}$$

$$= (x^{2+2+2})(y^{1+1+1}) \qquad \text{Add exponents of common bases.}$$

$$= x^6y^3 \qquad \text{Simplify.}$$

The result is exactly the same if each factor of the base is raised to the third power.

$$(x^2y)^3 = (x^2)^3 \cdot y^3 = x^6y^3$$

The definition of a^n applies as well when the base is a quotient. To simplify an expression like $\left(\dfrac{x}{7}\right)^4$,

$$\left(\frac{x}{7}\right)^4 = \underbrace{\left(\frac{x}{7}\right)\left(\frac{x}{7}\right)\left(\frac{x}{7}\right)\left(\frac{x}{7}\right)}_{4 \text{ factors of } \frac{x}{7}}$$

$$= \frac{x \cdot x \cdot x \cdot x}{7 \cdot 7 \cdot 7 \cdot 7} \qquad \text{Multiply fractions.}$$

$$= \frac{x^4}{7^4} \qquad \text{Simplify.}$$

Notice that the result is exactly the same if we raise both the numerator and the denominator of the base to the fourth power.

$$\left(\frac{x}{7}\right)^4 = \frac{x^4}{7^4}$$

In general, we have the following.

Power Rules for Products and Quotients

If n is a positive integer and a, b, and c are real numbers, then

$$(ab)^n = a^n b^n \quad \text{and} \quad \left(\frac{a}{c}\right)^n = \frac{a^n}{c^n}$$

as long as c is not 0.

In other words, to raise a product to a power, raise each factor of the product to the power. Also, to raise a quotient to a power, raise both the numerator and the denominator to the power.

EXAMPLE 5   Simplify each expression.

a. $(st)^4$ **b.** $\left(\dfrac{m}{n}\right)^7$ **c.** $(2a)^3$ **d.** $(-5x^2y^3z)^2$ **e.** $\left(\dfrac{2x^4}{3y^5}\right)^4$

Solution: **a.** $(st)^4 = s^4 \cdot t^4 = s^4 t^4$ Use the power rule for a product.

b. $\left(\dfrac{m}{n}\right)^7 = \dfrac{m^7}{n^7}, \ n \neq 0$ Use the power rule for a quotient.

c. $(2a)^3 = 2^3 \cdot a^3 = 8a^3$ Use the power rule for a product.

d. $(-5x^2y^3z)^2 = (-5)^2 \cdot (x^2)^2 \cdot (y^3)^2 \cdot (z^1)^2$ Use the power rule for a product.

$\qquad\qquad = 25x^4y^6z^2$

e. $\left(\dfrac{2x^4}{3y^5}\right)^4 = \dfrac{2^4 \cdot (x^4)^4}{3^4 \cdot (y^5)^4}$ Use the power rules for products and quotients.

$\qquad\qquad = \dfrac{16x^{16}}{81y^{20}}, \ y \neq 0$ Use the power rule for exponents.   ∎

---

5   The final pattern for simplifying exponential expressions relates to quotients. To simplify an expression like $\dfrac{x^5}{x^3}$, in which the numerator and the denominator have a common base, we can divide the numerator and the denominator by common factors. Assume for the remainder of this section, denominators are not 0.

$$\frac{x^5}{x^3} = \frac{x \cdot x \cdot x \cdot x \cdot x}{x \cdot x \cdot x}$$

$$= \frac{x \cdot x \cdot x \cdot x \cdot x}{x \cdot x \cdot x}$$

$$= x \cdot x = x^2$$

Notice that the result is exactly the same if we subtract exponents of the common base.

$$\frac{x^5}{x^3} = x^{5-3} = x^2$$

The quotient rule for exponents states this result in a general way.

Quotient Rule for Exponents

If m and n are positive integers and a is a real number, then

$$\frac{a^m}{a^n} = a^{m-n}$$

as long as a is not 0.

This rule says that to divide one exponential expression by another with a common base, keep the base and subtract exponents.

EXAMPLE 6 Simplify each quotient.

a. $\dfrac{x^5}{x^2}$ b. $\dfrac{4^7}{4^3}$ c. $\dfrac{(-3)^5}{(-3)^2}$ d. $\dfrac{2x^5y^2}{xy}$

Solution: a. $\dfrac{x^5}{x^2} = x^{5-2} = x^3$ Use the quotient rule.

b. $\dfrac{4^7}{4^3} = 4^{7-3} = 4^4 = 256$ Use the quotient rule.

c. $\dfrac{(-3)^5}{(-3)^2} = (-3)^3 = -27$

d. Begin by grouping common bases.

$$\frac{2x^5y^2}{xy} = 2 \cdot \frac{x^5}{x^1} \cdot \frac{y^2}{y^1}$$

$$= 2 \cdot (x^{5-1}) \cdot (y^{2-1}) \qquad \text{Use the quotient rule.}$$

$$= 2x^4y^1 \quad \text{or} \quad 2x^4y \quad \blacksquare$$

Let's look at one more case. To simplify $\dfrac{x^3}{x^3}$, we use the quotient rule and subtract exponents.

$$\frac{x^3}{x^3} = x^{3-3} = x^0$$

But our definition of a^n does not include the possibility that n might be 0. What is the meaning when 0 is an exponent? To find out, simplify $\dfrac{x^3}{x^3}$ by dividing the numerator and denominator by common factors.

$$\frac{x^3}{x^3} = \frac{x \cdot x \cdot x}{x \cdot x \cdot x} = 1$$

Since $\dfrac{x^3}{x^3} = x^0$ and $\dfrac{x^3}{x^3} = 1$, we conclude that $x^0 = 1$ as long as x is not 0.

> **Zero Exponent**
>
> $a^0 = 1$, as long as a is not 0.

EXAMPLE 7   Simplify the following expressions.
   a. 3^0 **b.** $(ab)^0$ **c.** $(-5)^0$ **d.** -5^0

Solution: **a.** $3^0 = 1$

b. Assume that neither a nor b is zero.

$$(ab)^0 = a^0 \cdot b^0 = 1 \cdot 1 = 1$$

c. $(-5)^0 = 1$ **d.** $-5^0 = -1 \cdot 5^0 = -1 \cdot 1 = -1$ ∎

HELPFUL HINT

These examples will remind you of the difference between adding and multiplying terms.

Addition	*Multiplication*
$5x^3 + 3x^3 = (5+3)x^3 = 8x^3$	$(5x^3)(3x^3) = 5 \cdot 3 \cdot x^3 \cdot x^3 = 15x^{3+3} = 15x^6$
$7x + 4x^2 = 7x + 4x^2$	$(7x)(4x^2) = 7 \cdot 4 \cdot x \cdot x^2 = 28x^{1+2} = 28x^3$

In the next example, exponential expressions are simplified using two or more of the exponent rules presented in this section.

EXAMPLE 8   Simplify the following.

a. $\left(\dfrac{-5x^2}{y^3}\right)^2$ **b.** $\dfrac{(x^3)^4 x}{x^7}$ **c.** $\dfrac{(2x)^5}{x^3}$ **d.** $\dfrac{(a^2 b)^3}{a^3 b^2}$

Solution: **a.** Use the power rules for products and quotients; then use the power rule for exponents.

$$\left(\frac{-5x^2}{y^3}\right)^2 = \frac{(-5)^2 (x^2)^2}{(y^3)^2} = \frac{25x^4}{y^6}$$

b. $\dfrac{(x^3)^4 x}{x^7} = \dfrac{x^{12} \cdot x}{x^7} = \dfrac{x^{12+1}}{x^7} = \dfrac{x^{13}}{x^7} = x^{13-7} = x^6$

c. Use the power rules for products and quotients; then use the quotient rule.

$$\frac{(2x)^5}{x^3} = \frac{2^5 \cdot x^5}{x^3} = 2^5 \cdot x^{5-3} = 32x^2$$

d. Begin by applying the power rule for a product to the numerator.

$$\frac{(a^2 b)^3}{a^3 b^2} = \frac{(a^2)^3 \cdot b^3}{a^3 \cdot b^2}$$

$$= \frac{a^6 b^3}{a^3 b^2} \qquad \text{Use the power rule for exponents.}$$

$$= a^{6-3} b^{3-2} \qquad \text{Use the quotient rule.}$$

$$= a^3 b^1 \quad \text{or} \quad a^3 b \quad ∎$$

MENTAL MATH

State the base and the exponent for each of the following expressions.

1. 3^2

2. 5^4

3. $(-3)^6$

4. -3^7

5. -4^2

6. $(-4)^3$

EXERCISE SET 3.1

Evaluate each expression. See Example 1.

1. 7^2

2. -3^2

3. $(-5)^1$

4. $(-3)^2$

5. -2^4

6. -4^3

7. $(-2)^4$

8. $(-4)^3$

Simplify each expression. See Examples 2 and 3.

9. $x^2 \cdot x^5$

10. $y^2 \cdot y$

11. $(5y^4)(3y)$

12. $(-2z^3)(-2z^2)$

13. $(4z^{10})(-6z^7)(z^3)$

14. $(12x^5)(-x^6)(x^4)$

Simplify each expression. See Examples 4 and 5.

15. $(pq)^7$

16. $(4s)^3$

17. $\left(\dfrac{m}{n}\right)^9$

18. $\left(\dfrac{xy}{7}\right)^2$

19. $(x^2y^3)^5$

20. $(a^4b)^7$

21. $\left(\dfrac{-2xz}{y^5}\right)^2$

22. $\left(\dfrac{y^4}{-3z^3}\right)^3$

Simplify each expression. See Example 6.

23. $\dfrac{x^3}{x}$

24. $\dfrac{y^{10}}{y^9}$

25. $\dfrac{(-2)^5}{(-2)^3}$

26. $\dfrac{(-5)^{14}}{(-5)^{11}}$

27. $\dfrac{p^7q^{20}}{pq^{15}}$

28. $\dfrac{x^8y^6}{y^5}$

29. $\dfrac{7x^2y^6}{14x^2y^3}$

30. $\dfrac{9a^4b^7}{3ab^2}$

Simplify the following. See Example 7.

31. $(2x)^0$

32. $-4x^0$

33. $-2x^0$

34. $(4y)^0$

35. $5^0 + y^0$

36. $-3^0 + 4^0$

Simplify the following. See Example 8.

37. $\dfrac{5p^3q^2}{pq}$

38. $\dfrac{4x^{11}y^8}{(2x^2y)^5}$

39. $\dfrac{(3x)^4yz^3}{zx^2}$

40. $\dfrac{4x(a^2b^3)^2}{3(a^2)^2}$

41. $\dfrac{(6mn)^5}{mn^2}$

42. $\dfrac{(6xy)^2x}{9x^2y^2}$

43. $\dfrac{a^5b^7}{a^4b^3}$

44. $\dfrac{10y^3x^3}{2xy}$

Simplify the following.

45. -5^1

46. $(6x)^1$

47. $(-4)^3$

48. -4^3

49. $(6b)^0$

50. $(5ab)^0$

51. $2^3 + 2^5$

52. $7^2 - 7^0$

53. b^4b^2

54. y^4y^1

55. $a^2a^3a^4$

56. $x^2x^{15}x^9$

57. $(2x^3)(-8x^4)$

58. $(3y^4)(-5y)$

59. $2x^3 - 8x^4$

60. $b^5 + 2b^3$

61. $(4a)^3$

62. $(2ab)^4$

63. $(-6xyz^3)^2$

64. $(-3xy^2a$

65. $\left(\dfrac{3y^5}{6x^4}\right)^3$

66. $\left(\dfrac{2ab}{6yz}\right)^4$

67. $\dfrac{x^5}{x^4}$

68. $\dfrac{5x^9}{x^3}$

69. $\dfrac{2x^3y^2z}{xyz}$

70. $\dfrac{(ab^2)^4}{a^3b^9}$

71. $\dfrac{10a^9y^{12}}{35a^4y^7}$

72. $\dfrac{x^{12}y^{13}}{x^5y^7}$

73. $\dfrac{3x^4y^6}{-2xy^2}$

74. $\dfrac{2x^3y^5}{5x^2y^4}$

75. $\left(\dfrac{a^3}{q^2}\right)^5$

76. $\left(\dfrac{9x}{7b}\right)^2$

77. $\left(\dfrac{y^3y^5}{xy^2}\right)^3$

78. $\left(\dfrac{a^9b^4}{a^3b^4}\right)^4$

79. $\left(\dfrac{5xy}{5^2x^2}\right)^2$

80. $\left(\dfrac{3x^4b^8}{3x}\right)^5$

81. $\dfrac{(3x^2)^5}{x^3}$

82. $\dfrac{(4a^2)^4}{a^4b}$

83. $\dfrac{7(xy)^3}{(7xy)^2}$

84. $\left(\dfrac{5}{-q}\right)^2$

85. $\dfrac{(-5)^2}{-q}$

86. $\left(\dfrac{w}{-v}\right)^5$

87. The following rectangle has width $4x^2$ feet and length $5x^3$ feet. Find its area.

$4x^2$ feet

$5x^3$ feet

88. Given the following circle with radius $5y$ centimeters, find its area. Do not approximate π.

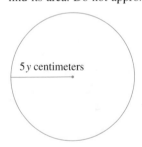

$5y$ centimeters

89. Given the following vault in the shape of a cube, if each side is $3y^4$ feet, find its volume.

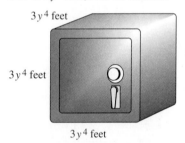

$3y^4$ feet

$3y^4$ feet

$3y^4$ feet

90. The silo shown is in the shape of a cylinder. If its radius is $4x$ meters and its height is $5x^3$ meters, find its volume.

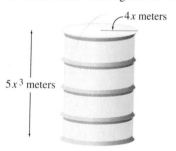

$4x$ meters

$5x^3$ meters

A Look Ahead

Simplify each expression. Assume that variables represent positive integers. See the following example.

EXAMPLE   Simplify $x^a \cdot x^{3a}$.

Solution: Like bases, so add exponents.

$$x^a \cdot x^{3a} = x^{a+3a} = x^{4a} \quad \blacksquare$$

91. $x^{5a}x^{4a}$

92. $b^{9a}b^{4a}$

93. $(a^b)^5$

94. $(2a^{4b})^4$

95. $\dfrac{x^{9a}}{x^{4a}}$

96. $\dfrac{y^{15b}}{y^{6b}}$

97. $(x^ay^bz^c)^{5a}$

98. $(9a^2b^3c^4d^5)^{ab}$

Writing in Mathematics

99. Explain why $(-5)^4 = 625$, while $-5^4 = -625$.

Skill Review

Evaluate each expression. See Section 1.6.

100. $6 + 7 \cdot 4 - 1$ **101.** $7 + 7 \cdot 2 - 3$ **102.** $2 + 9(6 + 1)$

103. $12 + 2(6 - 9)$ **104.** $(45 + 25) \div 7 - 14$ **105.** $(45 + 25) \div (14 - 7)$

3.2
Negative Exponents and Scientific Notation

OBJECTIVES

Tape 9

1   Evaluate numbers raised to negative integer powers.

2   Use all exponents rules and definitions to simplify exponential expressions.

3   Write numbers in scientific notation.

4   Convert numbers from scientific notation to standard form.

---

1   Our work with exponential expressions so far has been limited to exponents that are positive integers or 0. Here we expand to give meaning to an expression like x^{-3}.

Suppose that we wish to simplify the expression $\dfrac{x^2}{x^5}$. If we use the quotient rule for exponents, we subtract exponents:

$$\frac{x^2}{x^5} = x^{2-5} = x^{-3}, x \neq 0$$

But what does x^{-3} mean? Let's simplify $\dfrac{x^2}{x^5}$ using the definition of a^n.

$$\frac{x^2}{x^5} = \frac{x \cdot x}{x \cdot x \cdot x \cdot x \cdot x}$$

$$= \frac{x \cdot x}{x \cdot x \cdot x \cdot x \cdot x} \qquad \text{Divide numerator and denominator by common factors.}$$

$$= \frac{1}{x^3}$$

If the quotient rule is to hold true for negative exponents, then x^{-3} must equal $\dfrac{1}{x^3}$.

From this example, we state the definition for negative exponents.

Negative Exponents

If a is a real number other than 0, and n is an integer, then

$$a^{-n} = \frac{1}{a^n}$$

EXAMPLE 1   Simplify each expression. Write answers using positive exponents only.

a. 3^{-2} b. $2x^{-3}$ c. $\dfrac{1}{2^{-5}}$

Solution: a. $3^{-2} = \dfrac{1}{3^2} = \dfrac{1}{9}$ Use the definition of negative exponent.

b. $2x^{-3} = 2 \cdot \dfrac{1}{x^3} = \dfrac{2}{x^3}$ Use the definition of negative exponent.

Since there are no parentheses, notice that the exponent -3 applies only to the base of x.

c. $\dfrac{1}{2^{-5}} = \dfrac{1}{\dfrac{1}{2^5}}$ Use the definition of negative exponent.

$= 1 \div \dfrac{1}{2^5}$

$= 1 \cdot \dfrac{2^5}{1} = 2^5$ or   32   ■

HELPFUL HINT

It may help you to think about negative exponents in this way: When a **factor** is moved from numerator to denominator or denominator to numerator (as in the previous example), the **sign** of its **exponent** changes. For example,

$$x^{-2} = \frac{1}{x^2}, \qquad 2^{-3} = \frac{1}{2^3} \quad \text{or} \quad \frac{1}{8}$$

$$\frac{1}{y^{-4}} = \frac{1}{\dfrac{1}{y^4}} = y^4, \qquad \frac{1}{5^{-2}} = 5^2 \quad \text{or} \quad 25$$

EXAMPLE 2   Simplify each expression. Write answers with positive exponents.

a. $\left(\dfrac{2}{3}\right)^{-4}$ b. $2^{-1} + 4^{-1}$ c. $(-2)^{-4}$

Solution: a. $\left(\dfrac{2}{3}\right)^{-4} = \dfrac{2^{-4}}{3^{-4}} = \dfrac{3^4}{2^4} = \dfrac{81}{16}$ b. $2^{-1} + 4^{-1} = \dfrac{1}{2} + \dfrac{1}{4} = \dfrac{2}{4} + \dfrac{1}{4} = \dfrac{3}{4}$

c. $(-2)^{-4} = \dfrac{1}{(-2)^4} = \dfrac{1}{(-2)(-2)(-2)(-2)} = \dfrac{1}{16}$ ■

EXAMPLE 3 Simplify each expression. Write answers with positive exponents.

a. $\dfrac{y}{y^{-2}}$ b. $\dfrac{p^{-4}}{q^{-9}}$ c. $\dfrac{x^{-5}}{x^7}$

Solution: a. $\dfrac{y}{y^{-2}} = y^{1-(-2)} = y^3$ b. $\dfrac{p^{-4}}{q^{-9}} = \dfrac{q^9}{p^4}$ c. $\dfrac{x^{-5}}{x^7} = x^{-5-7} = x^{-12} = \dfrac{1}{x^{12}}$ ∎

2 The following is a summary of the rules and definitions for exponents.

Summary of Exponent Rules

If m and n are integers and a, b, and c are real numbers, then:

Product rule for exponents: $\qquad\qquad\qquad a^m \cdot a^n = a^{m+n}$

Power rule for exponents: $\qquad\qquad\qquad (a^m)^n = a^{m \cdot n}$

Power rules for products and quotients: $\quad (ab)^n = a^n b^n$ and

$$\left(\dfrac{a}{c}\right)^n = \dfrac{a^n}{c^n}, c \neq 0$$

Quotient rule for exponents: $\qquad\qquad \dfrac{a^m}{a^n} = a^{m-n}, a \neq 0$

Zero exponent: $\qquad\qquad\qquad\qquad\quad a^0 = 1, a \neq 0$

Negative exponent: $\qquad\qquad\qquad\quad a^{-n} = \dfrac{1}{a^n}, a \neq 0$

EXAMPLE 4 Simplify the following expressions. Write each answer using positive exponents only.

a. $(2x^3)(5x)^{-2}$ b. $\left(\dfrac{3a^2}{b}\right)^{-3}$ c. $\dfrac{4^{-1}x^{-3}y}{4^{-3}x^2y^{-6}}$ d. $(y^{-3}z^{-6})^{-6}$ e. $\left(\dfrac{-2x^3y}{xy^{-1}}\right)^3$

Solution: a. $(2x^3)(5x)^{-2} = 2x^3 \cdot 5^{-2}x^{-2}$ Use Power Rule.

$\qquad\qquad = \dfrac{2x^{3+(-2)}}{5^2}$ Use the Product and Quotient Rules, and Definition of Negative Exponent

$\qquad\qquad = \dfrac{2x}{25}$

b. $\left(\dfrac{3a^2}{b}\right)^{-3} = \dfrac{3^{-3}a^{-6}}{b^{-3}} = \dfrac{b^3}{3^3a^6} = \dfrac{b^3}{27a^6}$

c. $\dfrac{4^{-1}x^{-3}y}{4^{-3}x^2y^{-6}} = 4^{-1-(-3)}x^{-3-2}y^{1-(-6)} = 4^2x^{-5}y^7 = \dfrac{4^2y^7}{x^5}$

d. $(y^{-3}z^6)^{-6} = y^{18} \cdot z^{-36} = \dfrac{y^{18}}{z^{36}}$

e. $\left(\dfrac{-2x^3y}{xy^{-1}}\right)^3 = \dfrac{(-2)^3x^9y^3}{x^3y^{-3}} = \dfrac{-8x^9y^3}{x^3y^{-3}} = -8x^{9-3}y^{3-(-3)} = -8x^6y^6$ ∎

3   Both very large and very small numbers frequently occur in many fields of science. For example, the distance between the sun to the planet Pluto is approximately 5,906,000,000 kilometers, while the mass of a proton is approximately 0.000000000000000000000000165 gram. It takes too much time and paper space to write these numbers in standard notation like this, so **scientific notation** is used as a convenient shorthand for expressing very large and very small numbers.

Scientific Notation

A positive number is written in scientific notation if it is written as the product of a number a, where $1 \leq a < 10$, and an integer power r of 10:

$$a \times 10^r$$

The following numbers are written in scientific notation. The $\times$ sign for multiplication is used as part of the notation.

$$2.03 \times 10^2 \qquad 7.362 \times 10^7$$
$$1 \times 10^{-3} \qquad 8.1 \times 10^{-5}$$

To write the distance between the sun and Pluto in scientific notation, begin by moving the decimal point to the left until we have a number between 1 and 10.

$$5906000000.$$

Next, count the number of places the decimal point is moved.

$$5906000000.$$

9 decimal places

We moved the decimal point 9 places **to the left.** This count is used as the power of 10.

$$5{,}906{,}000{,}000 = 5.906 \times 10^9$$

To express the mass of a proton in scientific notation, move the decimal until the number is between 1 and 10.

$$0.000000000000000000000000165$$

The decimal point was moved 24 places **to the right,** so the exponent on 10 is negative 24.

$$0.000\ 000\ 000\ 000\ 000\ 000\ 000\ 001\ 65 = 1.65 \times 10^{-24}$$

To Write a Number in Scientific Notation

Step 1   Move the decimal point in the original number so that the new number has a value between 1 and 10.

Step 2   Count the number of decimal places the decimal point is moved in step 1. If the decimal point is moved to the left, the count is positive. If the decimal point is moved to the right, the count is negative.

Step 3   Multiply the new number in step 1 by 10 raised to an exponent equal to the count found in step 2.

EXAMPLE 5 Write the following numbers in scientific notation.
a. 367,000,000   b. 0.000003   c. 20,520,000,000   d. 0.00085

Solution: a. *Step 1*   Move the decimal point until the number is between 1 and 10.

$$367,000,000.$$

Step 2   The decimal point is moved to the left 8 places so the count is positive 8.

Step 3 $367,000,000 = 3.67 \times 10^8$.

b. *Step 1*   Move the decimal point until the number is between 1 and 10.

$$0.000003$$

Step 2   The decimal point is moved 6 places to the right so the count is -6.

Step 3 $0.000003 = 3.0 \times 10^{-6}$.

c. $20,520,000,000 = 2.052 \times 10^{10}$

d. $0.00085 = 8.5 \times 10^{-4}$ ■

4   A number written in scientific notation can be rewritten in standard form. To write 8.63×10^3 in standard form, recall that $10^3 = 1000$.

$$8.63 \times 10^3 = 8.63(1000) = 8630$$

Notice that the exponent on the 10 is positive three and we moved the decimal point three places to the right.

To write 8.63×10^{-3} in standard form, recall that $10^{-3} = \dfrac{1}{10^3} = \dfrac{1}{1000}$.

$$8.63 \times 10^{-3} = 8.63\left(\frac{1}{1000}\right) = \frac{8.63}{1000} = 0.00863$$

The exponent on the 10 is negative three, and we moved the decimal to the left three places.

In general, **to write a scientific notation number in standard form,** move the decimal point the same number of places as the exponent on 10. If the exponent is positive, move the decimal point to the right; if the exponent is negative, move the decimal point to the left.

EXAMPLE 6 Write the following numbers in standard notation, without exponents.
a. 1.02×10^5 b. 7.358×10^{-3} c. 8.4×10^7 d. 3.007×10^{-5}

Solution: a. Move the decimal point 5 places to the right and

$$1.02 \times 10^5 = 102,000.$$

b. Move the decimal point 3 places to the left and

$$7.358 \times 10^{-3} = 0.007358$$

c. $8.4 \times 10^7 = 84,000,000$

7 places to the right

d. $3.007 \times 10^{-5} = 0.00003007$

5 places to the left   ■

Performing operations on numbers written in scientific notation makes use of the rules and definitions of exponents.

EXAMPLE 7   Write each number without exponents.

a. $(8 \times 10^{-6})(7 \times 10^{3})$ **b.** $\dfrac{12 \times 10^{2}}{6 \times 10^{-3}}$

Solution: **a.** $(8 \times 10^{-6})(7 \times 10^{3}) = 8 \cdot 7 \cdot 10^{-6} \cdot 10^{3}$

$$= 56 \times 10^{-3}$$

$$= 0.056$$

b. $\dfrac{12 \times 10^{2}}{6 \times 10^{-3}} = \dfrac{12}{6} \times 10^{2-(-3)} = 2 \times 10^{5} = 200{,}000$ ∎

    CALCULATOR BOX

Scientific Notation

To enter a number written in scientific notation on a calculator, locate the key marked $\boxed{\text{EE}}$.

To enter 3.1×10^{7}, press $\boxed{3.1}\ \boxed{\text{EE}}\ \boxed{7}$. The display should read $\boxed{3.1 \qquad 07}$.

Enter the following numbers written in scientific notation on your calculator.

1. 5.31×10^{3}
2. -4.8×10^{14}
3. 6.6×10^{-9}
4. -9.9811×10^{-2}

Multiply the following on your calculator. Notice the form of the answer.

5. $3{,}000{,}000 \times 5{,}000{,}000$
6. $230{,}000 \times 1{,}000$

Multiply the following on your calculator.

7. $(3 \times 10^{6})(2 \times 10^{13})$
8. $(2 \times 10^{-4})(4 \times 10^{9})$

MENTAL MATH

State each expression using positive exponents.

1. $5x^{-2}$ **2.** $3x^{-3}$ **3.** $\dfrac{1}{y^{-6}}$

4. $\dfrac{1}{x^{-3}}$ **5.** $\dfrac{4}{y^{-3}}$ **6.** $\dfrac{16}{y^{-7}}$

EXERCISE SET 3.2

Simplify each expression. Write answers with positive exponents. See Examples 1 and 2.

1. 4^{-3}

2. 6^{-2}

3. $7x^{-3}$

4. $(7x)^{-3}$

5. $\left(\dfrac{1}{4}\right)^{-3}$

6. $\left(\dfrac{1}{8}\right)^{-2}$

7. $3^{-1} + 2^{-1}$

8. $4^{-1} + 4^{-2}$

9. $\dfrac{1}{p^{-3}}$

10. $\dfrac{1}{q^{-5}}$

Simplify each expression. Write answers with positive exponents. See Example 3.

11. $\dfrac{p^{-5}}{q^{-4}}$

12. $\dfrac{r^{-5}}{s^{-2}}$

13. $\dfrac{x^{-2}}{x}$

14. $\dfrac{y}{y^{-3}}$

15. $\dfrac{z^{-4}}{z^{-7}}$

16. $\dfrac{x^{-4}}{x^{-1}}$

Simplify the following. Write each answer with positive exponents only. See Example 4.

17. $(a^{-5})^{-6}$

18. $(4^{-1})^{-2}$

19. $\left(\dfrac{x^{-2}y^4}{x^3y^7}\right)^2$

20. $\left(\dfrac{a^5b}{a^7b^{-2}}\right)^{-3}$

21. $\dfrac{4^2z^{-3}}{4^3z^{-5}}$

22. $\dfrac{3^{-1}x^4}{3^3x^{-7}}$

Write each number in scientific notation. See Example 5.

23. 78,000

24. 9,300,000,000

25. 0.00000167

26. 0.00000017

27. The distance between Earth and the sun is 93,000,000 miles.

28. The population of the world is 4,800,000,000.

Write each number without using exponents. See Example 6.

29. 7.86×10^8

30. 1.43×10^7

31. 8.673×10^{-10}

32. 9.056×10^{-4}

33. One coulomb of electricity is 6.25×10^{18}.

34. The mass of a hydrogen atom is 1.7×10^{-24} grams.

Evaluate the following expressions using exponential rules. Write the answers without using exponents. See Example 7.

35. $(4 \times 10^4)(2 \times 10^6)$

36. $(1.8 \times 10^8)(3.4 \times 10^{-4})$

37. $(5x)^{-3}$

38. $5x^{-2}$

Simplify the following. Write answers with positive exponents.

39. $(-3)^{-2}$

40. $(-2)^{-4}$

41. $\dfrac{-1}{p^{-4}}$

42. $\dfrac{-1}{y^{-6}}$

43. $-2^0 - 3^0$

44. $5^0 + (-5)^0$

45. $\dfrac{r}{r^{-3}r^{-2}}$

46. $\dfrac{p}{p^{-3}q^{-5}}$

47. $(x^5y^3)^{-3}$

48. $(z^5x^5)^{-3}$

49. $2^0 + 3^{-1}$

50. $4^{-2} - 4^{-3}$

51. $\dfrac{2^{-3}x^{-4}}{2^2x}$

52. $\dfrac{5^{-1}z^7}{5^{-2}z^9}$

53. $\dfrac{7ab^{-4}}{7^{-1}a^{-3}b^2}$

54. $\dfrac{6^{-5}x^{-1}y^2}{6^{-2}x^{-4}y^4}$

55. $\left(\dfrac{a^{-5}b}{ab^3}\right)^{-4}$

56. $\left(\dfrac{r^{-2}s^{-3}}{r^{-4}s^{-3}}\right)^{-3}$

57. $\dfrac{(xy^3)^5}{(xy)^{-4}}$

58. $\dfrac{(rs)^{-3}}{(r^2s^3)^2}$

59. $\dfrac{(-2xy^{-3})^{-3}}{(xy^{-1})^{-1}}$

60. $\dfrac{(-3x^2y^2)^{-2}}{(xyz)^{-2}}$

Write each number in scientific notation.

61. 0.00635

62. 0.00194

63. 1,160,000

64. 700,000

65. The temperature at the interior of Earth is 20,000,000 degrees Celsius.

66. The half-life of a carbon isotope is 5000 years.

Write each number in standard notation.

67. 3.3×10^{-2}

68. 4.8×10^{-6}

69. 2.032×10^4

70. 9.07×10^{10}

71. The distance light travels in 1 year is 9.460×10^{12} kilometers.

72. The population of the United States is 2.34×10^8.

Evaluate the following expressions using exponential rules. Write the answers without exponents.

73. $(1.2 \times 10^{-3})(3 \times 10^{-2})$

74. $(2.5 \times 10^6)(2 \times 10^{-6})$

75. $(4 \times 10^{-10})(7 \times 10^{-9})$

76. $(5 \times 10^6)(4 \times 10^{-8})$

77. $\dfrac{8 \times 10^{-1}}{16 \times 10^5}$

78. $\dfrac{25 \times 10^{-4}}{5 \times 10^{-9}}$

79. $\dfrac{1.4 \times 10^{-2}}{7 \times 10^{-8}}$

80. $\dfrac{0.4 \times 10^5}{0.2 \times 10^{11}}$

81. The average amount of water flowing past the mouth of the Amazon River is 4.2×10^6 cubic feet per second. How much water flows past in an hour? (1 hour equals 3600 seconds.)

82. A beam of light travels 9.460×10^{12} kilometers per year. How far does light travel in 10,000 years?

83. The total force (F) against the face of a dam that is 100 feet long by 20 feet high is given by the following:

$$F = \frac{(6.24 \times 10)(4 \times 10^4)}{2}$$

Compute the force and express the answer in scientific notation.

84. Suppose $1000 is invested at a rate of 9% and compounded monthly. The amount of principal (P) after one year is given by

$$P = (1 \times 10^3)(1.09381)$$

Compute the amount of principal.

A Look Ahead

Simplify each expression. Assume that variables represent positive integers. See the following example.

EXAMPLE   Simplify the following expressions. Assume that the variable in the exponent represents an integer value.

a. $x^{m+1} \cdot x^m$ **b.** $(z^{2x+1})^x$ **c.** $\dfrac{y^{6a}}{y^{4a}}$

Solution: **a.** $x^{m+1} \cdot x^m = x^{(m+1)+m} = x^{2m+1}$ **b.** $(z^{2x+1})^x = z^{(2x+1)x} = z^{2x^2+x}$

c. $\dfrac{y^{6a}}{y^{4a}} = y^{6a-4a} = y^{2a}$ ∎

85. $a^{-4m} \cdot a^{5m}$

86. $(x^{-3s})^3$

87. $(3y^{2z})^3$

88. $a^{4m+1} \cdot a^4$

89. $\dfrac{y^{4a}}{y^{-a}}$

90. $\dfrac{y^{-6a}}{zy^{6a}}$

91. $(z^{3a+2})^{-2}$

92. $(a^{4x-1})^{-1}$

Writing in Mathematics

93. It was stated earlier that, for an integer n, $x^{-n} = \dfrac{1}{x^n}$, $x \neq 0$. Explain why x may not equal 0.

94. Explain why the following equation is true.
$$(a^{-1})^3 = (a^3)^{-1}$$

Skill Review

Solve the following. See Section 2.2.

95. $5x = 4x + 8$

96. $4x + 1 = 3x$

97. $2(3y - 1) = 20 + 7y$

98. $4(2x + 7) = 3(3x + 1)$

99. $9x + 1 = 8x - 3$

100. $8x + 2 = 7x + 4$

3.3
Adding and Subtracting Polynomials

OBJECTIVES

Tape 10

1 Define monomial, binomial, trinomial, polynomial, and degree.

2 Combine like terms.

3 Add and subtract polynomials.

1 We first review some definitions presented in Section 2.1. A **term** is a single number or a product of a number and one or more variables raised to powers.

Expression	Terms
$4x^2 + 3x$	$4x^2$, $3x$
$9x^4 - 7x - 1$	$9x^4$, $-7x$, -1

The **numerical coefficient** of a term, or simply the **coefficient,** is the numerical factor of each term. If no numerical factor appears in the term, then the coefficient is understood to be 1. If the term contains only a numerical factor, it is called a **constant** term, or simply a constant.

Term	Coefficient
x^5	1
$3x^2$	3
$-4x$	-4
$-x^2y$	-1
3 (constant)	3

A **polynomial** in x is a finite sum of terms of the form ax^n, where a is a real number and n is a whole number. For example,

$$x^5 - 3x^3 + 2x^2 - 5x + 1$$

is a polynomial. Notice that this polynomial is written in **descending powers** of x because the powers of x decrease from left to right. On the other hand,

$$x^{-5} + 2x - 3$$

is **not** a polynomial because it contains an exponent, -5, that is not a whole number.

A **monomial** is a polynomial with exactly one term.
A **binomial** is a polynomial with exactly two terms.
A **trinomial** is a polynomial with exactly three terms.

The following are examples of monomials, binomials, and trinomials. Each of these examples is also a polynomial.

Monomials	Binomials	Trinomials
ax^2	$x + y$	$x^2 + 4xy + y^2$
$-3z$	$3p + 2$	$x^5 + 7x^2 - x$
4	$4x^2 - 7$	$-q^4 + q^3r - 2q$

Each term in a polynomial has a **degree.**

Degree of a Term

The degree of a term is the sum of the exponents on the variables contained in the term.

EXAMPLE 1 Find the degree of each term.
a. $-3x^2$ **b.** $5x^3yz$ **c.** 2

Solution: **a.** The exponent on x is 2, so the degree of the term is 2.
b. $5x^3yz$ can be written as $5x^3y^1z^1$. The degree of the term is the sum of its exponents, so the degree is $3 + 1 + 1$ or 5.
c. The constant, 2, can be written as $2x^0$ (since $x^0 = 1$). The degree of 2 or $2x^0$ is 0. ∎

From the preceding, we can say that **the degree of a constant is 0.**
The polynomial as a whole can also be identified by degree.

Degree of a Polynomial

The degree of a polynomial is the largest degree of any term in the polynomial.

EXAMPLE 2 Find the degree of each polynomial and tell whether the polynomial is a monomial, binomial, trinomial, or none of these.
a. $-2t^2 + 3t + 6$ **b.** $15x - 10$ **c.** $7x + 3x^3 + 2x^2 - 1$

Solution: **a.** The degree of the trinomial $-2t^2 + 3t + 6$ is 2, the largest degree of any of its terms.
b. The degree of the binomial $15x - 10$ is 1.
c. The degree of the polynomial $7x + 3x^3 + 2x^2 - 1$ is 3. ∎

Polynomials have different values depending on replacement values for the variables.

EXAMPLE 3 Find the value of the polynomial $3x^2 - 2x + 1$ when $x = -2$.

Solution: Replace x with -2 and simplify.

$$3x^2 - 2x + 1 = 3(-2)^2 - 2(-2) + 1$$
$$= 3(4) + 4 + 1$$
$$= 12 + 4 + 1$$
$$= 17 \quad \blacksquare$$

2 Polynomials with like terms can be simplified by combining like terms. Recall that like terms are terms that contain exactly the same variables raised to exactly the same powers.

Like Terms

$$5x^2, -7x^2$$
$$y, 2y$$
$$\frac{1}{2}a^2b, -a^2b$$

Only like terms can be combined. Combine like terms by applying the distributive property.

EXAMPLE 4 Combine like terms.
   a. $-3x + 7x$ **b.** $11x^2 + 5 + 2x^2 - 7$

Solution: **a.** $-3x + 7x = (-3 + 7)x = 4x$
   b. $11x^2 + 5 + 2x^2 - 7 = 11x^2 + 2x^2 + 5 - 7$
   $$= 13x^2 - 2 \qquad \text{Combine like terms.} \quad \blacksquare$$

3 We now practice adding and subtracting polynomials.

> **To Add Polynomials**
>
> To add polynomials, combine all like terms.

EXAMPLE 5 Add $(-2x^2 + 5x - 1) + (-2x^2 + x + 3)$.

Solution: To add these polynomials, remove the parentheses and then group like terms.

$$(-2x^2 + 5x - 1) + (-2x^2 + x + 3) = -2x^2 + 5x - 1 - 2x^2 + x + 3$$
$$= (-2x^2 - 2x^2) + (5x + 1x) + (-1 + 3)$$
$$= -4x^2 + 6x + 2 \quad \blacksquare$$

EXAMPLE 6 Add $(4x^3 - 6x^2 + 2x + 7) + (5x^2 - 2x)$.

Solution:
$$(4x^3 - 6x^2 + 2x + 7) + (5x^2 - 2x) = 4x^3 - 6x^2 + 2x + 7 + 5x^2 - 2x$$
$$= 4x^3 + (-6x^2 + 5x^2) + (2x - 2x) + 7$$
$$= 4x^3 - x^2 + 7 \quad \blacksquare$$

Polynomials can be added vertically if we line up like terms underneath one another.

EXAMPLE 7 Add $(7y^3 - 2y^2 + 7)$ and $(6y^2 + 1)$ using the vertical format.

Solution: Vertically line up like terms and add.

$$\begin{array}{r} 7y^3 - 2y^2 + 7 \\ 6y^2 + 1 \\ \hline 7y^3 + 4y^2 + 8 \end{array} \quad \blacksquare$$

To subtract one polynomial from another, recall the definition of subtraction. To subtract a number, we add its opposite: $a - b = a + (-b)$. To subtract a polynomial, we also add its opposite. Just as $-b$ is the opposite of b, $-(x^2 + 5)$ is the opposite of $(x^2 + 5)$.

EXAMPLE 8 Subtract $(5x - 3) - (2x - 11)$.

Solution: From the definition of subtraction, we have

$$(5x - 3) - (2x - 11) = (5x - 3) + [-(2x - 11)] \qquad \text{Add the opposite.}$$
$$= (5x - 3) + (-2x + 11) \qquad \text{Apply the distributive property.}$$
$$= 3x + 8 \qquad \text{Combine like terms.}$$

$\blacksquare$

> **To Subtract Polynomials**
>
> To subtract two polynomials, change the signs of the second polynomial and then add.

EXAMPLE 9 Subtract $(2x^3 + 8x^2 - 6x) - (2x^3 - x^2 + 1)$.

Solution: First, change the sign of each term of the second polynomial and then add.

$$(2x^3 + 8x^2 - 6x) - (2x^3 - x^2 + 1) = (2x^3 + 8x^2 - 6x) + (-2x^3 + x^2 - 1)$$
$$= 2x^3 - 2x^3 + 8x^2 + x^2 - 6x - 1$$
$$= 9x^2 - 6x - 1 \qquad \text{Combine like terms.}$$

$\blacksquare$

EXAMPLE 10 Subtract $(5y^2 + 2y - 6)$ from $(-3y^2 - 2y + 11)$ using the vertical format.

Solution: Arrange the polynomials in vertical format, lining up like terms.

$$\begin{array}{r} -3y^2 - 2y + 11 \\ -(5y^2 + 2y - 6) \\ \hline \end{array} \qquad \begin{array}{r} -3y^2 - 2y + 11 \\ -5y^2 - 2y + 6 \\ \hline -8y^2 - 4y + 17 \end{array} \quad \blacksquare$$

EXAMPLE 11   Subtract $(5z - 7)$ from the sum of $(8z + 11)$ and $(9z - 2)$.

Solution: Notice that $(5z - 7)$ is to be subtracted **from** a sum. The translation is

$$[(8z + 11) + (9z - 2)] - (5z - 7)$$

$$= 8z + 11 + 9z - 2 - 5z + 7 \qquad \text{Remove grouping symbols.}$$

$$= 8z + 9z - 5z + 11 - 2 + 7 \qquad \text{Group like terms.}$$

$$= 12z + 16 \qquad \text{Combine like terms.} \quad \blacksquare$$

MENTAL MATH

Combine like terms.

1. $-9y - 5y$

2. $6m^5 + 7m^5$

3. $4y^3 + 3y^3$

4. $21y^5 - 19y^5$

5. $x + 6x$

6. $7z - z$

EXERCISE SET 3.3

Find the degree of each of the following polynomials. See Examples 1 and 2.

1. $x + 2$

2. $-6y + y^2 + 4$

3. $9m^3 - 5m^2 + 4m - 8$

4. $5a^2 + 3a^3 - 4a^4$

5. $12x^4 - x^2 - 12x^8$

6. $7r^2 + 2r - 3$

7. $3z$

8. $5y + 2$

Find the value of each polynomial when **(a)** $x = 0$ *and* **(b)** $x = -1$. *See Example 3.*

9. $x + 6$

10. $2x - 10$

11. $x^2 - 5x - 2$

12. $x^2 - 4$

Simplify each of the following by combining like terms. See Example 4.

13. $14x^2 + 9x^2$

14. $18x^3 - 4x^3$

15. $15x^2 - 3x^2 - y$

16. $12k^3 - 9k^3 + 11$

17. $8s - 5s + 4s$

18. $5y + 7y - 6y$

Add the following. See Examples 5 and 6.

19. $(3x + 7) + (9x + 5)$

20. $(3x^2 + 7) + (3x^2 + 9)$

21. $(-7x + 5) + (-3x^2 + 7x + 5)$

22. $(3x - 8) + (4x^2 - 3x + 3)$

23. $(-5x^2 + 3) + (2x^2 + 1)$

24. $(-y - 2) + (3y + 5)$

Subtract the following. See Examples 8 and 9. distributive property understood I make sure I change signs

25. $(2x + 5) - (3x - 9)$

26. $(5x^2 + 4) - (-2y^2 + 4)$

27. $3x - (5x - 9)$

28. $4 - (-y - 4)$

29. $(2x^2 + 3x - 9) - (-4x + 7)$

30. $(-7x^2 + 4x + 7) - (-8x + 2)$

Perform the indicated operations. See Examples 7 and 10.

31. $3t^2 + 4$
         $\underline{+ \ 5t^2 - 8}$

32. $7x^3 + 3$
         $\underline{+ \ 2x^3 + 1}$

33. $4z^2 - 8z + 3$
         $\underline{- \ (6z^2 + 8z - 3)}$

34. $5u^5 - 4u^2 + 3u - 7$
         $\underline{- \ (3u^5 + 6u^2 - 8u + 2)}$

35. $5x^3 - 4x^2 + 6x - 2$
         $\underline{- \ (3x^3 - 2x^2 - x - 4)}$

36. $7a^2 - 9a + 6$
         $\underline{- \ (11a^2 - 4a + 2)}$

37. $\begin{array}{r} 10a^3 - 8a^2 + 9 \\ +\ \ 5a^3 + 9a^2 + 7 \\ \hline \end{array}$

38. $\begin{array}{r} 2x^3 - 3x^2 + x - 4 \\ +\ 5x^3 + 2x^2 - 3x + 2 \\ \hline \end{array}$

Perform the indicated operations and simplify. See Example 11.

39. Subtract $(19x^2 + 5)$ from $(81x^2 + 10)$.

40. Subtract $(2x + xy)$ from $(3x - 9xy)$.

41. Subtract $(2x + 2)$ from the sum of $(8x + 1)$ and $(6x + 3)$.

42. Subtract $(-12x - 3)$ from the sum of $(-5x - 7)$ and $(12x + 3)$.

43. Subtract $(8x + 9)$ from $(9xy^2 + 7x - 18)$.

44. Subtract $(4x^2 + 7)$ from $(9x^3 + 9x^2 - 9)$.

Find the value of each polynomial when **(a)** $x = 3$ *and* **(b)** $x = -4$.

45. $3x + 7$

46. $-2x - 5$

47. $2x^2 - 6x + 1$

48. $3x^2 + 20$

49. $x^3 - 15$

50. $-4x^2 + 5x$

Simplify the following by performing the indicated operations.

51. $2x - 5 + 5x - 8$

52. $x - 3 + 8x + 10$

53. $(-3y^2 - 4y) + (2y^2 + y - 1)$

54. $(7x^2 + 2x - 9) + (-3^2 + 5)$

55. $(-7y^2 + 5) - (-8y^2 + 12)$

56. $(4 + 5a) - (-a - 5)$

57. $(5x + 8) - (-2x^2 - 6x + 8)$

58. $(-6y^2 + 3y - 4) - (9y^2 - 3y)$

59. $(-8x^2 + 7x) + (-8x^2 + x + 9)$

60. $(6y^2 - 6y + 4) + (-2y^2 - 8y - 7)$

61. $(3x^2 + 5x - 8) + (5x^2 + 9x + 12)$

62. $(-a^2 + 1) - (a^2 - 3)$

63. $-15x - (-4x)$

64. $16y - (-4y)$

65. $17w^2 - (-10w^2) + 3$

66. $-13w^2 - (-5w^2) - 15$

67. $(-4x^3 + 4x^2 + 3x) - (2x^3 + 7x^2 - 3)$

68. $(9a^3 - 2a^2 + 4a - 7) - (7a^3 - 8a^2 + 4a - 1)$

69. $(5x - 7) + (2x^2 + 3x - 12)$

70. $(9xy^2 + 4xy + 8) + (7xy^2 + 2xy - 15)$

71. Subtract $4x$ from $7x - 3$.

72. Subtract y from $y^2 - 4y + 1$.

73. Subtract $(5x + 7)$ from $(7x^2 + 3x + 9)$.

74. Subtract $(5y^2 + 8y + 2)$ from $(7y^2 + 9y - 8)$.

75. Subtract $(4y^2 - 6y - 3)$ from the sum of $(8y^2 + 7)$ and $(6y + 9)$.

76. Subtract $(5y + 7x^2)$ from the sum of $(8y - x)$ and $(3 + 8x^2)$.

77. Subtract $(-2x^2 + 4x - 12)$ from the sum of $(-x^2 - 2x)$ and $(5x^2 + x + 9)$.

78. Subtract $(4x^2 - 2x + 2)$ from the sum of $(x^2 + 7x + 1)$ and $(7x + 5)$.

Express each of the following as a polynomial. Simplify if possible.

79. Given the following triangle, find its perimeter.

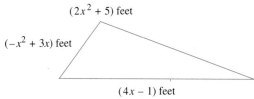

$(2x^2 + 5)$ feet

$(-x^2 + 3x)$ feet

$(4x - 1)$ feet

80. Given the following quadrilateral, find its perimeter.

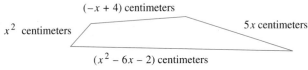

$(-x + 4)$ centimeters

x^2 centimeters

$5x$ centimeters

$(x^2 - 6x - 2)$ centimeters

81. A wooden beam is $(4y^2 + 4y + 1)$ meters long. If a piece $(y^2 - 10)$ meters is cut, express the length of the remaining piece of beam as a polynomial in y.

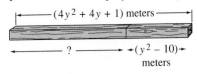

$(4y^2 + 4y + 1)$ meters

? $\qquad$ $(y^2 - 10)$

meters

82. A piece of taffy is $(13x - 7)$ inches long. If a piece $(2x + 2)$ inches is removed, express the length of the remaining piece of taffy as a polynomial in x.

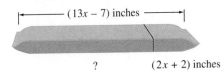

$(13x - 7)$ inches

? $\qquad$ $(2x + 2)$ inches

Writing in Mathematics

83. Simplify each expression, and explain the difference between the two problems.
  a. $(3x^2)(4x^2)$ **b.** $3x^2 + 4x^2$

84. Explain how to add polynomials.

Skill Review

Insert $<$, $>$, or $=$ in the appropriate space to make each of the following true. See Section 1.1.

85. 0   5

86. 45   54

87. $10 - 1$ 9

88. 100   10

Given the set of numbers $\left\{-2, 0, \frac{1}{2}, 25, \pi, -5\frac{2}{3}\right\}$, which of these also belong to the set of:

89. Integers

90. Real numbers

91. Whole numbers

92. Rational numbers

3.4
Multiplying Polynomials

Tape 10

OBJECTIVES

1 Use the distributive property to multiply polynomials.

2 Multiply polynomials vertically.

1 To multiply polynomials, we apply our knowledge of the rules and definitions of exponents.

To multiply two monomials such as $(-5x^3)$ and $(-2x^4)$, use the associative and commutative properties and regroup. Remember that to multiply exponential expressions with a common base we add exponents.

$$(-5x^3)(-2x^4) = (-5)(-2)(x^3)(x^4) = 10x^7$$

To multiply polynomials that are not monomials, use the distributive property.

EXAMPLE 1 Use the distributive property to find each product.
  a. $5x(2x^3 + 6)$ **b.** $-3x^2(5x^2 + 6x - 1)$ **c.** $(3n^2 - 5n + 4)(2n)$

Solution: **a.** $5x(2x^3 + 6) = 5x(2x^3) + 5x(6)$ Use the distributive property.

$$= 10x^4 + 30x$$ Multiply.

b. $-3x^2(5x^2 + 6x - 1)$

$$= (-3x^2)(5x^2) + (-3x^2)(6x) + (-3x^2)(-1)$$ Use the distributive property.

$$= -15x^4 - 18x^3 + 3x^2$$ Multiply.

c. $(3n^2 - 5n + 4)(2n)$

$$= (3n^2)(2n) + (-5n)(2n) + 4(2n)$$ Use the distributive property.

$$= 6n^3 - 10n^2 + 8n$$ Multiply.   ∎

We also use the distributive property to multiply two binomials. To multiply $(x + 3)$ by $(x + 1)$, distribute the factor $(x + 1)$ first.

$$(x + 3)(x + 1) = x(x + 1) + 3(x + 1) \qquad \text{Distribute } (x + 1).$$
$$= x(x) + x(1) + 3(x) + 3(1) \qquad \text{Apply distributive property a second time.}$$
$$= x^2 + x + 3x + 3 \qquad \text{Multiply.}$$
$$= x^2 + 4x + 3 \qquad \text{Combine like terms.}$$

This idea can be expanded so that we can multiply any two polynomials.

> **To Multiply Two Polynomials**
>
> Multiply each term of the first polynomial by each term of the second polynomial, and then combine like terms.

EXAMPLE 2   Find the product $(3x + 2)(2x - 5)$.

Solution: Multiply each term of the first binomial by each term of the second.

$$(3x + 2)(2x - 5) = 3x(2x) + 3x(-5) + 2(2x) + 2(-5)$$
$$= 6x^2 - 15x + 4x - 10 \qquad \text{Multiply.}$$
$$= 6x^2 - 11x - 10 \qquad \text{Combine like terms.} \qquad \blacksquare$$

EXAMPLE 3   Multiply $(2x - y)^2$.

Solution: Recall that $a^2 = a \cdot a$, so $(2x - y)^2 = (2x - y)(2x - y)$. Multiply each term of the first polynomial by each term of the second.

$$(2x - y)(2x - y) = 2x(2x) + 2x(-y) + (-y)(2x) + (-y)(-y)$$
$$= 4x^2 - 2xy - 2xy + y^2 \qquad \text{Multiply.}$$
$$= 4x^2 - 4xy + y^2 \qquad \text{Combine like terms.} \qquad \blacksquare$$

EXAMPLE 4   Multiply $(3a + b)^3$.

Solution: Write $(3a + b)^3$ as $(3a + b)(3a + b)(3a + b)$.

$$(3a + b)(3a + b)(3a + b) = (3a + b)(9a^2 + 3ab + 3ab + b^2)$$
$$= (3a + b)(9a^2 + 6ab + b^2)$$
$$= 3a(9a^2 + 6ab + b^2) + b(9a^2 + 6ab + b^2)$$
$$= 27a^3 + 18a^2b + 3ab^2 + 9a^2b + 6ab^2 + b^3$$
$$= 27a^3 + 27a^2b + 9ab^2 + b^3 \qquad \blacksquare$$

EXAMPLE 5   Multiply $(t + 2)$ by $(3t^2 - 4t + 2)$.

Solution: Multiply each term of the first polynomial by each term of the second.

$$(t + 2)(3t^2 - 4t + 2) = t(3t^2) + t(-4t) + t(2) + 2(3t^2) + 2(-4t) + 2(2)$$
$$= 3t^3 - 4t^2 + 2t + 6t^2 - 8t + 4$$
$$= 3t^3 + 2t^2 - 6t + 4 \qquad \text{Combine like terms.} \qquad \blacksquare$$

2 Another convenient method for multiplying polynomials is to use a vertical format similar to the format used to multiply real numbers. We demonstrate this method by multiplying $(3y^2 - 4y + 1)$ by $(y + 2)$.

Step 1 Write the polynomials in a vertical format.

$$\begin{array}{r} 3y^2 - 4y + 1 \\ \times \quad\quad y + 2 \\ \hline \end{array}$$

Step 2 Multiply 2 by each term of the top polynomial. Write the **partial product** below the line.

$$\begin{array}{r} \mathbf{3y^2 - 4y + 1} \\ \times \quad\quad y + 2 \\ \hline 6y^2 - 8y + 2 \end{array}$$

Step 3 Multiply y by each term of the top polynomial. Write this partial product underneath the previous one, being careful to line up like terms.

$$\begin{array}{r} 3y^2 - 4y + 1 \\ \times \quad\quad y + 2 \\ \hline 6y^2 - 8y + 2 \\ 3y^3 - 4y^2 + \quad y \quad\quad \end{array}$$

Step 4 Combine like terms of the partial products.

$$\begin{array}{r} 3y^2 - 4y + 1 \\ \times \quad\quad y + 2 \\ \hline 6y^2 - 8y + 2 \\ 3y^3 - 4y^2 + \quad y \quad\quad \\ \hline 3y^3 + 2y^2 - 7y + 2 \end{array}$$

Thus, $(y + 2)(3y^2 - 4y + 1) = 3y^3 + 2y^2 - 7y + 2$.

EXAMPLE 6 Find the product of $(2x^2 - 3x + 4)$ and $(x^2 + 5x - 2)$ using the vertical format.

Solution: Multiply each term of the second polynomial by each term of the first polynomial.

$$\begin{array}{r} 2x^2 - \quad 3x + 4 \\ \times \quad\quad x^2 + \quad 5x - 2 \\ \hline -4x^2 + \quad 6x - 8 \\ 10x^3 - 15x^2 + 20x \quad\quad \\ 2x^4 - \quad 3x^3 + \quad 4x^2 \quad\quad\quad\quad\quad \\ \hline 2x^4 + \quad 7x^3 - 15x^2 + 26x - 8 \end{array}$$

Multiply $2x^2 - 3x + 4$ by -2.
Multiply $2x^2 - 3x + 4$ by $5x$.
Multiply $2x^2 - 3x + 4$ by x^2.
Combine like terms. ■

MENTAL MATH

Find the following products mentally.

1. $5x(2y)$

2. $7a(4b)$

3. $x^2 \cdot x^5$

4. $z \cdot z^4$

5. $6x(3x^2)$

6. $5a^2(3a^2)$

EXERCISE SET 3.4

Find the following products. See Example 1.

1. $2a(2a - 4)$

2. $3a(2a + 7)$

3. $7x(x^2 + 2x - 1)$

4. $-5y(y^2 + y - 10)$

5. $3x^2(2x^2 - x)$

6. $-4y^2(5y - 6y^2)$

Find the following products. See Examples 2 and 3.

7. $(a + 7)(a - 2)$

8. $(y + 5)(y + 7)$

9. $(2y - 4)^2$

10. $(6x - 7)^2$

11. $(5x - 9y)(6x - 5y)$

12. $(3x - 7y)(7x + 2y)$

13. $(2x^2 - 5)^2$

14. $(x^2 - 4)^2$

Find the following products. See Example 4.

15. $(x + 2)^3$

16. $(y - 1)^3$

17. $(2y - 3)^3$

18. $(3x + 4)^3$

Find the following products. See Example 5.

19. $(x - 2)(x^2 - 3x + 7)$

20. $(x + 3)(x^2 + 5x - 8)$

21. $(x + 5)(x^3 - 3x + 4)$

22. $(a + 2)(a^3 - 3a^2 + 7)$

23. $(2a - 3)(5a^2 - 6a + 4)$

24. $(3 + b)(2 - 5b - 3b^2)$

Find the following products. Use the vertical multiplication method. See Example 5.

25. $(x + 3)(2x^2 + 4x - 1)$

26. $(2x - 5)(3x^2 - 4x + 7)$

27. $(x^2 + 5x - 7)(x^2 - 7x - 9)$

28. $(3x^2 - x + 2)(x^2 + 2x + 1)$

Find the following products.

29. $2a(a + 4)$

30. $-3a(2a + 7)$

31. $3x(2x^2 - 3x + 4)$

32. $-4x(5x^2 - 6x - 10)$

33. $(5x + 9y)(3x + 2y)$

34. $(5x - 5y)(2x - y)$

35. $(x + 2)(x^2 + 5x + 6)$

36. $(x - 7)(x^2 - 15x + 56)$

37. $(7x + 4)^2$

38. $(3x - 2)^2$

39. $-2a^2(3a^2 - 2a + 3)$

40. $-4b^2(3b^3 - 12b^2 - 6)$

41. $(x + 3)(x^2 + 7x + 12)$

42. $(n + 1)(n^2 - 7n - 9)$

43. $(a + 1)^3$

44. $(x - y)^3$

45. $(x + y)(x + y)$

46. $(x + 3)(7x + 1)$

47. $(x - 7)(x - 6)$

48. $(4x + 5)(-3x + 2)$

49. $3a(a^2 + 2)$

50. $x^3(x + 12)$

51. $-4y(y^2 + 3y - 11)$

52. $-2x(5x^2 - 6x + 1)$

53. $(5x + 1)(5x - 1)$

54. $(2x + y)(3x - y)$

55. $(5x + 4)(x^2 - x + 4)$

56. $(x - 2)(x^2 - x + 3)$

57. $(2x - 5)^3$

58. $(3y - 1)^3$

59. $(4x + 5)(8x^2 + 2x - 4)$

60. $(x + 7)(x^2 - 7x - 8)$

61. $(7xy - y)^2$

62. $(x + y)^2$

63. $(5y^2 - y + 3)(y^2 - 3y - 2)$

64. $(2x^2 + x - 1)(x^2 + 3x + 4)$

65. $(3x^2 + 2x - 4)(2x^2 - 4x + 3)$

66. $(a^2 + 3a - 2)(2a^2 - 5a - 1)$

Express each of the following as polynomials.

67. Find the area of the following rectangle.

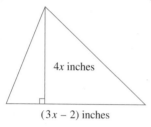

(2x + 5) yards

(2x − 5) yards

68. Find the area of the following triangle.

4x inches

(3x − 2) inches

69. Find the area of the square-shaped field.

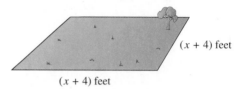

(x + 4) feet

(x + 4) feet

70. Find the volume of the cube-shaped glass block.

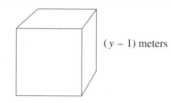

(y − 1) meters

Writing in Mathematics

71. Multiply the following polynomials.
  a. $(a + b)(a - b)$
  b. $(2x + 3y)(2x - 3y)$
  c. $(4x + 7)(4x - 7)$
  Can we make a general statement about all products of the form $(x + y)(x - y)$?

72. Simplify each of the following. Explain the difference between the two problems.
  a. $(3x + 5) + (3x + 7)$
  b. $(3x + 5)(3x + 7)$

Skill Review

Perform the indicated operation. See Section 3.3.

73. $(2x^2 + 5x) - (x + 7)$

74. $(3y^2 - 2y) - (7y + 3)$

75. $(9y^2 + 7y) + (5y - 1)$

76. $(6y^2 + 8y) + (5y + 8)$

77. $(x^2 + 4x - 4x) - (x + 7)$

78. $(x^2 + 3x - 3) + (x^2 - 9)$

3.5
Special Products

Tape 11

OBJECTIVES	
1	Multiply two binomials using the FOIL method.
2	Square a binomial.
3	Multiply the sum and difference of two terms.

1 In this section, we multiply binomials using special products. First, a method for multiplying binomials called the FOIL method is introduced. This method is demonstrated by multiplying $(3x + 1)$ by $(2x + 5)$.

F   stands for the product of the **First** terms. $(3x + 1)(2x + 5)$

$$(3x)(2x) = 6x^2 \quad \mathbf{F}$$

O   stands for the product of the **Outer** terms. $(3x + 1)(2x + 5)$

$$(3x)(5) = 15x \quad \mathbf{O}$$

I   stands for the product of the **Inner** terms. $(3x + 1)(2x + 5)$

$$(1)(2x) = 2x \quad \mathbf{I}$$

L   stands for the product of the **Last** terms. $(3x + 1)(3x + 5)$

$$(1)(5) = 5 \quad \mathbf{L}$$

$$(3x + 1)(2x + 5) = 6x^2 + 15x + 2x + 5$$

$$= 6x^2 + 17x + 5 \qquad \text{Combine like terms.}$$

EXAMPLE 1   Find $(x - 3)(x + 4)$ by the FOIL method.

Solution:
$$\overset{\text{F} \quad\quad \text{O} \quad\quad\quad \text{I} \quad\quad\quad \text{L}}{(x - 3)(x + 4) = (x)(x) + (x)(4) + (-3)(x) + (-3)(4)}$$

$$= x^2 + 4x - 3x - 12$$

$$= x^2 + x - 12 \qquad \text{Collect like terms.} \quad \blacksquare$$

EXAMPLE 2   Find $(5x - 7)(x - 2)$ by the FOIL method.

Solution:
$$\overset{\text{F} \quad\quad\quad \text{O} \quad\quad\quad \text{I} \quad\quad\quad \text{L}}{(5x - 7)(x - 2) = 5x(x) + 5x(-2) + (-7)(x) + (-7)(-2)}$$

$$= 5x^2 - 10x - 7x + 14$$

$$= 5x^2 - 17x + 14 \qquad \text{Collect like terms.} \quad \blacksquare$$

EXAMPLE 3   Multiply $(y + 6)(2y - 1)$.

Solution:
$$\overset{\text{F} \quad\;\; \text{O} \quad\;\; \text{I} \quad\;\; \text{L}}{(y + 6)(2y - 1) = 2y^2 - 1y + 12y - 6}$$

$$= 2y^2 + 11y - 6 \quad \blacksquare$$

2   Now, try squaring a binomial using the FOIL method.

EXAMPLE 4   Multiply $(3y + 1)^2$.

Solution: $(3y + 1)^2 = (3y + 1)(3y + 1)$

$$\overset{\text{F} \quad\quad\;\; \text{O} \quad\quad\; \text{I} \quad\quad \text{L}}{= (3y)(3y) + (3y)(1) + 1(3y) + 1(1)}$$

$$= 9y^2 + 3y + 3y + 1$$

$$= 9y^2 + 6y + 1 \quad \blacksquare$$

Notice the pattern that appears in Example 4.

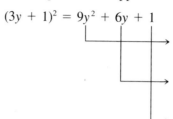

$9y^2$ is the first term of the binomial squared. $(3y)^2 = 9y^2$.

$6y$ is 2 times the product of both terms of the binomial. $(2)(3y)(1) = 6y$.

1 is the second term of the binomial squared. $(1)^2 = 1$.

This pattern leads to the following rule for squaring a binomial. We call the rules in this section **special products.**

Squaring a Binomial

$$(a + b)^2 = a^2 + 2ab + b^2$$

$$(a - b)^2 = a^2 - 2ab + b^2$$

This is, a binomial squared is equal to the square of the first term plus two times the product of the first and second terms plus the square of the second term.

HELPFUL HINT

Notice that $(x + y)^2 \neq x^2 + y^2$ and $(x - y)^2 \neq x^2 - y^2$.
$(x + y)^2 = (x + y)(x + y) = x^2 + 2xy + y^2$ and not $x^2 + y^2$.
Also, $(x - y)^2 = (x - y)(x - y) = x^2 - 2xy + y^2$ and not $x^2 - y^2$.

EXAMPLE 5   Use the preceding special products to square the following binomials.
   a. $(t + 2)^2$ **b.** $(p - q)^2$ **c.** $(2x + 3y)^2$ **d.** $(5r - 7s)^2$

Solution: **a.** The first term is t and the second term is 2.

$$(t + 2)^2 = t^2 + 2(t)(2) + 2^2 = t^2 + 4t + 4$$

b. $(p - q)^2 = p^2 - 2(p)(q) + q^2 = p^2 - 2pq + q^2$.
c. $2x$ is the first term and $3y$ is the second term.

$$(2x + 3y)^2 = (2x)^2 + 2(2x)(3y) + (3y)^2 = 4x^2 + 12xy + 9y^2$$

d. $(5r - 7s)^2 = (5r)^2 - 2(5r)(7s) + (7s)^2 = 25r^2 - 70rs + 49s^2$.   ∎

3   Another example of a special product is the sum or difference of two terms such as $(x + y)(x - y)$. Finding this product by the FOIL method, we see a pattern emerge.

$$(x + y)(x - y) = x^2 - xy + xy - y^2$$

$$= x^2 - y^2$$

Notice that middle two terms subtract out. This is because the **O**uter product is the opposite of the **I**nner product. Only the **difference of squares** remains.

> **Multiplying the Sum and Difference of Two Terms**
>
> $$(a + b)(a - b) = a^2 - b^2.$$

EXAMPLE 6   Find the following products using the preceding special product.

a. $(2x + y)(2x - y)$ **b.** $(6t + 7)(6t - 7)$ **c.** $\left(c - \dfrac{1}{4}\right)\left(c + \dfrac{1}{4}\right)$
d. $(2p - q)(2p + q)$

Solution: **a.** Using the preceding formula, $(2x + y)(2x - y) = (2x)^2 - y^2 = 4x^2 - y^2$.
b. $(6t + 7)(6t - 7) = (6t)^2 - 7^2 = 36t^2 - 49$.
c. $\left(c - \dfrac{1}{4}\right)\left(c + \dfrac{1}{4}\right) = c^2 - \left(\dfrac{1}{4}\right)^2 = c^2 - \dfrac{1}{16}$.
d. $(2p - q)(2p + q) = (2p)^2 - q^2 = 4p^2 - q^2$.   ■

EXAMPLE 7   Find $\left(x^2 - \dfrac{1}{3}y\right)\left(x^2 + \dfrac{1}{3}y\right)$.

Solution: $\left(x^2 - \dfrac{1}{3}y\right)\left(x^2 + \dfrac{1}{3}y\right) = (x^2)^2 - \left(\dfrac{1}{3}y\right)^2 = x^4 - \dfrac{1}{9}y^2$ ■

EXERCISE SET 3.5

Find each product using the FOIL method. See Examples 1 through 3.

1. $(x + 3)(x + 4)$ **2.** $(x + 5)(x - 1)$ **3.** $(x - 5)(x + 10)$
4. $(y - 12)(y + 4)$ **5.** $(5x - 6)(x + 2)$ **6.** $(3y - 5)(2y - 7)$
7. $(y - 6)(4y - 1)$ **8.** $(2x - 9)(x - 11)$ **9.** $(2x + 5)(3x - 1)$
10. $(6x + 2)(x - 2)$

Find each product. See Examples 4 and 5.

11. $(x - 2)^2$ **12.** $(x + 7)^2$ **13.** $(2x - 1)^2$
14. $(7x - 3)^2$ **15.** $(3a - 5)^2$ **16.** $(5a + 2)^2$
17. $(5x + 9)^2$ **18.** $(6s - 2)^2$

Find each product. See Examples 6 and 7.

19. $(a - 7)(a + 7)$ **20.** $(b + 3)(b - 3)$ **21.** $(3x - 1)(3x + 1)$

22. $(4x - 5)(4x + 5)$ **23.** $\left(3x - \dfrac{1}{2}\right)\left(3x + \dfrac{1}{2}\right)$ **24.** $\left(10x + \dfrac{2}{7}\right)\left(10x - \dfrac{2}{7}\right)$

25. $(9x + y)(9x - y)$ **26.** $(2x - y)(2x + y)$

Find each product.

27. $(a + 5)(a + 4)$ **28.** $(a - 5)(a - 7)$
29. $(a + 7)^2$ **30.** $(b - 2)^2$
31. $(4a + 1)(3a - 1)$ **32.** $(6a + 7)(6a + 5)$
33. $(x + 2)(x - 2)$ **34.** $(x - 10)(x + 10)$

35. $(3a + 1)^2$

36. $(4a - 2)^2$

37. $(x + y)(4x - y)$

38. $(3x + 2)(4x - 2)$

39. $(2a - 3)^2$

40. $(5b - 4x)^2$

41. $(5x - 6z)(5x + 6z)$

42. $(11x - 7y)(11x + 7y)$

43. $(x - 3)(x - 5)$

44. $(a + 5b)(a + 6b)$

45. $\left(x - \dfrac{1}{3}\right)\left(x + \dfrac{1}{3}\right)$

46. $\left(3x + \dfrac{1}{5}\right)\left(3x - \dfrac{1}{5}\right)$

47. $(a + 11)(a - 3)$

48. $(2x + 5)(x - 8)$

49. $(x - 2)^2$

50. $(3b + 7)^2$

51. $(3b + 7)(2b - 5)$

52. $(3y - 13)(y - 3)$

53. $(7p - 8)(7p + 8)$

54. $(3s - 4)(3s + 4)$

55. $\left(\dfrac{1}{3}a^2 - 7\right)\left(\dfrac{1}{3}a^2 + 7\right)$

56. $\left(\dfrac{2}{3}a - b^2\right)\left(\dfrac{2}{3}a - b^2\right)$

57. $(2r - 3s)(2r + 3s)$

58. $(6r - 2x)(6r + 2x)$

59. $(3x - 7y)^2$

60. $(4s - 2y)^2$

61. $(4x + 5)(4x - 5)$

62. $(3x + 5)(3x - 5)$

63. $(x + 4)(x + 4)$

64. $(3x + 2)(3x + 2)$

65. $\left(a - \dfrac{1}{2}y\right)\left(a + \dfrac{1}{2}y\right)$

66. $\left(\dfrac{a}{2} + 4y\right)\left(\dfrac{a}{2} - 4y\right)$

67. $\left(\dfrac{1}{5}x - y\right)\left(\dfrac{1}{5}x + y\right)$

68. $\left(\dfrac{y}{6} - 8\right)\left(\dfrac{y}{6} + 8\right)$

Express each of the following as a polynomial in x.

69. Find the area of the square-shaped rug shown if its side is $(2x + 1)$ feet.

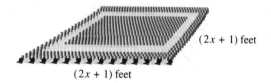

$(2x + 1)$ feet

$(2x + 1)$ feet

70. Find the area of the rectangularly shaped canvas if its length is $(3x - 2)$ inches and its width is $(x - 4)$ inches.

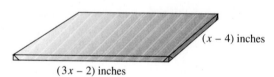

$(x - 4)$ inches

$(3x - 2)$ inches

A Look Ahead

Find each product. See the following example.

EXAMPLE   Find $[(a + b) - 2][(a + b) + 2]$.

Solution: If we think of $(a + b)$ as one term, we think of $[(a + b) - 2][(a + b) + 2]$ as the sum and difference of two terms.

$$[(a + b) - 2][(a + b) + 2] = (a + b)^2 - 2^2$$

Next, square $(a + b)$.

$$= a^2 + 2ab + b^2 - 4 \quad \blacksquare$$

71. $[(x + y) - 3][(x + y) + 3]$

72. $[(a + c) - 5][(a + c) + 5]$

73. $[(a - 3) + b][(a - 3) - b]$

74. $[(x - 2) + y][(x - 2) - y]$

75. $[(2x + 1) - y][(2x + 1) + y]$

76. $[(3x + 2) - z][(3x + 2) + z]$

Writing in Mathematics

77. Explain how to square a binomial.

78. Explain how to find the product of two binomials using the FOIL method.

Skill Review

Simplify each expression. See Sections 3.1 and 3.2.

79. $\dfrac{50b^{10}}{70b^5}$

80. $\dfrac{x^3 y^6}{x^6 y^2}$

81. $\dfrac{8a^{17}b^5}{-4a^7 b^{10}}$

82. $\dfrac{-6a^2 y}{3a^4 y}$

83. $\dfrac{2x^4 y^2}{3x^4 y^4}$

84. $\dfrac{-48ab}{32a^4 b^3}$

3.6
Division of Polynomials

Tape 11

OBJECTIVES

1 Divide a polynomial by a monomial.

2 Use long division to divide a polynomial by another polynomial.

1 Although we didn't use the word monomial, in Sections 3.1 and 3.2 we divided monomials in developing rules and definitions for exponents.

EXAMPLE 1 Simplify the following expressions. Write each answer with positive exponents.

a. $\dfrac{36a^2 b}{6ab^3}$ **b.** $\dfrac{3x^4 y^5}{9x^4 y^2}$

Solution: To simplify, use rules for exponents.

a. $\dfrac{36a^2 b}{6ab^3} = \dfrac{6a}{b^2}$ **b.** $\dfrac{3x^4 y^5}{9x^4 y^2} = \dfrac{y^3}{3}$ ■

To divide a polynomial by a monomial, recall addition of fractions. Fractions that have a common denominator are added by adding the numerators:

$$\frac{a}{c} + \frac{b}{c} = \frac{a+b}{c}$$

If we read this equation from right to left and let a, b, and c be monomials, $c \neq 0$, the following rule emerges.

> **To Divide a Polynomial by a Monomial**
>
> Divide each term of the polynomial by the monomial.
>
> $$\frac{a+b}{c} = \frac{a}{c} + \frac{b}{c}, \qquad c \neq 0$$

Throughout this section, we assume that denominators are not 0.

EXAMPLE 2 Divide $(6m^2 + 2m)$ by $2m$.

Solution: Begin by writing the quotient of the polynomials. Then divide each term of the polynomial $6m^2 + 2m$ by the monomial $2m$.

$$\frac{6m^2 + 2m}{2m} = \frac{6m^2}{2m} + \frac{2m}{2m} = 3m + 1 \quad \blacksquare$$

EXAMPLE 3 Simplify $\dfrac{8x^2y^2 - 16xy + 2x}{4xy}$.

Solution: $\dfrac{8x^2y^2 - 16xy + 2x}{4xy} = \dfrac{8x^2y^2}{4xy} - \dfrac{16xy}{4xy} + \dfrac{2x}{4xy}$ Divide each term by $4xy$.

$$= 2xy - 4 + \frac{1}{2y} \quad \blacksquare$$

EXAMPLE 4 Simplify $\dfrac{12x^5y^6 - 6x^2y^2 + 9x^3y^4 + 3}{6x^2y^2}$.

Solution: Divide each term by $6x^2y^2$.

$$\frac{12x^5y^6 - 6x^2y^2 + 9x^3y^4 + 3}{6x^2y^2} = \frac{12x^5y^6}{6x^2y^2} - \frac{6x^2y^2}{6x^2y^2} + \frac{9x^3y^4}{6x^2y^2} + \frac{3}{6x^2y^2}$$

$$= 2x^3y^4 - 1 + \frac{3xy^2}{2} + \frac{1}{2x^2y^2} \quad \blacksquare$$

2 To divide a polynomial by a polynomial other than a monomial, we use a process known as long division. Polynomial long division is similar to number long division, so we review long division by dividing 13 into 3660.

$$
\begin{array}{r}
281 \\
13\overline{)3660} \\
\underline{26}\!\downarrow\downarrow \\
106 \\
\underline{104}\!\downarrow \\
20 \\
\underline{13} \\
7
\end{array}
$$

$2 \cdot 13 = 26$

Subtract and bring down the next digit in the dividend.

$8 \cdot 13 = 104$

Subtract and bring down the next digit in the dividend.

$1 \cdot 13 = 13$

Subtract. There are no more digits to bring down, so the remainder is 7.

The quotient is 281 R 7, which can be written as $281\dfrac{7}{13}\left(\dfrac{\text{remainder}}{\text{divisor}}\right)$. Recall that division can be checked by multiplication. To check a division problem such as this one, we see that

$$13 \cdot 281 + 7 = 3660$$

Now we demonstrate long division of polynomials.

EXAMPLE 5 Divide $(x^2 + 7x + 12)$ by $(x + 3)$.

Solution: $(x^2 + 7x + 12)$ is the dividend polynomial and $(x + 3)$ is the divisor polynomial.

To subtract, change the signs of these terms and add.	$$\begin{array}{r} x \\ x + 3\overline{)x^2 + 7x + 12} \\ \underline{x^2 + 3x} \\ 4x + 12 \end{array}$$	How many times does x divide x^2? $\dfrac{x^2}{x} = \boxed{x}$.

Multiply: $x(x + 3)$.
Subtract and bring down the next term.

Next, repeat this process.

To subtract, change the signs of these terms and add.	$$\begin{array}{r} x + \boxed{4} \\ x + 3\overline{)x^2 + 7x + 12} \\ \underline{x^2 + 3x} \\ 4x + 12 \\ \underline{4x + 12} \\ 0 \end{array}$$	How many times does x divide $4x$? $\dfrac{4x}{x} = \boxed{4}$.

Multiply: $4(x + 3)$.
Subtract. The remainder is 0.

Then $(x^2 + 7x + 12)$ divided by $(x + 3)$ is $(x + 4)$, the quotient polynomial. To check, see that

$$\text{divisor} \cdot \text{quotient} + \text{remainder} = \text{dividend}$$

or

$(x + 3) \cdot (x + 4) + 0 = x^2 + 7x + 12$, the dividend, so the division checks.   ■

EXAMPLE 6   Divide $6x^2 + 10x - 5$ by $3x - 1$.

Solution: The **divisor polynomial** is $(3x - 1)$ and the **dividend polynomial** is $(6x^2 + 10x - 5)$.

$$\begin{array}{r} 2x + 4 \\ 3x - 1\overline{)6x^2 + 10x - 5} \\ \underline{6x^2 - 2x} \downarrow \\ 12x - 5 \\ \underline{12x - 4} \\ -1 \end{array}$$

$\dfrac{6x^2}{3x} = 2x$, so $\boxed{2x}$ is a term of the quotient.

$2x(3x - 1)$
Subtract and bring down the next term.

$\dfrac{12x}{3x} = 4, 4(3x - 1)$
Subtract. The remainder is -1.

Then $(6x^2 + 10x - 5)$ divided by $(3x - 1)$ is $(2x + 4)$ with a remainder of -1. This can be written as

$$\frac{6x^2 + 10x - 5}{3x - 1} = 2x + 4 + \frac{-1}{3x - 1} \quad \left(\frac{\text{remainder}}{\text{divisor}}\right).$$

We call $(2x + 4)$ the **quotient polynomial** and -1 the **remainder polynomial.** To check, see that divisor · quotient + remainder = dividend.

$$(3x - 1)(2x + 4) + (-1) = (6x^2 + 12x - 2x - 4) - 1$$
$$= 6x^2 + 10x - 5$$

The division checks.   ■

Notice that the division process is continued until the degree of the remainder polynomial is less than the degree of the divisor polynomial.

EXAMPLE 7 Find $\dfrac{4x^2 + 7 + 8x^3}{2x + 3}$.

Solution: Before we begin the division process, the dividend polynomial and the divisor polynomial should be written in descending order of exponents. Any missing powers are represented by a term whose coefficient is zero.

$$\frac{4x^2 + 7 + 8x^3}{2x + 3} = \frac{8x^3 + 4x^2 + \boxed{0x} + 7}{2x + 3}$$

There is no x term, so include $0x$ as the missing power.

$$
\require{enclose}
\begin{array}{r}
4x^2 - 4x + 6 \\
2x + 3 \enclose{longdiv}{8x^3 + 4x^2 + 0x + 7} \\
\underline{8x^3 + 12x^2} \\
-8x^2 + 0x \\
\underline{(-8x^2 - 12x)} \\
12x + 7 \\
\underline{(12x + 18)} \\
-11
\end{array}
$$

Remainder polynomial.

Thus, $\dfrac{4x^2 + 7 + 8x^3}{2x + 3} = 4x^2 - 4x + 6 + \dfrac{-11}{2x + 3}$. ■

EXAMPLE 8 Find $\dfrac{2x^4 - x^3 + 3x^2 + x - 1}{x^2 + 1}$.

Solution: Before dividing, rewrite the divisor polynomial $(x^2 + 1)$ as $(x^2 + 0x + 1)$. The $0x$ term represents the missing x^1 term in the divisor.

$$
\begin{array}{r}
2x^2 - x + 1 \\
x^2 + 0x + 1 \enclose{longdiv}{2x^4 - x^3 + 3x^2 + x - 1} \\
\underline{2x^4 + 0x^3 + 2x^2} \\
-x^3 + x^2 + x \\
\underline{-x^3 + 0x^2 - x} \\
x^2 + 2x - 1 \\
\underline{x^2 + 0x + 1} \\
2x - 2
\end{array}
$$

Remainder polynomial.

Thus, $\dfrac{2x^4 - x^3 + 3x^2 + x - 1}{x^2 + 1} =$

$2x^2 - x + 1 + \dfrac{2x - 2}{x^2 + 1}$. ■

MENTAL MATH

Simplify each expression mentally.

1. $\dfrac{a^6}{a^4}$

2. $\dfrac{y^2}{y}$

3. $\dfrac{a^3}{a}$

4. $\dfrac{p^8}{p^3}$

5. $\dfrac{k^5}{k^2}$

6. $\dfrac{k^7}{k^5}$

EXERCISE SET 3.6

Simplify each expression. See Example 1.

1. $\dfrac{8k^4}{2k}$

2. $\dfrac{27r^4}{3r^6}$

3. $\dfrac{-6m^4}{-2m^3}$

4. $\dfrac{15a^4}{-15a^5}$

5. $\dfrac{-24a^6b}{6ab^2}$

6. $\dfrac{-5x^4y^5}{15x^4y^2}$

7. $\dfrac{6x^2y^3}{-7xy^5}$

8. $\dfrac{-8xa^2b}{-5xa^5b}$

Perform each division. See Examples 2 through 4.

9. $\dfrac{15p^3 + 18p^2}{3p}$

10. $\dfrac{14m^2 - 27m^3}{7m}$

11. $\dfrac{-9x^4 + 18x^5}{6x^5}$

12. $\dfrac{6x^5 + 3x^4}{3x^4}$

13. $\dfrac{-9x^5 + 3x^4 - 12}{3x^3}$

14. $\dfrac{6a^2 - 4a + 12}{2a^2}$

15. $\dfrac{4x^4 - 6x^3 + 7}{-4x^4}$

16. $\dfrac{-12a^3 + 36a - 15}{3a}$

17. $\dfrac{25x^5 - 15x^3 + 5}{5x^2}$

18. $\dfrac{-4y^2 + 4y + 6}{2y}$

Perform each division. See Examples 5 through 8.

19. $\dfrac{x^2 + 4x + 3}{x + 3}$

20. $\dfrac{x^2 + 7x + 10}{x + 5}$

21. $\dfrac{2x^2 + 13x + 15}{x + 5}$

22. $\dfrac{3x^2 + 8x + 4}{x + 2}$

23. $\dfrac{2x^2 - 7x + 3}{x - 4}$

24. $\dfrac{3x^2 - x - 4}{x - 1}$

25. $\dfrac{8x^2 + 6x - 27}{2x - 3}$

26. $\dfrac{18w^2 + 18w - 8}{3w + 4}$

27. $\dfrac{9a^3 - 3a^2 - 3a + 4}{3a + 2}$

28. $\dfrac{-x^3 - 6x^2 + 2x - 3}{x - 1}$

29. $\dfrac{2b^3 + 9b^2 + 6b - 4}{b + 4}$

30. $\dfrac{2x^3 + 3x^2 - 3x + 4}{x + 2}$

Perform each division.

31. $\dfrac{20x^2 + 5x + 9}{5x^3}$

32. $\dfrac{8x^3 - 4x^2 + 6x + 2}{2x^2}$

33. $\dfrac{5x^2 + 28x - 10}{x + 6}$

34. $\dfrac{2x^2 + x - 15}{x + 3}$

35. $\dfrac{10x^3 - 24x^2 - 10x}{10x}$

36. $\dfrac{2x^3 + 12x^2 + 16}{4x^2}$

37. $\dfrac{6x^2 + 17x - 4}{x + 3}$

38. $\dfrac{2x^2 - 9x + 15}{x - 6}$

39. $\dfrac{12x^4 + 3x^2}{3x^2}$

40. $\dfrac{15x^2 - 9x^5}{9x^5}$

41. $\dfrac{2x^3 + 2x^2 - 17x + 8}{x - 2}$

42. $\dfrac{4x^3 + 11x^2 - 8x - 10}{x + 3}$

43. $\dfrac{30x^2 - 17x + 2}{5x - 2}$

44. $\dfrac{4x^2 - 13x - 12}{4x + 3}$

45. $\dfrac{3x^4 - 9x^3 + 12}{-3x}$

46. $\dfrac{8y^6 - 3y^2 - 4y}{4y}$

47. $\dfrac{8x^2 + 10x + 1}{2x + 1}$

48. $\dfrac{3x^2 + 17x + 7}{3x + 2}$

49. $\dfrac{4x^2 - 81}{2x - 9}$

50. $\dfrac{16x^2 - 36}{4x + 6}$

51. $\dfrac{4x^3 + 12x^2 + x - 12}{2x + 3}$

52. $\dfrac{6x^2 + 11x - 10}{3x - 2}$

53. $\dfrac{x^3 - 27}{x - 3}$

54. $\dfrac{x^3 + 64}{x + 4}$

55. $\dfrac{x^3 + 1}{x + 1}$

56. $\dfrac{x^5 + x^2}{x^2 + x}$

57. $\dfrac{1 - 3x^2}{x + 2}$

58. $\dfrac{7 - 5x^2}{x + 3}$

59. $\dfrac{-4b + 4b^2 - 5}{2b - 1}$

60. $\dfrac{-3y + 2y^2 - 15}{2y + 5}$

Express each of the following as a polynomial in x.

61. The perimeter of a square is $(12x^3 + 4x - 16)$ feet. Find the length of its side.

62. The area of the following parallelogram is $(10x^2 + 31x + 15)$ square meters. If its base is $(5x + 3)$ meters, find its height.

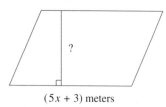

(5x + 3) meters

63. The area of the top of the ping pong table is $(49x^2 + 70x - 200)$ square inches. If its length is $(7x + 20)$ inches, find its width.

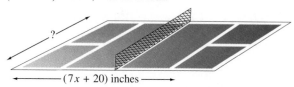

◄—————— (7x + 20) inches ——————►

64. The volume of the swimming pool shown is $(36x^5 - 12x^3 + 6x^2)$ cubic feet. If its height is $2x$ feet and its width is $3x$ feet, find its length.

Writing in Mathematics

65. Explain how to check a polynomial long division problem.

Skill Review

Multiply each expression. See Section 3.4.

66. $2a(a^2 + 1)$

67. $-4a(3a^2 - 4)$

68. $2x(x^2 + 7x - 5)$

69. $4y(y^2 - 8y - 4)$

70. $-3xy(xy^2 + 7x^2y + 8)$

71. $-9xy(4xyz + 7xy^2z + 2)$

72. $9ab(ab^2c + 4bc - 8)$

73. $-7sr(6s^2r + 9sr^2 + 9rs + 8)$

CRITICAL THINKING Chain letters have been around for years, yet many people do not realize that some chain letters violate U.S. postal regulations. The originator of a chain letter makes a list of, say, four person's names, including his own name at the bottom of the list. Then, to as many people as he can, he sends a letter and the list. The letter tells each person to send to the person at the top of the list a recipe, a postcard, a few dollars, or something else of seemingly insignificant value. The letter further tells the person to put his own name at the bottom of the list, take the top name off the list, and mail the letter and the list to his own friends. In just a few weeks, the letter promises, the person will be swamped with recipes, postcards, or dollars! Unfortunately, the chain usually dies out before many people receive anything at all, except perhaps the first few people in the chain.

Chain letters asking for money are against the law. Some people assume these letters are illegal because of a threatening tone or because they play on a person's superstitions. But these reasons are not the true reasons.

What justification can you give for the law against money chain letters? If the letter asks for $2.00, and the originator and each person who gets the letter sends it to 10 people, how much money does the originator hope to get if his name was on the bottom of his original list of 4 names? What if each person sent the letter to 15 people?

CHAPTER 3 GLOSSARY

A **binomial** is a polynomial with exactly two terms.

The **degree of a polynomial** is the highest degree of any term in the polynomial.

The **degree of a term** is the sum of the exponents of the variables contained in the term.

The polynomial $3x^7 - 2x^6 + 5x^3 - 2x + 7$ is a **polynomial in** x written in **descending powers** of x.

A **monomial** is a polynomial with exactly one term.

The **numerical coefficient** of a term, or simply the **coefficient,** is the numerical part of each term.

A **polynomial** is an algebraic expression that consists of a finite sum of terms.

A number is written in **scientific notation** if it is written as the product of a number x, where $1 \leq x < 10$, and a power of 10.

A **trinomial** is a polynomial with exactly three terms.

CHAPTER 3 SUMMARY

(3.1) and (3.2)

Summary of Exponent Rules

If m and n are integers and a, b, and c are real numbers, then:

Product rule for exponents:	$a^m \cdot a^n = a^{m+n}$
Power rule for exponents:	$(a^m)^n = a^{m \cdot n}$
Power rules for products and quotients:	$(ab)^n = a^n b^n$ and
	$\left(\dfrac{a}{c}\right)^n = \dfrac{a^n}{c^n}, c \neq 0$
Quotient rule for exponents:	$\dfrac{a^m}{a^n} = a^{m-n}, a \neq 0$
Zero exponent:	$a^0 = 1, a \neq 0$
Negative exponent:	$a^{-n} = \dfrac{1}{a^n}, a \neq 0$

(3.3)

To add polynomials, combine like terms.

To subtract polynomials, change all the signs of the second polynomial and then add like terms of the first and second polynomials.

(3.4)

To multiply two polynomials, multiply each term of the first polynomial by each term of the second polynomial and then combine products that are like terms.

FOIL METHOD OF MULTIPLYING TWO BINOMIALS   **(3.5)**

$$(a + b)(c + d) = \overset{F}{ac} + \overset{O}{ad} + \overset{I}{bc} + \overset{L}{bd}$$

SQUARING A BINOMIAL   **(3.5)**

$$(a + b)^2 = a^2 + 2ab + b^2$$

$$(a - b)^2 = a^2 - 2ab + b^2$$

TO MULTIPLY THE SUM AND DIFFERENCE OF TWO TERMS   **(3.5)**

$$(a + b)(a - b) = a^2 - b^2$$

(3.6)

To divide a polynomial by a monomial, divide each term of the polynomial by the monomial.

To divide a polynomial by a polynomial other than a monomial, use long division.

CHAPTER 3 REVIEW

(3.1)   *State the base and the exponent for each expression.*

1. 3^2

2. $(-5)^4$

3. -5^4

Evaluate each expression.

4. 8^3

5. $(-6)^2$

6. -6^2

7. $-4^3 - 4^0$

8. $(3b)^0$

9. $\dfrac{8b}{8b}$

Simplify each expression.

10. $5b^3b^5a^6$

11. $2^3 \cdot x^0$

12. $[(-3)^2]^3$

13. $(2x^3)(-5x^2)$

14. $\left(\dfrac{mn}{q}\right)^2 \cdot \left(\dfrac{mn}{q}\right)$

15. $\left(\dfrac{3ab^2}{6ab}\right)^4$

16. $\dfrac{x^9}{x^4}$

17. $\dfrac{2x^7y^8}{8xy^2}$

18. $\dfrac{12xy^6}{3x^4y^{10}}$

19. $5a^7(2a^4)^3$

20. $(2x)^2(9x)$

21. $\dfrac{(-4)^2(3^3)}{(4^5)(3^2)}$

22. $\dfrac{(-7)^2(3^5)}{(-7)^3(3^4)}$

23. $\dfrac{(2x)^0(-4)^2}{16x}$

24. $\dfrac{(8xy)(3xy)}{18x^2y^2}$

25. $m^0 + p^0 + 3q^0$

26. $(-5a)^0 + 7^0 + 8^0$

27. $(3xy^2 + 8x + 9)^0$

28. $8x^0 + 9^0$

29. $6(a^2b^3)^3$

30. $\dfrac{(x^3z)^a}{x^2z^2}$

(3.2)   *Simplify each expression.*

31. 7^{-2}

32. -7^{-2}

33. $2x^{-4}$

34. $(2x)^{-4}$

35. $\left(\dfrac{1}{5}\right)^{-3}$

36. $\left(\dfrac{-2}{3}\right)^{-2}$

37. $2^0 + 2^{-4}$

38. $6^{-1} - 7^{-1}$

Simplify each expression. Assume that variables in an exponent represent positive integers only. Write each answer using positive exponents.

39. $\dfrac{1}{(2q)^{-3}}$

40. $\dfrac{-1}{(qr)^{-3}}$

41. $\dfrac{r^{-3}}{s^{-4}}$

42. $\dfrac{rs^{-3}}{r^{-4}}$

43. $\dfrac{-6}{8x^{-3}r^4}$

44. $\dfrac{-4s}{16s^{-3}}$

45. $(2x^{-5})^{-3}$

46. $(3y^{-6})^{-1}$

47. $(3a^{-1}b^{-1}c^{-2})^{-2}$

48. $(4x^{-2}y^{-3}z)^{-3}$

49. $\dfrac{5^{-2}x^8}{5^{-3}x^{11}}$

50. $\dfrac{7^5y^{-2}}{7^7y^{-10}}$

51. $\left(\dfrac{bc^{-2}}{bc^{-3}}\right)^4$

52. $\left(\dfrac{x^{-3}y^{-4}}{x^{-2}y^{-5}}\right)^{-3}$

53. $\dfrac{x^{-4}y^{-6}}{x^2y^7}$

54. $\dfrac{a^5b^{-5}}{a^{-5}b^5}$

55. $-2^0 + 2^{-4}$

56. $-3^{-2} - 3^{-3}$

57. $a^{6m}a^{5m}$

58. $\dfrac{(x^{5+h})^3}{x^5}$

59. $(3xy^{2z})^3$

60. $a^{m+2}a^{m+3}$

Write each number in scientific notation.

61. 0.00027

62. 0.8868

63. $80,800,000$

64. $-868,000$

65. The population of the United States is 234,000,000.

66. The radius of Earth is 4000 miles.

Write each number in standard form.

67. 8.67×10^5

68. 3.86×10^{-3}

69. 8.6×10^{-4}

70. 8.936×10^5

71. The number of photons of light emitted by a 100-watt bulb every second is 1×10^{20}.

72. The real mass of all the galaxies in the constellation of Virgo is 3×10^{-25}.

Simplify. Express each answer in standard form.

73. $(8 \times 10^4)(2 \times 10^{-7})$

74. $\dfrac{8 \times 10^4}{2 \times 10^{-7}}$

(3.3) *Find the degree of each term.*

75. $-5x^4y^3$

76. $10x^3y^2z$

77. $35a^5bc^2$

78. $95xyz$

Find the degree of each polynomial.

79. $y^5 + 7x - 8x^4$

80. $9y^2 + 30y + 25$

81. $-14x^2yb - 28x^2y^3b - 42x^2y^2$

82. $6x^2y^2z^2 + 5x^2y^3 - 12xyz$

Combine like terms.

83. $6a^2b^2 + 4ab + 9a^2b^2$

84. $21x^2y^3 + 3xy + x^2y^3 + 6$

85. $4a^2b - 3b^2 - 8q^2 - 10a^2b + 7q^2$

86. $2s^{14} + 3s^{13} + 12s^{12} - s^{10}$

Add or subtract each polynomial.

87. $(3k^2 + 2k + 6) + (5k^2 + k)$

88. $(2s^5 + 3s^4 + 4s^3 + 5s^2) - (4s^2 + 7s + 6)$

89. $(2m^7 + 3x^4 + 7m^6) - (8m^7 + 4m^2 + 6x^4)$

90. Subtract $(4x^2 + 8x - 7)$ from the sum of $(x^2 + 7x + 9)$ and $(x^2 + 4)$.

(3.4) *Multiply each expression.*

91. $9x(x^2y)$

92. $-7(8xz^2)$

93. $(6xa^2)(xya^3)$

94. $(4xy)(-3xa^2y^3)$

95. $6(x + 5)$

96. $9(x - 7)$

97. $4(2a + 7)$

98. $9(6a - 3)$

99. $-7x(x^2 + 5)$

100. $-8y(4y^2 - 6)$

101. $-2(x^3 - 9x^2 + x)$

102. $-3a(a^2b + ab + b^2)$

103. $(3a^3 - 4a + 1)(-2a)$

104. $(6b^3 - 4b + 2)(7b)$

105. $(2x + 2)(x - 7)$

106. $(2x - 5)(3x + 2)$

107. $(4a - 1)(a + 7)$

108. $(6a - 1)(7a + 3)$

109. $(x + 7)(x^3 + 4x - 5)$

110. $(x + 2)(x^5 + x + 1)$

111. $(x^2 + 2x + 4)(x^2 + 2x - 4)$

112. $(x^3 + 4x + 4)(x^3 + 4x - 4)$

113. $(x + 7)^3$

114. $(2x - 5)^3$

(3.5)   *Use the special product rules to compute each expression.*

115. $(x + 7)^2$

116. $(x - 5)^2$

117. $(3x - 7)^2$

118. $(4x + 2)^2$

119. $(5x - 9)^2$

120. $(5x + 1)(5x - 1)$

121. $(7x + 4)(7x - 4)$

122. $(a + 2b)(a - 2b)$

123. $(2x - 6)(2x + 6)$

124. $(4a^2 - 2b)(4a^2 + 2b)$

(3.6)   *Perform each division.*

125. $\dfrac{4xy^2}{3xz^2y^3}$

126. $\dfrac{4xy^3}{32xy^2z}$

127. $\dfrac{x^2 + 21x + 49}{7x^2}$

128. $\dfrac{5a^3b - 15ab^2 + 20ab}{-5ab}$

129. $\dfrac{a^2 - a + 4}{a - 2}$

130. $\dfrac{4x^2 + 20x + 7}{x + 5}$

131. $\dfrac{a^3 + a^2 + 2a + 6}{a - 2}$

132. $\dfrac{9b^3 - 18b^2 + 8b - 1}{3b - 2}$

133. $\dfrac{4x^4 - 4x^3 + x^2 + 4x - 3}{2x - 1}$

134. $\dfrac{-10x^2 - x^3 - 21x + 18}{x - 6}$

CHAPTER 3 TEST

Evaluate each expression.

1. 2^5

2. $(-3)^4$

3. -3^4

4. 4^{-3}

Simplify each exponential expression.

5. $\left(\dfrac{5x^6y^3}{35x^7y}\right)^2$

6. $\dfrac{7(xy)^4}{(xy)^2}$

7. $4(x^2y^3)^{-3}$

Simplify each expression. Write the answer using only positive exponents.

8. $\left(\dfrac{x^2y^3}{x^3y^{-4}}\right)^{-2}$

9. $\dfrac{6^2x^{-4}y^{-1}}{6^3x^{-3}y^7}$

Express each number in scientific notation.

10. 563,000

11. 0.0000863

Write each expression without exponents.

12. 1.5×10^{-3}

13. 6.23×10^4

14. Simplify. Write the answer in standard form.

$(1.2 \times 10^5)(3 \times 10^{-7})$

15. Find the degree of the following polynomial.

$4xy^2 + 7xyz + 9x^3yz$

16. Simplify by combining like terms.

$6xyz + 9x^2y - 3xyz + 9x^2y$

Perform the indicated operations.

17. $(8x^3 + 7x^2 + 4x - 7) + (8x^3 - 7x - 6)$

18. $\begin{aligned} 5x^3 + \ x^2 + 5x - 2 \\ - (8x^3 - 4x^2 + x - 7) \end{aligned}$

19. Subtract $(4x + 2)$ from the sum of $(8x^2 + 7x + 5)$ and $(x^3 - 8)$.

20. Multiply $(3x + 7)(x^2 + 5x + 2)$.

21. Multiply
$$x^3 - x^2 + x + 1$$
$$\underline{2x^2 - 3x + 7}$$

22. Use the FOIL method to multiply $(x + 7)(3x - 5)$.

Use special products to multiply each of the following.

23. $(3x - 7)(3x + 7)$

24. $(4x - 2)^2$

25. $(8x + 3)^2$

26. $(x^2 - 9b)(x^2 + 9b)$

Divide.

27. $\dfrac{8xy^2}{4x^3y^3z}$

28. $\dfrac{4x^2 + 2xy - 7x}{8xy}$

29. $\dfrac{x^2 + 7x + 10}{x + 5}$

30. $\dfrac{27x^3 - 8}{3x + 2}$

CHAPTER 3 CUMULATIVE REVIEW

1. Insert $<$, $>$, or $=$ in the appropriate space to make each statement true;

   a. 2   3   **b.** 7   4   **c.** 72   27

2. Find the product of $\dfrac{2}{15}$ and $\dfrac{5}{13}$. Write the answer in lowest terms.

3. Add or subtract as indicated. Write each answer in lowest terms.

   a. $\dfrac{2}{5} + \dfrac{1}{4}$ **b.** $\dfrac{1}{2} + \dfrac{17}{22} - \dfrac{2}{11}$ **c.** $3\dfrac{1}{6} - 1\dfrac{11}{12}$

4. Simplify $\dfrac{8 + 2 \cdot 3}{2^2 - 1}$;

5. Simplify the following:

   a. $3 + (-7) + (-8)$

   b. $[7 + (-10)] + [-2 + (-4)]$

6. Find each difference:

   a. $-13 - (+4)$ **b.** $5 - (-6)$

   c. $3 - 6$ **d.** $-1 - (-7)$

7. Simplify the following if possible:

   a. $\dfrac{1}{0}$ **b.** $\dfrac{0}{-3}$ **c.** $\dfrac{0(-8)}{2}$

8. Find the additive inverse of each number:

   a. -3 **b.** 5   **c.** 0

9. Simplify the expressions by combining like terms:

   a. $2x + 3x + 5 + 2$

   b. $-5a - 3 + a + 2$

   c. $4y - 3y^2$

   d. $2.3x + 5x - 6$

10. Solve $2x + 3x - 5 + 7 = 10x + 3 - 6x - 4$ for x.

11. Solve $-z - 4 = 6$ for z.

12. Solve $-2(x - 5) + 10 = -3(x + 2) + x$

13. Find how long it takes for \$500 in a savings account that pays 5% simple interest to earn \$75.

14. Solve $F = \dfrac{9}{5}C + 32$ for C

15. Graph $2 < x \le 4$.

16. Solve $3 \le \dfrac{3x}{2} + 4 \le 5$, and graph the solution.

17. Simplify each expression. Write answers with positive exponents.

   a. $\dfrac{y}{y^{-2}}$ **b.** $\dfrac{p^{-4}}{q^{-9}}$ **c.** $\dfrac{x^{-5}}{x^7}$

18. Add $(-2x^2 + 5x - 1) + (-2x^2 + x + 3)$.

19. Multiply $(2x - y)^2$

20. Multiply the following

   a. $(2x + y)(2x - y)$

   b. $(6t + 7)(6t - 7)$

   c. $\left(c - \dfrac{1}{4}\right)\left(c + \dfrac{1}{4}\right)$

   d. $(2p - q)(2p + q)$

21. Divide $(x^2 + 7x + 12)$ by $(x + 3)$. Use long division

4.1 The Greatest Common Factor and Factoring by Grouping

4.2 Factoring Trinomials of the Form $x^2 + bx + c$

4.3 Factoring Trinomials of the Form $ax^2 + bx + c$

4.4 Factoring Binomials

4.5 Factoring Polynomials Completely

4.6 Solving Quadratic Equations by Factoring

4.7 Applications of Quadratic Equations

CHAPTER **4**

Factoring Polynomials

Peter yearns for the luxury of a swimming pool in his backyard, complete with surrounding cement patio. He knows how large a pool he can afford, but now he wonders how large the patio can be. (See Critical Thinking, page 196.)

INTRODUCTION

In Chapter 3, we learned how to multiply polynomials. This chapter deals with an operation that is the reverse process of multiplying, called factoring. Factoring is an important algebraic skill because it is used to simplify complicated expressions into simpler expressions. It is a process that allows us to change an addition problem to a multiplication problem.

4.1
The Greatest Common Factor and Factoring by Grouping

OBJECTIVES

Tape 12

1 Find the greatest common factor of a list of terms.

2 Factor out the greatest common factor from a polynomial.

3 Factor a polynomial by grouping.

When an integer is written as the product of two other integers, each integer in the product is called a **factor.** This is true for polynomials, also. When a polynomial is written as the product of two other polynomials, each polynomial in the product is called a factor. The process of writing a polynomial as a product is called **factoring.**

$$2 \cdot 3 = 6 \qquad x^2 \cdot x^3 = x^5 \qquad (x + 2)(x + 3) = x^2 + 5x + 6$$

factor factor product   factor factor product   factor   factor   product

Notice that factoring is the reverse process of multiplying.

$$\underset{\text{multiplying}}{\overset{\text{factoring}}{x^2 + 5x + 6 = (x + 2)(x + 3)}}$$

1   We begin our study of factoring by reviewing the greatest common factor (GCF). The GCF of a list of integers is the largest integer that is a factor of all the integers in the list. For example, the GCF of 12 and 20 is 4 because 4 is the largest integer that is a factor of both 12 and 20. With large integers, the GCF may not be easily found by inspection. When this happens, use the following steps.

To Find the GCF

Step 1   Write each number as a product of primes.

Step 2   Find the common prime factors.

Step 3   The product of all common prime factors found in step 2 is the greatest common factor. If no prime factors are found in step 2, the greatest common factor is 1.

EXAMPLE 1 Find the GCF of each list of numbers.
  a. 28 and 40   **b.** 55 and 21   **c.** 15, 18, and 66

Solution: **a.** Write each number as a product of primes.

$$28 = 2 \cdot 2 \cdot 7 = 2^2 \cdot 7$$

$$40 = 2 \cdot 2 \cdot 2 \cdot 5 = 2^3 \cdot 5$$

The common factors are two factors of 2, so the GCF is

$$\text{GCF} = 2 \cdot 2 = 4$$

  b. $55 = 5 \cdot 11$

  $21 = 3 \cdot 7$

There are no common prime factors; thus, the GCF is 1.

  c. $15 = 3 \cdot 5$

  $18 = 2 \cdot 3 \cdot 3 = 2 \cdot 3^2$

  $66 = 2 \cdot 3 \cdot 11$

The only factor common to all three numbers is one factor of 3, so the GCF is

$$\text{GCF} = 3 \quad \blacksquare$$

The greatest common factor of a list of variables is found in a similar way. For example, the GCF of x^2, x^3, and x^5 is x^2 because each term contains a factor of x^2.

$$x^2 = x^2$$

$$x^3 = x^2 \cdot x$$

$$x^5 = x^2 \cdot x^3$$

In other words, **the GCF of a list of common variables raised to powers is the variable raised to the smallest exponent in the list.**

EXAMPLE 2 Find the GCF of each list of terms.
  a. x^3, x^7, and x^2 **b.** x^3y, x^3y^4, and x^5y^7

Solution: **a.** The GCF is x^2 since 2 is the smallest exponent to which x is raised.
  b. The GCF is x^3y, since 3 is the smallest exponent on x and 1 is the smallest exponent on y. $\blacksquare$

The GCF of a list of terms is the product of all common factors.

EXAMPLE 3 Find the greatest common factor of each list of terms.
  a. $6x^2$, $10x^3$, and $-8x$ **b.** $8y^2$, y^3, and y^5

Solution: **a.** The GCF of the numerical coefficients 6, 10, and -8 is **2.**
  The GCF of variable factors x^2, x^3, and x is **x.**
  Thus, the GCF of the terms $6x^2$, $10x^3$, and $-8x$ is **$2x$.**

b. The GCF of the numerical coefficients 8, 1, and 1 is **1.**
The GCF of variable factors y^2, y^3, and y^5 is **y^2.**
Thus, the GCF of terms $8y^2$, y^3, and y^5 is **$1y^2$ or y^2.**   ∎

2   The first step in factoring a polynomial is to find the GCF of its terms. Once we do so, we can write the polynomial as a product by **factoring out** the GCF.

The polynomial $8x + 14$, for example, contains two terms: $8x$ and 14. The GCF of these terms is 2. We factor out 2 from each term by writing each term as a product of 2 and the term's remaining factors.

$$8x + 14 = 2 \cdot 4x + 2 \cdot 7$$

Using the distributive property, we can write

$$8x + 14 = 2 \cdot 4x + 2 \cdot 7$$
$$= 2(4x + 7)$$

Thus, a factored form of $8x + 14$ is $2(4x + 7)$.

EXAMPLE 4   Factor each polynomial by factoring out the GCF.
a. $6t + 18$ **b.** $y^5 - y^7$

Solution: **a.** The GCF of terms $6t$ and 18 is 6.

$$6t + 18 = \boxed{6} \cdot t + \boxed{6} \cdot 3$$
$$= \boxed{6} (t + 3) \qquad \text{Apply the distributive property.}$$

Our work can be checked by finding the product $6(t + 3)$.

$$6(t + 3) = 6t + 18, \text{ the original polynomial.}$$

b. The GCF of y^5 and y^7 is y^5. Thus,

$$y^5 - y^7 = (\boxed{y^5})\,1 - (\boxed{y^5})\,y^2$$
$$= \boxed{y^5} (1 - y^2) \quad ∎$$

EXAMPLE 5   Factor $-9a^2b + 18a^2b^2 - 3ab$.

Solution: $-9a^2b + 18a^2b^2 - 3ab = (3ab)\,(-3a) + (3ab)\,(6ab) + (3ab)\,(-1)$
$$= 3ab\,(-3a + 6ab - 1) \quad ∎$$

In Example 5 we could have chosen to factor out a $-3ab$ instead of $3ab$. If we factor out a $-3ab$, we have

$$-9a^2b + 18a^2b^2 - 3ab = (-3ab)\,(3a) + (-3ab)\,(-6ab) + (-3ab)\,(1)$$
$$= -3ab\,(3a - 6ab + 1)$$

EXAMPLE 6   Factor $25x^4z + 15x^3z + 5x^2z$.

Solution: The greatest common factor is $5x^2z$.

$$25x^4z + 15x^3z + 5x^2z = 5x^2z\,(5x^2 + 3x + 1) \quad ∎$$

HELPFUL HINT

Be careful when the GCF of the terms is the same as one of the terms in the polynomial. The greatest common factor of $8x^2 - 6x^3 + 2x$ is $2x$. When factoring out $2x$ from $8x^2 - 6x^3 + 2x$, don't forget a term of 1.

$$8x^2 - 6x^3 + 2x = 2x(4x) - 2x(3x^2) + 2x(1)$$
$$= 2x(4x - 3x^2 + 1)$$

Check by multiplying.

$$2x(4x - 3x^2 + 1) = 8x^2 - 6x^3 + 2x$$

EXAMPLE 7   Factor $5(x + 3) + y(x + 3)$.

Solution: The binomial $(x + 3)$ is the greatest common factor. Use the distributive property to factor out $(x + 3)$.

$$5(x + 3) + y(x + 3) = (x + 3)(5 + y) \quad \blacksquare$$

EXAMPLE 8   Factor $3m^2n(a + b) - (a + b)$.

Solution: The greatest common factor is $(a + b)$.

$$3m^2n(a + b) - 1(a + b) = (a + b)(3m^2n - 1) \quad \blacksquare$$

3   Once the GCF is factored out, we can often continue to factor the polynomial, using a variety of techniques. We discuss here a technique for factoring polynomials called **grouping.**

EXAMPLE 9   Factor $xy + 2x + 3y + 6$ by grouping. Check by multiplying.

Solution: The first two terms have a common factor of x, and the last two terms have a common factor of 3.

$$xy + 2x + 3y + 6 = x(y + 2) + 3(y + 2)$$

Next, factor out the common binomial factor of $(y + 2)$.

$$x(y + 2) + 3(y + 2) = (y + 2)(x + 3)$$

To check, multiply $(y + 2)$ by $(x + 3)$.

$$(y + 2)(x + 3) = xy + 2x + 3y + 6, \text{ the original polynomial.}$$

Thus, $xy + 2x + 3y + 6 = (y + 2)(x + 3)$. $\blacksquare$

To Factor a Four-term Polynomial by Grouping

Step 1   If there is a greatest common factor other than 1 for all four terms, factor it out.

Step 2   Arrange terms so that the first two terms have a common factor and the last two terms have a common factor.

Step 3   For each pair of terms, use the distributive property to factor out the pair's common factor.

> *Step 4*   If there is now a common binomial factor, factor it out.
>
> *Step 5*   If there is no common binomial factor in step 4, rearrange the terms in the polynomial and try steps 3 and 4 again.

EXAMPLE 10   Factor $3x^2 + 4xy - 3x - 4y$ by grouping.

Solution: The first two terms have a common factor of x. Factor -1 from the last two terms so that a common binomial factor of $(3x + 4y)$ appears.

$$3x^2 + 4xy - 3x - 4y = x(3x + 4y) - 1(3x + 4y)$$

Next, factor out the common factor of $(3x + 4y)$.

$$= (3x + 4y)(x - 1) \quad \blacksquare$$

HELPFUL HINT

When **factoring** a polynomial, make sure the polynomial is written as a **product.** For example, it is true that

$$xy + 2x + 3y + 6 = x(y + 2) + 3(y + 2)$$

but $x(y + 2) + 3(y + 2)$ is not a **factored form** of the original polynomial since it is a **sum,** not a **product.** The factored polynomial is

$$xy + 2x + 3y + 6 = (y + 2)(x + 3)$$

EXAMPLE 11   Factor $4ax - 4ab - 2bx + 2b^2$.

Solution: First, factor out a common factor of 2 from all four terms.

$$
\begin{aligned}
4ax &- 4ab - 2bx + 2b^2 \\
&= 2(2ax - 2ab - bx + b^2) \quad &&\text{Factor out common factors from each pair} \\
&= 2[2a(x - b) - b(x - b)] \quad &&\text{of terms.} \\
&= 2(x - b)(2a - b) \quad &&\text{Factor out the common binomial.}
\end{aligned}
$$

Notice that we factored out $-b$ instead of b from the second pair of terms so that the binomial factor of each pair is the same. $\blacksquare$

MENTAL MATH

Find the prime factorization of the following integers mentally.

1. 14   **2.** 15   **3.** 10   **4.** 70

Find the GCF of the following pairs of integers mentally.

5. 6, 15   **6.** 20, 15   **7.** 3, 18   **8.** 14, 35

EXERCISE SET 4.1

Find the GCF for each list. See Examples 1 and 2.

1. 32, 36

2. 36, 90

3. 12, 18, 36

4. 24, 14, 21

5. y^2, y^4, y^7

6. x^3, x^2, x^3

7. $x^{10}y^2, xy^2, x^3y^3$

8. p^7q, p^8q^2, p^9q^3

Find the GCF for each list. See Example 3.

9. $8x, 4$

10. $9y, y$

11. $12y^4, 20y^3$

12. $32x, 18x^2$

13. $12x^3, 6x^4, 3x^5$

14. $15y^2, 5y^7, 20y^3$

15. $18x^2y, 9x^3y^3, 36x^3y$

16. $7x, 21x^2y^2, 14xy$

or simplify

Factor the GCF from each polynomial. See Examples 4 through 6.

17. $3a + 6$

18. $18a - 12$

19. $30x - 15$

20. $42x - 7$

21. $24cd^3 - 18c^2d$

22. $25x^4y^3 - 15x^2y^2$

23. $-24a^4x + 18a^3x$

24. $-15a^2x + 9ax$

25. $12x^3 + 16x^2 - 8x$

26. $6x^3 - 9x^2 + 12x$

27. $5x^3y - 15x^2y + 10xy$

28. $14x^3y + 7x^2y - 7xy$

Factor the GCF from each polynomial. See Examples 7 and 8.

29. $y(x + 2) + 3(x + 2)$

30. $z(y + 4) + 3(y + 4)$

31. $x(y - 3) - 4(y - 3)$

32. $6(x + 2) - y(x + 2)$

33. $2x(x + y) - (x + y)$

34. $xy(y + 1) - (y + 1)$

Factor the following four-term polynomials by grouping. See Examples 9 through 11.

35. $5x + 15 + xy + 3y$

36. $xy + y + 2x + 2$

37. $2y - 8 + xy - 4x$

38. $6x - 42 + xy - 7y$

39. $3xy - 6x + 8y - 16$

40. $xy - 2yz + 5x - 10z$

41. $y^3 + 3y^2 + y + 3$

42. $x^3 + 4x + x^2 + 4$

Factor the following polynomials.

43. $3x - 6$

44. $4x - 16$

45. $-8x - 18$

46. $-6x - 40$

47. $32xy - 18x^2$

48. $10xy - 15x^2$

49. $4x - 8y + 4$

50. $7x + 21y - 7$

51. $8(x + 2) - y(x + 2)$

52. $x(y^2 + 1) - 3(y^2 + 1)$

53. $-40x^8y^6 - 16x^9y^5$

54. $-21x^3y - 49x^2y^2$

55. $5x + 10$

56. $7x + 35$

57. $-3x + 12$

58. $-10x + 20$

59. $18x^3y^3 - 12x^3y^2 + 6x^5y^2$

60. $32x^3y^3 - 24x^2y^3 + 8x^2y^4$

61. $-2a^3 - 6a^2b$

62. $-3b^3c - 9b^2$

63. $y^2(x - 2) + (x - 2)$

64. $x(y + 4) + (y + 4)$

65. $5xy + 15x + 6y + 18$

66. $2x^3 + x^2 + 8x + 4$

67. $4x^2 - 8xy - 3x + 6y$

68. $2x^3 - x^2 - 10x + 5$

69. $126x^3yz + 210y^4z^3$

70. $231x^3y^2z - 143yz^2$

71. $4y^2 - 12y + 4yz - 12z$

72. $5x^2 - 20x^2y + 5z - 20zy$

A Look Ahead

Factor by grouping. See the following example.

> **EXAMPLE**   Factor $a^3y + 10b^3 + 2b^3y + 5a^3$.
>
> **Solution:** $a^3y + 10b^3 + 2b^3y + 5a^3 = a^3y + 5a^3 + 10b^3 + 2b^3y$
>
> $= a^3(y + 5) + 2b^3(5 + y)$
>
> $= (y + 5)(a^3 + 2b^3)$ ■

73. $3y - 5x + 15 - xy$

74. $2x - 9y + 18 - xy$

75. $36x + 15y + 30 + 18xy$

76. $21y - 15x + 15xy - 21$

77. $12x^2y - 42x^2 - 4y + 14$

78. $90 + 15y^2 - 18x - 3xy^2$

Writing in Mathematics

79. Write a sentence that includes the words product, prime, and composite.

80. Describe the greatest common factor of a list of integers.

81. Describe what it means to factor a polynomial.

Skill Review

Solve the following. See Section 2.3.

82. $5x = 10$

83. $7x = 49$

84. $-3x = -9$

85. $-2x = 18$

86. $\frac{2}{3}x = \frac{5}{9}$

87. $-\frac{4}{9}x = 8$

4.2
Factoring Trinomials of the Form $x^2 + bx + c$

OBJECTIVES

Tape 12

1   Factor trinomials of the form $x^2 + bx + c$.

2   Factor out the greatest common factor before factoring a trinomial of the form $x^2 + bx + c$.

1   In this section, we factor trinomials of the form $x^2 + bx + c$, where the numerical coefficient of the squared variable is 1. Recall that factoring a polynomial is the process of writing the polynomial as a product. For example, since $(x + 3)(x + 1) = x^2 + 4x + 3$, we say that the factored form of $x^2 + 4x + 3$ is

$$x^2 + 4x + 3 = (x + 3)(x + 1)$$

Notice that the product of the first terms of the binomials is $x \cdot x = x^2$, the first term of the trinomial. Also, the product of the last two terms of the binomials is $3 \cdot 1 = 3$, the third term of the trinomial. The sum of these same terms is $3 + 1 = 4$, the coefficient of the x term of the trinomial.

$$\begin{array}{c} \overbrace{x^2 = x \cdot x} \\ x^2 + 4x + 3 = (x + 3)(x + 1) \\ \underbrace{3 = 3 \cdot 1} \\ 4 = 3 + 1 \end{array}$$

foil
add middle terms

Many trinomials, such as the preceding, factor into two binomials. To factor $x^2 + 7x + 10$, assume that it factors into two binomials and begin by writing two pairs of parentheses. The first term of the trinomial is x^2, so we will use x and x as first terms of the binomial factors.

$$(x + \quad)(x + \quad)$$

To determine the last term of each binomial factor, we look for two integers whose product is 10 and whose sum is 7. Since our numbers must have a positive product and a positive sum, we list positive integer factors of 10 only.

Postive Factors of 10	*Sum of Factors*
1, 10	$1 + 10 = 11$
2, 5	$2 + 5 = 7$

The correct pair of numbers is 2 and 5 because their product is 10 and their sum is 7. Now we can fill in the last terms of the binomial factors.

$$x^2 + 7x + 10 = (x + 5)(x + 2)$$

To see if we have factored correctly, multiply.

$$(x + 5)(x + 2) = x^2 + 2x + 5x + 10$$
$$= x^2 + 7x + 10 \qquad \text{Combine like terms.}$$

Since multiplication is commutative, the factored form of $x^2 + 7x + 10$ can also be written as $(x + 2)(x + 5)$.

In general, to factor a trinomial of the form $x^2 + bx + c$, look for two numbers whose product is c and whose sum is b. The factored form of $x^2 + bx + c$ is

$$(x + \text{ one number})(x + \text{ other number})$$

EXAMPLE 1   Factor $x^2 + 7x + 12$.

Solution: Begin by writing the first terms of the binomial factors.

$$(x + \quad)(x + \quad)$$

Next, look for two numbers whose product is 12 and whose sum is 7. Since our numbers must have a positive product and a positive sum, we look at positive factors of 12 only.

Positive Factors of 12	*Sum of Factors*
1, 12	$1 + 12 = 13$
2, 6	$2 + 6 = 8$
3, 4	$3 + 4 = 7$

The correct pair of numbers is 3 and 4 because their product is 12 and their sum is 7. Use these factors to fill in the last terms of the binomial factors.

$$x^2 + 7x + 12 = (x + 3)(x + 4)$$

To check, multiply $(x + 3)$ by $(x + 4)$.   ∎

EXAMPLE 2   Factor $x^2 - 8x + 15$.

Solution: Begin by writing the first terms of the binomials.

$$(x + \quad)(x + \quad)$$

Now look for two numbers whose product is 15 and whose sum is -8. Since our numbers must have a positive product and a negative sum, we look at negative factors of 15 only.

Negative Factors of 15	Sum of Factors
$-1, -15$	$-1 + (-15) = -16$
$-3, -5$	$-3 + (-5) = -8$

The correct pair of numbers is -3 and -5 because their product is 15 and their sum is -8. Then

$$x^2 - 8x + 15 = (x - 3)(x - 5) \quad \blacksquare$$

EXAMPLE 3   Factor $x^2 + 4x - 12$.

Solution: $$x^2 + 4x - 12 = (x + \quad)(x + \quad)$$

Look for two numbers whose product is -12 and whose sum is 4.

Factors of -12	Sum of Factors
$-1, 12$	$-1 + 12 = 11$
$1, -12$	$1 + (-12) = -11$
$-2, 6$	$-2 + 6 = 4$
$2, -6$	$2 + (-6) = -4$
$-3, 4$	$-3 + 4 = 1$
$3, -4$	$3 + (-4) = -1$

The correct pair of numbers is -2 and 6 since their product is -12 and their sum is 4. Hence

$$x^2 + 4x - 12 = (x - 2)(x + 6) \quad \blacksquare$$

EXAMPLE 4   Factor $r^2 - r - 42$.

Solution: Because the variable in this trinomial is r, the first term in each binomial factor is r.

$$r^2 - r - 42 = (r + \quad)(r + \quad)$$

Find two numbers whose product is -42 and whose sum is -1, the numerical coefficient of r. The numbers are 6 and -7. Therefore,

$$r^2 - r - 42 = (r + 6)(r - 7) \quad \blacksquare$$

EXAMPLE 5   Factor $a^2 + 2a + 10$.

Solution: Look for two numbers whose product is 10 and whose sum is 2. Neither 1 and 10 nor 2 and 5 give the required sum, 2. We conclude that $a^2 + 2a + 10$ is not factorable with integers. The polynomial $a^2 + 2a + 10$ is called a **prime polynomial.** $\blacksquare$

EXAMPLE 6   Factor $x^2 + 5xy + 6y^2$.

Solution:

$$x^2 + 5xy + 6y^2 = (x + \quad)(x + \quad)$$

Look for two numbers whose product is 6 and whose sum is 5. The numbers are 2 and 3. Since the last term of the trinomial is $6y^2$, we use $2y$ and $3y$. Notice that $2y \cdot 3y = 6y^2$ and $2y + 3y = 5y$. Therefore,

$$x^2 + 5xy + 6y^2 = (x + 2y)(x + 3y) \quad ■$$

2

EXAMPLE 7   Factor $3m^2 - 24m - 60$.

Solution: First factor out the greatest common factor, 3.

$$3m^2 - 24m - 60 = 3(m^2 - 8m - 20)$$

Next, factor $m^2 - 8m - 20$ by looking for two factors of -20 whose sum is -8. The factors are -10 and 2.

$$3m^2 - 24m - 60 = 3(m + 2)(m - 10)$$
$$\uparrow$$

Remember to write the common factor 3 as part of the answer.

Check by multiplying.

$$3(m + 2)(m - 10) = 3(m^2 - 8m - 20)$$
$$= 3m^2 - 24m - 60 \quad ■$$

HELPFUL HINT

When factoring a polynomial, remember that factored out common factors are part of the final factored form. For example,

$$5x^2 - 15x - 50 = 5(x^2 - 3x - 10)$$
$$= 5(x + 2)(x - 5)$$

Thus, $5x^2 - 15x - 50$ **factored completely** is $5(x + 2)(x - 5)$.

MENTAL MATH

Complete the following.

1. $x^2 + 9x + 20 = (x + 4)(x \quad)$

2. $x^2 + 12x + 35 = (x + 5)(x \quad)$

3. $x^2 - 7x + 12 = (x - 4)(x \quad)$

4. $x^2 - 13x + 22 = (x - 2)(x \quad)$

5. $x^2 + 4x + 4 = (x + 2)(x \quad)$

6. $x^2 + 10x + 24 = (x + 6)(x \quad)$

EXERCISE SET 4.2

Factor each trinomial. See Examples 1 through 5.

1. $x^2 + 7x + 6$ **2.** $x^2 + 6x + 8$ **3.** $x^2 + 9x + 20$

4. $x^2 + 13x + 30$ **5.** $x^2 - 8x + 15$ **6.** $x^2 - 9x + 14$

7. $x^2 - 10x + 9$ **8.** $x^2 - 6x + 9$ **9.** $x^2 - 15x + 5$

10. $x^2 - 13x + 30$ **11.** $x^2 - 3x - 18$ **12.** $x^2 - x - 30$

13. $x^2 + 5x + 2$ **14.** $x^2 - 7x + 5$

Factor each trinomial completely. See Example 6.

15. $x^2 + 8xy + 15y^2$ **16.** $x^2 + 6xy + 8y^2$ **17.** $x^2 - 2xy + y^2$

18. $x^2 - 11xy + 30y^2$ **19.** $x^2 - 3xy - 4y^2$ **20.** $x^2 - 4xy - 77y^2$

Factor each trinomial completely. See Example 7.

21. $2z^2 + 20z + 32$ **22.** $3x^2 + 30x + 63$ **23.** $2x^3 - 18x^2 + 40x$

24. $x^3 - x^2 - 56x$ **25.** $7x^2 + 14xy - 21y^2$ **26.** $6r^2 - 3rs - 3s^2$

Factor each trinomial completely.

27. $x^2 + 15x + 36$ **28.** $x^2 + 19x + 60$ **29.** $x^2 - x - 2$

30. $x^2 - 5x - 14$ **31.** $r^2 - 16r + 48$ **32.** $r^2 - 10r + 21$

33. $x^2 - 4x - 21$ **34.** $x^2 - 4x - 32$ **35.** $x^2 + 7xy + 10y^2$

36. $x^2 - 3xy - 4y^2$ **37.** $r^2 - 3r + 6$ **38.** $x^2 + 4x - 10$

39. $2t^2 + 24t + 64$ **40.** $2t^2 + 20t + 50$ **41.** $x^3 - 2x^2 - 24x$

42. $x^3 - 3x^2 - 28x$ **43.** $x^2 - 16x + 63$ ✓   **44.** $x^2 - 19x + 88$

45. $x^2 + xy - 2y^2$ **46.** $x^2 - xy - 6y^2$ **47.** $3x^2 + 9x - 30$

48. $4x^2 - 4x - 48$ **49.** $3x^2 - 60x + 108$ **50.** $2x^2 - 24x + 70$

51. $x^2 - 18x - 144$ **52.** $x^2 + x - 42$ **53.** $6x^3 + 54x^2 + 120x$

54. $3x^3 + 3x^2 - 126x$ **55.** $2t^5 - 14t^4 + 24t^3$ **56.** $3x^6 + 30x^5 + 72x^4$

57. $5x^3y - 25x^2y^2 - 120xy^3$ **58.** $3x^2 - 6x - 72$ **59.** $4x^2 + 4x - 12$

60. $3x^2 - 9x + 45$ **61.** $x^2 - 10x + 21$ **62.** $x^2 - 14x + 28$

A Look Ahead

Factor each trinomial completely.

        EXAMPLE   Factor $2t^5y^2 - 22t^4y^2 + 56t^3y^2$.

        Solution: First, factor out a greatest common factor of $2t^3y^2$.

$$2t^5y^2 - 22t^4y^2 + 56t^3y^2 = 2t^3y^2(t^2 - 11t + 28)$$
$$= 2t^3y^2(t - 4)(t - 7) \quad\blacksquare$$

63. $2x^2y + 30xy + 100y$ **64.** $3x^2z^2 + 9xz^2 + 6z^2$

65. $-12x^2y^3 - 24xy^3 - 36y^3$ **66.** $-4x^2t^4 + 4xt^4 + 24t^4$

67. $y^2(x + 1) - 2y(x + 1) - 15(x + 1)$ **68.** $z^2(x + 1) - 3z(x + 1) - 70(x + 1)$

Writing in Mathematics

Complete the following sentences in your own words.

69. If the sign of the constant term of a trinomial is negative, then the signs of the last term factors of the binomial are opposite because . . . .

70. If the sign of the constant term of a trinomial is positive, then the signs of the last term factors of the binomials are the same because . . . .

Skill Review

Simplify. See Sections 3.1 and 3.2.

71. 5^{-2}

72. 7^{-1}

73. $\dfrac{x^{-10}}{x^5}$

74. $\dfrac{y^{-7}}{y^4}$

Multiply the following. See Section 3.5.

75. $(x - 2)(x + 2)$

76. $(y + 5)^2$

77. $(y - 1)^2$

78. $(2x + 3)(2x - 3)$

4.3
Factoring Trinomials of the Form $ax^2 + bx + c$

Tape 13

OBJECTIVES

1 Factor trinomials of the form $ax^2 + bx + c$.

2 Factor out a GCF before factoring a trinomial of the form $ax^2 + bx + c$.

3 Factor perfect square trinomials.

4 Factor trinomials of the form $ax^2 + bx + c$ by an alternate method.

1 In this section, we factor trinomials of the form $ax^2 + bx + c$, where the numerical coefficient of the squared variable is any integer, not just 1.

To begin, let's review the relationship between the numerical coefficients of the trinomial and the numerical coefficients of its factored form. For example, since $(2x + 1)(x + 6) = 2x^2 + 13x + 6$, the factored form of $2x^2 + 13x + 6$ is

$$2x^2 + 13x + 6 = (2x + 1)(x + 6)$$

Notice that $2x$ and x are factors of $2x^2$. Also, 6 and 1 are factors of 6, as shown:

$$2x^2 = 2x \cdot x$$
$$2x^2 + 13x + 6 = (2x + 1)(x + 6)$$
$$6 = 1 \cdot 6$$

Also notice that $13x$ is the sum of the following products:

$$2x^2 + 13x + 6 = (2x + 1)(x + 6)$$
$$1x$$
$$+ 12x$$
$$13x$$

Use this information to factor $5x^2 + 7x + 2$. First, find factors of $5x^2$. Since all numerical coefficients in this trinomial are positive, we use factors with positive numerical coefficients only.

Factors of $5x^2$ are $5x$ and x. Try these factors as first terms of the binomials. Thus far, we have

$$5x^2 + 7x + 2 = (5x \qquad)(x \qquad)$$

Next, find factors of 2. Factors of 2 are 1 and 2. Try possible combinations of these factors as second terms of the binomials until a middle term of $7x$ is obtained.

$$(5x + \underbrace{1)(x}_{1x} + 2) = 5x^2 + 11x + 2$$

$$\frac{+\ 10x}{11x} \qquad \text{Incorrect middle term.}$$

Try switching factors 2 and 1.

$$(5x + \underbrace{2)(x}_{2x} + 1) = 5x^2 + 7x + 2$$

$$\frac{+\ 5x}{7x} \qquad \text{Correct middle term.}$$

The factored form of $5x^2 + 7x + 2$ is

$$5x^2 + 7x + 2 = (5x + 2)(x + 1)$$

EXAMPLE 1   Factor $3x^2 + 11x + 6$.

Solution: Since all numerical coefficients are positive, use factors with positive numerical coefficients. First find factors of $3x^2$.

$$\text{Factors of } 3x^2: \quad 3x^2 = 3x \cdot x$$

If factorable, the trinomial will be of the form:

$$3x^2 + 11x + 6 = (3x \qquad)(x \qquad)$$

Next, factor 6.

$$\text{Factors of 6:} \quad 6 = 1 \cdot 6, \quad\quad 6 = 2 \cdot 3$$

Try combinations of factors of 6 until a middle term of $11x$ is obtained. First, try 1 and 6.

$$(3x + \underbrace{1)(x}_{1x} + 6) = 3x^2 + 19x + 6$$

$$\frac{+\ 18x}{19x} \qquad \text{Incorrect middle term.}$$

Next, try 6 and 1.

$$(3x + 6)(x + 1)$$

Before multiplying, notice that the factor $3x + 6$ has a common factor of 3. The original trinomial $3x^2 + 11x + 6$ has no common factor other than 1, so the factored form of $3x^2 + 11x + 6$ will contain no common factor other than 1, also. This means that $(3x + 6)(x + 1)$ is not the factored form.

$(3x+3)(\quad)$

Try 2 and 3.

$$(3x + 2)(x + 3) = 3x^2 + 11x + 6$$

$$\underbrace{\qquad\qquad}_{2x}$$

$$\frac{+\ 9x}{11x} \quad \text{Correct middle term.}$$

The factored form of $3x^2 + 11x + 6$ is

$$3x^2 + 11x + 6 = (3x + 2)(x + 3) \quad \blacksquare$$

HELPFUL HINT

If a trinomial has no common factor (other than 1), then none of its binomial factors will contain a common factor (other than 1).

EXAMPLE 2 Factor $8x^2 - 22x + 5$.

Solution:
$$\text{Factors of } 8x^2\text{:} \quad 8x^2 = 8x \cdot x, \qquad 8x^2 = 4x \cdot 2x$$

Try $8x$ and x.

$$8x^2 - 22x + 5 = (8x \qquad)(x \qquad)$$

Since the middle term $-22x$ has a negative numerical coefficient, factor 5 into negative factors.

$$\text{Factors of } 5\text{:} \quad 5 = -1 \cdot -5$$

Try -1 and -5.

$$(8x - 1)(x - 5) = 8x^2 - 41x + 5$$

$$\underbrace{\qquad\qquad}_{-1x}$$

$$\frac{+\ (-40x)}{-41x} \quad \text{Incorrect middle term.}$$

Try -5 and -1.

$$(8x - 5)(x - 1) = 8x^2 - 13x + 5$$

$$\underbrace{\qquad\qquad}_{-5x}$$

$$\frac{+\ (-8x)}{-13x} \quad \text{Incorrect middle term.}$$

Don't give up yet. Try other factors of $8x^2$. Try $4x$ and $2x$ with -1 and -5.

$$(4x - 1)(2x - 5) = 8x^2 - 22x + 5$$

$$\underbrace{\qquad\qquad}_{-2x}$$

$$\frac{+\ (-20x)}{-22x} \quad \text{Correct middle term.}$$

The factored form of $8x^2 - 22x + 5$ is

$$8x^2 - 22x + 5 = (4x - 1)(2x - 5) \quad \blacksquare$$

EXAMPLE 3   Factor $2x^2 + 13x - 7$.

Solution: Factors of $2x^2$: $2x^2 = 2x \cdot x$

Factors of -7: $-7 = -1 \cdot 7$, $-7 = 1 \cdot -7$

Try possible combinations. The combination that yields the middle term of $13x$ is

$$(2x - \underbrace{1)(x}_{-1x} + 7) = 2x^2 + 13x - 7$$

$$\begin{array}{r} -1x \\ + \ 14x \\ \hline 13x \end{array} \quad \text{Correct middle term.}$$

The factored form of $2x^2 + 13x - 7$ is

$$2x^2 + 13x - 7 = (2x - 1)(x + 7) \quad \blacksquare$$

EXAMPLE 4   Factor $10x^2 - 13xy - 3y^2$.

Solution: Factors of $10x^2$: $10x^2 = 2x \cdot 5x$, $10x^2 = 10x \cdot x$

Factors of $-3y^2$: $-3y^2 = -3y \cdot y$, $-3y^2 = 3y \cdot -y$

Try possible combinations. The combination that yields the correct middle term is

$$(2x - \underbrace{3y)(5x}_{-15xy} + y) = 10x^2 - 13xy - 3y^2$$

$$\begin{array}{r} -15xy \\ + \ \ 2xy \\ \hline -13xy \end{array} \quad \text{Correct middle term.}$$

The factored form of $10x^2 - 13xy - 3y^2$ is

$$10x^2 - 13xy - 3y^2 = (2x - 3y)(5x + y) \quad \blacksquare$$

2   Don't forget that the first step in factoring any polynomial is to look for a common factor to factor out.

EXAMPLE 5   Factor $24x^4 + 40x^3 + 6x^2$.

Solution: Notice that all three terms have a common factor of $2x^2$. First, factor out $2x^2$.

$$24x^4 + 40x^3 + 6x^2 = 2x^2(12x^2 + 20x + 3)$$

Next, factor $12x^2 + 20x + 3$.

Factors of $12x^2$: $12x^2 = 6x \cdot 2x$, $12x^2 = 4x \cdot 3x$, $12x^2 = 12x \cdot x$

Since all terms in the trinomial have positive numerical coefficients, factor 3 using positive factors only.

Factors of 3: $3 = 1 \cdot 3$

The correct combination of factors is

$$2x^2(2x + \underbrace{3)(6x}_{18x} + 1) = 2x^2(12x^2 + 20x + 3)$$

$$\begin{array}{r} 18x \\ + \ \ 2x \\ \hline 20x \end{array} \quad \text{Correct middle term.}$$

The factored form of $24x^4 + 40x^3 + 6x^2$ is

$$24x^4 + 40x^3 + 6x^2 = 2x^2(2x + 3)(6x + 1) \quad \blacksquare$$

EXAMPLE 6 Factor $4x^2 - 12x + 9$.

Solution: $\qquad$ Factors of $4x^2$:$\quad 4x^2 = 2x \cdot 2x, \qquad 4x^2 = 4x \cdot x$

Since the middle term $-12x$ has a negative numerical coefficient, factor 9 into negative factors only.

$$\text{Factors of 9:}\quad 9 = -3 \cdot -3, \qquad 9 = -1 \cdot -9$$

The correct combination is

$$(2x - \underbrace{3)(2x}_{-6x} - 3) = 4x^2 - 12x + 9$$

$$\underline{+ \ (-6x)}$$
$$-12x \qquad \text{Correct middle term.}$$

Thus, $4x^2 - 12x + 9 = (2x - 3)(2x - 3)$, which can be written as $(2x - 3)^2$. ■

3 Notice in Example 6 that $4x^2 - 12x + 9 = (2x - 3)^2$. The trinomial $4x^2 - 12x + 9$ is called a **perfect square trinomial** since it is the square of the binomial $2x - 3$.

In the last chapter, we learned a special product for squaring a binomial.

$$(a + b)^2 = a^2 + 2ab + b^2$$

The trinomial $a^2 + 2ab + b^2$ is also a perfect square trinomial, since it is the square of the binomial $a + b$. We can use this pattern to help us factor perfect square trinomials. To use this pattern, we must first be able to recognize a perfect square trinomial. A trinomial is a perfect square when its first term is the square of some quantity a, its last term is the square of some quantity b, and its middle term is twice the product of the quantities a and b.

Perfect Square Trinomials

$$a^2 + 2ab + b^2 = (a + b)^2$$
$$a^2 - 2ab + b^2 = (a - b)^2$$

EXAMPLE 7 Factor $x^2 + 12x + 36$.

Solution: This trinomial is a perfect square trinomial since:

1. The first term is a square:$\quad x^2 = (x)^2$.
2. The last term is a square:$\quad 36 = (6)^2$.
3. The middle term is twice the product of x and 6:$\quad 12x = 2 \cdot x \cdot 6$.

Thus, $x^2 + 12x + 36 = (x + 6)^2$. ■

EXAMPLE 8 Factor $25x^2 + 25xy + 4y^2$.

Solution: Determine whether or not this trinomial is a perfect square by considering the same three questions.

1. Is the first term a square? Yes, $25x^2 = (5x)^2$.

2. Is the last term a square? Yes, $4y^2 = (2y)^2$.

3. Is the middle term twice the product of $5x$ and $2y$? **No.** $2(5x)(2y) = 20xy$, not $25xy$.

Therefore, try to factor $25x^2 + 25xy + 4y^2$ by other methods. It is factorable, and $25x^2 + 25xy + 4y^2 = (5x + 4y)(5x + y)$.   ■

HELPFUL HINT

A perfect square trinomial that is not recognized as such can be factored by other methods.

EXAMPLE 9   Factor $4m^2 - 4m + 1$.

Solution: This is a perfect square trinomial since $4m^2 = (2m)^2$, $1 = (1)^2$, and $4m = 2(2m)(1)$.

$$4m^2 - 4m + 1 = (2m - 1)^2$$ ■

4   There is another method for factoring trinomials of the form $ax^2 + bx + c$. This method is described next.

To factor $2x^2 + 11x + 12$, find two numbers whose product is $2 \cdot 12 = 24$ and whose sum is 11. Since we want a positive product and a positive sum, we consider positive factors of 24 only.

Factors of 24	Sum of Factors
1, 24	$1 + 24 = 25$
2, 12	$2 + 12 = 14$
3, 8	$3 + 8 = 11$

The factors are 3 and 8. Use these factors to write the middle term $11x$ as $3x + 8x$. Replace $11x$ with $3x + 8x$ in the original trinomial and factor by grouping.

$$2x^2 + \boxed{11x} + 12 = 2x^2 + \boxed{3x + 8x} + 12$$
$$= x(2x + 3) + 4(2x + 3)$$
$$= (2x + 3)(x + 4)$$

In general, we have the following:
To factor trinomials of the form $ax^2 + bx + c$ by grouping,

Step 1   Find two numbers whose product is $a \cdot c$ and whose sum is b.
Step 2   Write the middle term, bx, using the factors found in step 1.
Step 3   Factor by grouping.

EXAMPLE 10   Factor $8x^2 - 14x + 5$.

Solution: In this trinomial, $a = 8$, $b = -14$, and $c = 5$.

Step 1   Find two numbers whose product is $a \cdot c$ or $8 \cdot 5 = 40$ and whose sum is b or -14. The numbers are -4 and -10.

Step 2   Write $-14x$ as $-4x - 10x$ so that

$$8x^2 \; \boxed{-14x} \; + 5 = 8x^2 \; \boxed{-4x - 10x} \; + 5$$

Step 3   Factor by grouping.

$$8x^2 - 4x - 10x + 5 = 4x(2x - 1) - 5(2x - 1)$$
$$= (2x - 1)(4x - 5) \quad \blacksquare$$

EXAMPLE 11   Factor $3x^2 - x - 10$ by the method just described.

Solution: In $3x^2 - x - 10$, $a = 3$, $b = -1$, and $c = -10$.

Step 1   Find two numbers whose product is $a \cdot c$ or $3(-10) = -30$ and whose sum is b, -1. The numbers are -6 and 5.

Step 2 $3x^2 \; \boxed{-x} \; - 10 = 3x^2 \; \boxed{-6x + 5x} \; - 10$

Step 3 $\qquad\qquad\qquad = 3x(x - 2) + 5(x - 2)$
$\qquad\qquad\qquad = (x - 2)(3x + 5) \quad \blacksquare$

MENTAL MATH

State whether or not each trinomial is a perfect trinomial square.

1. $x^2 + 14x + 49$
2. $9x^2 - 12x + 4$ ✓
3. $y^2 + 2y + 4$ ✓
4. $x^2 - 4x + 2$ ⌣
5. $9y^2 + 6y + 1$
6. $y^2 - 16y + 64$ ✓

EXERCISE SET 4.3

Factor completely. See Examples 1 through 4. (See Example 10 for alternate method.)

1. $2x^2 + 13x + 15$
2. $3x^2 + 8x + 4$
3. $2x^2 - 9x - 5$
4. $3x^2 + 20x - 63$
5. $2y^2 - y - 6$
6. $8y^2 - 17y + 9$
7. $16a^2 - 24a + 9$
8. $25x^2 + 20x + 4$
9. $36r^2 - 5r - 24$
10. $20r^2 + 27r - 8$
11. $10x^2 + 17x + 3$
12. $21x^2 - 41x + 10$

Factor completely. See Example 5. See Example 11 for alternate method.

13. $21x^2 - 48x - 45$
14. $12x^2 - 14x - 10$
15. $12x^2 - 14x - 6$
16. $20x^2 - 2x + 6$
17. $4x^3 - 9x^2 - 9x$
18. $6x^3 - 31x^2 + 5x$

Factor the following perfect square trinomials. See Examples 6 through 9.

19. $x^2 + 22x + 121$
20. $x^2 + 18x + 81$
21. $x^2 - 16x + 64$
22. $x^2 - 12x + 36$
23. $16y^2 - 40y + 25$
24. $9y^2 + 48y + 64$
25. $x^2y^2 - 10xy + 25$
26. $4x^2y^2 - 28xy + 49$

Factor the following completely.

27. $2x^2 - 7x - 99$

28. $2x^2 + 7x - 72$

29. $4x^2 - 8x - 21$

30. $6x^2 - 11x - 10$

31. $30x^2 - 53x + 21$

32. $21x^3 - 6x - 30$

33. $24x^2 - 58x + 9$

34. $36x^2 + 55x - 14$

35. $9x^2 - 24xy + 16y^2$

36. $25x^2 + 60xy + 36y^2$

37. $x^2 - 14xy + 49y^2$

38. $x^2 + 10xy + 25y^2$

39. $2x^2 + 7x + 5$

40. $2x^2 + 7x + 3$

41. $3x^2 - 5x + 1$

42. $3x^2 - 7x + 6$

43. $-2y^2 + y + 10$

44. $-4x^2 - 23x + 6$

45. $16x^2 + 24xy + 9y^2$

46. $4x^2 - 36xy + 81y^2$

47. $8x^2y + 34xy - 84y$

48. $6x^2y^2 - 2xy^2 - 60y^2$

49. $3x^2 + x - 2$

50. $8y^2 + y - 9$

51. $x^2y^2 + 4xy + 4$

52. $x^2y^2 - 6xy + 9$

53. $49y^2 + 42xy + 9x^2$

54. $16x^2 - 8xy + y^2$

55. $3x^2 - 42x + 63$

56. $5x^2 - 75x + 60$

57. $42a^2 - 43a + 6$

58. $54a^2 + 39ab - 8b^2$

59. $18x^2 - 9x - 14$

60. $8x^2 + 6x - 27$

61. $25p^2 - 70pq + 49q^2$

62. $36p^2 - 18pq + 9q^2$

63. $15x^2 - 16x - 15$

64. $12x^2 + 7x - 12$

65. $-27t + 7t^2 - 4$

66. $4t^2 - 7 - 3t$

A Look Ahead

Factor completely. See the following example.

EXAMPLE   Factor $10x^4y - 5x^3y^2 - 15x^2y^3$.

Solution: Start by factoring out a greatest common factor of $5x^2y$.

$$10x^4y - 5x^3y^2 - 15x^2y^3 = 5x^2y(2x^2 - xy - 3y^2)$$

Next, factor $2x^2 - xy - 3y^2$. Try $2x$ and x as factors of $2x^2$ and $-3y$ and y as factors of $-3y^2$.

$$(2x - 3y)(x + y)$$

$$\underbrace{}_{-3xy}$$

$$\begin{array}{r} + \quad 2xy \\ \hline -xy \end{array}$$ The middle term is correct.

Hence $10x^4y - 5x^3y^2 - 15x^2y^3 = 5x^2y(2x - 3y)(x + y)$.   ■

67. $-12x^3y^2 + 3x^2y^2 + 15xy^2$

68. $-12r^3x^2 + 38r^2x^2 + 14rx^2$

69. $-30p^3q + 88p^2q^2 + 6pq^3$

70. $3x^3y^2 + 3x^2y^3 - 18xy^4$

71. $4x^2(y - 1)^2 + 10x(y - 1)^2 + 25(y - 1)^2$

72. $3x^2(a + 3)^3 - 28x(a + 3)^3 + 25(a + 3)^3$

Writing in Mathematics

73. Describe a perfect square trinomial, and then give one example.

Skill Review

Solve the following. See Section 2.4.

74. $2x + 7 = 4x - 5$

75. $15x - 10 = 12x - 8$

76. $5(x - 3) = x - 3$

77. $2(3x + 4) = -2(2x + 1)$

Multiply the following. See Section 3.4.

78. $(x + 3)(x^2 - 3x + 9)$

79. $(x - 2)(x^2 + 2x + 4)$

4.4
Factoring Binomials

OBJECTIVES

Tape 14

1 Factor the difference of two squares.

2 Factor the sum or difference of two cubes.

1 When learning to multiply binomials in Chapter 3, we studied a special product: the product of the sum and difference of two terms, a and b.

$$(a + b)(a - b) = a^2 - b^2$$

For example, the product of $x + 3$ and $x - 3$ is

$$(x + 3)(x - 3) = x^2 - 9$$

The binomial $x^2 - 9$ is called a **difference of squares.** In this section, we use the pattern for the product of a sum and difference to factor the binomial difference of squares.

To use this pattern to help us factor, we must be able to recognize a difference of squares. A binomial is a difference of squares when it is the difference of the square of some quantity a and the square of some quantity b.

Difference of Two Squares

$$a^2 - b^2 = (a + b)(a - b)$$

EXAMPLE 1 Factor $4x^2 - 1$.

Solution: $4x^2 - 1$ is the difference of two squares since $4x^2 = (2x)^2$ and $1 = (1)^2$; therefore,

$$4x^2 - 1 = (2x)^2 - 1^2 = (2x + 1)(2x - 1)$$

Multiply to check. ■

EXAMPLE 2 Factor $25a^2 - 9b^2$.

Solution: $25a^2 - 9b^2 = (5a)^2 - (3b)^2 = (5a + 3b)(5a - 3b)$ ■

EXAMPLE 3   Factor $9x^2 - 36$.

Solution: Remember when factoring to always check first for common factors. If there are common factors, factor out the GCF, and then factor the resulting polynomial.

$$9x^2 - 36 = 9(x^2 - 4) \qquad \text{Factor out the GCF 9.}$$
$$= 9(x^2 - 2^2)$$
$$= 9(x + 2)(x - 2)$$

In this example, if we forget to factor out the GCF first, we still have the difference of two squares.

$$9x^2 - 36 = (3x)^2 - (6)^2 = (3x + 6)(3x - 6)$$

This binomial has not been factored completely since both terms of both binomial factors have a common factor of 3.

$$3x + 6 = 3(x + 2) \quad \text{and} \quad 3x - 6 = 3(x - 2)$$

Then

$$9x^2 - 36 = (3x + 6)(3x - 6) = 3(x + 2)3(x - 2) = 9(x + 2)(x - 2)$$

Factoring is easier if the GCF is factored out first before using other methods.   ■

EXAMPLE 4   Factor $x^2 + 4$.

Solution: The binomial $x^2 + 4$ is the **sum** of squares since we can write $x^2 + 4 = x^2 + 2^2$. We might try to factor using $(x + 2)(x + 2)$ or $(x - 2)(x - 2)$. But when multiplying to check, neither factoring is correct.

$$(x + 2)(x + 2) = x^2 + 4x + 4$$
$$(x - 2)(x - 2) = x^2 - 4x + 4$$

In both cases, the product is a trinomial, not the required binomial. Thus, $x^2 + 4$ is a prime polynomial.   ■

2   Although the sum of two squares usually does not factor, the sum or difference of two cubes can be factored and reveals factoring patterns. The pattern for the sum of cubes is illustrated by multiplying the binomial $x + y$ and the trinomial $x^2 - xy + y^2$.

$$
\begin{array}{r}
x^2 - xy + y^2 \\
x + y \\
\hline
x^2y - xy^2 + y^3 \\
x^3 - x^2y + xy^2 \quad\quad\quad \\
\hline
x^3 \quad\quad\quad\quad\quad\quad + y^3
\end{array}
$$

$$(x + y)(x^2 - xy + y^2) = x^3 + y^3 \qquad \text{Sum of cubes.}$$

The pattern for the differences of two cubes is illustrated by multiplying the binomial $x - y$ by the trinomial $x^2 + xy + y^2$. The result is

$$(x - y)(x^2 + xy + y^2) = x^3 - y^3 \qquad \text{Difference of cubes.}$$

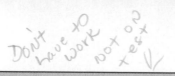

Sum or Difference of Two Cubes

$$a^3 + b^3 = (a + b)(a^2 - ab + b^2)$$

$$a^3 - b^3 = (a - b)(a^2 + ab + b^2)$$

EXAMPLE 5   Factor $x^3 + 8$.

Solution: First, write the binomial in the form $a^3 + b^3$.

$$x^3 + 8 = x^3 + 2^3 \qquad \text{Write in the form } a^3 + b^3.$$

If we replace a with x and b with 2 in the formula above, we have

$$x^3 + 2^3 = (x + 2)(x^2 - (x)(2) + 2^2)$$

$$= (x + 2)(x^2 - 2x + 4) \quad \blacksquare$$

HELPFUL HINT

When factoring sums or differences of cubes, notice the sign patterns.

$$\overset{\text{same sign}}{\underset{\text{opposite sign}}{x^3 + y^3 = (x + y)(x^2} - xy + \underset{\text{always positive}}{y^2})}$$

$$\overset{\text{same sign}}{\underset{\text{opposite sign}}{x^3 - y^3 = (x - y)(x^2} + xy + \underset{\text{always positive}}{y^2})}$$

EXAMPLE 6   Factor $y^3 - 27$.

Solution: $y^3 - 27 = y^3 - 3^3$ $\qquad\qquad$ Write in the form $a^3 - b^3$.

$$= (y - 3)[y^2 + (y)(3) + 3^2]$$

$$= (y - 3)(y^2 + 3y + 9) \quad \blacksquare$$

EXAMPLE 7   Factor $64x^3 + 1$.

Solution: $64x^3 + 1 = (4x)^3 + 1^3$

$$= (4x + 1)[(4x)^2 - (4x)(1) + 1^2]$$

$$= (4x + 1)(16x^2 - 4x + 1) \quad \blacksquare$$

EXAMPLE 8   Factor $54a^3 - 16b^3$.

Solution: Remember to factor out common factors first before using other factoring methods.

$$54a^3 - 16b^3 = 2(27a^3 - 8b^3) \qquad\qquad\qquad \text{Factor out the GCF 2.}$$

$$= 2[(3a)^3 - (2b)^3] \qquad\qquad\qquad \text{Difference of two cubes.}$$

$$= 2(3a - 2b)[(3a)^2 + (3a)(2b) + (2b)^2]$$

$$= 2(3a - 2b)(9a^2 + 6ab + 4b^2) \quad \blacksquare$$

MENTAL MATH

State each number as a square.

1. 1   **2.** 25   **3.** 81   **4.** 64
5. 9   **6.** 100

State each number as a cube.

7. 1   **8.** 64   **9.** 8   **10.** 27

EXERCISE SET 4.4

Factor the difference of two squares. See Examples 1 through 3.

1. $25y^2 - 9$ **2.** $49a^2 - 16$ **3.** $121 - 100x^2$
4. $144 - 81x^2$ **5.** $12x^2 - 27$ **6.** $36x^2 - 64$
7. $169a^2 - 49b^2$ **8.** $225a^2 - 81b^2$ **9.** $x^2y^2 - 1$
10. $16 - a^2b^2$

Factor the sum or difference of two cubes. See Examples 5 through 8.

11. $a^3 + 27$ **12.** $b^3 - 8$ **13.** $8a^3 + 1$
14. $64x^3 - 1$ **15.** $5k^3 + 40$ **16.** $6r^3 - 162$
17. $x^3y^3 - 64$ **18.** $8x^3 - y^3$ **19.** $x^3 + 125$
20. $a^3 - 216$ **21.** $24x^4 - 81xy^3$ **22.** $375y^6 - 24y^3$

Factor the binomials completely.

23. $x^2 - 4$ **24.** $x^2 - 36$ **25.** $81 - p^2$
26. $100 - t^2$ **27.** $4r^2 - 1$ **28.** $9t^2 - 1$
29. $9x^2 - 16$ **30.** $36y^2 - 25$ **31.** $16r^2 + 1$
32. $49y^2 + 1$ **33.** $27 - t^3$ **34.** $125 + r^3$
35. $8r^3 - 64$ **36.** $54r^3 + 2$ **37.** $t^3 - 343$
38. $s^3 + 216$ **39.** $x^2 - 169y^2$ **40.** $x^2 - 225y^2$
41. $x^2y^2 - z^2$ **42.** $x^3y^3 - z^3$ **43.** $x^3y^3 + 1$
44. $x^2y^2 + z^2$ **45.** $s^3 - 64t^3$ **46.** $8t^3 + s^3$
47. $18r^2 - 8$ **48.** $32t^2 - 50$ **49.** $9xy^2 - 4x$
50. $16xy^2 - 64x$ **51.** $25y^4 - 100y^2$ **52.** $xy^3 - 9xyz^2$
53. $x^3y - 4xy^3$ **54.** $12s^3t^3 + 192s^5t$ **55.** $8s^6t^3 + 100s^3t^6$
56. $25x^5y + 121x^3y$ **57.** $27x^2y^3 - xy^2$ **58.** $8x^3y^3 + x^3y$

A Look Ahead

Factor each difference of squares. See the following example.

EXAMPLE   Factor $(x + y)^2 - (x - y)^2$.

Solution: Use the method for factoring the difference of squares.

$$(x + y)^2 - (x - y)^2 = [(x + y) + (x - y)][(x + y) - (x - y)]$$
$$= (x + y + x - y)(x + y - x + y)$$
$$= (2x)(2y)$$
$$= 4xy \quad \blacksquare$$

59. $x^4 - 16$

60. $x^6 - 1$

61. $a^2 - (2 + b)^2$

62. $(x + 3)^2 - y^2$

63. $(x^2 - 4)^2 - (x - 2)^2$

64. $(x^2 - 9) - (3 - x)$

65. What binomial multiplied by $(x - 6)$ gives the difference of two squares?

66. What binomial multiplied by $(5 + y)$ gives the difference of two squares?

67. What binomial multiplied by $(4x^2 - 2xy + y^2)$ gives the sum or difference of two cubes?

68. What binomial multiplied by $(1 + 4y + 16y^2)$ gives the sum or difference of two cubes?

Skill Review

Divide the following. See Section 3.6.

69. $\dfrac{8x^4 + 4x^3 - 2x + 6}{2x}$

70. $\dfrac{3y^4 + 9y^2 - 6y + 1}{3y^2}$

Use long division to divide the following. See Section 3.6.

71. $\dfrac{2x^2 - 3x - 2}{x - 2}$

72. $\dfrac{4x^2 - 21x + 21}{x - 3}$

73. $\dfrac{3x^2 + 13x + 10}{x + 3}$

74. $\dfrac{5x^2 + 14x + 12}{x + 2}$

4.5
Factoring Polynomials Completely

OBJECTIVE   **1**   Factor polynomials completely.

Tape 14

1   A polynomial is factored completely when it is written as the product of prime polynomials. This section uses the various methods of factoring polynomials that have been discussed in earlier sections. Since these methods are applied throughout the remainder of this text, as well as in later courses, it is important to master the skills of factoring. The following is a set of guidelines for factoring polynomials.

> **To Factor a Polynomial**
>
> *Step 1*   Are there any common factors? If so, factor them out.
>
> *Step 2*   How many terms are in the polynomial?
>   **a.** If there are **two** terms, decide if one of the following can be applied.
>   **i.** Difference of two squares: $a^2 - b^2 = (a - b)(a + b)$.
>   **ii.** Difference of two cubes: $a^3 - b^3 = (a - b)(a^2 + ab + b^2)$.
>   **iii.** Sum of two cubes: $a^3 + b^3 = (a + b)(a^2 - ab + b^2)$.

> **b.** If there are **three** terms, try one of the following.
>   **i.** Perfect square trinomial: $a^2 + 2ab + b^2 = (a + b)^2$.
>   **ii.** If not a perfect square trinomial, factor using the methods presented in Sections 4.2 and 4.3.
> **c.** If there are **four** or more terms, try factoring by grouping.
> *Step 3*   Last, see if any factors in the factored polynomial can be factored further.

EXAMPLE 1   Factor $10t^2 - 17t + 3$.

Solution: This polynomial contains no greatest common factor (other than 1). There are three terms, so this polynomial is a trinomial. This trinomial is not a perfect square trinomial, so factor using methods from earlier sections.

$$\text{Factors of } 10t^2: \quad 10t^2 = 2t \cdot 5t, \qquad 10t^2 = t \cdot 10t$$

Since the middle term, $-17t$, has a negative numerical coefficient, find negative factors of 3.

$$\text{Factors of 3:} \quad 3 = -1 \cdot -3$$

The correct combination is

$$(2t - 3)(5t - 1) = 10t^2 - 17t + 3$$

$$\begin{array}{c} -15t \\ \underline{-2t} \\ -17t \end{array} \quad \text{Correct middle term.}$$

The factored form of $10t^2 - 17t + 3$ is

$$10t^2 - 17t + 3 = (2t - 3)(5t - 1)$$

To check, multiply $2t - 3$ and $5t - 1$.   ■

EXAMPLE 2   Factor $2x^3 + 3x^2 - 2x - 3$.

Solution: There are no factors common to all terms. Try factoring by grouping since this polynomial has four terms.

$$2x^3 + 3x^2 - 2x - 3 = x^2(2x + 3) - (2x + 3) \qquad \text{Factor out a common factor from each pair of terms.}$$

$$= (2x + 3)(x^2 - 1) \qquad \text{Factor out } 2x + 3.$$

The binomial $x^2 - 1$ can be factored further.

$$= (2x + 3)(x + 1)(x - 1) \qquad \text{Factor } x^2 - 1 \text{ as a difference of squares.}$$

Check by finding the product of the three binomials.   ■

EXAMPLE 3   Factor $12m^2 - 3n^2$.

Solution: $12m^2 - 3n^2 = 3(4m^2 - n^2) \qquad \text{Factor out the greatest common factor.}$

$$= 3(2m + n)(2m - n) \qquad \text{Factor the difference of squares.} \quad ■$$

EXAMPLE 4   Factor $x^3 + 27y^3$.

Solution: This binomial is the sum of two cubes.

$$x^3 + 27y^3 = (x)^3 + (3y)^3$$
$$= (x + 3y)[x^2 - 3xy + (3y)^2]$$
$$= (x + 3y)(x^2 - 3xy + 9y^2) \quad \blacksquare$$

EXAMPLE 5   Factor $30a^2b^3 + 55a^2b^2 - 35a^2b$.

Solution: $30a^2b^3 + 55a^2b^2 - 35a^2b = 5a^2b(6b^2 + 11b - 7)$ Factor out the GCF.

$$= 5a^2b(2b - 1)(3b + 7)$$ Factor the resulting trinomial.

Check by multiplying. $\blacksquare$

EXERCISE SET 4.5

Factor the following completely. See Examples 1 through 5.

1. $a^2 + 2ab + b^2$ **2.** $a^2 - 2ab + b^2$ **3.** $a^2 + a - 12$

4. $a^2 - 7a + 10$ **5.** $a^2 - a - 6$ **6.** $a^2 + 2a + 1$

7. $x^2 + 2x + 1$ **8.** $x^2 + x - 2$ **9.** $x^2 + 4x + 3$

10. $x^2 + x - 6$ **11.** $x^2 + 7x + 12$ **12.** $x^2 + x - 12$

13. $x^2 + 3x - 4$ **14.** $x^2 - 7x + 10$ **15.** $x^2 + 2x - 15$

16. $x^2 + 11 + 30$ **17.** $x^2 - x - 30$ **18.** $x^2 + 11x + 24$

19. $2x^2 - 98$ **20.** $3x^2 - 75$ **21.** $x^2 + 3x + xy + 3y$

22. $3y - 21 + xy - 7x$ **23.** $x^2 + 6x - 16$ **24.** $x^2 - 3x - 28$

25. $4x^3 + 20x^2 - 56x$ **26.** $6x^3 - 6x^2 - 120x$ **27.** $12x^2 + 34x + 24$

28. $8a^2 + 6ab - 5b^2$ **29.** $4a^2 - b^2$ **30.** $28 - 13x - 6x^2$

31. $20 - 3x - 2x^2$ **32.** $x^2 - 2x + 4$ **33.** $a^2 + a - 3$

34. $6y^2 + y - 15$ **35.** $4x^2 - x - 5$ **36.** $x^2y - y^3$

37. $4t^2 + 36$ **38.** $x^2 + x + xy + y$ **39.** $ax + 2x + a + 2$

40. $18x^3 - 63x^2 + 9x$ **41.** $12a^3 - 24a^2 + 4a$ **42.** $x^2 + 14x - 32$

43. $x^2 - 14x - 48$ **44.** $16a^2 - 56ab + 49b^2$ **45.** $25p^2 - 70pq + 49q^2$

46. $7x^2 + 24xy + 9y^2$ **47.** $125 - 8y^3$ **48.** $64x^3 + 27$

49. $-x^2 - x + 30$ **50.** $-x^2 + 6x - 8$ **51.** $14 + 5x - x^2$

52. $3 - 2x - x^2$ **53.** $3x^4y + 6x^3y - 72x^2y$ **54.** $2x^3y + 8x^2y^2 - 10xy^3$

55. $5x^3y^2 - 40x^2y^3 + 35xy^4$ **56.** $4x^4y - 8x^3y - 60x^2y$ **57.** $12x^3y + 243xy$

58. $6x^3y^2 + 8xy^2$ **59.** $(x - y)^2 - z^2$ **60.** $(x + 2)^2 - 9$

61. $3rs - s + 12r - 4$ **62.** $x^3 - 2x^2 + 3x - 6$ **63.** $4x^2 - 8xy - 3x + 6y$

64. $4x^2 - 2xy - 7yz + 14xz$ **65.** $6x^2 + 18xy + 12y^2$ **66.** $12x^2 + 46xy - 8y^2$

67. $xy^2 - 4x + 3y^2 - 12$ **68.** $x^2y^2 - 9x^2 + 3y^2 - 27$ **69.** $5(x + y) + x(x + y)$

70. $7(x - y) + y(x - y)$ **71.** $14t^2 - 9t + 1$ **72.** $3t^2 - 5t + 1$

73. $3x^2 + 2x - 5$ **74.** $7x^2 + 19x - 6$ **75.** $x^2 + 9xy - 36y^2$

76. $3x^2 + 10xy - 8y^2$ **77.** $1 - 8ab - 20a^2b^2$ **78.** $1 - 7ab - 60a^2b^2$

79. $x^4 - 10x^2 + 9$ **80.** $x^4 - 13x^2 + 36$ **81.** $x^4 - 14x^2 - 32$

82. $x^4 - 22x^2 - 75$

83. $x^2 - 23x + 120$

84. $y^2 + 22y + 96$

85. $6x^3 - 28x^2 + 16x$

86. $6y^3 - 8y^2 - 30y$

87. $27x^3 - 125y^3$

88. $216y^3 - z^3$

89. $x^3y^3 + 8z^3$

90. $27a^3b^3 + 8$

91. $2xy - 72x^3y$

92. $2x^3 - 18x$

93. $x^3 + 6x^2 - 4x - 24$

94. $x^3 - 2x^2 - 36x + 72$

95. $6a^3 + 10a^2$

96. $4n^2 - 6n$

97. $a^2(a + 2) + 2(a + 2)$

98. $a - b + x(a - b)$

99. $x^3 - 28 + 7x^2 - 4x$

100. $a^3 - 45 - 9a + 5a^2$

Writing in Mathematics

101. Explain why it makes good sense to factor out the GCF first, before using other methods of factoring.

Skill Review

Perform indicated operations. See Sections 1.6 and 1.7.

102. $8 - (-3)$

103. $4 + (-5)$

104. $-2 + (-6)$

105. $-7 - 10$

Solve the following. See Section 2.5.

106. The following suitcase has a volume of 960 cubic inches. Find its length.

10 inches

12 inches

x inches

107. The sail shown has an area of 25 square feet. Find its height.

x feet

10 feet

4.6
Solving Quadratic Equations by Factoring

OBJECTIVES

Tape 15

1   Define quadratic equation.

2   Solve quadratic equations by factoring.

3   Solve equations with degree greater than 2 by factoring.

1   In this section, we learn how to solve **quadratic equations.**

> **Quadratic Equation**
>
> A quadratic equation is one that can be written in the form $ax^2 + bx + c = 0$, where a, b, and c are real numbers, and $a \neq 0$.

Notice that the degree of the polynomial $ax^2 + bx + c$ is 2. Here are some examples of quadratic equations.

Quadratic Equations

$$3x^2 + 5x + 6 = 0 \qquad x^2 = 9 \qquad y^2 + y = 1$$

The form $ax^2 + bx + c = 0$ is called the **standard form** of a quadratic equation. The quadratic equation $3x^2 + 5x + 6 = 0$ is in standard form. One side of the equation is 0 and the other side is a polynomial of degree 2 written in descending powers of x.

2   Some quadratic equations can be solved by making use of factoring and **the zero factor theorem.**

Zero Factor Theorem

If a and b are real numbers and if $ab = 0$, then $a = 0$ or $b = 0$.

This theorem states that if the product of two numbers is 0 then at least one of the numbers must be 0. If the equation

$$(x - 3)(x + 1) = 0$$

is a true statement, then either the factor $x - 3$ must be 0 or the factor $x + 1$ must be 0. In other words,

$$x - 3 = 0 \quad \text{or} \quad x + 1 = 0$$
$$x = 3 \quad \text{or} \qquad x = -1$$

Thus, 3 and -1 are both solutions of the equation $(x - 3)(x + 1) = 0$. To check, replace x with 3 in the original equation. Then replace x with -1 in the original equation.

$$(x - 3)(x + 1) = 0$$
$$(3 - 3)(3 + 1) = 0 \qquad \text{Replace } x \text{ with 3.}$$
$$0(4) = 0 \qquad \text{True.}$$
$$(x - 3)(x + 1) = 0$$
$$(-1 - 3)(-1 + 1) = 0 \qquad \text{Replace } x \text{ with } -1.$$
$$(-4)(0) = 0 \qquad \text{True.}$$

EXAMPLE 1   Solve $(x - 5)(2x + 7) = 0$.

Solution: Use the zero factor theorem; set each factor equal to 0 and solve the resulting linear equations.

$$(x - 5)(2x + 7) = 0$$
$$x - 5 = 0 \quad \text{or} \quad 2x + 7 = 0$$
$$x = 5 \quad \text{or} \qquad 2x = -7$$
$$x = -\frac{7}{2}$$

If x is either 5 or $-\dfrac{7}{2}$ the product $(x - 5)(2x + 7)$ is 0. Check by replacing x with 5 in the original equation; then replace x with $-\dfrac{7}{2}$ in the original equation. Both 5 and $-\dfrac{7}{2}$ are solutions. ■

$2, \left(\dfrac{-7}{2}\right)$

EXAMPLE 2 Solve $x^2 - 9x = -20$.

Solution: First, write the equation in standard form; then factor.

$$x^2 - 9x = -20$$

$$x^2 - 9x + 20 = 0 \qquad \text{Write in standard form by adding 20 to both sides.}$$

$$(x - 4)(x - 5) = 0 \qquad \text{Factor.}$$

Next, use the zero factor theorem and set each factor equal to 0.

$$x - 4 = 0 \quad \text{or} \quad x - 5 = 0 \qquad \text{Set each factor equal to 0.}$$

$$x = 4 \quad \text{or} \qquad x = 5 \qquad \text{Solve.}$$

Check the solutions by replacing x with each value in the original equation. ■

The following steps may be used to solve a quadratic equation by factoring.

To Solve Quadratic Equations by Factoring

Step 1 Write the equation in standard form: $ax^2 + bx + c = 0$.
Step 2 Factor the quadratic completely.
Step 3 Set each factor containing a variable equal to 0.
Step 4 Solve the resulting equations.
Step 5 Check each solution in the original equation.

Since it is not always possible to factor a quadratic polynomial, not all quadratic equations can be solved by factoring. Other methods of solving quadratic equations are presented in Chapter 9.

EXAMPLE 3 Solve $x(2x - 7) = 4$.

Solution: First, write the equation in standard form; then factor.

$$x(2x - 7) = 4$$

$$2x^2 - 7x = 4 \qquad \qquad \text{Multiply.}$$

$$2x^2 - 7x - 4 = 0 \qquad \qquad \text{Write in standard form.}$$

$$(2x + 1)(x - 4) = 0 \qquad \qquad \text{Factor.}$$

$$2x + 1 = 0 \quad \text{or} \quad x - 4 = 0 \qquad \text{Set each factor equal to zero.}$$

$$2x = -1 \quad \text{or} \quad x = 4 \qquad \text{Solve.}$$

$$x = -\frac{1}{2}$$

Check both solutions $-\dfrac{1}{2}$ and 4. ■

> **HELPFUL HINT**
>
> To apply the zero factor theorem, one side of the equation must be 0 and the other side of the equation must be factored. To solve the equation $x(2x - 7) = 4$, for example, you may **not** set each factor equal to 4.

EXAMPLE 4 Solve $-2x^2 - 4x + 30 = 0$.

Solution: Begin by factoring out a common factor of -2.

$$-2x^2 - 4x + 30 = 0$$

$$-2(x^2 + 2x - 15) = 0 \qquad \text{Factor out } -2.$$

$$-2(x + 5)(x - 3) = 0 \qquad \text{Factor the quadratic.}$$

Next, set each factor **containing a variable** equal to 0.

$$x + 5 = 0 \quad \text{or} \quad x - 3 = 0 \qquad \text{Set each factor containing a variable equal to 0.}$$

$$x = -5 \quad \text{or} \qquad x = 3 \qquad \text{Solve.}$$

Note that the factor -2 is a constant term containing no variables and can never equal 0. The solutions are -5 and 3. ■

3 Some equations involving polynomials of degree higher than 2 may also be solved by factoring and then applying the zero factor theorem.

EXAMPLE 5 Solve $3x^3 - 12x = 0$.

Solution: Factor the left side of the equation. Begin by factoring out the common factor of $3x$.

$$3x^3 - 12x = 0$$

$$3x(x^2 - 4) = 0 \qquad \text{Factor out the GCF } 3x.$$

$$3x(x + 2)(x - 2) = 0 \qquad \text{Factor } x^2 - 4, \text{ a difference of squares.}$$

$$3x = 0 \text{ or } x + 2 = 0 \quad \text{or } x - 2 = 0 \qquad \text{Set each factor equal to 0.}$$

$$x = 0 \text{ or} \qquad x = -2 \text{ or} \qquad x = 2 \qquad \text{Solve.}$$

Thus, the equation $3x^3 - 12x = 0$ has three solutions: 0, -2, and 2. To check, replace x with each solution in the original equation.

Let $x = 0$	Let $x = -2$	Let $x = 2$
$3(0)^3 - 12(0) = 0$	$3(-2)^3 - 12(-2) = 0$	$3(2)^3 - 12(2) = 0$
$0 = 0$	$3(-8) + 24 = 0$	$3(8) - 24 = 0$
or	$0 = 0$	$0 = 0$

Substituting 0, -2, or 2 into the original equation results each time in a true equation. ■

EXAMPLE 6 Solve $(5x - 1)(2x^2 + 15x + 18) = 0$.

Solution:
$$(5x - 1)(2x^2 + 15x + 18) = 0$$

$$(5x - 1)(2x + 3)(x + 6) = 0 \qquad \text{Factor the trinomial.}$$

$$5x - 1 = 0 \quad \text{or} \quad 2x + 3 = 0 \quad \text{or} \quad x + 6 = 0 \qquad \text{Set each factor equal to 0.}$$

$$5x = 1 \quad \text{or} \qquad 2x = -3 \, \text{or} \qquad x = -6 \quad \text{Solve.}$$

$$x = \frac{1}{5} \quad \text{or} \qquad x = -\frac{3}{2}$$

The solutions are $\frac{1}{5}$, $-\frac{3}{2}$, and -6. Check by replacing x with each solution in the original equation. ■

EXAMPLE 7   Solve $2x^3 - 4x^2 - 30x = 0$.

Solution: Begin by factoring out the GCF $2x$.

$$2x^3 - 4x^2 - 30x = 0$$

$$2x(x^2 - 2x - 15) = 0 \qquad \text{Factor out the GCF } 2x.$$

$$2x(x - 5)(x + 3) = 0 \qquad \text{Factor the quadratic.}$$

$$2x = 0 \quad \text{or} \quad x - 5 = 0 \quad \text{or} \quad x + 3 = 0 \qquad \begin{array}{l} \text{Set each factor containing a} \\ \text{variable equal to 0.} \end{array}$$

$$x = 0 \quad \text{or} \qquad x = 5 \quad \text{or} \qquad x = -3 \quad \text{Solve.}$$

The solutions are 0, 5, and -3. Check by replacing x with each solution in the cubic equation. ■

MENTAL MATH

Solve each equation by inspection.

1. $(a - 3)(a - 7) = 0$ **2.** $(a - 5)(a - 2) = 0$ **3.** $(x + 8)(x + 6) = 0$

4. $(x + 2)(x + 3) = 0$ **5.** $(x + 1)(x - 3) = 0$ **6.** $(x - 1)(x + 2) = 0$

EXERCISE SET 4.6

Solve each equation. See Example 1.

1. $(x - 2)(x + 1) = 0$ **2.** $(x + 3)(x + 2) = 0$ **3.** $x(x + 6) = 0$

4. $2x(x - 7) = 0$ **5.** $(2x + 3)(4x - 5) = 0$ **6.** $(3x - 2)(5x + 1) = 0$

7. $(2x - 7)(7x + 2) = 0$ **8.** $(9x + 1)(4x - 3) = 0$

Solve each equation. See Examples 2 through 4.

9. $x^2 - 13x + 36 = 0$ **10.** $x^2 + 2x - 63 = 0$ **11.** $x^2 + 2x - 8 = 0$

12. $x^2 - 5x + 6 = 0$ **13.** $x^2 - 4x = 32$ **14.** $x^2 - 5x = 24$

15. $x(3x - 1) = 14$ **16.** $x(4x - 11) = 3$ **17.** $3x^2 + 19x - 72 = 0$

18. $36x^2 + x - 21 = 0$

Solve each equation. See Examples 5 through 7.

19. $x^3 - 12x^2 + 32x = 0$

20. $x^3 - 14x^2 + 49x = 0$

21. $(4x - 3)(16x^2 - 24x + 9) = 0$

22. $(2x + 5)(4x^2 - 10x + 25) = 0$

23. $4x^3 - x = 0$

24. $4y^3 - 36y = 0$

25. $32x^3 - 4x^2 - 6x = 0$

26. $15x^3 + 24x^2 - 63x = 0$

Solve each equation. Be careful. Some of the equations are linear, not quadratic.

27. $x(x + 7) = 0$

28. $y(6 - y) = 0$

29. $(x + 5)(x - 4) = 0$

30. $(x - 8)(x - 1) = 0$

31. $x^2 - x = 30$

32. $x^2 + 13x = -36$

33. $6y^2 - 22y - 40 = 0$

34. $3x^2 - 6x - 9 = 0$

35. $(2x + 3)(2x^2 - 5x - 3) = 0$

36. $(2x - 9)(x^2 + 5x - 36) = 0$

37. $x^2 - 15 = -2x$

38. $x^2 - 26 = -11x$

39. $x^2 - 16x = 0$

40. $x^2 + 5x = 0$

41. $-18y^2 - 33y + 216 = 0$

42. $-20y^2 + 145y - 35 = 0$

43. $12x^2 - 59x + 55 = 0$

44. $30x^2 - 97x + 60 = 0$

45. $18x^2 + 9x - 2 = 0$

46. $28x^2 - 27x - 10 = 0$

47. $x(6x + 7) = 5$

48. $4x(8x + 9) = 5$

49. $4(x - 7) = 6$

50. $5(3 - 4x) = 9$

51. $5x^2 - 6x - 8 = 0$

52. $9x^2 + 6x + 2 = 0$

53. $(y - 2)(y + 3) = 6$

54. $(y - 5)(y - 2) = 28$

55. $4y^2 - 1 = 0$

56. $4y^2 - 81 = 0$

57. $t^2 + 13t + 22 = 0$

58. $x^2 - 9x + 18 = 0$

59. $5t - 3 = 12$

60. $9 - t = -1$

61. $x^2 + 6x - 17 = -26$

62. $x^2 - 8x - 4 = -20$

63. $12x^2 + 7x - 12 = 0$

64. $30x^2 - 11x - 30 = 0$

65. $10t^3 - 25t - 15t^2 = 0$

66. $36t^3 - 48t - 12t^2 = 0$

67. The sum of a number and its square is 132. Find the number.

68. The sum of a number and its square is 72. Find the number.

69. The sum of a positive number and its square is 90. Find the number.

70. The difference of a positive number and its square is 12. Find the number.

71. Write a quadratic equation in standard form that has two solutions, 5 and 7.

72. Write an equation that has three solutions, 0, 1, and 2.

A Look Ahead

Solve each equation. See the following example.

EXAMPLE Solve $(x - 6)(2x - 3) = (x + 2)(x + 9)$.

Solution:
$$(x - 6)(2x - 3) = (x + 2)(x + 9)$$
$$2x^2 - 15x + 18 = x^2 + 11x + 18$$
$$x^2 - 26x = 0$$
$$x(x - 26) = 0$$
$$x = 0 \quad \text{or} \quad x - 26 = 0$$
$$x = 26 \quad \blacksquare$$

73. $(x - 3)(3x + 4) = (x + 2)(x - 6)$

74. $(2x - 3)(x + 6) = (x - 9)(x + 2)$

75. $(2x - 3)(x + 8) = (x - 6)(x + 4)$

76. $(x + 6)(x - 6) = (2x - 9)(x + 4)$

77. $(4x - 1)(x - 8) = (x + 2)(x + 4)$

78. $(5x - 2)(x + 3) = (2x - 3)(x + 2)$

Skill Review

Perform the following operations. Write all answers in lowest terms. See Section 1.3.

79. $\dfrac{3}{5} + \dfrac{4}{9}$

80. $\dfrac{2}{3} + \dfrac{3}{7}$

81. $\dfrac{7}{10} - \dfrac{5}{12}$

82. $\dfrac{5}{9} - \dfrac{5}{12}$

83. $\dfrac{7}{8} \div \dfrac{7}{15}$

84. $\dfrac{5}{12} - \dfrac{3}{10}$

85. $\dfrac{4}{5} \cdot \dfrac{7}{8}$

86. $\dfrac{3}{7} \cdot \dfrac{12}{17}$

4.7
Applications of Quadratic Equations

OBJECTIVE	**1**	Translate word problems into equations and solve.

Tape 15

1 The skills gained in factoring and solving equations allow us to solve a wider variety of problems. Use the same steps for solving a word problem as in Section 2.6.

> **Solving a Word Problem**
>
> *Step 1* Read and reread the problem. Choose a variable to represent one unknown quantity to be found.
>
> *Step 2* Use this variable to represent any other unknown quantities.
>
> *Step 3* Draw a diagram if possible to visualize the known facts.
>
> *Step 4* Translate the problem into an equation.
>
> *Step 5* Solve the equation.
>
> *Step 6* Answer the question asked and check to see if the answer is a reasonable one.
>
> *Step 7* Check the possible solution in the originally stated problem.

Keep in mind that a solution of an equation may not always be a solution of the word problem. For example, a person's age or the length of a rectangle is always a positive number. Discard solutions that do not make sense as solutions of the word problem.

EXAMPLE 1 The length of a rectangle is 1 centimeter more than twice the width. The area of the rectangle is 36 square centimeters. Find the dimensions of the rectangle.

Solution: *Step 1* Let a variable represent one unknown quantity.
Let x be the width.

Step 2 Use this variable to represent other unknown quantities.
Since the length is one more than twice the width, let $2x + 1$ represent the length.

Step 3 Draw a diagram.

x

$2x + 1$

Step 4 Write an equation from the stated problem.
Since the area of a rectangle is the product of the length and width, the equation to solve is

$$\text{length} \cdot \text{width} = \text{area}$$

or

$$(2x + 1)x = 36$$

Step 5 Solve the equation.

$$x(2x + 1) = 36$$

$$2x^2 + x = 36 \qquad \text{Multiply.}$$

$$2x^2 + x - 36 = 0 \qquad \text{Write in standard form.}$$

$$(2x + 9)(x - 4) = 0 \qquad \text{Factor.}$$

$$2x + 9 = 0 \quad \text{or} \quad x - 4 = 0 \qquad \text{Set each factor equal to 0.}$$

$$2x = -9 \text{ or} \qquad x = 4 \qquad \text{Solve.}$$

$$x = -\frac{9}{2}$$

Step 6 Answer the question and see if it is reasonable.
The negative solution $-\frac{9}{2}$ can be discarded because width must be positive.

The other solution, 4, is the width of the rectangle, and the length is $2(4) + 1 = 9$. Thus, the dimensions of the rectangle are 4 centimeters by 9 centimeters.

Step 7 Check the solution.
To check, notice that $\text{length} \cdot \text{width} = 9 \cdot 4 = 36$, the required area. ∎

EXAMPLE 2 The height of a triangle is 2 feet less than twice the length of the base. If the triangle has an area of 30 square feet, find the length of its base and height.

Solution: *Step 1* Let a variable represent one unknown quantity.
Let x represent the length of the base of the triangle.

Step 2 Use this variable to represent other unknown quantities.
The height is 2 feet less than twice the base, so let $2x - 2$ represent the height.

Step 3   Draw a diagram.

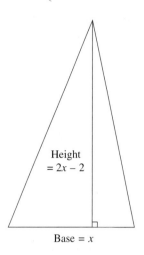

Height
= $2x - 2$

Base = x

Step 4   Write an equation from the stated problem.
We are given the area of the triangle, so use the formula:

$$\text{area of triangle} = \frac{1}{2}\,\text{base} \cdot \text{height}$$

or

$$30 = \frac{1}{2}x(2x - 2)$$

Step 5   Solve the equation.

$$30 = \frac{1}{2}x(2x - 2)$$

$30 = x^2 - x$	Multiply.
$x^2 - x - 30 = 0$	Write in standard form.
$(x - 6)(x + 5) = 0$	Factor.
$x - 6 = 0 \quad \text{or} \quad x + 5 = 0$	Set each factor equal to 0.
$x = 6 \quad \text{or} \qquad x = -5$	

Step 6   Answer the question and see if it is reasonable.
Since x represents the length of the base, discard the solution -5. The base of a triangle cannot be negative. The base is then 6 feet and the height is $2(6) - 2 = 10$ feet.

Step 7   Check.

To check this problem, recall that $\frac{1}{2}$ base $\cdot$ height = area, or

$$\frac{1}{2}(6)(10) = 30, \quad \text{the required area.} \quad \blacksquare$$

The next examples involve consecutive integers.

Consecutive integers:

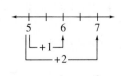

$x \quad x + 1 \quad x + 2$

Consecutive even integers:

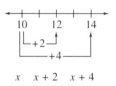

$$x \quad x + 2 \quad x + 4$$

Consecutive odd integers:

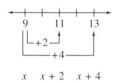

$$x \quad x + 2 \quad x + 4$$

EXAMPLE 3 Find two consecutive even integers whose product is 120.

Solution: Let x represent the first even integer. Then $x + 2$ represents the next consecutive even integer. Since their product is 120, we have the equation

$$x(x + 2) = 120$$

Now solve.

$$x(x + 2) = 120$$
$$x^2 + 2x = 120 \qquad \text{Multiply.}$$
$$x^2 + 2x - 120 = 0 \qquad \text{Write in standard form.}$$
$$(x + 12)(x - 10) = 0 \qquad \text{Factor.}$$
$$x + 12 = 0 \quad \text{or} \quad x - 10 = 0 \qquad \text{Set each factor equal to 0.}$$
$$x = -12 \quad \text{or} \qquad x = 10 \qquad \text{Solve.}$$

If $x = -12$, then $x + 2 = -10$ and their product $(-12)(-10) = 120$, the required product.

If $x = 10$ then $x + 2 = 12$ and their product $10(12) = 120$, the required product. Both pairs of solutions are reasonable and meet the requirements of the problem. ■

The next example makes use of the **Pythagorean theorem.** Before we review this theorem, recall that a **right triangle** is a triangle that contains a 90° or right angle. The **hypotenuse** of a right triangle is the side opposite the right angle and is the longest side of the triangle. The **legs** of a right triangle are the other sides of the triangle.

Hypotenuse
c

Leg b

Leg a

Pythagorean Theorem

In a right triangle, the sum of the squares of the lengths of the two legs is equal to the square of the length of the hypotenuse.

$$(\text{leg})^2 + (\text{leg})^2 = (\text{hypotenuse})^2 \quad \text{or} \quad a^2 + b^2 = c^2$$

EXAMPLE 4   Find the lengths of the sides of a right triangle if the lengths can be expressed by three consecutive even integers.

Solution: Let x, $x + 2$, and $x + 4$ be three consecutive even integers. Since these integers represent lengths of the sides of a right triangle, we have

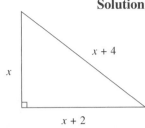

$$x = \text{one leg}$$

$$x + 2 = \text{other leg}$$

$$x + 4 = \text{hypotenuse (longest side)}$$

The Pythagorean theorem says

$$(\text{hypotenuse})^2 = (\text{leg})^2 + (\text{leg})^2$$

or

$$(x + 4)^2 = (x)^2 + (x + 2)^2$$

Now solve the equation.

$$(x + 4)^2 = x^2 + (x + 2)^2$$

$$x^2 + 8x + 16 = x^2 + x^2 + 4x + 4 \qquad \text{Multiply.}$$

$$x^2 + 8x + 16 = 2x^2 + 4x + 4$$

$$x^2 - 4x - 12 = 0 \qquad\qquad\qquad \text{Write in standard form.}$$

$$(x - 6)(x + 2) = 0$$

$$x - 6 = 0 \quad \text{or} \quad x + 2 = 0$$

$$x = 6 \quad \text{or} \qquad x = -2$$

Discard $x = -2$ since length cannot be negative. If $x = 6$, then $x + 2 = 8$ and $x + 4 = 10$.

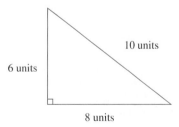

To check, see that $(\text{hypotenuse})^2 = (\text{leg})^2 + (\text{leg})^2$,

$$10^2 = 6^2 + 8^2$$

or

$$100 = 36 + 64 \qquad \text{True.} \quad \blacksquare$$

EXERCISE SET 4.7

Represent the given conditions using a single variable, x.

1. Two numbers, if one number is two times another.

2. The age of a man and the age of his son if the man is 5 years older than twice his son's age.

3. Two numbers whose sum is 36.

4. Two consecutive odd integers.

5. The length and width of a rectangle whose length is 4 centimeters less than three times the width.

6. The length and width of a rectangle whose length is twice its width.

7. A woman's age now and her age 10 years ago.

8. Three consecutive integers.

9. The value in cents of dimes and quarters if there are four more dimes than quarters.

10. The dollar amount of two investments if $10,000 was invested all together.

11. The three sides of a triangle if the first side is 2 inches less than twice the second side. The third side is 10 inches longer than the second side.

Solve each exercise. See Examples 1 and 2.

12. The length of a rectangle is 7 centimeters less than twice the width. Its area is 30 square centimeters. Find the dimensions of the rectangle.

13. The length of a rectangle is 9 inches more than its width. Its area is 112 square inches. Find the dimensions of the rectangle.

14. The altitude of a triangle is 8 centimeters more than twice the length of the base. If the triangle has an area of 96 square centimeters, find the length of its base and altitude.

15. The base of a triangle is 4 meters less than twice the length of the altitude. If the triangle has an area of 15 square meters, find the length of its base and altitude.

16. If the sides of a square are increased by 3 inches, the area becomes 64 square inches. Find the length of the sides of the original square.

17. If the sides of a square are increased by 5 meters, the area becomes 100 square meters. Find the length of the sides of the original square.

Solve the following. See Example 3.

18. The sum of a number and its square is 132. Find the number.

19. The sum of a number and its square is 182. Find the number.

20. The sum of two numbers is 20, and the sum of their squares is 218. Find the numbers.

21. The sum of two numbers is 25, and the sum of their squares is 325. Find the numbers.

22. The sum of the squares of two consecutive integers is 9 greater than 8 times the smaller integer. Find the integers.

23. The square of the largest of three consecutive integers is equal to the sum of the squares of the other two. Find the integers.

Solve the following. See Example 4.

24. Find the lengths of the sides of a right triangle if the hypotenuse is 10 centimeters longer than the short leg and 5 centimeters longer than the long leg.

25. Find the lengths of the sides of a right triangle if the length of the hypotenuse is 12 kilometers longer than the short leg and 6 kilometers longer than the long leg.

26. Find the length of the short leg of a right triangle if the long leg is 12 feet more than the short leg and the hypotenuse is 12 feet less than twice the short leg.

27. Find the length of the short leg of a right triangle if the long leg is 10 miles more than the short leg and the hypotenuse is 10 miles less than twice the short leg.

Solve the following.

28. A rectangle has a perimeter of 42 miles and an area of 104 square miles. Find the dimensions of the rectangle.

29. A rectangle has a perimeter of 60 meters and an area of 209 square meters. Find the dimensions of the rectangle.

30. Find two consecutive even numbers whose product is 624.

31. Find two consecutive odd numbers whose product is 399.

32. A rectangular pool is surrounded by a walk 4 meters wide.

The pool is 6 meters longer than its width. If the total area is 576 square meters more than the area of the pool, find the dimensions of the pool.

33. A rectangular garden is surrounded by a walk of uniform width. The area of the garden is 180 square yards. If the dimensions of the garden plus the walk are 16 yards by 24 yards, find the width of the walk.

34. If $D = \dfrac{n(n-3)}{2}$ is the formula for the number of diagonals N of a polygon with n sides, find the number of sides for a polygon with five diagonals.

35. If $D = \dfrac{n(n-3)}{2}$ is the formula for the number of diagonals D of a polygon with n sides, find the number of sides for a polygon with 35 diagonals.

36. One leg of a right triangle is 4 millimeters more than the smaller leg and the hypotenuse is 8 millimeters more than the smaller leg. Find the lengths of the sides of the triangle.

37. One leg of a right triangle is 9 centimeters longer than the other leg and the hypotenuse is 45 centimeters. Find the lengths of the legs of the triangle.

38. The length of the base of a triangle is twice its altitude. If the area of the triangle is 100 square kilometers, find the altitude.

39. The altitude of a triangle is 2 millimeters less than the base. If the area is 60 square millimeters, find the base.

40. The sum of the squares of two consecutive negative integers is 221. Find the integers.

41. The sum of the squares of two consecutive even positive integers is 100. Find the integers.

42. Find the dimensions of a rectangle whose length is 2 yards more than twice its width and whose area is 60 square yards.

43. Find the dimensions of a rectangle whose length is 9

centimeters more than its width and whose area is 112 square centimeters.

44. Find the dimensions of a rectangle whose width is 7 miles less than its length and whose area is 120 square miles.

45. Find the dimensions of a rectangle whose width is 2 inches less than half its length and whose area is 160 square inches.

46. At the end of 2 years, P dollars invested at r percent interest compounded annually increases to an amount, A dollars, given by $A = P(1 + r)^2$. Find the interest rate if $100 increased to $144 in 2 years.

47. At the end of 2 years, P dollars invested at r percent interest compounded annually increases to an amount, A dollars, given by $A = P(1 + r)^2$. Find the interest rate if $2000 increased to $2420 in 2 years.

48. If the cost, C, for manufacturing x units of a certain product is given by $C = x^2 - 15x + 50$, find the number of units manufactured at a cost of $9500.

49. If a switchboard handles n telephones, the number C of telephone connections it can make simultaneously is given by the equation $C = \dfrac{n(n-1)}{2}$. Find how many telephones are handled by a switchboard making 120 telephone connections simultaneously.

50. Two boats travel at a right angle to each other after leaving the same dock at the same time. One hour later the boats are 17 miles apart. If one boat travels 7 miles per hour faster than the other boat, find the rate of each boat.

51. The side of a square equals the width of a rectangle. The length of the rectangle is 6 meters longer than its width. The sum of the areas of the square and the rectangle is 176 square meters. Find the side of the square.

52. A rectangle has a perimeter of 42 yards and an area of 104 square yards. Find the dimensions of the rectangle.

Writing in Mathematics

53. Describe the kind of applied problem you find easiest to do and why.

54. Describe the kind of applied problem you find most difficult and why.

Skill Review

Write each fraction in simplest form. See Section 1.3.

55. $\dfrac{20}{35}$

56. $\dfrac{24}{32}$

57. $\dfrac{27}{18}$

58. $\dfrac{15}{27}$

59. $\dfrac{84}{120}$

60. $\dfrac{25}{120}$

61. $\dfrac{24}{64}$

62. $\dfrac{35}{56}$

CRITICAL
THINKING

Peter has reached the pleasant point in his life when he is convinced he can afford a swimming pool for his backyard. He's talked to local suppliers, and has settled on a pool 10 feet by 15 feet. Peter's pool will be surrounded by a cement patio, equally wide on all sides. Constructing the patio costs $8.00 for every square foot of patio, and Peter can afford to pay no more than $2400 for the patio.

How wide can Peter's patio be? What if Peter chose a smaller pool costing $300 less and used the savings for a bigger patio? How wide can Peter's patio be now?

CHAPTER 4 GLOSSARY

Factoring a polynomial is the process of writing a polynomial as a product.

A **perfect square trinomial** is a trinomial that is the square of some binomial.

A **quadratic equation** is an equation that can be written in the form $ax^2 + bx + c = 0$ with a not 0.

CHAPTER 4 SUMMARY

(4.5)

To Factor a Polynomial

Step 1   Are there any common factors? If so, factor them out.

Step 2   How many terms are in the polynomial?
  a. If there are **two** terms, decide if one of the following can be applied.
   i. Difference of two squares: $a^2 - b^2 = (a - b)(a + b)$.
   ii. Difference of two cubes: $a^3 - b^3 = (a - b)(a^2 + ab + b^2)$.
   iii. Sum of two cubes: $a^3 + b^3 = (a + b)(a^2 - ab + b^2)$.
  b. If there are **three** terms, try one of the following:
   i. Perfect square trinomial: $a^2 + 2ab + b^2 = (a + b)^2$.
   ii. If not a perfect square trinomial, factor using the methods presented in Sections 4.2 and 4.3.
  c. If there are **four** or more terms, try factoring by grouping.

Step 3   Last, see if any factors in the factored polynomial can be factored further.

(4.6)

> **To Solve Quadratic Equations by Factoring**
>
> *Step 1*   Write the equation in standard form: $ax^2 + bx + c = 0$.
> *Step 2*   Factor the quadratic completely.
> *Step 3*   Set each factor containing a variable equal to 0.
> *Step 4*   Solve the resulting equations.
> *Step 5*   Check each solution in the original equation.

CHAPTER 4 REVIEW

(4.1)   *Complete the factoring.*

1. $6x^2 - 15x = 3x(\quad)$

2. $2x^3y - 6x^2y^2 - 8xy^3 = 2xy(\quad)$

Factor the GCF from each polynomial.

3. $20x^2 + 12x$

4. $6x^2y^2 - 3xy^3$

5. $-8x^3y + 6x^2y^2$

6. $3x(2x + 3) - 5(2x + 3)$

7. $5x(x + 1) - (x + 1)$

Factor.

8. $3x^2 - 3x + 2x - 2$

9. $6x^2 + 10x - 3x - 5$

10. $3a^2 + 9ab + 3b^2 + ab$

(4.2)   *Factor each trinomial.*

11. $x^2 + 6x + 8$

12. $x^2 - 11x + 24$

13. $x^2 + x + 2$

14. $x^2 - 5x - 6$

15. $x^2 + 2x - 8$

16. $x^2 + 4x - 12y^2$

17. $x^2 + 8xy + 15y^2$

18. $3x^2y + 6xy^2 + 3y^3$

19. $72 - 18x - 2x^2$

20. $32 + 12x - 4x^2$

(4.3)   *Factor each trinomial.*

21. $2x^2 + 11x - 6$

22. $4x^2 - 7x + 4$

23. $4x^2 + 4x - 3$

24. $6x^2 + 5xy - 4y^2$

25. $6x^2 - 25xy + 4y^2$

26. $18x^2 - 60x + 50$

27. $2x^2 - 23xy - 39y^2$

28. $4x^2 - 28xy + 49y^2$

29. $18x^2 - 9xy - 20y^2$

30. $36x^3y + 24x^2y^2 - 45xy^3$

(4.4)   *Factor each binomial.*

31. $4x^2 - 9$

32. $9t^2 - 25s^2$

33. $16x^2 + y^2$

34. $x^3 - 8y^3$

35. $8x^3 + 27$

36. $2x^3 + 8x$

37. $54 - 2x^3y^3$

38. $9x^2 - 4y^2$

39. $16x^4 - 1$

40. $x^4 + 16$

(4.5) *Factor.*

41. $2x^2 + 5x - 12$

42. $3x^2 - 12$

43. $x(x - 1) + 3(x - 1)$

44. $x^2 + xy - 3x - 3y$

45. $4x^2y - 6xy^2$

46. $8x^2 - 15x - x^3$

47. $125x^3 + 27$

48. $24x^2 - 3x - 18$

49. $(x + 7)^2 - y^2$

50. $x^2(x + 3) - 4(x + 3)$

(4.6) *Solve the following equations.*

51. $(x + 6)(x - 2) = 0$

52. $3x(x + 1)(7x - 2) = 0$

53. $4(5x + 1)(x + 3) = 0$

54. $x^2 + 8x + 7 = 0$

55. $x^2 - 2x - 24 = 0$

56. $x^2 + 10x = -25$

57. $x(x - 10) = -16$

58. $(3x - 1)(9x^2 + 3x + 1) = 0$

59. $56x^2 - 5x - 6 = 0$

60. $20x^2 - 7x - 6 = 0$

61. $5(3x + 2) = 4$

62. $6x^2 - 3x + 8 = 0$

63. $12 - 5t = -3$

64. $5x^3 + 20x^2 + 20x = 0$

65. $4t^3 - 5t^2 - 21$

(4.7) *Solve the following problems.*

66. A girl is 12 years older than her sister. If the product of their ages is 540, find the age of each.

67. A number is one more than twice another. Their squares differ by 176. Find the numbers.

68. A rectangular pool is surrounded by a walk 5 feet wide. The pool is 10 feet longer than it is wide. If the total area is 480 square feet more than the area of the pool, what are the dimensions of the pool?

69. The length of a rectangle is 15 centimeters less than twice the width. The area is 500 square centimeters. Find the dimensions of the rectangle.

70. The base of a triangle is four times its altitude. If the area of the triangle is 162 square feet, find the base.

71. The sum of a number and twice its square is 105. Find the number.

72. Find two consecutive positive integers whose product is 380.

73. Find the length of the long leg of a right triangle if the hypotenuse is 8 feet longer than the long leg and the short leg is 8 feet shorter than the long leg.

CHAPTER 4 TEST

Factor each polynomial completely. If a polynomial cannot be factored, write "prime."

1. $9x^3 + 39x^2 + 12x$

2. $x^2 + x - 10$

3. $x^2 + 4$

4. $y^2 - 8y - 48$

5. $3a^2 + 3ab - 7a - 7b$

6. $3x^2 - 5x + 2$

7. $x^2 + 20x + 90$

8. $x^2 + 14xy + 24y^2$

9. $26x^6 - x^4$

10. $50x^3 + 10x^2 - 35x$

11. $180 - 5x^2$

12. $64x^3 - 1$

13. $6t^2 - t - 5$

14. $xy^2 - 7y^2 - 4x + 28$

15. $x - x^5$

16. $-xy^3 - x^3y$

Solve each equation.

17. $x^2 + 5x = 14$

18. $(x + 3)^2 = 16$

19. $3x(2x - 3)(3x + 4) = 0$

20. $5t^3 - 45t = 0$

21. $3x^2 = -12x$

22. $t^2 - 2t - 15 = 0$

23. $7x^2 = 168 + 35x$

24. $6x^2 = 15x$

Solve each problem.

25. Find the dimensions of a rectangle whose length is 5 feet longer than its width and whose area is 66 square feet.

26. The sum of two numbers is 17, and the sum of their squares is 145. Find the numbers.

27. The length of the base of a triangle is 9 centimeters longer than its altitude. If the area of the triangle is 68 square centimeters, find the length of the base.

28. The sum of the squares of two consecutive odd integers is 202. Find the numbers.

29. A rectangle is twice as long as it is wide. If the length and the width are both increased by 3 inches, the area is increased by 72 square inches. Find the dimensions of the original rectangle.

CHAPTER 4 CUMULATIVE REVIEW

1. Translate each sentence into symbols.
   a. The product of 2 and 3 is 6.
   b. The difference of 8 and 4 is less than or equal to 4.
   c. The quotient of 10 and 2 is not equal to 6.

2. Find each quotient and write all answers in lowest terms.
   a. $\dfrac{4}{5} \div \dfrac{5}{16}$ **b.** $\dfrac{7}{10} \div 14$ **c.** $\dfrac{3}{8} \div \dfrac{3}{10}$

3. Simplify each expression.
   a. $6 \div 3 + 5^2$ **b.** $\dfrac{2(12 + 3)}{|-15|}$
   c. $3 \cdot 10 - 7 \div 7$
   d. $3 \cdot 4^2$ **e.** $\dfrac{3}{2} \cdot \dfrac{1}{2} - \dfrac{1}{2}$

4. Find the opposite of the following numbers.
   a. 5   **b.** 0   **c.** -6

5. Find the following quotients.
   a. $-\dfrac{18}{3}$ **b.** $\dfrac{-14}{-2}$ **c.** $\dfrac{20}{-4}$

6. Subtract $4x - 2$ from $2x - 3$.

7. Solve $x - 7 = 10$ for x.

8. Solve $-\dfrac{2}{3}x = -5$ for x.

9. Solve $\dfrac{2(a + 3)}{3} = 6a + 2$.

10. A jogger starts running at 6 miles per hour. One hour later a biker starts riding along the same path at 15 mph. How long will it take the biker to catch up to the jogger?

11. Solve $2x - 4$, and graph the solution.

12. Evaluate the following expressions.
   a. 2^3 **b.** 3^1 **c.** $(-4)^2$ **d.** -4^2

13. Simplify each expression. Write answers using positive exponents only.
   a. 3^{-2} **b.** $2x^{-3}$ **c.** $\dfrac{1}{2^{-5}}$

14. Write the following numbers in standard notation, without exponents
   a. 1.02×10^5
   b. 7.358×10^{-3}
   c. 8.4×10^7
   d. 3.007×10^{-5}

15. Multiply $(t + 2)$ by $(3t^2 - 4t + 2)$

16. Multiply $(y + 6)(2y - 1)$.

17. Simplify $\dfrac{8x^2y^2 - 16xy + 2x}{4xy}$.

18. Factor $25x^4z + 15x^3z + 5x^2z$.

19. Factor $x^2 + 4x - 12$.

20. Factor $2x^2 + 13x - 7$.

21. Factor $12m^2 - 3n^2$.

22. Solve $x^2 - 9x = -20$.

5.1 Simplifying Rational Expressions

5.2 Multiplying and Dividing Rational Expressions

5.3 Adding and Subtracting Rational Expressions with Common Denominators and Least Common Denominator

5.4 Adding and Subtracting Rational Expressions with Unlike Denominators

5.5 Complex Fractions

5.6 Solving Equations Containing Rational Expressions

5.7 Ratio and Proportion

5.8 Applications of Equations Containing Rational Equations

CHAPTER 5

Rational Expressions

Depending on the medicine, too large a dose can be extremely dangerous. Mathematical models predicting the correct dose are available, but are, at best, approximations. (See Critical Thinking, page 249.)

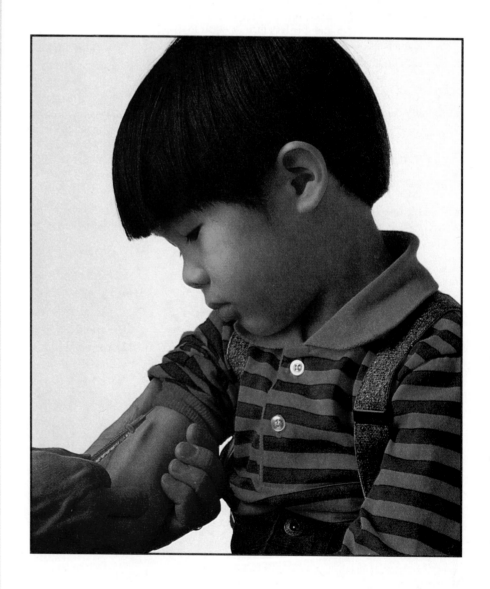

INTRODUCTION

In this chapter, we expand our knowledge of algebraic expressions to include another category called **rational expressions,** such as $\dfrac{x + 1}{x}$. We explore the operations of addition, subtraction, multiplication, and division for these special fractions, using principles similar to the principles for number fractions. Thus, the material in this chapter will make full use of your knowledge of number fractions.

5.1
Simplifying Rational Expressions

OBJECTIVES

Tape 16

1   Define rational expressions.

2   Find the value of a rational expression given a replacement number.

3   Identify when a rational expression is undefined.

4   Write rational expressions in lowest terms.

1   As we reviewed in Chapter 1, a rational number is a number that can be written as a quotient of integers. A **rational expression** is also a quotient; it is a quotient of polynomials.

> **Rational Expression**
>
> A rational expression is an expression that can be written in the form $\dfrac{P}{Q}$, where P and Q are polynomials and Q does not equal 0.

Rational Expressions

$$\frac{3y^3}{8} \qquad \frac{-4p}{p^3 + 2p + 1} \qquad \frac{5x^2 - 3x + 2}{3x + 7}$$

2   Rational expressions have different values depending on what value replaces the variable. Next, we review the standard order of operations by finding the value of rational expressions at given replacement values of the variable.

EXAMPLE 1 Find the value of $\dfrac{x + 4}{2x - 3}$ at the given replacement values.

a. $x = 5$ **b.** $x = -2$

Solution: **a.** Replace each x in the expression with 5 and then simplify.

$$\frac{x + 4}{2x - 3} = \frac{5 + 4}{2(5) - 3} = \frac{9}{10 - 3} = \frac{9}{7}.$$

b. Replace each x in the expression with -2 and then simplify.

$$\frac{x + 4}{2x - 3} = \frac{-2 + 4}{2(-2) - 3} = \frac{2}{-7} \quad \text{or} \quad -\frac{2}{7}. \quad \blacksquare$$

For a fraction such as $\dfrac{2}{-7}$, recall from Chapter 1 that

$$\frac{2}{-7} = \frac{-2}{7} = -\frac{2}{7}$$

In general, for any fraction

$$\frac{-a}{b} = \frac{a}{-b} = -\frac{a}{b}, \qquad b \neq 0$$

3   In the preceding box, notice that we wrote $b \neq 0$ for the denominator b. This is because the denominator of a rational expression must not equal 0 since division by 0 is not defined. This means we must be careful when replacing the variable in a rational expression by a number. For example, suppose we replace x with 5 in the rational expression $\dfrac{2 + x}{x - 5}$. The expression becomes

$$\frac{2 + x}{x - 5} = \frac{2 + 5}{5 - 5} = \frac{7}{0}$$

which is undefined. Therefore, in this expression we can allow x to be any real number except 5. A rational expression is undefined at values that make the denominator 0.

EXAMPLE 2 Are there any values for which each expression is undefined?

a. $\dfrac{x}{x - 3}$ **b.** $\dfrac{x^2 + 2}{x^2 - 3x + 2}$ **c.** $\dfrac{x^3 - 6x^2 - 10x}{3}$ **d.** $\dfrac{2}{x^2 + 1}$

Solution: To find values for which each expression is undefined, find values that make the denominator 0.

a. The denominator of $\dfrac{x}{x - 3}$ is 0 when $x - 3 = 0$ or when $x = 3$. Thus, when $x = 3$, the expression $\dfrac{x}{x - 3}$ is undefined.

b. Set the denominator equal to zero.

$$x^2 - 3x + 2 = 0$$

$$(x - 2)(x - 1) = 0 \qquad \text{Factor.}$$

$$x - 2 = 0 \quad \text{or} \quad x - 1 = 0 \qquad \text{Set each factor equal to zero.}$$
$$x = 2 \quad \text{or} \qquad x = 1 \qquad \text{Solve.}$$

The expression $\dfrac{x^2 + 2}{x^2 - 3x + 2}$ is undefined when $x = 2$ or when $x = 1$.

c. The denominator of $\dfrac{x^3 - 6x^2 - 10x}{3}$ is never zero, so there are no values of x that make this expression undefined.

d. No matter what real number x is replaced by, the denominator $x^2 + 1$ does not equal 0, so there are no real numbers that make this expression undefined.   ∎

4   A fraction is said to be written in lowest terms or simplest form when the numerator and denominator have no common factors other than 1 (or -1). For example, the fraction $\dfrac{7}{10}$ is in lowest terms since the numerator and denominator have no common factors other than 1 (or -1).

The process of writing a rational expression in lowest terms or simplest form is called **simplifying** a rational expression. The following fundamental principle of rational expressions is used to simplify a rational expression.

Fundamental Principle of Rational Expressions

If $\dfrac{P}{Q}$ is a rational expression and R is a polynomial, $R \neq 0$, then

$$\frac{PR}{QR} = \frac{P}{Q} \quad \text{and} \quad \frac{P \div R}{Q \div R} = \frac{P}{Q}$$

Simplifying a rational expression is similar to simplifying a fraction. To simplify the fraction $\dfrac{15}{20}$, factor the numerator and the denominator and use the fundamental principle to divide both numerator and denominator by any common factors.

$$\frac{15}{20} = \frac{3 \cdot 5}{2 \cdot 2 \cdot 5} = \frac{3}{4}$$

To simplify the rational expression $\dfrac{x^2 - 9}{x^2 + x - 6}$, we also factor the numerator and denominator and then use the fundamental principle of rational expressions to divide both numerator and denominator by any common factors.

$$\frac{x^2 - 9}{x^2 + x - 6} = \frac{(x - 3)(x + 3)}{(x - 2)(x + 3)} = \frac{x - 3}{x - 2}$$

The following steps may be used to simplify rational expressions.

To Simplify a Rational Expression

Step 1   Completely factor the numerator and denominator.

Step 2   Divide both the numerator and denominator by all common factors.

EXAMPLE 3 Write $\dfrac{21a^2b}{3a^5b}$ in simplest form.

Solution: Factor the numerator and denominator. Then divide the numerator and denominator by common factors.

$$\frac{21a^2b}{3a^5b} = \frac{7 \cdot 3 \cdot a^2 \cdot b}{3 \cdot a^3 \cdot a^2 \cdot b} = \frac{7}{a^3} \quad \blacksquare$$

EXAMPLE 4 Simplify $\dfrac{5x - 5}{x^3 - x^2}$.

Solution: Factor the numerator and denominator and then divide both by any common factors.

$$\frac{5x - 5}{x^3 - x^2} = \frac{5(x - 1)}{x^2(x - 1)} = \frac{5}{x^2} \quad \blacksquare$$

EXAMPLE 5 Write $\dfrac{x^2 + 8x + 7}{x^2 - 4x - 5}$ in simplest form.

Solution: Factor the numerator and denominator and divide by common factors.

$$\frac{x^2 + 8x + 7}{x^2 - 4x - 5} = \frac{(x + 7)(x + 1)}{(x - 5)(x + 1)} = \frac{x + 7}{x - 5} \quad \blacksquare$$

EXAMPLE 6 Simplify $\dfrac{x^2 + 4x + 4}{x^2 + 2x}$.

Solution: Factor the numerator and denominator and divide by common factors.

$$\frac{x^2 + 4x + 4}{x^2 + 2x} = \frac{(x + 2)(x + 2)}{x(x + 2)} = \frac{x + 2}{x} \quad \blacksquare$$

HELPFUL HINT

When simplifying a rational expression, **common factors** are divided out, **not common terms.** For example, $\dfrac{x + 2}{x}$ cannot be simplified any further because the numerator and denominator have no **common factors.**

EXAMPLE 7 Write each rational expression in lowest terms.

a. $\dfrac{x + y}{y + x}$ **b.** $\dfrac{x - y}{y - x}$

Solution: **a.** The expression $\dfrac{x + y}{y + x}$ can be simplified by using the commutative property of addition and rewriting the denominator $y + x$ as $x + y$.

$$\frac{x + y}{y + x} = \frac{x + y}{x + y} = 1$$

b. The expression $\dfrac{x - y}{y - x}$ can be simplified by recognizing that $y - x$ and $x - y$ are opposites. In other words, $y - x = -1(x - y)$. Proceed as follows:

$$\frac{x - y}{y - x} = \frac{(x - y)}{(-1)(x - y)} = \frac{1}{-1} = -1 \quad \blacksquare$$

EXAMPLE 8 Simplify $\dfrac{4 - x^2}{3x^2 - 5x - 2}$.

Solution: $\dfrac{4 - x^2}{3x^2 - 5x - 2} = \dfrac{(2 - x)(2 + x)}{(x - 2)(3x + 1)}$ \qquad Factor.

$$= \frac{(-1)(x - 2)(2 + x)}{(x - 2)(3x + 1)} \qquad \text{Write } 2 - x \text{ as } -1(x - 2).$$

$$= \frac{(-1)(2 + x)}{3x + 1} \quad \text{or} \quad \frac{-2 - x}{3x + 1} \qquad \text{Simplify.} \quad \blacksquare$$

EXAMPLE 9 Simplify $\dfrac{2x^2 - 2xy + 3x - 3y}{2x + 3}$.

Solution: First, factor the four-term numerator by grouping.

$$\frac{2x^2 - 2xy + 3x - 3y}{2x + 3} = \frac{2x(x - y) + 3(x - y)}{2x + 3}$$

$$= \frac{(2x + 3)(x - y)}{2x + 3} \qquad \text{Factor.}$$

$$= \frac{x - y}{1} \quad \text{or} \quad x - y \qquad \text{Simplify.} \quad \blacksquare$$

MENTAL MATH

Find any real numbers for which each rational expression is undefined. See Example 2.

1. $\dfrac{x + 5}{x}$

2. $\dfrac{x^2 - 5x}{x - 3}$

3. $\dfrac{x^2 + 4x - 2}{x(x - 1)}$

4. $\dfrac{x + 2}{(x - 5)(x - 6)}$

EXERCISE SET 5.1

Find the value of the following expressions when $x = 2$, $y = -2$, and $z = -5$. See Example 1.

1. $\dfrac{x + 5}{x + 2}$

2. $\dfrac{x + 8}{2x + 5}$

3. $\dfrac{z - 8}{z + 2}$

4. $\dfrac{y - 2}{-5 + y}$

5. $\dfrac{x^2 + 8x + 2}{x^2 - x - 6}$

6. $\dfrac{z^2 + 8}{z^3 - 25z}$

7. $\dfrac{x + 5}{x^2 + 4x - 8}$

8. $\dfrac{z^3 + 1}{z^2 + 1}$

9. $\dfrac{y^3}{y^2 - 1}$

10. $\dfrac{z}{z^2 - 5}$

Find any real numbers for which each rational expression is undefined. See Example 2.

11. $\dfrac{x + 3}{x + 2}$

12. $\dfrac{5x + 1}{x - 3}$

13. $\dfrac{4x^2 + 9}{2x - 8}$

14. $\dfrac{9x^3 + 4x}{15x + 45}$

15. $\dfrac{9x^3 + 4x}{15x + 30}$

16. $\dfrac{19x^3 + 2}{x^3 - x}$

17. $\dfrac{x^2 - 5x - 2}{x^2 + 4}$

18. $\dfrac{9y^5 + y^3}{x^2 + 9}$

Write each expression in lowest terms. See Examples 3 and 4.

19. $\dfrac{8x^5}{4x^9}$

20. $\dfrac{12y^7}{-2y^6}$

21. $\dfrac{5(x - 2)}{(x - 2)(x + 1)}$

22. $\dfrac{9(x - 7)(x + 7)}{3(x - 7)}$

23. $\dfrac{-5a - 5b}{a + b}$

24. $\dfrac{7x + 35}{x^2 + 5x}$

Simplify each expression. See Examples 5 and 6.

25. $\dfrac{x + 5}{x^2 - 4x - 45}$

26. $\dfrac{x - 3}{x^2 - 6x + 9}$

27. $\dfrac{5x^2 + 11x + 2}{x + 2}$

28. $\dfrac{12x^2 + 4x - 1}{2x + 1}$

29. $\dfrac{x^2 - x - 12}{2x^2 - 5x - 3}$

30. $\dfrac{x^2 + 3x - 4}{x^2 - x - 20}$

Simplify each expression. See Examples 7 and 8.

31. $\dfrac{x - 7}{7 - x}$

32. $\dfrac{y - z}{z - y}$

33. $\dfrac{y^2 - 2y}{4 - 2y}$

34. $\dfrac{x^2 + 5x}{20 + 4x}$

35. $\dfrac{x^2 - 4x + 4}{4 - x^2}$

36. $\dfrac{x^2 + 10x + 21}{-2x - 14}$

Simplify each expression. See Example 9.

37. $\dfrac{x^2 + xy + 2x + 2y}{x + 2}$

38. $\dfrac{ab + ac + b^2 + bc}{b + c}$

39. $\dfrac{5x + 15 - xy - 3y}{2x + 6}$

40. $\dfrac{xy - 6x + 2y - 12}{y^2 - 6y}$

Write each expression in lowest terms.

41. $\dfrac{15x^4 y^8}{-5x^8 y^3}$

42. $\dfrac{24a^3 b^3}{6a^2 b^4}$

43. $\dfrac{(x - 2)(x + 3)}{5(x + 3)}$

44. $\dfrac{-2(y - 9)}{(y - 9)^2}$

45. $\dfrac{-6a - 6b}{a + b}$

46. $\dfrac{4a - 4y}{4y - 4a}$

47. $\dfrac{2x^2 - 8}{4x - 8}$

48. $\dfrac{5x^2 - 500}{35x + 350}$

49. $\dfrac{11x^2 - 22x^3}{6x - 12x^2}$

50. $\dfrac{16r^2 - 4s^2}{4r - 2s}$

51. $\dfrac{x + 7}{x^2 + 5x - 14}$

52. $\dfrac{x - 10}{x^2 - 17x + 70}$

53. $\dfrac{2x^2 + 3x - 2}{2x - 1}$

54. $\dfrac{4x^2 + 24x}{x + 6}$

55. $\dfrac{x^2 - 1}{x^2 - 2x + 1}$

56. $\dfrac{x^2 - 16}{x^2 - 8x + 16}$

57. $\dfrac{m^2 - 6m + 9}{m^2 - 9}$

58. $\dfrac{m^2 - 4m + 4}{m^2 + m - 6}$

59. $\dfrac{-2a^2 + 12a - 18}{9 - a^2}$

60. $\dfrac{-4a^2 + 8a - 4}{2a^2 - 2}$

61. $\dfrac{2 - x}{x - 2}$

62. $\dfrac{7 - y}{y - 7}$

63. $\dfrac{x^2 - 1}{1 - x}$

64. $\dfrac{x^2 - xy}{2y - 2x}$

65. $\dfrac{x^2 + 7x + 10}{x^2 - 3x - 10}$

66. $\dfrac{2x^2 + 7x - 4}{x^2 + 3x - 4}$

67. $\dfrac{3x^2 + 7x + 2}{3x^2 + 13x + 4}$

68. $\dfrac{4x^2 - 4x + 1}{2x^2 + 9x - 5}$

69. $\dfrac{x^2 + 3x - 2x - 6}{x^2 - 2x}$ **70.** $\dfrac{ax + ay - bx - by}{x^2 - y^2}$ **71.** $\dfrac{x^3 + 8}{x + 2}$ **72.** $\dfrac{x^2 + 64}{x + 4}$

73. $\dfrac{x^2 + xy + 5x + 5y}{3x + 3y}$ **74.** $\dfrac{x^2 + 2x + xy + 2y}{x^2 - 4}$ **75.** $\dfrac{x^3 - 1}{1 - x}$ **76.** $\dfrac{3 - x}{x^3 - 27}$

Writing in Mathematics

77. Why can't the denominator of a fraction or a rational expression equal zero?

78. Explain how to write a rational expression in lowest terms.

Skill Review

Perform the indicated operations. See Section 3.3.

79. $(x^2 + 2x - 6) + (x^2 + 5)$

80. $(5x^2 - x) + (-2x^2 + 7)$

81. $(2x - 3) - (5x + 4)$

82. $(-8x - 10) - (x + 6)$

83. Subtract $(3x - 1)$ from $(5x + 2)$

84. Subtract $(9x + 3)$ from $(7x + 7)$

5.2
Multiplying and Dividing Rational Expressions

OBJECTIVES

Tape 16

1 Multiply rational expressions.

2 Divide rational expressions.

1   Just as simplifying rational expressions is similar to simplifying number fractions, multiplying and dividing rational expressions is similar to multiplying and dividing number fractions. To find the product of $\dfrac{3}{5}$ and $\dfrac{1}{4}$, multiply the numerators and then multiply the denominators of both fractions.

$$\frac{3}{5} \cdot \frac{1}{4} = \frac{3 \cdot 1}{5 \cdot 4} = \frac{3}{20}$$

Use this same procedure to multiply rational expressions.

Multiplying Rational Expressions

Let P, Q, R, and S be polynomials. Then

$$\frac{P}{Q} \cdot \frac{R}{S} = \frac{PR}{QS}$$

as long as $Q \neq 0$ and $S \neq 0$.

EXAMPLE 1 Find the following products.

a. $\dfrac{25x}{2} \cdot \dfrac{1}{y^3}$ **b.** $\dfrac{-7x^2}{5y} \cdot \dfrac{3y^5}{14x^2}$

Solution: To multiply rational expressions, multiply the numerators and then multiply the denominators of both expressions. Then simplify if possible.

a. $\dfrac{25x}{2} \cdot \dfrac{1}{y^3} = \dfrac{25x \cdot 1}{2 \cdot y^3} = \dfrac{25x}{2y^3}$

b. $\dfrac{-7x^2}{5y} \cdot \dfrac{3y^5}{14x^2} = \dfrac{-7x^2 \cdot 3y^5}{5y \cdot 14x^2}$ Multiply.

Next, we simplify.

$$= \dfrac{-1 \cdot \boxed{7} \cdot 3 \cdot \boxed{x^2} \cdot \boxed{y} \cdot y^4}{5 \cdot 2 \cdot \boxed{7} \cdot \boxed{x^2} \cdot \boxed{y}}$$

$$= -\dfrac{3y^4}{10} \quad \blacksquare$$

EXAMPLE 2 Multiply $\dfrac{x^2 + x}{3x} \cdot \dfrac{6}{5x + 5}$.

Solution: $\dfrac{x^2 + x}{3x} \cdot \dfrac{6}{5x + 5} = \dfrac{x(x + 1)}{3x} \cdot \dfrac{2 \cdot 3}{5(x + 1)}$ Factor numerators and denominators.

$$= \dfrac{\boxed{x}\,(x + 1) \cdot 2 \cdot \boxed{3}}{\boxed{3x}\, \cdot 5\,\boxed{(x + 1)}}$$ Multiply.

$$= \dfrac{2}{5}$$ Simplify. $\blacksquare$

The following steps may be used to multiply rational expressions.

To Multiply Rational Expressions

Step 1 Completely factor numerators and denominators.

Step 2 Multiply numerators and multiply denominators.

Step 3 Write the product in lowest terms by dividing the numerator and the denominator by all common factors.

EXAMPLE 3 Multiply $\dfrac{3x + 3}{5x - 5x^2} \cdot \dfrac{2x^2 + x - 3}{4x^2 - 9}$.

Solution: $\dfrac{3x + 3}{5x - 5x^2} \cdot \dfrac{2x^2 + x - 3}{4x^2 - 9} = \dfrac{3(x + 1)}{5x(1 - x)} \cdot \dfrac{(2x + 3)(x - 1)}{(2x - 3)(2x + 3)}$ Factor.

$$= \dfrac{3(x + 1)\,\boxed{(2x + 3)}\,(x - 1)}{5x(1 - x)(2x - 3)\,\boxed{(2x + 3)}}$$ Multiply.

$$= \frac{3(x + 1)(x - 1)}{5x(1 - x)(2x - 3)}$$ Divide out common factors.

Next, recall that $x - 1$ and $1 - x$ are opposites so that $x - 1 = -1(1 - x)$.

$$= \frac{3(x + 1)(-1)(1 - x)}{5x(1 - x)(2x - 3)}$$ Write $x - 1$ as $-1(1 - x)$.

$$= \frac{-3(x + 1)}{5x(2x - 3)}$$ Simplify.   ■

2   Rational expressions are divided in the same way as fractions. To divide two fractions, multiply the first fraction by the reciprocal of the second fraction.

For example to divide $\frac{3}{2}$ by $\frac{7}{8}$, multiply $\frac{3}{2}$ by $\frac{8}{7}$.

$$\frac{3}{2} \div \frac{7}{8} = \frac{3}{2} \cdot \frac{8}{7} = \frac{3 \cdot 4 \cdot 2}{2 \cdot 7} = \frac{12}{7}$$

Dividing Rational Expressions

Let P, Q, R, and S be polynomials. Then,

$$\frac{P}{Q} \div \frac{R}{S} = \frac{P}{Q} \cdot \frac{S}{R} = \frac{PS}{QR}$$

as long as $Q \neq 0$, $S \neq 0$, and $R \neq 0$.

EXAMPLE 4   Divide $\dfrac{3x^3y^7}{40} \div \dfrac{4x^3}{y^2}$.

Solution: $\dfrac{3x^3y^7}{40} \div \dfrac{4x^3}{y^2} = \dfrac{3x^3y^7}{40} \cdot \dfrac{y^2}{4x^3}$ Multiply by the reciprocal of $\dfrac{4x^3}{y^2}$.

$$= \frac{3\,x^3\,y^9}{160\,x^3}$$

$$= \frac{3y^9}{160}$$ Simplify.   ■

EXAMPLE 5   Divide $\dfrac{(x - 1)(x + 2)}{10} \div \dfrac{2x + 4}{5}$.

Solution: $\dfrac{(x - 1)(x + 2)}{10} \div \dfrac{2x + 4}{5} = \dfrac{(x - 1)(x + 2)}{10} \cdot \dfrac{5}{2x + 4}$ Multiply by the reciprocal of $\dfrac{2x + 4}{5}$.

$$= \frac{(x - 1)(x + 2) \cdot 5}{5 \cdot 2 \cdot 2(x + 2)}$$ Factor and multiply.

$$= \frac{x - 1}{4}$$ Simplify.   ■

The following may be used to divide rational expressions.

> **To Divide Rational Expressions**
>
> Multiply the first rational expression (the dividend) by the reciprocal of the second rational expression (the divisor).

EXAMPLE 6   Divide $\dfrac{6x + 2}{x^2 - 1} \div \dfrac{3x^2 + x}{x - 1}$.

Solution: $\dfrac{6x + 2}{x^2 - 1} \div \dfrac{3x^2 + x}{x - 1} = \dfrac{6x + 2}{x^2 - 1} \cdot \dfrac{x - 1}{3x^2 + x}$ Multiply by the reciprocal.

$$= \frac{2\,(3x + 1)(x - 1)}{(x + 1)\,(x - 1) \cdot x\,(3x + 1)}$$ Factor and multiply.

$$= \frac{2}{x(x + 1)}$$ Simplify.   ■

EXAMPLE 7   Divide $\dfrac{2x^2 - 11x + 5}{5x - 25} \div \dfrac{4x - 2}{10}$.

Solution:

$$\frac{2x^2 - 11x + 5}{5x - 25} \div \frac{4x - 2}{10} = \frac{2x^2 - 11x + 5}{5x - 25} \cdot \frac{10}{4x - 2}$$ Multiply by the reciprocal.

$$= \frac{(2x - 1)(x - 5) \cdot 2 \cdot 5}{5(x - 5) \cdot 2(2x - 1)}$$ Factor and multiply.

$$= \frac{1}{1} \quad \text{or} \quad 1$$ Simplify.   ■

MENTAL MATH

Find the following products. See Example 1.

1. $\dfrac{2}{y} \cdot \dfrac{x}{3}$

2. $\dfrac{3x}{4} \cdot \dfrac{1}{y}$

3. $\dfrac{5}{7} \cdot \dfrac{y^2}{x^2}$

4. $\dfrac{x^5}{11} \cdot \dfrac{4}{z^3}$

5. $\dfrac{9}{x} \cdot \dfrac{x}{5}$

6. $\dfrac{y}{7} \cdot \dfrac{3}{y}$

EXERCISE SET 5.2

Find each product and simplify if possible. See Example 1.

1. $\dfrac{3x}{y^2} \cdot \dfrac{7y}{4x}$

2. $\dfrac{9x^2}{y} \cdot \dfrac{4y}{3x^2}$

3. $\dfrac{8x}{2} \cdot \dfrac{x^5}{4x^2}$

4. $\dfrac{6x^2}{10x^3} \cdot \dfrac{5x}{12}$

5. $-\dfrac{5a^2b}{30a^2b^2} \cdot b^3$

6. $-\dfrac{9x^3y^2}{18xy^5} \cdot y^3$

Find each product and simplify if possible. See Example 2.

7. $\dfrac{x}{2x - 14} \cdot \dfrac{x^2 - 7x}{5}$

8. $\dfrac{4x - 24}{20x} \cdot \dfrac{5}{x - 6}$

9. $\dfrac{6x + 6}{5} \cdot \dfrac{10}{36x + 36}$

10. $\dfrac{x^2 + x}{8} \cdot \dfrac{16}{x + 1}$

Multiply the following. See Example 3.

11. $\dfrac{m^2 - n^2}{m + n} \cdot \dfrac{m}{m^2 - mn}$

12. $\dfrac{(m - n)^2}{m + n} \cdot \dfrac{m}{m^2 - mn}$

13. $\dfrac{x^2 - 25}{x^2 - 3x - 10} \cdot \dfrac{x + 2}{x}$

14. $\dfrac{a^2 + 6a + 9}{a^2 - 4} \cdot \dfrac{a + 3}{a - 2}$

Find each quotient and simplify. See Example 4.

15. $\dfrac{5x^7}{2x^5} \div \dfrac{10x}{4x^3}$

16. $\dfrac{9y^4}{6y} \div \dfrac{y^2}{3}$

17. $\dfrac{8x^2}{y^3} \div \dfrac{4x^2y^3}{6}$

18. $\dfrac{7a^2b}{3ab^2} \div \dfrac{21a^2b^2}{14ab}$

Divide the following. See Example 5.

19. $\dfrac{(x - 6)(x + 4)}{4x} \div \dfrac{2x - 12}{8x^2}$

20. $\dfrac{(x + 3)^2}{5} \div \dfrac{5x + 15}{25}$

21. $\dfrac{3x^2}{x^2 - 1} \div \dfrac{x^5}{(x + 1)^2}$

22. $\dfrac{(x + 1)}{(x + 1)(2x + 3)} \div \dfrac{20}{2x + 3}$

Divide the following. See Examples 6 and 7.

23. $\dfrac{m^2 - n^2}{m + n} \div \dfrac{m}{m^2 + nm}$

24. $\dfrac{(m - n)^2}{m + n} \div \dfrac{m^2 - mn}{m}$

25. $\dfrac{x + 2}{7 - x} \div \dfrac{x^2 - 5x + 6}{x^2 - 9x + 14}$

26. $x - 3 \div \dfrac{x^2 + 3x - 18}{x}$

27. $\dfrac{x^2 + 7x + 10}{1 - x} \div \dfrac{x^2 + 2x - 15}{x - 1}$

28. $\dfrac{a^2 - b^2}{9} \cdot \dfrac{27x^2}{3b - 3a}$

Perform the indicated operations.

29. $\dfrac{5a^2b}{30a^2b^2} \cdot \dfrac{1}{b^3}$

30. $\dfrac{9x^3y^2}{42xy^5} \cdot \dfrac{6}{x^5}$

31. $\dfrac{12x^3y}{8xy^7} \div \dfrac{7x^5y}{6x}$

32. $\dfrac{4y^2z}{3y^7z^7} \div \dfrac{12y}{6z}$

33. $\dfrac{5x - 10}{12} \div \dfrac{4x - 8}{8}$

34. $\dfrac{6x + 6}{5} \div \dfrac{3x + 3}{10}$

35. $\dfrac{x + 5}{8} \cdot \dfrac{9}{3x + 15}$

36. $\dfrac{3x + 12}{6} \cdot \dfrac{9}{2x + 8}$

37. $\dfrac{7}{6p^2 + q} \div \dfrac{14}{18p^2 + 3q}$

38. $\dfrac{5x - 10}{12} \div \dfrac{4x - 8}{8}$

39. $\dfrac{3x + 4y}{x^2 + 4xy + 4y^2} \cdot \dfrac{x + 2y}{2}$

40. $\dfrac{2a + 2b}{3} \div \dfrac{a^2 - b^2}{a - b}$

41. $\dfrac{x^2 - 9}{x^2 + 8} \div \dfrac{3 - x}{2x^2 + 16}$

42. $\dfrac{x^2 - y^2}{3x^2 + 3xy} \cdot \dfrac{3x^2 + 6x}{3x^2 - 2xy - y^2}$

43. $\dfrac{(x + 2)^2}{x - 2} \div \dfrac{x^2 - 4}{2x - 4}$

44. $\dfrac{x^2 - 4}{2y} \div \dfrac{2 - x}{6xy}$

45. $\dfrac{a^2 + 7a + 12}{a^2 + 5a + 6} \cdot \dfrac{a^2 + 8a + 15}{a^2 + 5a + 4}$

46. $\dfrac{b^2 + 2b - 3}{b^2 + b - 2} \cdot \dfrac{b^2 - 4}{b^2 + 6b + 8}$

47. $\dfrac{1}{-x - 4} \div \dfrac{x^2 - 7x}{x^2 - 3x - 28}$

48. $\dfrac{x^2 - 10x + 21}{7 - x} \div x + 3$

49. $\dfrac{x^2 - 5x - 24}{2x^2 - 2x - 24} \cdot \dfrac{4x^2 + 4x - 24}{x^2 - 10x + 16}$

50. $\dfrac{a^2 - b^2}{a} \cdot \dfrac{a + b}{a^2 + ab}$

51. $x - 5 \div \dfrac{5 - x}{x^2 + 2}$

52. $\dfrac{2x^2 + 3xy + y^2}{x^2 - y^2} \div \dfrac{1}{2x + 2y}$

53. $\dfrac{x^2 - y^2}{x^2 - 2xy + y^2} \cdot \dfrac{y - x}{x + y}$

54. $\dfrac{x + 3}{x^2 - 9} \cdot \dfrac{x^2 - 8x + 15}{5x}$

55. $\dfrac{a^2 + ac + ba + bc}{a - b} \div \dfrac{a + c}{a + b}$

56. $\dfrac{x^2 + 2x - xy - 2y}{x^2 - y^2} \div \dfrac{2x + 4}{x + y}$

57. $\dfrac{3x^2 + 8x + 5}{x^2 + 8x + 7} \cdot \dfrac{x + 7}{x^2 + 4}$

58. $\dfrac{16x^2 + 2x}{16x^2 + 10x + 1} \cdot \dfrac{1}{4x^2 + 2x}$

59. Find the area of the following rectangle.

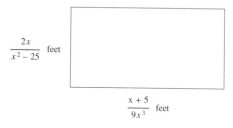

60. Find the area of the following square.

61. Find the quotient of $\dfrac{x^2 - 9}{2x}$ and $\dfrac{x + 3}{8x^4}$.

62. Find the quotient of $\dfrac{4x^2 + 4x + 1}{4x + 2}$ and $\dfrac{4x + 2}{16}$.

A Look Ahead

Perform the following operations. See the following example.

  EXAMPLE Perform the indicated operations.

$$\frac{15x^2 - x - 6}{12x^3} \cdot \frac{4x}{9 - 25x^2} \div \frac{x}{3x - 2}$$

  Solution: $\dfrac{15x^2 - x - 6}{12x^3} \cdot \dfrac{4x}{9 - 25x^2} \div \dfrac{x}{3x - 2} = \left(\dfrac{15x^2 - x - 6}{12x^3} \cdot \dfrac{4x}{9 - 25x^2}\right) \cdot \dfrac{3x - 2}{x}$

$$= \frac{(3x - 2)(5x + 3) \cdot 4x(3x - 2)}{12x^3(3 - 5x)(3 + 5x) \cdot x}$$

$$= \frac{(3x - 2)^2}{3x^3(3 - 5x)} \quad \blacksquare$$

63. $\left(\dfrac{x^2 - y^2}{x^2 + y^2} \div \dfrac{x^2 - y^2}{3x}\right) \cdot \dfrac{x^2 + y^2}{6}$

64. $\left(\dfrac{x^2 - 9}{x^2 - 1} \cdot \dfrac{x^2 + 2x + 1}{2x^2 + 9x + 9}\right) \div \dfrac{2x + 3}{1 - x}$

65. $\left(\dfrac{2a + b}{b^2} \cdot \dfrac{3a^2 - 2ab}{ab + 2b^2}\right) \div \dfrac{a^2 - 3ab + 2b^2}{5ab - 10b^2}$

66. $\left(\dfrac{x^2y^2 - xy}{4x - 4y} \div \dfrac{3y - 3x}{8x - 8y}\right) \cdot \dfrac{y - x}{8}$

Writing in Mathematics

67. List the basic steps necessary to multiply two rational expressions.

68. List the basic steps necessary to divide two rational expressions.

Skill Review

Perform each operation. See Section 1.3.

69. $\dfrac{1}{5} + \dfrac{4}{5}$

70. $\dfrac{3}{15} + \dfrac{6}{15}$

71. $\dfrac{9}{9} - \dfrac{19}{9}$

72. $\dfrac{4}{3} - \dfrac{8}{3}$

73. $\dfrac{6}{5} + \left(\dfrac{1}{5} - \dfrac{8}{5}\right)$

74. $-\dfrac{3}{2} + \left(\dfrac{1}{2} - \dfrac{3}{2}\right)$

5.3
Adding and Subtracting Rational Expressions with Common Denominators and Least Common Denominator

OBJECTIVES

Tape 17

1 Add and subtract rational expressions with common denominators.

2 Find the least common denominator of a list of rational expressions.

3 Write a rational expression as an equivalent rational expression whose denominator is given.

---

1 Like multiplication and division, addition and subtraction of rational expressions is similar to addition and subtraction of rational numbers. For example, to add fractions with common denominators such as $\dfrac{6}{5}$ and $\dfrac{2}{5}$, add the numerators and write the sum over the common denominator 5.

$$\frac{6}{5} + \frac{2}{5} = \frac{6 + 2}{5} = \frac{8}{5}$$

To subtract two fractions with common denominators, subtract the numerators and write the result over the common denominator.

$$\frac{6}{13} - \frac{5}{13} = \frac{6 - 5}{13} = \frac{1}{13}$$

Rational expressions with common denominators are added and subtracted in the same manner.

Adding and Subtracting Rational Expressions with Common Denominators

If P, Q, and R are polynomials, then

$$\frac{P}{R} + \frac{Q}{R} = \frac{P + Q}{R} \quad \text{and} \quad \frac{P}{R} - \frac{Q}{R} = \frac{P - Q}{R}$$

as long as $R \neq 0$.

EXAMPLE 1   Add $\dfrac{5m}{2n} + \dfrac{m}{2n}$.

Solution: $\dfrac{5m}{2n} + \dfrac{m}{2n} = \dfrac{5m + m}{2n}$ Add the numerators.

$= \dfrac{6m}{2n}$ Write the sum over the common denominator.

$= \dfrac{3m}{n}$ Simplify.   ∎

EXAMPLE 2   Subtract $\dfrac{2y}{2y - 7} - \dfrac{7}{2y - 7}$.

Solution: $\dfrac{2y}{2y - 7} - \dfrac{7}{2y - 7} = \dfrac{2y - 7}{2y - 7}$ Subtract the numerators.

$= \dfrac{1}{1}$ or   1   Simplify.   ∎

EXAMPLE 3   Subtract $\dfrac{3x^2 + 2x}{x - 1} - \dfrac{10x - 5}{x - 1}$.

Solution: $\dfrac{3x^2 + 2x}{x - 1} - \dfrac{10x - 5}{x - 1} = \dfrac{3x^2 + 2x - (10x - 5)}{x - 1}$ Subtract the numerators. Notice the parentheses.

$= \dfrac{3x^2 + 2x - 10x + 5}{x - 1}$ Use the distributive property.

$= \dfrac{3x^2 - 8x + 5}{x - 1}$ Combine like terms.

$= \dfrac{(x - 1)(3x - 5)}{x - 1}$ Factor.

$= 3x - 5$ Simplify by dividing common factors.   ∎

HELPFUL HINT

Notice how the numerators have been subtracted in Example 3.

This − sign applies to the entire numerator of $10x - 5$.   Parentheses are inserted here to indicate this.

$$\dfrac{3x^2 + 2x}{x - 1} - \dfrac{10x - 5}{x - 1} = \dfrac{3x^2 + 2x - (10x - 5)}{x - 1}$$

---

2   To add and subtract fractions with **unlike** denominators, we first find a lowest common denominator LCD and write all fractions as equivalent fractions with the common denominator.

For example, suppose we add $\frac{8}{3}$ and $\frac{2}{5}$. The LCD of 3 and 5 is 15, since 15 is the smallest number that both 3 and 5 divide into evenly. Rewrite each fraction so that its denominator is 15.

$$\frac{8}{3} + \frac{2}{5} = \frac{8(5)}{3(5)} + \frac{2(3)}{5(3)} = \frac{40}{15} + \frac{6}{15} = \frac{40 + 6}{15} = \frac{46}{15}$$

To add or subtract rational expressions, the expressions must also have a common denominator. The **least common denominator LCD of a list of rational expressions** is a polynomial of least degree whose factors include all the factors in the denominators.

To Find the Least Common Denominator (LCD)

Step 1   Factor each denominator completely.

Step 2   The least common denominator LCD is the product of all unique factors formed in step 1, each raised to a power equal to the greatest number of times that the factor appears in any one factored denominator.

EXAMPLE 4   Find the LCD for each pair.

a. $\frac{1}{8}, \frac{3}{22}$ **b.** $\frac{7}{5x}, \frac{6}{15x^2}$

Solution: **a.** Start by finding the prime factorization of each denominator.

$$8 = 2 \cdot 2 \cdot 2 = 2^3 \quad \text{and} \quad 22 = 2 \cdot 11$$

Next, write the product of all the unique factors, each raised to a power equal to the greatest number of times that the factor appears.

The greatest number of times that the factor 2 appears is 3.
The greatest number of times that the factor 11 appears is 1.

$$\text{LCD} = 2^3 \cdot 11^1 = 8 \cdot 11 = 88$$

b. Factor each denominator.

$$5x = 5 \cdot x \quad \text{and} \quad 15x^2 = 3 \cdot 5 \cdot x^2$$

The greatest number of times that the factor 5 appears is 1.
The greatest number of times that the factor 3 appears is 1.
The greatest number of times that the factor x appears is 2.

$$\text{LCD} = 3^1 \cdot 5^1 \cdot x^2 = 15x^2 \quad \blacksquare$$

EXAMPLE 5   Find the LCD of $\frac{7x}{x + 2}$ and $\frac{5x^2}{x - 2}$.

Solution: The denominators $x + 2$ and $x - 2$ do not factor any further. The factor $x + 2$ appears once and the factor $x - 2$ appears once.

$$\text{LCD} = (x + 2)(x - 2) \quad \blacksquare$$

EXAMPLE 6 Find the LCD of $\dfrac{6m^2}{3m + 15}$ and $\dfrac{2}{(m + 5)^2}$.

Solution: Factor each denominator.

$$3m + 15 = 3(m + 5)$$

$(m + 5)^2$ is already factored

The greatest number of times that the factor 3 appears is 1.

The greatest number of times that the factor $m + 5$ appears in any one denominator is 2.

$$LCD = 3(m + 5)^2 \quad \blacksquare$$

EXAMPLE 7 Find the LCD of $\dfrac{t - 10}{t^2 - t - 6}$ and $\dfrac{t + 5}{t^2 + 3t + 2}$.

Solution: Start by factoring each denominator.

$$t^2 - t - 6 = (t - 3)(t + 2)$$
$$t^2 + 3t + 2 = (t + 1)(t + 2)$$
$$LCD = (t - 3)(t + 2)(t + 1) \quad \blacksquare$$

EXAMPLE 8 Find the LCD for $\dfrac{2}{x - 2}$ and $\dfrac{10}{2 - x}$.

Solution: The denominators $x - 2$ and $2 - x$ are opposites. In other words, $2 - x = -1(x - 2)$. Use $x - 2$ or $2 - x$ as the LCD.

$$LCD = x - 2 \quad \text{or} \quad LCD = 2 - x \quad \blacksquare$$

3 Next we practice writing a rational expression as an equivalent rational expression with a given indicated denominator. To do this, we multiply the numerator and the denominator of a rational expression by the same appropriate expression. This is equivalent to multiplying by 1.

EXAMPLE 9 Write $\dfrac{4b}{9a}$ as an equivalent fraction with the given denominator.

$$\frac{4b}{9a} = \frac{}{27a^2b}$$

Solution: Ask yourself: "What do we multiply $9a$ by to get $27a^2b$?" The answer is $3ab$, since $9a(3ab) = 27a^2b$. Multiply the numerator and denominator by $3ab$.

$$\frac{4b}{9a} = \frac{4b\ (3ab)}{9a\ (3ab)} = \frac{12ab^2}{27a^2b} \quad \blacksquare$$

EXAMPLE 10 Write the rational expression as an equivalent rational expression with the given denominator.

$$\frac{5}{x^2 - 4} = \frac{}{(x - 2)(x + 2)(x - 4)}$$

Solution: First, factor the denominator $x^2 - 4$.

$$x^2 - 4 = (x + 2)(x - 2)$$

If we multiply the original denominator $(x + 2)(x - 2)$ by $x - 4$, the result is the new denominator $(x + 2)(x - 2)(x - 4)$. Thus, multiply the numerator and the denominator by $x - 4$.

$$\frac{5}{x^2 - 4} = \frac{5}{(x - 2)(x + 2)} = \frac{5 \cdot x - 4}{(x - 2)(x + 2) \cdot (x - 4)}$$

$$= \frac{5x - 20}{(x - 2)(x + 2)(x - 4)} \quad \blacksquare$$

MENTAL MATH

Perform the indicated operation.

1. $\dfrac{2}{3} + \dfrac{1}{3}$

2. $\dfrac{5}{11} + \dfrac{1}{11}$

3. $\dfrac{3x}{9} + \dfrac{4x}{9}$

4. $\dfrac{3y}{8} + \dfrac{2y}{8}$

5. $\dfrac{8}{9} - \dfrac{7}{9}$

6. $-\dfrac{4}{12} - \dfrac{3}{12}$

7. $\dfrac{7}{5} - \dfrac{10y}{5}$

8. $\dfrac{12x}{7} - \dfrac{4x}{7}$

EXERCISE SET 5.3

Add or subtract as indicated. Write the result in lowest terms. See Examples 1 and 2.

1. $\dfrac{a}{13} + \dfrac{9}{13}$

2. $\dfrac{x + 1}{7} + \dfrac{6}{7}$

3. $\dfrac{9}{3 + y} + \dfrac{y + 1}{3 + y}$

4. $\dfrac{9}{y + 9} + \dfrac{y}{y + 9}$

5. $\dfrac{4m}{3n} + \dfrac{5m}{3n}$

6. $\dfrac{3p}{2} + \dfrac{11p}{2}$

7. $\dfrac{2x + 1}{x - 3} + \dfrac{3x + 6}{x - 3}$

8. $\dfrac{4p - 3}{2p + 7} + \dfrac{3p + 8}{2p + 7}$

9. $\dfrac{7}{8} - \dfrac{3}{8}$

10. $\dfrac{4}{5} - \dfrac{13}{5}$

11. $\dfrac{4m}{m - 6} - \dfrac{24}{m - 6}$

12. $\dfrac{8y}{y - 2} - \dfrac{16}{y - 2}$

Add or subtract as indicated. Write the result in lowest terms. See Example 3.

13. $\dfrac{2x^2}{x - 5} - \dfrac{25 + x^2}{x - 5}$

14. $\dfrac{6x^2}{2x - 5} - \dfrac{25 + 2x^2}{2x - 5}$

15. $\dfrac{-3x^2 - 4}{x - 4} - \dfrac{12 - 4x^2}{x - 4}$

16. $\dfrac{7x^2 - 9}{2x - 5} - \dfrac{16 + 3x^2}{2x - 5}$

17. $\dfrac{2x + 3}{x + 1} - \dfrac{x + 2}{x + 1}$

18. $\dfrac{1}{x^2 - 2x - 15} - \dfrac{4 - x}{x^2 - 2x - 15}$

Find the LCD for the following lists of rational expressions. See Examples 4 through 8.

19. $\dfrac{2}{3}, \dfrac{4}{33}$

20. $\dfrac{8}{20}, \dfrac{4}{15}$

21. $\dfrac{19}{2x}, \dfrac{5}{4x^3}$

22. $\dfrac{17x}{4y^5}, \dfrac{2}{8y}$

23. $\dfrac{9}{8x}, \dfrac{3}{2x + 4}$

24. $\dfrac{1}{6y}, \dfrac{3x}{4y + 12}$

25. $\dfrac{1}{3x + 3}, \dfrac{8}{2x^2 + 4x + 2}$

26. $\dfrac{19x + 5}{4x - 12}, \dfrac{3}{2x^2 - 12x + 18}$

27. $\dfrac{5}{x - 8}, \dfrac{3}{8 - x}$

28. $\dfrac{2x + 5}{3x - 7}, \dfrac{5}{7 - 3x}$

Rewrite each rational expression as an equivalent rational expression whose denominator is the given polynomial. See Examples 9 and 10.

29. $\dfrac{3}{2x}$; $4x^2$

30. $\dfrac{3}{9y^5}$; $72y^9$

31. $\dfrac{9a + 2}{5a + 10}$; $5b(a + 2)$

32. $\dfrac{5 + y}{2x^2 + 10}$; $4(x^2 + 5)$

33. $\dfrac{x}{x^3 + 6x^2 + 8x}$; $x(x + 4)(x + 2)(x + 1)$

34. $\dfrac{5x}{x^2 + 2x - 3}$; $(x - 1)(x - 5)(x + 3)$

Perform the indicated operations. Simplify the result if possible.

35. $\dfrac{3}{x^3} + \dfrac{9}{x^3}$

36. $\dfrac{5}{xy} + \dfrac{8}{xy}$

37. $\dfrac{5}{x + 4} - \dfrac{10}{x + 4}$

38. $\dfrac{4}{2x + 1} - \dfrac{8}{2x + 1}$

39. $\dfrac{x}{x + y} - \dfrac{2}{x + y}$

40. $\dfrac{y + 1}{y + 2} - \dfrac{3}{y + 2}$

41. $\dfrac{8x}{2x + 5} + \dfrac{20}{2x + 5}$

42. $\dfrac{12y - 5}{3y - 1} + \dfrac{1}{3y - 1}$

43. $\dfrac{5x + 4}{x - 1} - \dfrac{2x + 7}{x - 1}$

44. $\dfrac{x^2 + 9x}{x + 7} - \dfrac{4x + 14}{x + 7}$

45. $\dfrac{a}{a^2 + 2a - 15} - \dfrac{3}{a^2 + 2a - 15}$

46. $\dfrac{3y}{y^2 + 3y - 10} - \dfrac{6}{y^2 + 3y - 10}$

47. $\dfrac{2x + 3}{x^2 - x - 30} - \dfrac{x - 2}{x^2 - x - 30}$

48. $\dfrac{3x - 1}{x^2 + 5x - 6} - \dfrac{2x - 7}{x^2 + 5x - 6}$

Find the LCD for the following lists of rational expressions.

49. $\dfrac{7}{2x}, \dfrac{8}{5x}$

50. $\dfrac{13}{7x}, \dfrac{12}{11x}$

51. $\dfrac{1}{9x^3y}, \dfrac{5}{12xy}$

52. $\dfrac{13a^3}{6ab^2}, \dfrac{9a^5}{18ab^3}$

53. $\dfrac{4 + x}{8x^2(x - 1)^2}, \dfrac{17}{10x^3(x - 1)}$

54. $\dfrac{2x + 3}{9x(x + 2)}, \dfrac{9x + 5}{12(x + 2)^2}$

55. $\dfrac{9x + 1}{2x + 1}, \dfrac{3x - 5}{2x - 1}$

56. $\dfrac{5}{4x - 2}, \dfrac{7}{4x + 2}$

57. $\dfrac{5x + 1}{2x^2 + 7x - 4}, \dfrac{3x}{2x^2 + 5x - 3}$

58. $\dfrac{4}{x^2 + 4x + 3}, \dfrac{4x - 2}{x^2 + 10x + 21}$

Write each rational expression with a denominator of $x - 2$.

59. $\dfrac{5}{2 - x}$

60. $\dfrac{8y}{2 - x}$

61. $-\dfrac{7 + x}{2 - x}$

62. $\dfrac{x - 3}{-(x - 2)}$

Write each rational expression as an equivalent expression whose denominator is the given polynomial.

63. $\dfrac{6}{3a}$; $12ab^2$

64. $\dfrac{17a}{4y^2x}$; $32y^3x^2z$

65. $\dfrac{9}{x + 3}$; $2(x + 3)$

66. $\dfrac{4x + 1}{3x + 6}$; $3y(x + 2)$

67. $\dfrac{9y - 1}{15x^2 - 30}$; $30x^2 - 60$

68. $\dfrac{6}{x^2 - 9}$; $(x + 3)(x - 3)(x + 2)$

69. $\dfrac{1}{x^2 - 16}$; $x(x - 4)^2(x + 4)$

70. $\dfrac{-3}{x^2 - 9}$; $(x - 3)(x + 3)^2$

71. $\dfrac{5}{2x^2 - 9x - 5}$; $3x(2x + 1)(x - 7)(x - 5)$

72. $\dfrac{x - 9}{3x^2 + 10x + 3}$; $x(x + 3)(x + 5)(3x + 1)$

73. A square-shaped pasture has a side of length $\dfrac{5}{x - 2}$ meters. Express its perimeter as a rational expression.

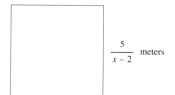

$\dfrac{5}{x - 2}$ meters

74. The following trapezoid has sides of indicated length. Find its perimeter.

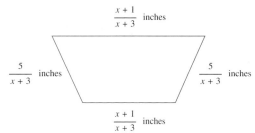

$\dfrac{x + 1}{x + 3}$ inches

$\dfrac{5}{x + 3}$ inches

$\dfrac{5}{x + 3}$ inches

$\dfrac{x + 1}{x + 3}$ inches

Writing in Mathematics

75. Describe the process for adding and subtracting two rational expressions with common denominators.

76. Write some instructions to help a friend who is having difficulty finding the LCD of two rational expressions.

77. Explain the similarities between subtracting $\dfrac{3}{8}$ from $\dfrac{7}{8}$ and subtracting $\dfrac{6}{x + 3}$ from $\dfrac{9}{x + 3}$.

Skill Review

Solve the following quadratic equations by factoring. See Section 4.6.

78. $x(x - 3) = 0$

79. $2x(x + 5) = 0$

80. $(2x + 1)(x - 6) = 0$

81. $(4x - 1)(2x + 3) = 0$

82. $x^2 + 6x + 5 = 0$

83. $x^2 - 6x + 5 = 0$

5.4
Adding and Subtracting Rational Expressions with Unlike Denominators

Tape 17

| **OBJECTIVE** | | |

1   Add and subtract rational expressions with unlike denominators.

1   We add and subtract rational expressions with unlike denominators the same way we add and subtract rational numbers with unlike denominators. The steps are as follows:

> **Adding and Subtracting Rational Expressions with Unlike Denominators**
>
> *Step 1*   Find the LCD of all denominators.
>
> *Step 2*   Rewrite each rational expression as an equivalent expression whose denominator is the LCD found in step 1.
>
> *Step 3*   Add or subtract numerators and write the sum or difference over the common denominator.
>
> *Step 4*   Write the rational expression in lowest terms.

EXAMPLE 1   Perform the indicated operation.

$$\textbf{a. } \frac{a}{4} - \frac{2a}{8} \quad \textbf{b. } \frac{3}{10x^2} + \frac{7}{25x}$$

Solution: **a.** Since $4 = 2^2$ and $8 = 2^3$, the LCD $= 2^3 = 8$. Write each fraction as an equivalent fraction with the denominator 8; then subtract.

$$\frac{a}{4} - \frac{2a}{8} = \frac{a\,(2)}{4\,(2)} - \frac{2a}{8} = \frac{2a}{8} - \frac{2a}{8} = \frac{2a - 2a}{8} = \frac{0}{8} = 0$$

b. Since $10x^2 = 2 \cdot 5 \cdot x \cdot x$ and $25x = 5 \cdot 5 \cdot x$, the LCD $= 2 \cdot 5^2 \cdot x^2 = 50x^2$. Write each fraction as an equivalent fraction with a denominator of $50x^2$.

$$\frac{3}{10x^2} + \frac{7}{25x} = \frac{3\,(5)}{10x^2\,(5)} + \frac{7\,(2x)}{25x\,(2x)}$$

$$= \frac{15}{50x^2} + \frac{14x}{50x^2}$$

$$= \frac{15 + 14x}{50x^2} \qquad \text{Add.} \quad \blacksquare$$

EXAMPLE 2   Subtract $\dfrac{6x}{x^2 - 4} - \dfrac{3}{x + 2}$.

Solution: Since $x^2 - 4 = (x + 2)(x - 2)$,

$$\text{LCD} = (x - 2)(x + 2)$$

Write equivalent expressions with the LCD as denominators.

$$\frac{6x}{x^2 - 4} - \frac{3}{x + 2} = \frac{6x}{(x - 2)(x + 2)} - \frac{3\,(x - 2)}{(x + 2)\,(x - 2)}$$

$$= \frac{6x - 3(x - 2)}{(x + 2)(x - 2)} \qquad \text{Subtract.}$$

$$= \frac{6x - 3x + 6}{(x + 2)(x - 2)} \qquad \text{Use the distributive property.}$$

$$= \frac{3x + 6}{(x + 2)(x - 2)} \qquad \text{Combine like terms in the numerator.}$$

Next, factor the numerator to see if this rational expression can be simplified.

$$= \frac{3x + 6}{(x + 2)\,(x - 2)}$$

$$= \frac{3\,(x + 2)}{(x + 2)\,(x - 2)} \qquad \text{Factor.}$$

$$= \frac{3}{x - 2} \qquad \text{Simplify.} \quad \blacksquare$$

EXAMPLE 3   Add $\dfrac{2}{3t} + \dfrac{5}{t+1}$.

Solution: The LCD is $3t(t+1)$. Write each rational expression as an equivalent rational expression with a denominator of $3t(t+1)$.

$$\frac{2}{3t} + \frac{5}{t+1} = \frac{2\,(t+1)}{3t\,(t+1)} + \frac{5\,(3t)}{(t+1)\,(3t)}$$

$$= \frac{2(t+1) + 5(3t)}{3t(t+1)} \qquad \text{Add.}$$

$$= \frac{2t + 2 + 15t}{3t(t+1)} \qquad \text{Apply the distributive property.}$$

$$= \frac{17t + 2}{3t(t+1)} \qquad \text{Combine like terms in the numerator.} \qquad\blacksquare$$

EXAMPLE 4   Find $\dfrac{7}{x-3} - \dfrac{9}{3-x}$.

Solution: To find a common denominator, notice that $x-3$ and $3-x$ are opposites. In other words, $3-x = -1(x-3)$. Write the denominator $3-x$ as $-1(x-3)$ and simplify.

$$\frac{7}{x-3} - \frac{9}{3-x} = \frac{7}{x-3} - \frac{9}{-1(x-3)}$$

$$= \frac{7}{x-3} + \frac{9}{x-3} \qquad \text{Simplify.}$$

$$= \frac{16}{x-3} \qquad \text{Add.} \qquad\blacksquare$$

EXAMPLE 5   Add $1 + \dfrac{m}{m+1}$.

Solution: Recall that 1 is the same as $\dfrac{1}{1}$. The LCD of 1 and $m+1$ is $m+1$.

$$1 + \frac{m}{m+1} = \frac{1}{1} + \frac{m}{m+1} \qquad \text{Write 1 as } \frac{1}{1}.$$

$$= \frac{1\,(m+1)}{1\,(m+1)} + \frac{m}{m+1} \qquad \begin{array}{l}\text{Multiply the numerator and the} \\ \text{denominator of } \frac{1}{1} \text{ by } m+1.\end{array}$$

$$= \frac{m + 1 + m}{m+1} \qquad \text{Add.}$$

$$= \frac{2m + 1}{m+1} \qquad \text{Combine like terms.} \qquad\blacksquare$$

EXAMPLE 6 Subtract $\dfrac{3}{2x^2 + x} - \dfrac{2x}{6x + 3}$.

Solution: First, factor the denominators.

$$\frac{3}{2x^2 + x} - \frac{2x}{6x + 3} = \frac{3}{x(2x + 1)} - \frac{2x}{3(2x + 1)}$$

The LCD is $3x(2x + 1)$. Write equivalent expressions with denominators of $3x(2x + 1)$.

$$= \frac{3\,(3)}{x(2x + 1)\,(3)} - \frac{2x\,(x)}{3(2x + 1)\,(x)}$$

$$= \frac{9 - 2x^2}{3x(2x + 1)} \qquad \text{Subtract.} \qquad \blacksquare$$

EXAMPLE 7 Add $\dfrac{2x}{x^2 + 2x + 1} + \dfrac{x}{x^2 - 1}$.

Solution: First, factor the denominators.

$$\frac{2x}{x^2 + 2x + 1} + \frac{x}{x^2 - 1} = \frac{2x}{(x + 1)(x + 1)} + \frac{x}{(x + 1)(x - 1)}$$

Write the rational expressions as equivalent expressions with denominators of $(x + 1)(x + 1)(x - 1)$, the LCD.

$$= \frac{2x\,(x - 1)}{(x + 1)(x + 1)\,(x - 1)} + \frac{x\,(x + 1)}{(x + 1)(x - 1)\,(x + 1)}$$

$$= \frac{2x(x - 1) + x(x + 1)}{(x + 1)^2(x - 1)} \qquad \text{Add.}$$

$$= \frac{2x^2 - 2x + x^2 + x}{(x + 1)^2(x - 1)} \qquad \text{Apply the distributive property.}$$

$$= \frac{3x^2 - x}{(x + 1)^2(x - 1)} \quad \text{or} \quad \frac{x(3x - 1)}{(x + 1)^2(x - 1)}$$

The numerator was factored as a last step to see if the rational expression could be simplified. $\blacksquare$

EXERCISE SET 5.4

Perform the indicated operations. See Example 1.

1. $\dfrac{4}{2x} + \dfrac{9}{3x}$

2. $\dfrac{15}{7a} + \dfrac{8}{6a}$

3. $\dfrac{15a}{b} + \dfrac{6b}{5}$

4. $\dfrac{4c}{d} - \dfrac{8x}{5}$

5. $\dfrac{3}{x} + \dfrac{5}{2x^2}$

6. $\dfrac{14}{3x^2} + \dfrac{6}{x}$

Perform the indicated operations. See Examples 2 and 3.

7. $\dfrac{6}{x+1} + \dfrac{9}{2x+2}$

8. $\dfrac{8}{x+4} - \dfrac{3}{3x+12}$

9. $\dfrac{15}{2x-4} + \dfrac{x}{x^2-4}$

10. $\dfrac{3}{x+2} - \dfrac{1}{x^2-4}$

11. $\dfrac{3}{4x} + \dfrac{8}{x-2}$

12. $\dfrac{x}{x+1} + \dfrac{3}{x-1}$

13. $\dfrac{5}{y^2} - \dfrac{y}{2y+1}$

14. $\dfrac{x}{4x-3} - \dfrac{3}{8x-6}$

Add or subtract as indicated. See Example 4.

15. $\dfrac{6}{x-3} + \dfrac{8}{3-x}$

16. $\dfrac{9}{x-3} + \dfrac{9}{3-x}$

17. $\dfrac{8}{x^2-1} - \dfrac{7}{1-x^2}$

18. $\dfrac{9}{25x^2-1} + \dfrac{7}{1-25x^2}$

19. $\dfrac{x}{x^2-4} - \dfrac{2}{4-x^2}$

20. $\dfrac{5}{2x-6} - \dfrac{3}{6-2x}$

Add or subtract as indicated. See Example 5.

21. $\dfrac{5}{x} + 2$

22. $\dfrac{7}{x^2} - 5x$

23. $\dfrac{5}{x-2} + 6$

24. $\dfrac{6y}{y+5} + 1$

25. $\dfrac{y+2}{y+3} - 2$

26. $\dfrac{7}{2x-3} - 3$

Perform indicated operations. See Examples 6 and 7.

27. $\dfrac{5x}{x+2} - \dfrac{3x-4}{x+2}$

28. $\dfrac{7x}{x-3} - \dfrac{4x+9}{x-3}$

29. $\dfrac{3x^4}{x} - \dfrac{4x^2}{x^2}$

30. $\dfrac{5x}{6} + \dfrac{15x^2}{2}$

31. $\dfrac{1}{x+3} - \dfrac{1}{(x+3)^2}$

32. $\dfrac{5x}{(x-2)^2} - \dfrac{3}{x-2}$

33. $\dfrac{4}{5b} + \dfrac{1}{b-1}$

34. $\dfrac{1}{y+5} + \dfrac{2}{3y}$

35. $\dfrac{2}{m} + 1$

36. $\dfrac{6}{x} - 1$

37. $\dfrac{6}{1-2x} - \dfrac{4}{2x-1}$

38. $\dfrac{10}{3n-4} - \dfrac{5}{4-3n}$

39. $\dfrac{7}{(x+1)(x-1)} + \dfrac{8}{(x+1)^2}$

40. $\dfrac{5x+2}{(x+1)(x+5)} - \dfrac{2}{x+5}$

41. $\dfrac{x}{x^2-1} - \dfrac{2}{x^2-2x+1}$

42. $\dfrac{x}{x^2-4} - \dfrac{5}{x^2-4x+4}$

43. $\dfrac{3a}{2a+6} - \dfrac{a-1}{a+3}$

44. $\dfrac{1}{x+y} - \dfrac{y}{x^2-y^2}$

45. $\dfrac{5}{2-x} + \dfrac{x}{2x-4}$

46. $\dfrac{-1}{a-2} + \dfrac{4}{4-2a}$

47. $\dfrac{-7}{y^2-3y+2} - \dfrac{2}{y-1}$

48. $\dfrac{2}{x^2+4x+4} + \dfrac{1}{x+2}$

49. $\dfrac{13}{x^2-5x+6} - \dfrac{5}{x-3}$

50. $\dfrac{27}{y^2-81} + \dfrac{3}{2(y+9)}$

51. $\dfrac{8}{(x+2)(x-2)} + \dfrac{4}{(x+2)(x-3)}$

52. $\dfrac{5}{6x^2(x+2)} + \dfrac{4x}{x(x+2)^2}$

53. $\dfrac{5}{9x^2-4} + \dfrac{2}{3x-2}$

54. $\dfrac{4}{x^2-x-6} + \dfrac{x}{x^2+5x+6}$

55. $\dfrac{x+8}{x^2-5x-6} + \dfrac{x+1}{x^2-4x-5}$

56. $\dfrac{x}{x^2+12x+20} - \dfrac{1}{x^2+8x-20}$

Perform the indicated operations. Addition, subtraction, multiplication, and division of rational expressions are included here.

57. $\dfrac{15x}{x+8} \cdot \dfrac{2x+16}{3x}$

58. $\dfrac{9z+5}{15} \cdot \dfrac{5z}{81z^2-25}$

59. $\dfrac{8x+7}{3x+5} - \dfrac{2x-3}{3x+5}$

60. $\dfrac{2z^2}{4z-1} - \dfrac{z-2z^2}{4z-1}$

61. $\dfrac{5a+10}{18} \div \dfrac{a^2-4}{10a}$

62. $\dfrac{9}{x^2-1} \div \dfrac{12}{3x+3}$

63. $\dfrac{5}{x^2-3x+2} + \dfrac{1}{x-2}$

64. $\dfrac{4}{2x^2+5x-3} + \dfrac{2}{x+3}$

65. A board of length $\dfrac{3}{x+4}$ inches was cut into two pieces. If one piece is $\dfrac{1}{x-4}$ inches, express the length of the other board as a rational expression.

66. Two angles are said to be complementary if their sum is $90°$. If one angle measures $\dfrac{40}{x}$ degrees, find the measure of its complement.

67. The length of a rectangle is $\dfrac{3}{y-5}$ feet, while its width is $\dfrac{2}{y}$ feet. Find its perimeter.

68. A triangle has sides of the indicated lengths. Express the perimeter of the triangle as a rational expression.

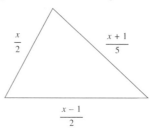

A Look Ahead

Add or subtract as indicated. See the following example.

EXAMPLE Perform the indicated operations:

$$\frac{3}{x^2-16} + \frac{2}{x^2-9x+20} - \frac{4}{x^2-x-20}$$

Solution: Factor the denominators.

$$= \frac{3}{(x+4)(x-4)} + \frac{2}{(x-4)(x-5)} - \frac{4}{(x-5)(x+4)}$$

Write each expression with a LCD of $(x-4)(x+4)(x-5)$.

$$= \frac{3(x-5)}{(x-4)(x+4)(x-5)} + \frac{2(x+4)}{(x-4)(x+4)(x-5)} - \frac{4(x-4)}{(x-4)(x+4)(x-5)}$$

$$= \frac{3(x-5) + 2(x+4) - 4(x-4)}{(x-4)(x+4)(x-5)}$$

$$= \frac{3x-15+2x+8-4x+16}{(x-4)(x+4)(x-5)}$$

$$= \frac{x+9}{(x-4)(x+4)(x-5)} \quad ■$$

69. $\dfrac{5}{x^2-4} + \dfrac{2}{x^2-4x+4} - \dfrac{3}{x^2-x-6}$

70. $\dfrac{8}{x^2+6x+5} - \dfrac{3x}{x^2+4x-5} + \dfrac{2}{x^2-1}$

71. $\dfrac{9}{x^2+9x+14} - \dfrac{3x}{x^2+10x+21} + \dfrac{4}{x^2+5x+6}$

72. $\dfrac{10}{x^2-3x-4} - \dfrac{8}{x^2+6x+5} - \dfrac{9}{x^2+x-20}$

73. $\dfrac{5 + x}{x^3 - 27} + \dfrac{x}{x^3 + 3x^2 + 9x}$

74. $\dfrac{x + 5}{x^3 + 1} - \dfrac{3}{2x^2 - 2x + 2}$

Skill Review

Factor the following. See Sections 4.1 and 4.4.

75. $x^3 - 1$

76. $8y^3 + 1$

77. $125z^3 + 8$

78. $a^3 - 27$

79. $xy + 2x + 3y + 6$

80. $x^2 - x + xy - y$

5.5
Complex Fractions

OBJECTIVES

Tape 18

1   Define complex fractions.

2   Simplify complex fractions using method 1.

3   Simplify complex fractions using method 2.

1   A fraction whose numerator or denominator or both numerator and denominator contain fractions is called a **complex fraction.**

Examples of Complex Fractions

$$\frac{4}{2 - \dfrac{1}{2}} \qquad \frac{\dfrac{3}{2}}{\dfrac{4}{7} - x} \qquad \frac{\dfrac{1}{x + 2}}{x + 2 - \dfrac{1}{x}}$$

The parts of a complex fraction are

$$\left.\begin{array}{c} \dfrac{x}{y + 2} \\[2ex] 7 + \dfrac{1}{y} \end{array}\right\}$$

← Numerator of complex fraction.
← Main fraction bar.
← Denominator of complex fraction.

2   In this section, two methods of simplifying complex fractions are presented. The first method presented uses the fact that the main fraction bar indicates division.

Method 1: To Simplify a Complex Fraction

Step 1   Add or subtract fractions in the numerator or denominator so that the numerator is a single fraction and the denominator is a single fraction.

Step 2   Perform the indicated division by multiplying the numerator of the complex fraction by the reciprocal of the denominator of the complex fraction.

Step 3   Simplify if possible.

EXAMPLE 1 Simplify the complex fraction $\dfrac{\frac{5}{8}}{\frac{2}{3}}$.

Solution: Since the numerator and denominator of the complex fraction are already single fractions, proceed to step 2: multiply the numerator $\dfrac{5}{8}$ by the reciprocal of the denominator $\dfrac{2}{3}$.

$$\frac{\frac{5}{8}}{\frac{2}{3}} = \frac{5}{8} \cdot \frac{3}{2} = \frac{5 \cdot 3}{8 \cdot 2} = \frac{15}{16} \quad \blacksquare$$

EXAMPLE 2 Simplify $\dfrac{\frac{2}{3} + \frac{1}{5}}{\frac{2}{3} - \frac{2}{9}}$.

Solution: Simplify above and below the main fraction bar separately. First, add $\dfrac{2}{3}$ and $\dfrac{1}{5}$ to obtain a single fraction in the numerator; then subtract $\dfrac{2}{9}$ from $\dfrac{2}{3}$ to obtain a single fraction in the denominator.

$$\frac{\frac{2}{3} + \frac{1}{5}}{\frac{2}{3} - \frac{2}{9}} = \frac{\frac{2(5)}{3(5)} + \frac{1(3)}{5(3)}}{\frac{2(3)}{3(3)} - \frac{2}{9}}$$

The LCD of the numerator's fractions is 15.

The LCD of the denominator's fractions is 9.

$$= \frac{\frac{10}{15} + \frac{3}{15}}{\frac{6}{9} - \frac{2}{9}}$$

Simplify.

$$= \frac{\frac{13}{15}}{\frac{4}{9}}$$

Add the numerator's fractions.

Subtract the denominator's fractions.

Next, perform division by multiplying the numerator of the complex fraction by the reciprocal of the denominator of the complex fraction.

$$\frac{\frac{13}{15}}{\frac{4}{9}} = \frac{13}{15} \cdot \frac{9}{4}$$

$$= \frac{13 \cdot 3 \cdot 3}{3 \cdot 5 \cdot 4} = \frac{39}{20} \quad \blacksquare$$

EXAMPLE 3   Simplify $\dfrac{\dfrac{x+1}{y}}{\dfrac{x}{y}+2}$.

Solution: The numerator of this complex fraction is already a single fraction. Write 2 as $\dfrac{2}{1}$ and

add $\dfrac{x}{y}$ and $\dfrac{2}{1}$ to get a single fraction in the denominator.

$$\frac{\dfrac{x+1}{y}}{\dfrac{x}{y}+\dfrac{2}{1}} = \frac{\dfrac{x+1}{y}}{\dfrac{x}{y}+\dfrac{2(y)}{1(y)}} \qquad \text{The LCD of } y \text{ and } 1 \text{ is } y.$$

$$= \frac{\dfrac{x+1}{y}}{\dfrac{x+2y}{y}}$$

$$= \frac{x+1}{y} \cdot \frac{y}{x+2y} \qquad \text{Multiply by the reciprocal of } \dfrac{x+2y}{y}.$$

$$= \frac{x+1}{x+2y} \qquad \blacksquare$$

EXAMPLE 4   Simplify $\dfrac{\dfrac{1}{z}-\dfrac{1}{2}}{\dfrac{1}{3}-\dfrac{z}{6}}$.

Solution: Subtract to get a single fraction in the numerator and a single fraction in the denominator of the complex fraction.

$$\frac{\dfrac{1}{z}-\dfrac{1}{2}}{\dfrac{1}{3}-\dfrac{z}{6}} = \frac{\dfrac{2}{2z}-\dfrac{z}{2z}}{\dfrac{2}{6}-\dfrac{z}{6}} \qquad \begin{array}{l}\text{The LCD of the numerator's fractions is } 2z.\\[6pt]\text{The LCD of the denominator's fractions is } 6.\end{array}$$

$$= \frac{\dfrac{2-z}{2z}}{\dfrac{2-z}{6}}$$

$$= \frac{2-z}{2z} \cdot \frac{6}{2-z} \qquad \text{Multiply by the reciprocal of } \dfrac{2-z}{6}.$$

$$= \frac{3}{z} \qquad \blacksquare$$

3   Next we study a second method for simplifying complex fractions. In this method, we multiply the numerator and the denominator of the complex fraction by the LCD of all fractions in the complex fraction.

> **Method 2: To Simplify a Complex Fraction**
>
> *Step 1* Multiply the numerator and the denominator of the complex fraction by the LCD of fractions in both the numerator and the denominator.
>
> *Step 2* Simplify.

We use method 2 to rework Examples 2 and 3.

EXAMPLE 5 Simplify $\dfrac{\frac{2}{3} + \frac{1}{5}}{\frac{2}{3} - \frac{2}{9}}$.

Solution: The LCD of $\frac{2}{3}, \frac{1}{5}, \frac{2}{3}$, and $\frac{2}{9}$ is 45. Multiply the numerator and the denominator of the complex fraction by 45 and simplify.

$$\frac{\frac{2}{3} + \frac{1}{5}}{\frac{2}{3} - \frac{2}{9}} = \frac{45\left(\frac{2}{3} + \frac{1}{5}\right)}{45\left(\frac{2}{3} - \frac{2}{9}\right)}$$

$$= \frac{45\left(\frac{2}{3}\right) + 45\left(\frac{1}{5}\right)}{45\left(\frac{2}{3}\right) - 45\left(\frac{2}{9}\right)} \qquad \text{Apply the distributive property.}$$

$$= \frac{30 + 9}{30 - 10} = \frac{39}{20} \qquad \text{Simplify.} \qquad ■$$

EXAMPLE 6 Simplify $\dfrac{\frac{x+1}{y}}{\frac{x}{y} + 2}$.

Solution: The LCD of $\frac{x+1}{y}, \frac{x}{y}$, and $\frac{2}{1}$ is y. Multiply both the numerator and denominator of the complex fraction by y and simplify.

$$\frac{\frac{x+1}{y}}{\frac{x}{y} + 2} = \frac{y\left(\frac{x+1}{y}\right)}{y\left(\frac{x}{y} + 2\right)}$$

$$= \frac{y\left(\frac{x+1}{y}\right)}{y\left(\frac{x}{y}\right) + y(2)} \qquad \text{Apply the distributive property.}$$

$$= \frac{x+1}{x + 2y} \qquad \text{Simplify.} \qquad ■$$

EXERCISE SET 5.5

Simplify each complex fraction. See Example 1.

1. $\dfrac{\dfrac{1}{2}}{\dfrac{3}{4}}$

2. $\dfrac{\dfrac{1}{8}}{-\dfrac{5}{12}}$

3. $\dfrac{-\dfrac{4x}{9}}{-\dfrac{2x}{3}}$

4. $\dfrac{-\dfrac{6y}{11}}{\dfrac{4y}{9}}$

5. $\dfrac{\dfrac{1+x}{6}}{\dfrac{1+x}{3}}$

6. $\dfrac{\dfrac{6x-3}{5x^2}}{\dfrac{2x-1}{10x}}$

7. $\dfrac{\dfrac{(y+1)(y-1)}{6}}{\dfrac{(y+1)(y+2)}{8}}$

8. $\dfrac{\dfrac{(y+4)(2y-1)}{9}}{\dfrac{(y+3)(2y-1)}{18}}$

Simplify each complex fraction. See Examples 2 and 5.

9. $\dfrac{\dfrac{1}{2}+\dfrac{2}{3}}{\dfrac{5}{9}-\dfrac{5}{6}}$

10. $\dfrac{\dfrac{3}{4}-\dfrac{1}{2}}{\dfrac{3}{8}+\dfrac{1}{6}}$

11. $\dfrac{2+\dfrac{7}{10}}{1+\dfrac{3}{5}}$

12. $\dfrac{4-\dfrac{11}{12}}{5+\dfrac{1}{4}}$

13. $\dfrac{\dfrac{1}{3}}{\dfrac{1}{2}-\dfrac{1}{4}}$

14. $\dfrac{\dfrac{7}{10}-\dfrac{3}{5}}{\dfrac{1}{2}}$

Simplify each complex fraction. See Examples 3, 4, and 6.

15. $\dfrac{\dfrac{m}{n}-1}{\dfrac{m}{n}+1}$

16. $\dfrac{\dfrac{x}{2}+2}{\dfrac{x}{2}-2}$

17. $\dfrac{\dfrac{1}{5}-\dfrac{1}{x}}{\dfrac{7}{10}+\dfrac{1}{x^2}}$

18. $\dfrac{\dfrac{1}{y^2}+\dfrac{2}{3}}{\dfrac{1}{y}-\dfrac{5}{6}}$

19. $\dfrac{1+\dfrac{1}{y-2}}{y+\dfrac{1}{y-2}}$

20. $\dfrac{x-\dfrac{1}{2x+1}}{1-\dfrac{x}{2x+1}}$

Simplify each complex fraction.

21. $\dfrac{-\dfrac{2}{9}}{-\dfrac{14}{3}}$

22. $\dfrac{\dfrac{3}{8}}{\dfrac{4}{15}}$

23. $\dfrac{-\dfrac{5}{12x^2}}{\dfrac{25}{16x^3}}$

24. $\dfrac{-\dfrac{7}{8y}}{\dfrac{21}{4y}}$

25. $\dfrac{\dfrac{4y-8}{16}}{\dfrac{6y-12}{4}}$

26. $\dfrac{\dfrac{7y+21}{3}}{\dfrac{3y+9}{8}}$

27. $\dfrac{\dfrac{x}{y}+1}{\dfrac{x}{y}-1}$

28. $\dfrac{\dfrac{3}{5y}+8}{\dfrac{3}{5y}-8}$

29. $\dfrac{1}{2+\dfrac{1}{3}}$

30. $\dfrac{3}{1-\dfrac{4}{3}}$

31. $\dfrac{\dfrac{ax+ab}{x^2-b^2}}{\dfrac{x+b}{x-b}}$

32. $\dfrac{\dfrac{m+2}{m-2}}{\dfrac{2m+4}{m^2-4}}$

33. $\dfrac{\dfrac{-3+y}{4}}{\dfrac{8+y}{28}}$

34. $\dfrac{\dfrac{-x+2}{18}}{\dfrac{8}{9}}$

35. $\dfrac{3+\dfrac{12}{x}}{1-\dfrac{16}{x^2}}$

36. $\dfrac{\dfrac{5}{6}+\dfrac{1}{2}}{\dfrac{1}{5}-4}$

37. $\dfrac{2+\dfrac{6}{x}}{1-\dfrac{9}{x^2}}$

38. $\dfrac{\dfrac{3}{8}+2}{\dfrac{5}{4}-\dfrac{1}{3}}$

39. $\dfrac{\dfrac{8}{x+4}+2}{\dfrac{12}{x+4}-2}$

40. $\dfrac{\dfrac{25}{x+5}+5}{\dfrac{3}{x+5}-5}$

41. $\dfrac{\dfrac{s}{r}+\dfrac{r}{s}}{\dfrac{s}{r}-\dfrac{r}{s}}$

42. $\dfrac{\dfrac{2}{x}+\dfrac{x}{2}}{\dfrac{2}{x}-\dfrac{x}{2}}$

43. Astronomers occasionally need to know the day of the week a particular date fell on. The complex fraction

$$\dfrac{J+\dfrac{3}{2}}{7},$$

where J, the *Julian day number,* is used to make this calculation. Find the day of the week February 17, 1985, fell on if its Julian day number is 2,446,113.5 by carrying out parts (a) and (b):

a. Evaluate the complex fraction with $J = 2{,}446{,}113.5$.

b. Consider only the decimal portion of the answer found in step 1. Multiply this value by 7 and round to the ones place. If this number equals 0, the answer is Sunday. If the number is 1, the answer is Monday. If this number is 2, then the answer is Tuesday, and so forth. What day of the week did February 17, 1985, fall on?

A Look Ahead

Simplify the following. See the following example.

EXAMPLE Simplify $\dfrac{1+x^{-1}}{x^{-1}}$.

Solution: $\quad \dfrac{1+x^{-1}}{x^{-1}} = \dfrac{1+\dfrac{1}{x}}{\dfrac{1}{x}}$

$$= \dfrac{x\left(1+\dfrac{1}{x}\right)}{x\left(\dfrac{1}{x}\right)}$$

$$= \dfrac{x+1}{1} \quad \text{or} \quad x+1 \quad \blacksquare$$

44. $\dfrac{x^{-1}+2^{-1}}{x^{-2}-4^{-1}}$

45. $\dfrac{3^{-1}-x^{-1}}{9^{-2}-x^{-2}}$

46. $\dfrac{x+y^{-1}}{\dfrac{x}{y}}$

47. $\dfrac{x-xy^{-1}}{\dfrac{1+x}{y}}$

48. $\dfrac{y^{-2}}{1-y^{-2}}$

49. $\dfrac{4+x^{-1}}{3+x^{-1}}$

Writing in Mathematics

50. Explain how to simplify a complex fraction using method 1.

51. Explain how to simplify a complex fraction using method 2.

Skill Review

Solve the following linear and quadratic equations. See Sections 2.4 and 4.6.

52. $3x + 5 = 7$

53. $5x - 1 = 8$

54. $2x^2 - x - 1 = 0$

55. $4x^2 - 9 = 0$

Simplify the following rational expressions. See Section 5.1.

56. $\dfrac{2 + x}{x + 2}$

57. $\dfrac{2 - x}{x - 2}$

5.6
Solving Equations Containing Rational Expressions

Tape 18

| **OBJECTIVES** | **1** | Solve a linear or quadratic equation containing rational expressions. |
| | **2** | Solve an equation containing rational expressions for a specified variable. |

1   In this section we solve equations that contain rational expressions. Two examples of such equations are

$$\frac{x}{5} + \frac{x + 2}{9} = 8 \quad \text{and} \quad \frac{x + 1}{9x - 5} = \frac{2}{3x}$$

To solve, use the multiplication property of equality to clear the equation of fractions by multiplying both sides of the equation by the LCD.

> **To Solve an Equation Containing Rational Expressions**
>
> *Step 1*   Multiply both sides of the equation by the LCD of all rational expressions in the equation.
> *Step 2*   Remove any grouping symbols.
> *Step 3*   Determine whether the equation is linear or quadratic and solve accordingly.
> *Step 4*   Check the solution in the original equation.

EXAMPLE 1   Solve $\dfrac{t-4}{2} - \dfrac{t-3}{9} = \dfrac{5}{18}$.

Solution: The LCD of denominators 2, 9, and 18 is 18. Multiply both sides of the equation by 18.

$$18\left(\dfrac{t-4}{2} - \dfrac{t-3}{9}\right) = 18\left(\dfrac{5}{18}\right)$$

$$18\left(\dfrac{t-4}{2}\right) - 18\left(\dfrac{t-3}{9}\right) = 18\left(\dfrac{5}{18}\right) \qquad \text{Apply the distributive property.}$$

$$9(t-4) - 2(t-3) = 5 \qquad \text{Simplify.}$$

$$9t - 36 - 2t + 6 = 5 \qquad \text{Use the distributive property.}$$

$$7t - 30 = 5 \qquad \text{Combine like terms.}$$

$$7t = 35 \text{ or } t = 5 \qquad \text{Solve for } t.$$

To check, substitute 5 for t in the original equation.

$$\dfrac{t-4}{2} - \dfrac{t-3}{9} = \dfrac{5}{18}$$

$$\dfrac{5-4}{2} - \dfrac{5-3}{9} = \dfrac{5}{18} \qquad \text{Let } t = 5.$$

$$\dfrac{1}{2} - \dfrac{2}{9} = \dfrac{5}{18} \qquad \text{Simplify.}$$

Next, subtract the fractions on the left by writing each with a denominator of 18.

$$\dfrac{1(9)}{2(9)} - \dfrac{2(2)}{9(2)} = \dfrac{5}{18}$$

$$\dfrac{9-4}{18} = \dfrac{5}{18}$$

$$\dfrac{5}{18} = \dfrac{5}{18} \qquad \text{True.}$$

Since the statement is true, 5 is the solution.   ■

EXAMPLE 2   Solve the equation $3 - \dfrac{6}{x} = x + 8$.

Solution: The LCD is x. Multiply both sides of the equation by x.

$$x\left(3 - \dfrac{6}{x}\right) = x(x+8)$$

$$x(3) - x\left(\dfrac{6}{x}\right) = x^2 + 8x \qquad \text{Apply the distributive property.}$$

$$3x - 6 = x^2 + 8x \qquad \text{Simplify.}$$

Write the quadratic equation in standard form and solve for x.

$$0 = x^2 + 5x + 6$$

$$0 = (x+3)(x+2) \qquad \text{Factor.}$$

$$x + 3 = 0 \quad \text{or} \quad x + 2 = 0 \qquad \text{Set each factor equal to 0 and solve.}$$

$$x = -3 \quad \text{or} \qquad x = -2$$

To check, replace x in the original equation by -3, and then by -2. This will verify that both -3 and -2 are solutions. ∎

EXAMPLE 3 Solve the equation $\dfrac{x + 1}{9x - 5} = \dfrac{1}{2x}$.

Solution: Multiply both sides of the equation by the LCD $2x(9x - 5)$.

$$2x(9x - 5)\left(\frac{x + 1}{9x - 5}\right) = 2x(9x - 5)\left(\frac{1}{2x}\right)$$

$$2x(x + 1) = (9x - 5)(1) \qquad \text{Simplify.}$$

$$2x^2 + 2x = 9x - 5 \qquad \text{Use the distributive property.}$$

$$2x^2 - 7x + 5 = 0 \qquad \text{Write the quadratic equation in the standard form.}$$

$$(2x - 5)(x - 1) = 0 \qquad \text{Factor.}$$

$$2x - 5 = 0 \quad \text{or} \quad x - 1 = 0 \qquad \text{Set each factor equal to zero.}$$

$$x = \frac{5}{2} \quad \text{or} \qquad x = 1 \qquad \text{Solve.}$$

Check to see that $\dfrac{5}{2}$ and 1 are both solutions. ∎

If a denominator in the original equation contains variables, we must be certain that the proposed solution does not make the denominator 0. If replacing the variable with the proposed solution makes the denominator 0, this proposed solution must be rejected. It is called an **extraneous solution.**

EXAMPLE 4 Solve the equation $x + \dfrac{14}{x - 2} = \dfrac{7x}{x - 2} + 1$.

Solution: The LCD is $x - 2$. Multiply both sides of the equation by $x - 2$.

$$(x - 2)\left(x + \frac{14}{x - 2}\right) = (x - 2)\left(\frac{7x}{x - 2} + 1\right)$$

$$(x - 2)(x) + (x - 2)\left(\frac{14}{x - 2}\right) = (x - 2)\left(\frac{7x}{x - 2}\right) + (x - 2)(1)$$

$$x^2 - 2x + 14 = 7x + x - 2 \qquad \text{Simplify.}$$

$$x^2 - 2x + 14 = 8x - 2 \qquad \text{Combine like terms.}$$

$$x^2 - 10x + 16 = 0 \qquad \text{Write the quadratic equation in standard form.}$$

$$(x - 8)(x - 2) = 0 \qquad \text{Factor.}$$

$$x - 8 = 0 \quad \text{or} \quad x - 2 = 0 \qquad \text{Set each factor equal to 0.}$$

$$x = 8 \quad \text{or} \qquad x = 2 \qquad \text{Solve.}$$

Check both 8 and 2. Since 2 makes the denominators in the original equation 0, then 2 is an extraneous solution. Replacing x with 8 in the original equation, we find that 8 is a solution. The only solution is 8. ■

If an equation contains rational expressions with variables in the denominator, make sure that proposed solutions are checked in the original equation.

EXAMPLE 5   Solve the equation $\dfrac{3a}{3a - 2} - \dfrac{5}{3a^2 + 7a - 6} = 1$.

Solution: Since $3a^2 + 7a - 6 = (3a - 2)(a + 3)$, the LCD is $(3a - 2)(a + 3)$. Multiply both sides of the equation by the LCD.

$$(3a - 2)(a + 3) \left(\frac{3a}{3a - 2} - \frac{5}{(3a - 2)(a + 3)} \right) = (3a - 2)(a + 3) \quad (1)$$

$$(3a - 2)(a + 3)\left(\frac{3a}{3a - 2}\right) - (3a - 2)(a + 3)\left(\frac{5}{(3a - 2)(a + 3)}\right)$$
$$= (3a - 2)(a + 3)(1)$$

$$3a(a + 3) - 5 = (3a - 2)(a + 3) \qquad \text{Simplify.}$$

$$3a^2 + 9a - 5 = 3a^2 + 7a - 6 \qquad \text{Apply the distributive property.}$$

$$9a - 5 = 7a - 6 \qquad \text{Subtract } 3a^2 \text{ from both sides.}$$

$$2a = -1 \qquad \text{Subtract } 7a \text{ from both sides and add 5 to both sides.}$$

$$a = -\frac{1}{2} \qquad \text{Divide both sides by 2.}$$

Check that $-\dfrac{1}{2}$ is the solution. ■

At this point, let's make sure we understand the difference between solving an equation containing rational expressions and adding or subtracting rational expressions.

EXAMPLE 6   **a.** Solve for x: $\dfrac{x}{4} + 2x = 9$.   **b.** Add $\dfrac{x}{4} + 2x$.

Solution: **a.** Since this is an equation, we can multiply both sides by the LCD, 4.

$$4 \left(\frac{x}{4} + 2x \right) = 4 \, (9)$$

$$4\left(\frac{x}{4}\right) + 4(2x) = 4(9) \qquad \text{Apply the distributive property.}$$

$$x + 8x = 36 \qquad \text{Simplify.}$$

$$9x = 36 \qquad \text{Combine like terms.}$$

$$x = 4 \qquad \text{Solve.}$$

Check to see that 4 is the solution.

b. This example is **not an equation** to solve; it is an addition to perform. To add these rational expressions, we find the LCD and write each rational expression as an equivalent expression whose denominator is the LCD. The LCD is 4.

$$\frac{x}{4} + 2x = \frac{x}{4} + \frac{2x(4)}{4}$$

$$= \frac{x + 8x}{4} \qquad \text{Add.}$$

$$= \frac{9x}{4} \qquad \text{Combine like terms.} \quad \blacksquare$$

2   The last example in this section is an equation containing several variables. We are directed to solve for one of them. The steps used in the preceding examples can be applied to solve equations for a specified variable as well.

EXAMPLE 7   Solve for x: $\dfrac{2x}{a} - 5 = \dfrac{3x}{b} + a$.

Solution: The LCD is ab. Multiply both sides of the equation by ab.

$$ab\left(\frac{2x}{a} - 5\right) = ab\left(\frac{3x}{b} + a\right)$$

$$ab\left(\frac{2x}{a}\right) - ab(5) = ab\left(\frac{3x}{b}\right) + ab(a) \qquad \text{Apply the distributive property.}$$

$$2xb - 5ab = 3xa + a^2b \qquad \text{Simplify.}$$

Next, write the equation so that all terms containing the variable x appear on one side of the equation. To do this, subtract $3xa$ from both sides and add $5ab$ to both sides.

$$2xb - 3xa = a^2b + 5ab$$

$$x(2b - 3a) = a^2b + 5ab \qquad \text{Factor out } x \text{ from each term.}$$

$$\frac{x(2b - 3a)}{2b - 3a} = \frac{a^2b + 5ab}{2b - 3a} \qquad \text{Divide both sides of the equation by } 2b - 3a.$$

$$x = \frac{a^2b + 5ab}{2b - 3a} \qquad \text{Simplify.} \quad \blacksquare$$

MENTAL MATH

Solve each equation for the variable.

1. $\dfrac{x}{5} = 2$ 　　　　**2.** $\dfrac{x}{8} = 4$ 　　　　**3.** $\dfrac{z}{6} = 6$ 　　　　**4.** $\dfrac{y}{7} = 8$

EXERCISE SET 5.6

Solve each equation. See Examples 1 and 2.

1. $\dfrac{x}{5} + 3 = 9$ 　　**2.** $\dfrac{x}{5} - 2 = 9$ 　　**3.** $\dfrac{x}{2} + \dfrac{5x}{4} = \dfrac{x}{12}$ 　　**4.** $\dfrac{x}{6} + \dfrac{4x}{3} = \dfrac{x}{18}$

5. $\dfrac{2}{9} = \dfrac{x}{15}$

6. $\dfrac{k}{3} = \dfrac{6}{9}$

7. $\dfrac{a}{5} = \dfrac{a-3}{2}$

8. $\dfrac{2b}{5} = \dfrac{b+2}{6}$

9. $\dfrac{x-3}{5} + \dfrac{x-2}{2} = \dfrac{1}{2}$

10. $\dfrac{a+5}{4} + \dfrac{a+5}{2} = \dfrac{a}{8}$

Solve each equation. See Examples 3 and 4.

11. $\dfrac{9}{2a-5} = -2$

12. $\dfrac{6}{4-3x} = 3$

13. $\dfrac{y}{y+4} + \dfrac{4}{y+4} = 3$

14. $\dfrac{5y}{y+1} - \dfrac{3}{y+1} = 4$

15. $\dfrac{2x}{x+2} - 2 = \dfrac{x-8}{x-2}$

16. $\dfrac{4y}{y-3} - 3 = \dfrac{3y-1}{y+3}$

17. $\dfrac{4y}{y-4} + 5 = \dfrac{5y}{y-4}$

18. $\dfrac{2a}{a+2} - 5 = \dfrac{7a}{a+2}$

19. $\dfrac{7}{x-2} + 1 = \dfrac{x}{x+2}$

20. $1 + \dfrac{3}{x+1} = \dfrac{x}{x-1}$

Solve each equation. See Example 5.

21. $\dfrac{x+1}{x+3} = \dfrac{2x^2 - 15x}{x^2 + x - 6} - \dfrac{x-3}{x-2}$

22. $\dfrac{3}{x+3} = \dfrac{12x+19}{x^2 + 7x + 12} - \dfrac{5}{x+4}$

23. $\dfrac{y}{2y+2} + \dfrac{2y-16}{4y+4} = \dfrac{2y-3}{y+1}$

24. $\dfrac{1}{x+2} = \dfrac{4}{x^2 - 4} - \dfrac{1}{x-2}$

Perform the indicated operation. See Example 6.

25. $\dfrac{1}{x} + \dfrac{2}{3}$

26. $\dfrac{3}{a} + \dfrac{5}{6}$

27. $\dfrac{2}{x+1} - \dfrac{1}{x}$

28. $\dfrac{4}{x-3} - \dfrac{1}{x}$

29. $\dfrac{t}{3t+6} - \dfrac{7}{5t+10}$

30. $\dfrac{m}{m^2 - 1} + \dfrac{1}{m+1}$

Solve each equation for the indicated variable. See Example 7.

31. $\dfrac{D}{R} = T$; for R

32. $\dfrac{A}{W} = L$; for W

33. $\dfrac{3}{x} = \dfrac{5y}{x+2}$; for y

34. $\dfrac{7x-1}{2x} = \dfrac{5}{y}$; for y

35. $\dfrac{3a+2}{3b-2} = -\dfrac{4}{2a}$; for b

36. $\dfrac{6x+y}{7x} = \dfrac{3x}{h}$; for h

Solve each equation.

37. $\dfrac{2x}{7} - 5x = 9$

38. $\dfrac{4x}{8} - 5x = 10$

39. $\dfrac{2}{y} + \dfrac{1}{2} = \dfrac{5}{2y}$

40. $\dfrac{6}{3y} + \dfrac{3}{y} = 1$

41. $\dfrac{4x+10}{7} = \dfrac{8}{2}$

42. $\dfrac{1}{2} = \dfrac{x+1}{8}$

43. $2 + \dfrac{3}{a-3} = \dfrac{a}{a-3} = 0$

44. $\dfrac{2y}{y-2} - \dfrac{4}{y-2} = 4$

45. $\dfrac{5}{x} + \dfrac{2}{3} = \dfrac{7}{2x}$

46. $\dfrac{5}{3} - \dfrac{3}{2x} = \dfrac{5}{4}$

47. $\dfrac{2a}{a+4} = \dfrac{3}{a-1}$

48. $\dfrac{5}{3x-8} = \dfrac{x}{x-2}$

49. $\dfrac{x+1}{3} - \dfrac{x-1}{6} = \dfrac{1}{6}$

50. $\dfrac{3x}{5} - \dfrac{x-6}{3} = \dfrac{1}{5}$

51. $\dfrac{4r-1}{r^2 + 5r - 14} + \dfrac{2}{r+7} = \dfrac{1}{r-2}$

52. $\dfrac{2t+3}{t-1} - \dfrac{2}{t+3} = \dfrac{5-6t}{t^2 + 2t - 3}$

53. $\dfrac{t}{t-4} = \dfrac{t+4}{6}$

54. $\dfrac{15}{x+4} = \dfrac{x-4}{x}$

55. $\dfrac{x}{2x + 6} + \dfrac{x + 1}{3x + 9} = \dfrac{2}{4x + 12}$

56. $\dfrac{a}{5a - 5} - \dfrac{a - 2}{2a - 2} = \dfrac{5}{4a - 4}$

57. $\dfrac{5}{a^2 + 4a + 3} + \dfrac{2}{a^2 + a - 6} - \dfrac{3}{a^2 - a - 2} = 0$

58. $-\dfrac{2}{a^2 + 2a - 8} + \dfrac{1}{a^2 + 9a + 20} = \dfrac{-4}{a^2 + 3a - 10}$

Solve each equation for the indicated variable.

59. $\dfrac{A}{BH} = \dfrac{1}{2}$; for B

60. $\dfrac{V}{\pi r^2 h} = 1$; for h

61. $\dfrac{C}{\pi r} = 2$; for r

62. $\dfrac{3V}{A} = H$; for V

63. $\dfrac{1}{a} = \dfrac{1}{b} + \dfrac{1}{c}$; for a

64. $\dfrac{1}{2} - \dfrac{1}{x} = \dfrac{1}{y}$; for x

65. $\dfrac{m^2}{6} - \dfrac{n}{3} = \dfrac{p}{2}$; for n

66. $\dfrac{x^2}{r} + \dfrac{y^2}{t} = 1$; for r

Writing in Mathematics

67. Explain the difference between solving an equation such as $\dfrac{x}{2} + \dfrac{3}{4} = \dfrac{x}{4}$ for x and performing an operation such as adding $\dfrac{x}{2} + \dfrac{3}{4}$.

Skill Review

Use the distributive property to multiply the following. See Section 1.9.

68. $-8(2x^2 + 3x + 4)$

69. $-2(a^2 - 5)$

70. $5a(a - 6)$

71. $3m(m^2 - 1)$

Factor the following. See Section 4.2.

72. $x^2 - 7x + 10$

73. $x^2 - 2x - 3$

5.7
Ratio and Proportion

OBJECTIVES

Tape 19

1 Rewrite word phrases as ratios using fractional notation.

2 Solve proportions.

3 Use proportions to solve word problems.

1 A **ratio** is the quotient of two numbers or two quantities.

Ratio

If a and b are two numbers or quantities, $b \neq 0$, the **ratio of a to b** is

$$\frac{a}{b} \quad \text{or} \quad a : b$$

EXAMPLE 1   Write a ratio for each phrase. Use fractional notation.

a. The ratio of 2 parts salt to 5 parts water

b. The ratio of 12 almonds to 16 pecans

Solution: **a.** The ratio of 2 parts salt to 5 parts water is $\dfrac{2}{5}$.

b. The ratio of 12 almonds to 16 pecans is $\dfrac{12}{16} = \dfrac{3}{4}$ in lowest terms.   ■

2   If two ratios are equal, we say the ratios are **in proportion** to each other. A **proportion** is a mathematical statement that two ratios are equal.

For example, $\dfrac{1}{2} = \dfrac{4}{8}$ is a proportion, as is $\dfrac{x}{5} = \dfrac{8}{10}$. When we want to emphasize the equation as a proportion, we

> **read the proportion $\dfrac{1}{2} = \dfrac{4}{8}$ as "one is to two as four is to eight"**

In the proportion $\dfrac{1}{2} = \dfrac{4}{8}$, the 1 and the 8 are called the **extremes;** the 2 and the 4 are called the **means.**

> **For any proportion, the product of the means equals the product of the extremes.**

To see this, multiply both sides of the proportion $\dfrac{a}{b} = \dfrac{c}{d}$ by the LCD bd.

$$\frac{a}{b} = \frac{c}{d}$$

$$bd\left(\frac{a}{b}\right) = bd\left(\frac{c}{d}\right) \qquad \text{Multiply both sides by } bd.$$

$$\underbrace{ad}_{} = \underbrace{bc}_{} \qquad \text{Simplify.}$$
product   product
of extremes   of means

The products ad and bc are also called **cross products** and can be found by

$$\frac{a}{b} \underset{\nearrow}{\overset{\searrow}{=}} \frac{c}{d} \quad \longrightarrow \quad \begin{array}{c} bc \\ ad \end{array}$$

Equating these cross products is called **cross multiplication.**

Cross Multiplication

For any ratios $\dfrac{a}{b}$ and $\dfrac{c}{d}$, if

$$\frac{a}{b} = \frac{c}{d}, \quad \text{then } ad = bc.$$

Cross multiplication can be used to solve a proportion for a variable.

EXAMPLE 2   Solve for x: $\dfrac{45}{x} = \dfrac{5}{7}$.

Solution: To solve, cross multiply.

$$\dfrac{45}{x} = \dfrac{5}{7}$$

$$45 \cdot 7 = 5 \cdot x \qquad \text{Cross multiply.}$$

$$\dfrac{315}{5} = \dfrac{5x}{5} \qquad \text{Divide both sides by 5.}$$

$$63 = x \qquad \text{Simplify.}$$

To check, substitute 63 for x in the original proportion.   ■

EXAMPLE 3   Solve for x: $\dfrac{x}{x-2} = \dfrac{x-3}{x+1}$.

Solution: To solve, cross multiply.

$$\dfrac{x}{x-2} = \dfrac{x-3}{x+1}$$

$$x(x+1) = (x-3)(x-2) \qquad \text{Cross multiply.}$$

$$x^2 + x = x^2 - 5x + 6 \qquad \text{Multiply.}$$

$$x = -5x + 6 \qquad \text{Subtract } x^2 \text{ from both sides.}$$

$$6x = 6 \qquad \text{Add } 5x \text{ to both sides.}$$

$$x = 1 \qquad \text{Divide both sides by 6.}$$

To check, substitute 1 for x in the original proportion.   ■

3   Proportions can be used to model and solve many real-life problems. When using proportions in this way, it is important to judge whether the solution is reasonable. Doing so helps us to decide if the proportion has been formed correctly. In these word problems, assume that ratios are constant.

EXAMPLE 4   If 3 pens cost \$0.87, how much will 5 pens cost?

Solution: Assume that the ratio of pens to money stays the same. Since 3 pens cost \$0.87, one ratio we might form is

$$\dfrac{\text{pens}}{\text{price}} \quad \text{or} \quad \dfrac{3}{0.87}$$

Let x represent the cost of 5 pens and set up a second $\dfrac{\text{pens}}{\text{price}}$ ratio. If the first ratio is $\dfrac{\text{pens}}{\text{price}}$, the second ratio also should be $\dfrac{\text{pens}}{\text{price}}$ or $\dfrac{5}{x}$.

The proportion is then $\dfrac{\text{pens}}{\text{price}} = \dfrac{\text{pens}}{\text{price}}$ or

$$\begin{array}{c} \text{pens} \rightarrow \\ \text{price} \rightarrow \end{array} \dfrac{3}{0.87} = \dfrac{5}{x} \begin{array}{c} \leftarrow \text{pens} \\ \leftarrow \text{price} \end{array}$$

$$3x = 0.87(5) \qquad \text{Cross multiply.}$$
$$3x = 4.35$$
$$x = 1.45 \qquad \text{Divide both sides by 3.}$$

The 5 pens cost $1.45. This answer is reasonable since we know that the cost of 5 pens is more than the cost of 3 pens, $0.87, and less than the cost of 6 pens, which is double the cost of 3 pens or 2($0.87) or $1.74. The proportion $\dfrac{\text{price}}{\text{pens}} = \dfrac{\text{price}}{\text{pens}}$ could also have been used to solve this problem. ■

When using proportions to solve problems, make sure the units in the numerators of the proportion are the same and the units in the denominators are the same.

EXAMPLE 5 In a chemistry experiment, the ratio of salt to water must be constant. If the ratio is 2 pounds of salt to 8 gallons of water, how many pounds of salt must be in 5 gallons of water?

Solution: Let x be the pounds of salt in 5 gallons of water and use the proportion

$$\frac{x \text{ pounds of salt}}{5 \text{ gallons of water}} = \frac{2 \text{ pounds of salt}}{8 \text{ gallons of water}}$$

or

$$\frac{x}{5} = \frac{2}{8}$$
$$8(x) = 2(5) \qquad \text{Cross multiply.}$$
$$8x = 10 \qquad \text{Multiply.}$$
$$x = \frac{10}{8} \quad \text{or} \quad \frac{5}{4}$$

There are $\dfrac{5}{4}$ or $1\dfrac{1}{4}$ pounds of salt in 5 gallons of water. Notice that this amount is less than 2 pounds, which is reasonable since there are 2 pounds of salt in 8 gallons of water and we have 5 gallons. ■

EXERCISE SET 5.7

Rewrite the following phrases as ratios in fractional notation. See Example 1.

1. 2 parts flea dip to 15 parts water.
2. 4 parts plant food to 18 parts water.
3. $120 to 8 hours of work.
4. $40.00 to 8 hours of work.
5. 50 chips to every 1 cup of salsa.
6. 3 hot dogs to every 1 hamburger.

Solve each proportion. See Examples 2 and 3.

7. $\dfrac{2}{3} = \dfrac{x}{6}$

8. $\dfrac{x}{2} = \dfrac{16}{6}$

9. $\dfrac{x}{10} = \dfrac{5}{9}$

10. $\dfrac{9}{4x} = \dfrac{6}{2}$

11. $\dfrac{9}{5} = \dfrac{12}{3x + 2}$

12. $\dfrac{6}{11} = \dfrac{27}{3x - 2}$

13. $\dfrac{3}{x + 1} = \dfrac{5}{2x}$

14. $\dfrac{7}{x - 3} = \dfrac{8}{2x}$

15. $\dfrac{x+1}{x} = \dfrac{x+2}{x-2}$

16. $\dfrac{x-1}{x+3} = \dfrac{x+3}{x}$

Solve each problem by forming a proportion and solving. See Examples 4 and 5.

17. The manufacturers of cans of salted mixed nuts state that the ratio of peanuts to other nuts is 3 to 2. If 324 peanuts are in a can, find how many other nuts should also be in the can.

18. If Sam can travel 343 miles in 7 hours, find how far he can travel if he maintains the same speed for 5 hours.

19. Simon is paid $53.20 for 4 hours of work. If he worked 8 hours, find his pay.

20. The instructions on a bottle of plant food read as follows: "Use four tablespoons plant food per 3 gallons of water."

Find how many tablespoons of plant food should be mixed into 6 gallons of water.

21. To correctly mix weed killer with water, it is necessary to mix 8 teaspoons of weed killer with 2 gallons of water. Find how many gallons of water are needed to mix with the entire box of weed killer if the box contains 36 teaspoons.

22. There are 290 milligrams of sodium per 1-ounce serving of Rice Crispies. Find how many milligrams of sodium are in three 1-ounce servings of the cereal.

Solve the following proportions.

23. $\dfrac{4x}{6} = \dfrac{7}{2}$

24. $\dfrac{a}{5} = \dfrac{3}{2}$

25. $\dfrac{a}{25} = \dfrac{12}{10}$

26. $\dfrac{n}{10} = 9$

27. $\dfrac{x-3}{x} = \dfrac{4}{7}$

28. $\dfrac{y}{y-16} = \dfrac{5}{3}$

29. $\dfrac{5x+1}{x} = \dfrac{6}{3}$

30. $\dfrac{3x-2}{5} = \dfrac{4x}{1}$

31. $\dfrac{x+1}{2x+3} = \dfrac{2}{3}$

32. $\dfrac{x+1}{x+2} = \dfrac{5}{3}$

Solve the following.

33. There are 110 calories per 28.4 grams of Crispy Rice cereal. Find how many calories are in 42.6 grams of this cereal.

34. There are 1280 calories in a 14-ounce portion of Eagle Brand Milk. Find how many calories are in 2 ounces of Eagle Brand Milk.

35. A box of flea and tick powder instructs the user to mix 4 tablespoons of powder with 1 gallon of water. Find how much powder shold be mixed with 5 gallons of water.

36. Miss Babola's new Mazda gets 35 miles per gallon. Find how far she can drive if the tank contains 13.5 gallons of gas.

37. In a week of city driving, Miss Babola noticed that she was able to drive 418.5 miles on a tank of gas (13.5 gallons). Find how many miles per gallon Miss Babola got in city traffic.

38. Sara Jane was able to travel 420 miles in 6 hours. If she maintains the same speed, find how far she could travel in 1 hour. Find how fast she was traveling in miles per hour.

39. Mr. Lin's contract states that he will be paid $153 per 8-hour day to teach mathematics. Find how much he earns per hour.

40. Mr. Gonzales, a pool contractor, bases the cost of labor on the volume of the pool to be constructed. The cost of labor on a wading pool of 803 cubic feet is $750.00. If the customer decided to cut the volume by a third, find the cost of labor for the smaller pool.

41. An accountant finds that Country Collections earned $35,063 during its first 6 months. Find how much on the average, the business earned each week.

Skill Review

Evaluate $\dfrac{y-b}{x-a}$ for the given values. See Section 1.5.

42. $x = 2, y = 3, a = 5,$ and $b = 2$

43. $x = -3, y = 5, a = 4,$ and $b = 5$

Factor the following. See Section 4.3.

44. $4x^2 - 22x + 10$

45. $9x^2 - 3x - 12$

5.8
Applications of Equations Containing Rational Expressions

OBJECTIVES

Tape 19

1   Solve direct translation word problems.

2   Solve word problems involving work.

3   Solve word problems involving distance.

4   Solve word problems involving similar triangles.

1   In this section, word problems are presented that will lead to equations with rational expressions. We begin this section with a direct translation.

EXAMPLE 1   The quotient of a number and 6 minus $\frac{5}{3}$ is the quotient of the number and 2. Find the number.

Solution: Carefully read the problem. Let x represent the unknown number we are asked to find and translate the problem into an equation.

The quotient of x and 6	minus	$\frac{5}{3}$	is	the quotient of x and 2.
$\frac{x}{6}$	$-$	$\frac{5}{3}$	$=$	$\frac{x}{2}$

Begin solving this equation by multiplying both sides of the equation by the LCD 6.

$$6\left(\frac{x}{6} - \frac{5}{3}\right) = 6\left(\frac{x}{2}\right)$$

$$6\left(\frac{x}{6}\right) - 6\left(\frac{5}{3}\right) = 6\left(\frac{x}{2}\right) \qquad \text{Apply the distributive property.}$$

$$x - 10 = 3x \qquad \text{Simplify.}$$

$$-10 = 2x \qquad \text{Subtract } x \text{ from both sides.}$$

$$-\frac{10}{2} = \frac{2x}{2} \qquad \text{Divide both sides by 2.}$$

$$-5 = x \qquad \text{Simplify.}$$

The unknown number is -5. To check, verify that "the quotient of -5 and 6 minus $\frac{5}{3}$ is the quotient of -5 and 2, or $-\frac{5}{6} - \frac{5}{3} = -\frac{5}{2}$.   ■

Before we attempt another word problem, let's first review steps for solving word problems. The same steps we presented in earlier sections can be used to solve this section's problems.

> **Solving a Word Problem**
>
> *Step 1* Read and reread the problem. Choose a variable to represent an unknown quantity that we are asked to find.
> *Step 2* Use this variable to represent any other unknown quantities.
> *Step 3* If possible, draw a diagram.
> *Step 4* Translate the problem into an equation.
> *Step 5* Solve the equation.
> *Step 6* Answer the question asked and check to see if the answer is a reasonable one.
> *Step 7* Check the solution in the originally stated problem.

2   The next example is a word problem known as a work problem. Work problems usually involve people or machines doing a certain task.

EXAMPLE 2   Sam can clean the pump house in 3 hours while Frank needs 7 hours to complete the job. The regional manager is coming to inspect the plant facilities, so the operations supervisor directs Sam and Frank to clean the pump house together. How long will it take the two of them to do the job if they work together?

Solution: *Step 1*   Let a variable represent one unknown quantity.
Let x = number of hours for them to do the job together.

Step 2   Use this variable to represent other unknown quantities.
Sam can clean the pump house alone in 3 hours. Therefore,

$$\text{in 1 hour, Sam can clean } \frac{1}{3} \text{ of the pump house.}$$

Frank can clean the pump house alone in 7 hours. Therefore,

$$\text{in 1 hour he can clean } \frac{1}{7} \text{ of the pump house.}$$

The two men can clean the pump house together in x hours. Therefore,

$$\text{in 1 hour, they can clean } \frac{1}{x} \text{ of the pump house.}$$

Step 4 Translate the problem into an equation.
Use the fact that the part of the pump house cleaned by Sam in 1 hour plus the part cleaned by Frank in 1 hour must equal the part cleaned by both in 1 hour.

part cleaned by Sam in 1 hour	+	part cleaned by Frank in 1 hour	=	part cleaned by both in 1 hour
$\dfrac{1}{3}$	$+$	$\dfrac{1}{7}$	$=$	$\dfrac{1}{x}$

Step 5 Solve the equation.

Begin solving the equation by multiplying both sides of the equation by the LCD 21x.

$$21x\left(\frac{1}{3}\right) + 21x\left(\frac{1}{7}\right) = 21x\left(\frac{1}{x}\right)$$

$$7x + 3x = 21 \qquad \text{Simplify.}$$

$$10x = 21$$

$$x = \frac{21}{10} \quad \text{or} \quad 2\frac{1}{10} \text{ hours}$$

Step 6 Answer the question asked.

Frank and Sam can clean the pump house together in $2\frac{1}{10}$ hours. Since Sam can clean the pump house alone in 3 hours, we expect x to be less than 3 hours. It is, and we conclude that our answer is reasonable. ∎

3   Next we look at a problem solved by the distance formula.

EXAMPLE 3   A car travels 180 miles in the same time that a semi-truck travels 120 miles. If the car's speed is 20 miles per hour faster than the truck's, find the car's speed and the truck's speed.

Solution: *Step 1*   Choose a variable to represent the unknown quantity you are asked to find. Let $x =$ the speed of the truck.

Step 2   Use this variable to represent any other unknown quantities.

Since the car's speed is 20 miles per hour faster than the truck's, then $x + 20 =$ the speed of the car.

Step 3   Draw a diagram.

Use the formula $d = r \cdot t$ or **distance** = **rate** (speed) · **time**. Prepare a chart to organize the information in the problem. Recall that if $d = r \cdot t$, then $t = \frac{d}{r}$.

	distance	=	rate	·	time
Truck	120		x		$\dfrac{120}{x}\left(\dfrac{\text{distance}}{\text{rate}}\right)$
Car	180		$x + 20$		$\dfrac{180}{x + 20}\left(\dfrac{\text{distance}}{\text{rate}}\right)$

Step 4   Write an equation.

Since the car and the truck traveled the same amount of time, set up an equation by letting

$$\text{car's time} = \text{truck's time}$$

$$\frac{180}{x + 20} = \frac{120}{x}$$

Step 5   Solve the equation.

Begin solving the equation by cross multiplying.

$$180x = 120(x + 20)$$ Cross multiplying.

$$180x = 120x + 2400$$ Use the distributive property.

$$60x = 2400$$ Subtract $120x$ from both sides.

$$x = 40$$ Divide both sides by 60.

Step 6   Answer the question.

The speed of the truck is 40 miles per hour. The speed of the car must then be $x + 20$ or 60 miles per hour.

Step 7   Check.

To check the solution, notice that the car traveled 180 miles at a rate of 60 mph. This takes how long?

$$d = r \cdot t$$

$$180 = 60 \cdot t$$ Let $d = 180$ miles and $r = 60$ mph.

$$3 \text{ hours} = t$$ Solve for t.

The truck traveled 120 miles at a rate of 40 mph. How long does this take?

$$d = r \cdot t$$

$$120 = 40 \cdot t$$ Let $d = 120$ miles and $r = 40$ mph.

$$3 \text{ hours} = t$$ Solve for t.

The times are the same as the original problem states, so the solutions check.   ∎

4   **Similar triangles** have the same shape but not necessarily the same size. It can be shown that corresponding sides of similar triangles are in proportion. For example, the following two triangles are similar.

Side *a* corresponds to side *d*.

Side *b* corresponds to side *e*.

Side *c* corresponds to side *f*.

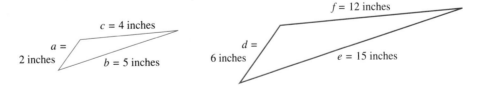

In this section, we will position similar triangles so that they are oriented the same.

To show that corresponding sides are in proportion, write the ratios of the corresponding sides.

$$\frac{a}{d} = \frac{2}{6} = \frac{1}{3}, \qquad \frac{b}{e} = \frac{5}{15} = \frac{1}{3}, \qquad \frac{c}{f} = \frac{4}{12} = \frac{1}{3}$$

EXAMPLE 4 If the following two triangles are similar, find the missing length x.

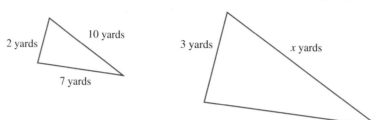

Solution: Since the triangles are similar, their corresponding sides are in proportion and we have

$$\frac{2}{3} = \frac{10}{x}$$

To solve, cross multiply.

$$2x = 30 \qquad \text{Cross multiply.}$$

$$x = 15 \qquad \text{Divide both sides by 2.}$$

The missing length is 15 yards. ∎

EXERCISE SET 5.8

Solve the following. See Example 1.

1. Three times the reciprocal of a number equals nine times the reciprocal of six. Find the number.

2. Twelve divided by the sum of x and 2 equals the quotient of 4 and the difference of x and 2. Find x.

3. If twice a number added to 3 is divided by the number plus 1, the result is three halves. Find the number.

4. A number added to the product of 6 and the reciprocal of the number equals -5. Find the number.

See Example 2.

5. Smith Engineering found that an experienced surveyor surveys a roadbed in four hours. An apprentice surveyor needs five hours to survey the same stretch of road. If the two work together, find how long it takes them to complete the job.

6. An experienced bricklayer constructs a small wall in three hours. The apprentice completes the job in six hours. Find how long it takes if they work together.

7. In 2 minutes, a conveyor belt moves 300 pounds of recyclable aluminum from the delivery truck to a storage area. A smaller belt moves the same quantity of cans the same distance in 6 minutes. If both belts are used, find how long it takes to move the cans to the storage area.

8. Find how long it takes the conveyor belts described in Exercise 7 to move 1200 pounds of cans. (*Hint:* Think of 1200 pounds as four jobs.)

See Example 3.

9. A jogger begins her workout by walking to the park. This takes her $1\frac{1}{4}$ hours. On the return trip she jogs one mile per hour faster than she walked and is home in an hour. Complete the chart at the right and use it to find the rate for each trip.

	distance	=	rate	·	time
Trip			r		$\frac{5}{4}$
Return trip			$r + 1$		1

10. A boat can travel 9 miles upstream in the same amount of time it takes to travel 11 miles downstream. If the current of the river is 3 miles per hour, complete the chart below and use it to find the speed of the boat in still water.

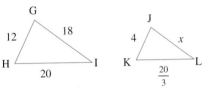

	distance	rate	time = $\dfrac{\text{distance}}{\text{rate}}$
Upstream	9	$r - 3$	
Downstream	11	$r + 3$	

11. A cyclist rode the first 20-mile portion of his workout at a constant rate. For the 16-mile cooldown portion of his workout, he reduced his speed by 2 miles per hour. Each portion of the workout took equal time. Find the cyclist's rate during the first portion and find his rate during the cooldown portion.

12. A semitruck travels 300 miles through the flatland in the same amount of time that it travels 180 miles through mountains. The rate of the truck is 20 miles per hour slower in the mountains than in the flatland. Find both the flatland rate and mountain rate.

Given that the following pairs of triangles are similar, find the missing lengths. See Example 4.

13.

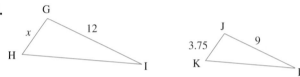

14.

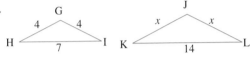

15.

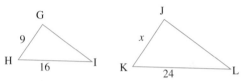

16.

Solve the following.

17. One-fourth equals the quotient of a number and eight. Find the number.

18. Four times a number added to 5 is divided by six. The result is $\dfrac{7}{2}$. Find the number.

19. Marcus and Tony work for Lombardo's Pipe and Concrete. Mr. Lombardo is preparing an estimate for a customer. He knows that Marcus lays a slab of concrete in six hours. Tony lays the same size slab in four hours. If both work on the job and the cost of labor is $45.00 per hour, decide what the labor estimate should be.

20. Mr. Dodson can paint his house by himself in four days.

His son needs an additional day to complete the job if he works by himself. If they work together, find how long it takes to paint the house.

21. While road testing a new make of car, the editor of a consumer magazine finds that he can go 10 miles into a 3-mile-per-hour wind in the same amount of time he can go 11 miles with a 3 miles per hour wind behind him. Find the speed of the car in still air.

22. A fisherman rows 9 miles downstream in the same amount of time he rows 3 miles upstream. If the current is 6 miles per hour, find how long it takes him to cover the 12 miles.

Find the unknown length in the following pairs of similar triangles.

23.

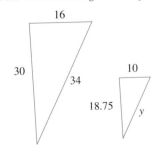

24.

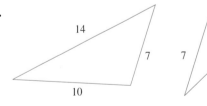

Solve the following.

25. Two divided by the difference of a number and three, minus four divided by a number plus three, equals 8 times the reciprocal of the difference of the number squared and 9. What is the number?

26. If fifteen times the reciprocal of a number is added to the ratio of nine times a number minus seven and the number plus two, the result is 9. What is the number?

27. A pilot flies 630 miles with a tail wind of 35 miles per hour. Against the wind, he flies only 455 miles. Find the rate of the plane in still air.

28. A marketing manager travels 1080 miles in a corporate jet and then an additional 240 miles by car. If the car ride takes one hour longer, and if the rate of the jet is 6 times the rate of the car, find the time the manager travels by jet and find the time the manager travels by car.

29. A cyclist rides 16 miles per hour on level ground on a still day. He finds that he rides 48 miles with the wind behind him in the same amount of time that he rides 16 miles into the wind. Find the rate of the wind.

30. The current on a large river is 3 miles per hour. A barge can go 6 miles upstream in the same amount of time it takes to go 10 miles downstream. Find the speed of the boat in still water.

31. One custodian cleans a suite of offices in three hours. When a second worker is asked to join the regular custodian, the job only takes an hour and half. How long does it take the second worker to do the same job?

32. One person proofreads copy for a small newspaper in four hours. If a second proofreader also is employed, the job can be down in two and a half hours. How long does it take for the second proofreader to do the same job?

33. One pipe fills a storage pool in 20 hours. A second pipe fills the same pool in 15 hours. When a third pipe is added and all three are used to fill the pond, it takes only 6 hours. Find how long it takes the third pipe to do the job.

34. Mr. Jamison can do an audit in 400 hours. Mr. Ling can do the same audit in 300 hours with the help of a new computer program. How long will it take if both are assigned to the audit?

35. A toy maker wishes to make a triangular mainsail for a toy sailboat that will be the same shape as a regular-size sailboat's mainsail. Use the following diagram to find the missing dimensions.

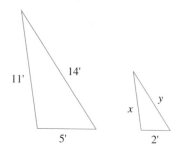

36. A mother wishes to make a doll's triangular diaper that will have the same shape as a full-size diaper. Use the following diagram to find the missing dimensions.

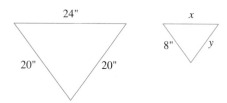

37. In 6 hours, an experienced cook prepares enough pies to supply a local restaurant's daily order. Another cook prepares the same number of pies in seven hours. Together with a third cook, they prepare the pies in two hours. Find the work rate of the third cook.

38. Mrs. Smith balances the company books in 8 hours. It takes her assistant half again as long to do the same job. If they work together, find how how long it takes them to balance the books.

39. One pump fills a tank in three times the time it takes a faster pump. If the pumps work together, they fill the tank in 21 minutes. How long does it take for each pump to fill the tank?

Writing in Mathematics

One of the great algebraists of ancient times was a man named Diophantus. Little is known of his life other than that he lived and worked in Alexandria. Some historians believe he lived during the first century of the Christian era, about the time of Nero. The only clue to his personal life is the following epigram found in a collection called the Palatine Anthology.

God granted him youth for a sixth of his life and added a twelfth part to this. He clothed his cheeks in down. He lit him the light of wedlock after a seventh part and five years after his marriage, He granted him a son. Alas, lateborn wretched child. After attaining the measure of half his father's life, cruel fate overtook him, thus leaving Diophantus during the last four years of his life only such consolation as the science of numbers. How old was Diophantus at his death?*

*From *The Nature and Growth of Modern Mathematics,* Edna Kramer, 1970, Fawcett Premier Books, Vol. 1, pages 107–108.

We are looking for Diophantus' age when he died, so let x represent that age. If we sum the parts of his life, we should get the total age.

Parts of his life
$$\begin{cases} \dfrac{1}{6} \cdot x + \dfrac{1}{12} \cdot x \text{ is the time of his youth.} \\[2mm] \dfrac{1}{7} \cdot x \text{ is the time between his youth and when he married.} \\[2mm] 5 \text{ years is the time between his marriage and the birth of his son.} \\[2mm] \dfrac{1}{2}x \text{ is the time Diophantus had with his son.} \\[2mm] 4 \text{ years is the time between his son's death and his own.} \end{cases}$$

The sum of these parts should equal Diophantus' age when he died.

$$\frac{1}{6} \cdot x + \frac{1}{12} \cdot x + \frac{1}{7} \cdot x + 5 + \frac{1}{2} \cdot x + 4 = x$$

40. Solve the epigram.

41. How old was Diophantus when his son was born? How old was the son when he died?

42. Solve the following epigram:

I was four when my mother packed my lunch and sent me off to school. Half my life was spent in school and another sixth was spent on a farm. Alas, hard times befell me. My crops and cattle fared poorly and my land was sold. I returned to school for 3 years and have spent one tenth of my life teaching. How old am I?

43. Write an epigram describing your life. Be sure that none of the time periods in your epigram overlap.

Skill Review

Solve the following equations. See Section 4.6.

44. $4z^2 - 11z + 6 = 0$

45. $6z^2 - 25z + 14 = 0$

CRITICAL THINKING   Doctors must make careful judgments when prescribing drugs, not only to prescribe the proper kind but to prescribe the proper amount. Too little has no effect; too much may have an unwanted, even fatal, effect. Particularly for children, gauging the proper dose is critical. Two dose formulas for children are well known among doctors. Each formula relates a child's age A and an adult dose D to the proper child's dose C:

$$\text{Young's rule:} \quad C = \frac{DA}{A + 12}$$

$$\text{Cowling's rule:} \quad C = \frac{D(A + 1)}{24}$$

Unlike formulas for, say, area or distance, these dose formulas describe only an approximate relationship. The formulas are most accurate when applied to children between the ages of 2 and 13.

Since the two formulas are not identical, they predict different doses. Will either formula consistently predict a larger dose than the other? For what age is the difference of the predicted doses greatest? For what age does Cowling's rule predict exactly an adult dose?

Many doctors prefer to use formulas that relate doses to factors other than a child's age. Why is age not necessarily the most important factor when predicting a child's dose? What other factors might be used?

CHAPTER 5 GLOSSARY

A fraction whose numerator or denominator or both numerator and denominator contain fractions is called a **complex fraction**

In the proportion $\dfrac{a}{b} = \dfrac{c}{d}$ the products ad and bc are called **cross products. Cross multiplication** is the process of setting the cross products equal: $ad = bc$.

A proposed solution that is not a solution of the original equation is called an **extraneous solution.**

In the proportion $\dfrac{1}{2} = \dfrac{4}{8}$, the 1 and the 8 are called the **extremes;** the 2 and the 4 are called the **means.**

A **proportion** is a mathematical statement that two ratios are equal.

A **ratio** is the quotient of two numbers or quantities.

A **rational expression** is a quotient of two polynomials.

CHAPTER 5 SUMMARY

If P, Q, R, and S are polynomials, then:

MULTIPLYING RATIONAL EXPRESSIONS   (5.2)

$$\frac{P}{Q} \cdot \frac{R}{S} = \frac{PR}{QS}, \qquad Q \neq 0, \quad S \neq 0$$

DIVIDING RATIONAL EXPRESSIONS

$$\frac{P}{Q} \div \frac{R}{S} = \frac{P}{Q} \cdot \frac{S}{R} = \frac{PS}{QR}, \qquad Q \neq 0, \quad S \neq 0, \quad R \neq 0$$

ADDING RATIONAL EXPRESSIONS   (5.3)

$$\frac{P}{R} + \frac{Q}{R} = \frac{P + Q}{R}, \qquad R \neq 0$$

SUBTRACTING RATIONAL EXPRESSIONS

$$\frac{P}{R} - \frac{Q}{R} = \frac{P - Q}{R}, \qquad R \neq 0$$

CHAPTER 5 REVIEW

(5.1)   *Find any real number for which each rational expression is undefined.*

1. $\dfrac{x + 5}{x^2 - 4}$

2. $\dfrac{5x + 9}{4x^2 - 4x - 15}$

Find the value of each rational expression when $x = 5$, $y = 7$, and $z = -2$.

3. $\dfrac{z^2 - z}{z + xy}$

4. $\dfrac{x^2 + xy - z^2}{x + y + z}$

Simplify each rational expression.

5. $\dfrac{x + 2}{x^2 - 3x - 10}$

6. $\dfrac{x + 4}{x^2 + 5x + 4}$

7. $\dfrac{x^3 - 4x}{x^2 + 3x + 2}$

8. $\dfrac{5x^2 - 125}{x^2 + 2x - 15}$

9. $\dfrac{x^2 - x - 6}{x^2 - 3x - 10}$

10. $\dfrac{x^2 - 2x}{x^2 + 2x - 8}$

11. $\dfrac{x^2 + 6x + 5}{2x^2 + 11x + 5}$

12. $\dfrac{x^2 + xa + xb + ab}{x^2 - xc + bx - bc}$

13. $\dfrac{x^2 + xb - 2x - 2b}{x^2 - xc - 2x + 2c}$

14. $\dfrac{x^2 - 9}{9 - x^2}$

15. $\dfrac{4 - x}{x^3 - 64}$

(5.2) *Perform the indicated operations and simplify.*

16. $\dfrac{15x^3 y^2}{z} \cdot \dfrac{z}{5xy^3}$

17. $\dfrac{-y^3}{8} \cdot \dfrac{9x^2}{y^3}$

18. $\dfrac{x^2 - 9}{x^2 - 4} \cdot \dfrac{x - 2}{x + 3}$

19. $\dfrac{2x + 5}{x - 6} \cdot \dfrac{2x}{-x + 6}$

20. $\dfrac{x^2 - 5x - 24}{x^2 - x - 12} \div \dfrac{x^2 - 10x + 16}{x^2 + x - 6}$

21. $\dfrac{4x + 4y}{xy^2} \div \dfrac{3x + 3y}{x^2 y}$

22. $\dfrac{x^2 + x - 42}{x - 3} \cdot \dfrac{(x - 3)^2}{x + 7}$

23. $\dfrac{2a + 2b}{3} \cdot \dfrac{a - b}{a^2 - b^2}$

24. $\dfrac{x^2 - 9x + 14}{x^2 - 5x + 6} \cdot \dfrac{x + 2}{x^2 - 5x - 14}$

25. $(x - 3) \cdot \dfrac{x}{x^2 + 3x - 18}$

26. $\dfrac{2x^2 - 9x + 9}{8x - 12} \div \dfrac{x^2 - 3x}{2x}$

27. $\dfrac{x^2 - y^2}{x^2 + xy} \div \dfrac{3x^2 - 2xy - y^2}{3x^2 + 6x}$

28. $\dfrac{x^2 - y^2}{8x^2 - 16xy + 8y^2} \div \dfrac{x + y}{4x - y}$

29. $\dfrac{x - y}{4} \div \dfrac{y^2 - 2y - xy + 2x}{16x + 24}$

30. $\dfrac{y - 3}{4x + 3} \div \dfrac{9 - y^2}{4x^2 - x - 3}$

(5.3, 5.4) *Perform the indicated operations and simplify.*

31. $\dfrac{5x - 4}{3x - 1} + \dfrac{6}{3x - 1}$

32. $\dfrac{4x - 5}{3x^2} - \dfrac{2x + 5}{3x^2}$

33. $\dfrac{9x + 7}{6x^2} - \dfrac{3x + 4}{6x^2}$

Find the LCD of each pair of rational expressions.

34. $\dfrac{x + 4}{2x}, \dfrac{3}{7x}$

35. $\dfrac{x - 2}{x^2 - 5x - 24}, \dfrac{3}{x^2 + 11x + 24}$

Rewrite the following rational expressions as equivalent expressions whose denominator is the given polynomial.

36. $\dfrac{x + 2}{x^2 + 11x + 18}, (x + 2)(x - 5)(x + 9)$

37. $\dfrac{3x - 5}{x^2 + 4x + 4}, (x + 2^2)^2 \cdot (x + 3)$

Perform the indicated operations and simplify.

38. $\dfrac{4}{5x^2} - \dfrac{6}{y}$

39. $\dfrac{2}{x - 3} - \dfrac{4}{x - 1}$

40. $\dfrac{x + 7}{x + 3} - \dfrac{x - 3}{x + 7}$

41. $\dfrac{x}{x^2 - 9} + \dfrac{4}{x + 3}$

42. $\dfrac{3}{x^2 + 2x - 8} + \dfrac{2}{x^2 - 3x + 2}$

43. $\dfrac{2x - 5}{6x + 9} - \dfrac{4}{2x^2 + 3x}$

44. $\dfrac{x - 1}{x^2 - 2x + 1} - \dfrac{x + 1}{x - 1}$

45. $\dfrac{x - 1}{x^2 + 4x + 4} + \dfrac{x - 1}{x + 2}$

(5.5) *Simplify each complex fraction.*

46. $\dfrac{1 - \dfrac{9}{16}}{3 + \dfrac{4}{5}}$

47. $\dfrac{\dfrac{3}{5} + \dfrac{2}{7}}{\dfrac{1}{5} + \dfrac{5}{6}}$

48. $\dfrac{\dfrac{2}{a} + \dfrac{1}{2a}}{a + \dfrac{a}{2}}$

49. $\dfrac{3 - \dfrac{1}{y}}{2 - \dfrac{1}{y}}$

50. $\dfrac{\dfrac{b^2}{a} - a}{\dfrac{a^2}{b} - b}$

51. $\dfrac{2 + \dfrac{1}{x^2}}{\dfrac{1}{x} + \dfrac{2}{x^2}}$

52. $\dfrac{\dfrac{1}{a} + \dfrac{1}{b}}{\dfrac{1}{ab}}$

53. $\dfrac{\dfrac{1}{a} + 1}{\dfrac{1}{b} - 1}$

54. $\dfrac{2a^{-1} + (2a)^{-1}}{a + 2^{-1}a}$

55. $\dfrac{x^{-1} + 2x^2}{2 + x^{-2}}$

(5.6) *Solve each equation for the variable or perform the indicated operation.*

56. $\dfrac{x + 4}{9} = \dfrac{5}{9}$

57. $\dfrac{n}{10} = 9 - \dfrac{n}{5}$

58. $\dfrac{5y - 3}{7} = \dfrac{15y - 2}{28}$

59. $\dfrac{2}{x + 1} - \dfrac{1}{x - 2} = -\dfrac{1}{2}$

60. $\dfrac{1}{a + 3} + \dfrac{1}{a - 3} = -\dfrac{5}{a^2 - 9}$

61. $\dfrac{y}{2y + 2} + \dfrac{2y - 16}{4y + 4} = \dfrac{y - 3}{y + 1}$

62. $\dfrac{4}{x + 3} + \dfrac{8}{x^2 - 9} = 0$

63. $\dfrac{2}{x - 3} - \dfrac{4}{x + 3} = \dfrac{8}{x^2 - 9}$

64. $\dfrac{x - 3}{x + 1} - \dfrac{x - 6}{x + 5} = 0$

65. $x + 5 = \dfrac{6}{x}$

Solve the equation for the indicated variable.

66. $\dfrac{4A}{5b} = x^2$, for b

67. $\dfrac{x}{7} + \dfrac{y}{8} = 10$, for y

(5.7) *Write each phrase as a ratio in fractional notation.*

68. 20 francs to 1 dollar

69. four parts red to six parts white

Solve each proportion.

70. $\dfrac{x}{2} = \dfrac{12}{4}$

71. $\dfrac{20}{1} = \dfrac{x}{25}$

72. $\dfrac{32}{100} = \dfrac{100}{x}$

73. $\dfrac{20}{2} = \dfrac{c}{5}$

74. $\dfrac{2}{x - 1} = \dfrac{3}{x + 3}$

75. $\dfrac{4}{y - 3} = \dfrac{2}{y - 3}$

76. $\dfrac{y + 2}{y} = \dfrac{5}{3}$

77. $\dfrac{x - 3}{3x + 2} = \dfrac{2}{6}$

Solve.

78. A machine can process 300 parts in 20 minutes. Find how many parts can be processed in 45 minutes.

79. As his consulting fee, Mr. Visconti charges $90.00 per day. Find how much he charges for 3 hours of consulting.

Assume an 8-hour work day.

80. One fund raiser can address 100 letters in 35 minutes. Find how many he can address in 55 minutes.

(5.8) *Solve each problem.*

81. A car travels 90 miles in the same time that a car traveling 10 miles per hour slower travels 60 miles. Find the speed of each car.

82. The speed of a stream is 4 miles per hour. A paddle boat travels 48 miles upstream in the same amount of time it takes to travel 72 miles downstream. Find the speed of the boat in still water.

83. When Mark and Maria manicure Mr. Stergeon's lawn, it takes them 5 hours. If Mark works alone, it takes 7 hours. Find how long it takes Maria alone.

84. It takes pipe 1 20 days to fill a pond. Pipe 2 takes 15 days. Find how long it takes both pipes together to fill the pond.

85. Five times the reciprocal of a number equals the sum of $\frac{3}{2}$ times the reciprocal of the number and $\frac{7}{6}$. What is the number?

86. The reciprocal of a number equals the reciprocal of the difference of 4 and the number. Find the number.

CHAPTER 5 TEST

1. Find any real numbers for which the following expression is undefined.

$$\frac{x + 5}{x^2 + 4x + 3}$$

Simplify each rational expression.

2. $\dfrac{4}{8x - 4}$

3. $\dfrac{3x - 6}{5x - 10}$

4. $\dfrac{x + 10}{x^2 - 100}$

5. $\dfrac{x + 6}{x^2 + 12x + 36}$

6. $\dfrac{x + 3}{x^3 + 27}$

7. $\dfrac{2m^3 - 2m^2 - 12m}{m^2 - 5m + 6}$

8. $\dfrac{ay + 3a + 2y + 6}{ay + 3a + 5y + 15}$

9. $\dfrac{y - x}{x^2 - y^2}$

Perform the indicated operation and simplify if possible.

10. $\dfrac{a^2 + 10a + 25}{a + 5} \cdot \dfrac{a - 5}{a^2 - 25}$

11. $\dfrac{x^2 - 13x + 42}{x^2 + 10x + 21} \div \dfrac{x^2 - 4}{x^2 + x - 6}$

12. $\dfrac{3}{x - 1} \cdot (5x - 5)$

13. $\dfrac{y^2 - 5y + 6}{2y + 4} \cdot \dfrac{y + 2}{2y - 6}$

14. $\dfrac{5}{2x + 5} - \dfrac{6}{2x + 5}$

15. $\dfrac{x - 1}{x^2 - x - 6} + \dfrac{x^2 - 1}{x^2 + 5x + 6}$

16. $\dfrac{5a}{a^2 - a - 6} - \dfrac{2}{a - 3}$

17. $\dfrac{6}{x^2 - 1} + \dfrac{3}{x + 1}$

18. $\dfrac{x^2 - 9}{x^2 - 3x} \div \dfrac{xy + 5x + 3y + 15}{2x + 10}$

19. $\dfrac{x + 2}{x^2 + 11x + 18} + \dfrac{5}{x^2 - 3x - 10}$

20. $\dfrac{4y}{y^2 + 6y + 5} - \dfrac{3}{y^2 + 5y + 4}$

Solve each equation.

21. $\dfrac{4}{y} - \dfrac{5}{3} = \dfrac{-1}{5}$

22. $\dfrac{5}{y+1} = \dfrac{4}{y+2}$

23. $\dfrac{a}{a-3} = \dfrac{3}{a-3} - \dfrac{3}{2}$

24. $\dfrac{10}{x^2-25} = \dfrac{3}{x+5} + \dfrac{1}{x-5}$

Simplify each complex fraction.

25. $\dfrac{\dfrac{b}{a} - \dfrac{a}{b}}{\dfrac{b}{a} + \dfrac{b}{a}}$

26. $\dfrac{3 - \dfrac{1}{6}}{2 - \dfrac{1}{6}}$

27. A car wash can wash 5 cars in 30 minutes. Find how many cars can be washed in two hours. (*Hint:* Make sure your units are consistent.)

28. One number plus five times its reciprocal is equal to six. Find the number.

29. A pleasure boat traveling down the Red River takes the same time to go 14 miles upstream as it takes to go 16 miles downstream. If the current of the river is 2 miles per hour, find the speed of the boat in still water.

30. An inlet pipe can fill a tank in 12 hours. A second pipe can fill the tank in 15 hours. If both pipes are used, find how long it takes to fill the tank.

CHAPTER 5 CUMULATIVE REVIEW

1. Write $\dfrac{2}{5}$ as an equivalent fraction with a denominator of 20

2. Evaluate the following
a. 3^2 **b.** 5^3 **c.** 2^4 **d.** 7^1 **e.** $\left(\dfrac{3}{7}\right)^2$

3. Find the following products
a. $(-1.2)(0.05)$
b. $\dfrac{2}{3} \cdot -\dfrac{7}{10}$

4. Write the following phrases as algebraic expressions. Let x represent the unknown number
a. Twice a number, added to six.
b. The difference of a number and four, divided by 7.

5. Solve $-5(2a-1) - (-11a+6) = 7$ for a.

6. Solve $4(2x-3) + 7 = 3x + 5$.

7. If the current Fahrenheit temperature is 59°, find the equivalent temperature in degrees Celsius.

8. Solve $x + 4 \le -6$ for x. Graph the solution.

9. Simplify each product.
a. $4^2 \cdot 4^5$ **b.** $x^2 \cdot x^5$ **c.** $y^3 \cdot y$
d. $y^3 \cdot y^2 \cdot y^7$ **e.** $(-5)^7 \cdot (-5)^8$

10. Simplify each quotient.
a. $\dfrac{x^5}{x^2}$ **b.** $\dfrac{4^7}{4^3}$ **c.** $\dfrac{(-3)^5}{(-3)^2}$ **d.** $\dfrac{2x^5y^2}{xy}$

11. Subtract $(5x-3) - (2x-11)$.

12. Multiply $(x-3)(x+4)$.

13. Factor $3m^2n(a+b) - (a+b)$.

14. Factor $x^2 + 7x + 12$.

15. Factor $24x^4 + 40x^3 + 6x^2$.

16. Factor $4x^2 - 1$.

17. Factor $y^3 - 27$.

18. Write $\dfrac{x^2+8x+7}{x^2-4x-5}$ in simplest form.

19. Multiply $\dfrac{x^2+x}{3x} \cdot \dfrac{6}{5x+5}$.

20. Subtract $\dfrac{3}{2x^2+x} - \dfrac{2x}{6x+3}$.

21. Simplify $\dfrac{\dfrac{2}{3} + \dfrac{1}{5}}{\dfrac{2}{3} - \dfrac{2}{9}}$.

22. Solve the equation $3 - \dfrac{6}{x} = x + 8$.

23. Write a ratio for each phrase. Use fractional notation.
  a. The ratio of 2 parts salt to 5 parts water.
  b. The ratio of 12 almonds to 16 pecans.

6.1 The Cartesian Coordinate System

6.2 Graphing Linear Equations

6.3 Slope

6.4 Equations of Lines

6.5 Graphing Linear Inequalities

6.6 Functions

CHAPTER 6

Graphing Linear Equations and Inequalities

You are responsible for arranging the food to be served at the upcoming company picnic. The thought of preparing all the food on a few barbecue grills is overwhelming, so you decide to hire a caterer. (See Critical Thinking page 303.)

INTRODUCTION

In Chapter 2 we learned to solve and graph the solutions of linear equations and inequalities in one variable. Now we define and present techniques for solving and graphing linear equations and inequalities in two variables. Two-variable equations lead directly to the concept of function, perhaps the most important concept in all mathematics. We introduce you to functions in the concluding section of this chapter.

6.1
The Cartesian Coordinate System

Tape 20

OBJECTIVES

1   Define an ordered pair and linear equations in two variables.

2   Determine whether an ordered pair is a solution of a linear equation in two variables.

3   Find the missing coordinate of an ordered pair, given one coordinate of the pair.

4   Define the Cartesian coordinate system.

5   Plot ordered pairs.

---

1   An equation in one variable such as $x + 1 = 5$ has one solution, which is 4: the number 4 is the value of the variable x that makes the equation true. We can graph this solution on a number line.

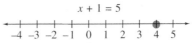

$$x + 1 = 5$$

An equation in two variables, such as $2x + y = 8$, has solutions consisting of two values, one for x and one for y. For example, $x = 3$ and $y = 2$ is a solution of $2x + y = 8$ because, if x is replaced with 3 and y with 2, we get a true statement.

$$2x + y = 8$$
$$2(3) + 2 = 8$$
$$8 = 8 \qquad \text{True.}$$

The solution $x = 3$ and $y = 2$ can be written as $(3, 2)$, an **ordered pair** of numbers. The first number 3 is the x-value and the second number 2 is the y-value. Equations such as $2x + y = 8$ are called **linear equations in two variables.**

> **Linear Equation in Two Variables**
>
> A linear equation in two variables is an equation that can be written in the form
>
> $$Ax + By = C$$
>
> where A, B, and C are real numbers, and A and B are not both 0.

257

The form $Ax + By = C$ is called the **standard form** of a linear equation in two variables.

Examples of Linear Equations in Two Variables

$$2x + y = 8 \qquad -2x = 7y \qquad y = \frac{1}{3}x + 2$$

The equation $2x + y = 8$ is written in standard form.

2   An ordered pair is a **solution** of an equation in two variables if replacing the variables by the values of the ordered pair results in a true statement. For example, $(1, 4)$ is a solution of $x + y = 5$ because, if we substitute 1 for x and 4 for y, we obtain the true statement $1 + 4 = 5$.

EXAMPLE 1   Determine whether each ordered pair is a solution to the equation $x - 2y = 6$.
  a. $(6, 0)$ **b.** $(0, 3)$ **c.** $(2, -2)$

Solution: **a.** Let $x = 6$ and $y = 0$ in the equation $x - 2y = 6$.

$$x - 2y = 6$$
$$6 - 2(0) = 6 \qquad \text{Replace } x \text{ with 6 and } y \text{ with 0.}$$
$$6 - 0 = 6 \qquad \text{Simplify.}$$
$$6 = 6 \qquad \text{True.}$$

$(6, 0)$ is a solution, since $6 = 6$ is a true statement.
  b. Let $x = 0$ and $y = 3$.

$$x - 2y = 6$$
$$0 - 2(3) = 6 \qquad \text{Replace } x \text{ with 0 and } y \text{ with 3.}$$
$$0 - 6 = 6$$
$$-6 = 6 \qquad \text{False.}$$

$(0, 3)$ is **not** a solution, since $-6 = 6$ is a false statement.
  c. Let $x = 2$ and $y = -2$ in the equation.

$$x - 2y = 6$$
$$2 - 2(-2) = 6 \qquad \text{Replace } x \text{ with 2 and } y \text{ with } -2.$$
$$2 + 4 = 6$$
$$6 = 6 \qquad \text{True.}$$

$(2, -2)$ is a solution, since $6 = 6$ is a true statement.   ∎

3   If one value of an ordered pair solution of a linear equation is known, the other value can be determined. To find the unknown value, replace one variable in the equation by the known value. Doing so will result in an equation with just one variable that can be solved for the variable using the methods of Chapter 2.

EXAMPLE 2   Complete the following ordered pairs for the equation $3x + y = 12$.
  a. $(0, \ )$ **b.** $(\ , 6)$ **c.** $(-1, \ )$

Solution: **a.** In the ordered pair $(0, \ )$, the x-value is 0. Let $x = 0$ in the equation and solve for y.

$$3x + y = 12$$
$$3(0) + y = 12 \qquad \text{Replace } x \text{ with 0.}$$
$$0 + y = 12 \qquad \text{Solve for } y.$$
$$y = 12$$

The completed ordered pair is $(0, 12)$.

b. In the ordered pair $(\ , 6)$, the y-value is 6. Let $y = 6$ in the equation and solve for x.

$$3x + y = 12$$
$$3x + 6 = 12 \qquad \text{Replace } y \text{ with 6.}$$
$$3x = 6 \qquad \text{Subtract 6 from both sides.}$$
$$x = 2 \qquad \text{Divide both sides by 3.}$$

The ordered pair is $(2, 6)$.

c. In the ordered pair $(-1, \ )$, the x-value is -1. Let $x = -1$ in the equation and solve for y.

$$3x + y = 12$$
$$3(-1) + y = 12 \qquad \text{Replace } x \text{ with } -1.$$
$$-3 + y = 12$$
$$y = 15 \qquad \text{Add 3 to both sides.}$$

The ordered pair is $(-1, 15)$.   ■

EXAMPLE 3   Complete the following ordered pairs for the equation $y = 3x$.

a. $(-1, \ )$ **b.** $(\ , 0)$ **c.** $(\ , -9)$

Solution: **a.** Replace x with -1 in the equation and solve for y.

$$y = 3x$$
$$y = 3(-1) \qquad \text{Let } x = -1.$$
$$y = -3$$

The ordered pair is $(-1, -3)$.

b. Replace y with 0 in the equation and solve for x.

$$y = 3x$$
$$0 = 3x \qquad \text{Let } y = 0.$$
$$0 = x \qquad \text{Divide both sides by 3.}$$

The completed ordered pair is $(0, 0)$.

c. Replace y with -9 in the equation and solve for x.

$$y = 3x$$
$$-9 = 3x \qquad \text{Let } y = -9.$$
$$-3 = x \qquad \text{Divide both sides by 3.}$$

The completed ordered pair is $(-3, -9)$.   ■

EXAMPLE 4 Complete each ordered pair so that it is a solution of the equation $y = 3$.
a. (2,)   **b.** (0,)   **c.** (−5,)

Solution: The equation $y = 3$ is the same as $0x + y = 3$. No matter what value we replace x by, y always equals 3. The completed ordered pairs are:
a. (2, 3)   **b.** (0, 3)   **c.** (−5, 3)   ∎

By now, you have probably noticed that linear equations in two variables have more than one solution. We will discuss this more in the next section. First, let's find out how to graph solutions of linear equations in two variables.

To graph these ordered pair solutions, a system is needed that can represent two-valued solutions.

4 It took the genius of the seventeenth-century French mathematician Rene Descartes to devise a method of graphing solutions of two-variable equations. This method uses a system called the **Cartesian coordinate system,** in honor of Descartes, or the **rectangular coordinate system.** The rectangular coordinate system consists of a pair of number lines perpendicular to each other, intersecting at the point 0 on each. This point of intersection is called the **origin.** We call the horizontal x-value number line the **x-axis,** and the vertical y-value number line the **y-axis.** These two number lines or axes separate the plane into four regions called **quadrants.** The quadrants are usually numbered with Roman numerals as shown. The axes are not considered to be in any quadrant.

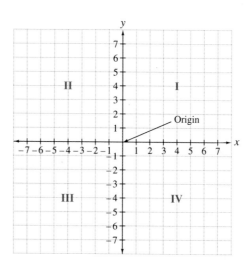

5 Every point on the rectangular coordinate system corresponds to one ordered pair, such as (3, 2). The x-value 3 is also called the **x-coordinate** of an ordered pair and it is associated with the x-axis. The y-value 2 is also called the **y-coordinate** of an ordered pair and it is associated with the y-axis. The origin is associated with the ordered pair (0, 0). To find the point on the rectangular coordinate system that corresponds to the ordered pair (3, 2), start at the origin. Move 3 units in the positive x direction; from there, move 2 units in the positive y direction. Finding the point on the coordinate system that corresponds to a particular ordered pair is called **plotting** the point.

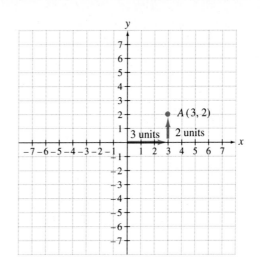

Here are some more points that have been plotted, along with their coordinates.

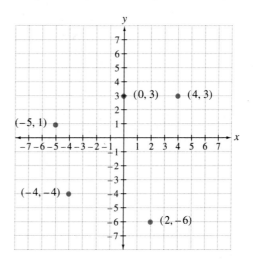

EXAMPLE 5   On a single coordinate system, plot the points corresponding to the following ordered pairs. State in which quadrant, if any, each point lies.

a. $(3, 2)$ **b.** $(-2, -4)$ **c.** $(1, -2)$ **d.** $(-5, 3)$ **e.** $(0, 0)$ **f.** $(0, 2)$ **g.** $(-5, 0)$
h. $\left(0, -1\frac{1}{2}\right)$

Solution:

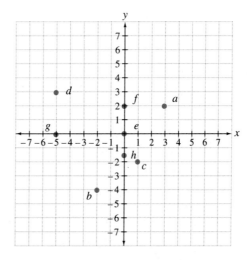

261

Solution: Point **a** lies in quadrant I.
Point **d** lies in quadrant II.
Point **b** lies in quadrant III.
Point **c** lies in quadrant IV.
Points **e, f, g,** and **h** lie on an axis, so they are not in any quadrant. ∎

MENTAL MATH

Give two ordered pair solutions for each of the following linear equations.

1. $x + y = 10$

2. $x + y = 6$

3. $x = 3$

4. $y = -2$

EXERCISE SET 6.1

Determine whether each ordered pair is a solution of the given linear equation. See Example 1.

1. $2x + y = 7$; $(3, 1)$, $(7, 0)$, $(0, 7)$

2. $x - y = 6$; $(5, -1)$, $(7, 1)$, $(0, -6)$

3. $y = -5x$; $(-1, -5)$, $(0, 0)$, $(2, -10)$

4. $x = 2y$; $(0, 0)$, $(2, 1)$, $(-2, -1)$

5. $x = 5$; $(4, 5)$, $(5, 4)$, $(5, 0)$

6. $y = 2$; $(-2, 2)$, $(2, 2)$, $(0, 2)$

Complete each ordered pair so that it is a solution of the given linear equation. See Examples 2 through 4.

7. $x - 4y = 4$; $(\ \ , -2)$, $(4, \ \ )$

8. $x - 5y = -1$; $(\ \ , -2)$, $(4, \ \ )$

9. $3x + y = 9$; $(0, \ \ )$, $(\ \ , 0)$

10. $x + 5y = 15$; $(0, \ \ )$, $(\ \ , 0)$

11. $y = -7$; $(11, \ \ )$, $(\ \ , -7)$

12. $x = \dfrac{1}{2}$; $(\ \ , 0)$, $\left(\dfrac{1}{2}, \ \ \right)$

Plot the following points. State in which quadrant if any that each point lies. See Example 5.

13. $(1, 5)$

14. $(-5, -2)$

15. $(-6, 0)$

16. $(0, -1)$

17. $(2, -4)$

18. $(-1, 4)$

19. $\left(4\dfrac{3}{4}, 0\right)$

20. $\left(0, \dfrac{7}{8}\right)$

Determine whether each ordered pair is a solution of the given linear equation.

21. $x + 2y = 9$; $(5, 2)$, $(0, 9)$

22. $3x + y = 8$; $(2, 3)$, $(0, 8)$

23. $2x - y = 11$; $(3, -4)$, $(9, 8)$

24. $x - 4y = 14$; $(2, -3)$, $(14, 6)$

25. $x = \frac{1}{3}y$; $(0, 0)$, $(3, 9)$

26. $y = -\frac{1}{2}x$; $(0, 0)$, $(4, 2)$

27. $y = -2$; $(-2, -2)$, $(5, -2)$

28. $x = 4$; $(4, 0)$, $(4, 4)$

Complete the ordered pair so that it is a solution of the given linear equation.

29. $2x + 5y = 11$; $(, 1)$, $(0,)$

30. $4x + 7y = 29$; $(2,)$, $(, 0)$

31. $y = 7$; $(2,)$

32. $x = -4$; $(, 2)$

33. $x = -5y$; $(, 0)$

34. $y = \frac{1}{4}x$; $(8,)$

35. $\frac{1}{2}x + y = 10$; $(, 4)$

36. $x + \frac{3}{4}y = 12$; $(, 8)$

37. $-x - 2y = 5$; $(0,)$

38. $-x - 4y = 15$; $(0,)$

Complete the ordered pairs so that each is a solution of the given linear equation; then plot each solution. Use a single coordinate system for each equation.

39. $x + 3y = 6$
  a. $(0,)$
  b. $(, 0)$
  c. $(, 1)$

40. $2x + y = 4$
  a. $(0,)$
  b. $(, 0)$
  c. $(, 2)$

41. $2x - y = 12$
  a. $(0,)$
  b. $(, -2)$
  c. $(-3,)$

42. $-5x + y = 10$
  a. $(, 0)$
  b. $(, 5)$
  c. $(2,)$

43. $2x + 7y = 5$
  a. $(0,)$
  b. $(, 0)$
  c. $(, 1)$

44. $x - 6y = 3$
  a. $(0,)$
  b. $(1,)$
  c. $(, -1)$

45. $x = 3$
  a. $(, 0)$
  b. $(, -0.5)$
  c. $\left(, \frac{1}{4}\right)$

46. $y = -1$
  a. $(-2,)$
  b. $(0,)$
  c. $(-1,)$

47. $x = -5y$
  a. $(, 0)$
  b. $(, 1)$
  c. $(10,)$

48. $y = -3x$
  a. $(0,)$
  b. $(-2,)$
  c. $(, 9)$

Find the x- and y-coordinates of the following labeled points.

49.

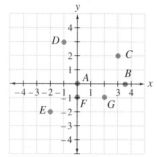

50.

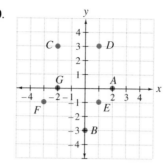

See the following examples of graphs found in newspapers. Study the graphs for a moment and then use the graphs as indicated.

51. Use this graph to estimate the number of members of Ochsner's HMO in January 1986. *(Times/Picayune Newspaper)*

52. Use this graph to estimate the stock market activity on Tuesday. (Sept. 8, 1991 *Times/Picayune*)

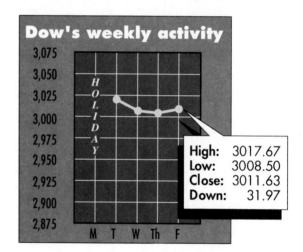

Writing in Mathematics

53. Explain why linear equations in two variables have more than one solution.

54. Explain why order is important in ordered pairs.

55. According to legend, Rene Descartes watched a fly move about the ceiling while lying in bed and tried to think of a way to describe the location of the fly as it moved from place to place. Describe the system he designed.

56. Explain how you decide if an ordered pair is a solution of a linear equation.

57. For a given linear equation, if one coordinate of an ordered pair solution is known, explain how to find the other coordinate.

Skill Review

Factor the following expressions. See Section 4.5.

58. $2x^2 - 5x - 3$

59. $2x^2 + 2x - 20$

60. $ax + am + x + m$

61. $x^2 + x + xy + y$

Find the perimeter of each figure. See Section 5.3.

62.

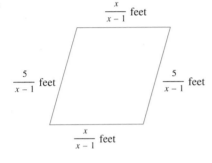

$\dfrac{x}{x-1}$ feet

$\dfrac{5}{x-1}$ feet

$\dfrac{5}{x-1}$ feet

$\dfrac{x}{x-1}$ feet

63.

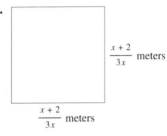

$\dfrac{x+2}{3x}$ meters

$\dfrac{x+2}{3x}$ meters

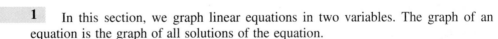

6.2
Graphing Linear Equations

OBJECTIVES

Tape 20

1 Graph a linear equation by finding and plotting ordered pair solutions.

2 Graph a linear equation by using the intercept method.

3 Graph vertical and horizontal lines.

1   In this section, we graph linear equations in two variables. The graph of an equation is the graph of all solutions of the equation.

In the previous section, we mentioned that linear equations in two variables have more than one solution. For example, the ordered pairs $(-1, 6)$, $(0, 5)$, $(1, 4)$, $(2, 3)$ and $(5, 0)$ are all solutions of the linear equation $x + y = 5$. How many solutions does this equation have? In fact, there are an infinite number of solutions to $x + y = 5$ since there are an infinite number of pairs of numbers whose sum is 5.

In general, a linear equation in two variables has an infinite number of ordered pair solutions. It is impossible to list every solution to such an equation. Instead, graph the solutions to $x + y = 5$ listed previously and notice the pattern formed. All points graphed lie on the same straight line. It can be shown that **the graph of a linear equation in two variables is a straight line.**

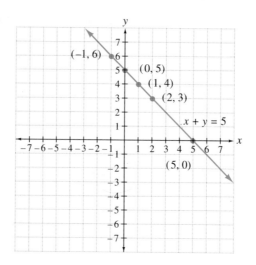

Therefore, the graph of the $x + y = 5$ is the straight line passing through points $(-1, 6)$, $(0, 5)$, $(1, 4)$, $(2, 3)$, and $(5, 0)$ as shown. All points on the line correspond to ordered pair solutions of $x + y = 5$, and all solutions of $x + y = 5$ correspond to points on the line.

From geometry, we know that a straight line is determined by just two points. Graphing a linear equation in two variables, then, requires that we find just two of its infinitely many solutions. Once we do so, we plot the solution points and draw the line connecting the points. Usually, we find a third solution as well, as a check.

EXAMPLE 1 Graph the linear equation $2x + y = 5$.

Solution: Find three ordered pair solutions of $2x + y = 5$. To do this, choose a value for one variable, x or y, and solve for the other variable. For example, let $x = 1$. Then $2x + y = 5$ becomes

$$2(1) + y = 5$$

$$2 + y = 5 \qquad \text{Multiply.}$$

$$y = 3 \qquad \text{Subtract 2 from both sides.}$$

Since $y = 3$ when $x = 1$, the ordered pair $(1, 3)$ is a solution of $2x + y = 5$. Next, let $x = 0$.

$$2x + y = 5$$

$$2(0) + y = 5 \qquad \text{Replace } x \text{ with 0.}$$

$$0 + y = 5$$

$$y = 5$$

The ordered pair $(0, 5)$ is a second solution.

The two solutions found so far will allow us to draw the straight line that is the graph of all solutions of $2x + y = 5$. However, we will find a third ordered pair as a check. Let $y = -1$. Then $2x + y = 5$ becomes

$$2x + (-1) = 5$$

$$2x - 1 = 5$$

$$2x = 6 \qquad \text{Add 1 to both sides.}$$

$$x = 3 \qquad \text{Divide both sides by 2.}$$

The third solution is $(3, -1)$. These three ordered pair solutions can be listed in table form as shown. The graph of $2x + y = 5$ is the line through the three points.

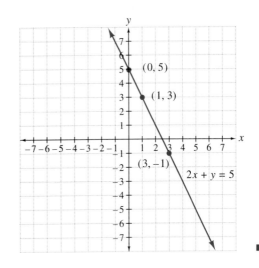

x	y
1	3
0	5
3	-1

EXAMPLE 2 Graph $y = 4x - 8$.

Solution: Since this equation is solved for y, we will choose three x-values.

If $x = 2$, then	If $x = 0$, then	If $x = 3$, then
$y = 4(2) - 8$ or	$y = 4(0) - 8$ or	$y = 4(3) - 8$ or
$y = 8 - 8$ or	$y = 0 - 8$ or	$y = 12 - 8$ or
$y = 0$	$y = -8$	$y = 4$

Three ordered pair solutions of $y = 4x - 8$ are $(2, 0)$, $(0, -8)$, and $(3, 4)$. These solutions are shown in a table along with the graph of $y = 4x - 8$.

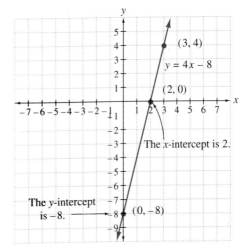

x	y
3	4
2	0
0	-8

The x-intercept is 2.

The y-intercept is -8.

2   The graph of $y = 4x - 8$ in the preceding figure crosses the y-axis at the point $(0, -8)$. The y-coordinate of this point, -8, is called the **y-intercept.** Likewise, the graph crosses the x-axis at $(2, 0)$ and the x-coordinate 2 is called the **x-intercept.**

To find the y-intercept of a graph, let $x = 0$ since a point on the y-axis has an x-coordinate of 0. To find the x-intercept of a line, let $y = 0$ since a point on the x-axis has a y-coordinate of 0.

> **Finding x- and y-Intercepts**
>
> To find the x-intercept, let $y = 0$ and solve for x.
> To find the y-intercept, let $x = 0$ and solve for y.

Intercept points are usually easy to find and plot since one coordinate is 0.

EXAMPLE 3   Graph $x - 3y = 6$ by plotting intercept points.

Solution: Let $y = 0$ to find the x-intercept and $x = 0$ to find the y-intercept.

$$\begin{array}{cc} \text{If } y = 0 \text{ then} & \text{If } x = 0 \text{ then} \\ x - 3(0) = 6 & 0 - 3y = 6 \\ x - 0 = 6 & -3y = 6 \\ x = 6 & y = -2 \end{array}$$

The x-intercept is 6 and the y-intercept is -2. We find a third ordered pair solution to check our work. If we let $y = -1$, then $x = 3$. Plot the points $(6, 0)$, $(0, -2)$, and $(3, -1)$. The graph of $x - 3y = 6$ is the line drawn through these points, as shown.

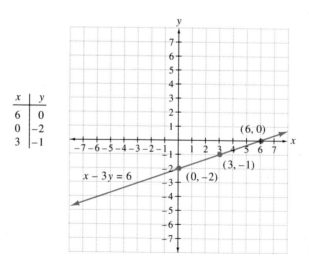

x	y
6	0
0	-2
3	-1

EXAMPLE 4   Graph $x = -2y$ by plotting intercept points.

Solution: Let $y = 0$ to find the x-intercept and $x = 0$ to find the y-intercept.

$$\text{If } y = 0 \quad \text{then} \qquad \text{If } x = 0 \quad \text{then}$$

$$x = -2(0) \quad \text{or} \qquad 0 = -2y \quad \text{or}$$

$$x = 0 \qquad\qquad\qquad 0 = y$$

Both the x-intercept and y-intercept are 0. In other words, when $x = 0$, then $y = 0$, which gives the ordered pair $(0, 0)$. Also, when $y = 0$, then $x = 0$, which gives the same ordered pair $(0, 0)$. This happens when the graph passes through the origin. Since two points are needed to determine a line, we must find at least one more ordered pair that satisfies $x = -2y$. Let $y = -1$ to find a second ordered pair solution and let $y = 1$ as a checkpoint.

If $y = -1$ then	If $y = 1$ then
$x = -2(-1)$ or	$x = -2(1)$ or
$x = 2$	$x = -2$

The ordered pairs are $(0, 0)$, $(2, -1)$, and $(-2, 1)$. Plot these points to graph $x = -2y$.

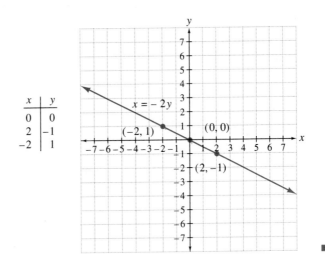

x	y
0	0
2	-1
-2	1

EXAMPLE 5   Graph $4x = 3y - 9$.

Solution: Find the x- and y-intercepts, and then choose $x = 2$ to find a third checkpoint.

If $y = 0$ then	If $x = 0$ then	If $x = 2$ then
$4x = 3(0) - 9$ or	$0 = 3y - 9$ or	$4(2) = 3y - 9$ or
$4x = -9$	$9 = 3y$	$8 = 3y - 9$
Solve for x.	Solve for y.	Solve for y.
$x = -\dfrac{9}{4}$ or $-2\dfrac{1}{4}$	$3 = y$	$17 = 3y$
		$\dfrac{17}{3} = y$ or $y = 5\dfrac{2}{3}$

The ordered pairs are $\left(-2\dfrac{1}{4}, 0\right)$, $(0, 3)$, $\left(2, 5\dfrac{2}{3}\right)$. The equation $4x = 3y - 9$ is graphed as follows.

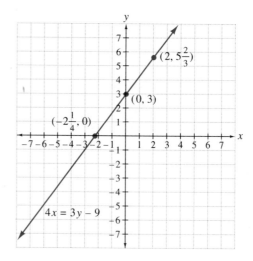

x	y
$-2\frac{1}{4}$	0
0	3
2	$5\frac{2}{3}$

3   The equation $x = c$, where c is a real number constant, is a linear equation in two variables because it can be written in the form $x + 0y = c$. The graph of this equation is a vertical line as shown in the next example.

EXAMPLE 6   Graph $x = 2$.

Solution: The equation $x = 2$ can be written as $x + 0y = 2$. For any y-value chosen, notice that x is 2. No other value for x satisfies $x + 0y = 2$. Any ordered pair whose x-coordinate is 2 is a solution to $x + 0y = 2$. We will use the ordered pairs $(2, 3)$, $(2, 0)$ and $(2, -3)$ to graph $x = 2$.

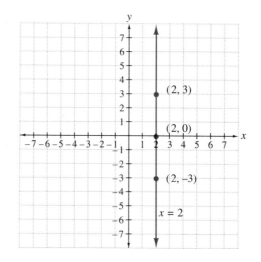

x	y
2	3
2	0
2	-3

The graph is a vertical line with x-intercept 2. Note that this graph has no y-intercept because x is never 0.   ■

Vertical Lines

The graph of $x = c$, where c is a real number, is a vertical line with x-intercept c.

EXAMPLE 7   Graph $y = -3$.

Solution: The equation $y = -3$ can be written as $0x + y = -3$. For any x-value chosen, y is -3. If we choose 4, 1, and -2 as x-values, the ordered pair solutions are $(4, -3)$, $(1, -3)$, and $(-2, -3)$. Use these ordered pairs to graph $y = -3$. The graph is a horizontal line with y-intercept -3 and no x-intercept.

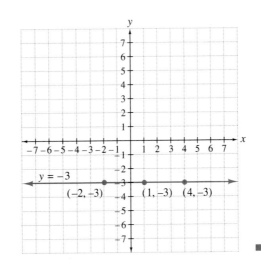

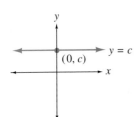

Horizontal Lines

The graph of $y = c$, where c is a real number, is a horizontal line with y-intercept c.

EXERCISE SET 6.2

Graph each linear equation by finding three ordered pair solutions that satisfy it. See Examples 1 and 2.

1. $x + y = 4$

2. $x + y = 7$

3. $x - y = -2$

4. $-x + y = 6$

5. $x - 2y = 4$

6. $-x + 5y = 5$

7. $y = 6x + 3$

8. $y = -2x + 7$

Graph each linear equation by finding x- and y- intercepts. See Examples 3 and 4.

9. $x - y = 3$ **10.** $x - y = -4$ **11.** $x = 5y$ **12.** $2x = y$

13. $-x + 2y = 6$ **14.** $x - 2y = -8$ **15.** $2x - 4y = 8$ **16.** $2x + 3y = 6$

Graph each linear equation. See Example 5.

17. $x - 2y = -6$ **18.** $-x + 2y = 5$ **19.** $y = 6x$

20. $x = -2y$ **21.** $3y - 10 = 5x$ **22.** $-2x + 7 = 2y$

Graph each linear equation. See Examples 6 and 7.

23. $x = -1$ **24.** $y = 5$ **25.** $y = 0$

26. $x = 0$

27. $y + 7 = 0$

28. $x - 2 = 0$

Graph each linear equation.

29. $x + 2y = 8$

30. $x - 3y = 3$

31. $x - 7 = 3y$

32. $y - 3x = 2$

33. $x = -3$

34. $y = 3$

35. $3x + 5y = 7$

36. $3x - 2y = 5$

37. $x = y$

38. $x = -y$

39. $x + 8y = 8$

40. $x - 3y = 9$

41. $5 = 6x - y$

42. $4 = x - 3y$

43. $-x + 10y = 11$

44. $-x + 9 = -y$

45. $y = 1$

46. $x = 1$

47. $x = 2y$

48. $y = -2x$

49. $x + 3 = 0$

50. $y - 6 = 0$

51. $x = 4y - \dfrac{1}{3}$

52. $y = -3x + \dfrac{3}{4}$

53. $2x + 3y = 6$

54. $4x + y = 5$

Writing in Mathematics

55. Discuss whether a vertical line ever has a y-intercept.

56. Discuss whether a horizontal line ever has an x-intercept.

57. Explain why it is a good idea to use three points to graph a linear equation.

58. Explain how to find intercepts.

Skill Review

Simplify the following complex fractions. See Section 5.5.

59. $\dfrac{\dfrac{x}{2}}{\dfrac{y}{6}}$

60. $\dfrac{\dfrac{x^2}{7}}{\dfrac{x^5}{14}}$

61. $\dfrac{\dfrac{1}{2} + \dfrac{2}{3}}{\dfrac{1}{6}}$

62. $\dfrac{\dfrac{3}{4} + \dfrac{1}{6}}{\dfrac{5}{12}}$

63. $\dfrac{\dfrac{x}{8}}{\dfrac{x}{2} + \dfrac{1}{4}}$

64. $\dfrac{\dfrac{y}{10}}{\dfrac{y}{5} + 2}$

6.3
Slope

Tape 21

OBJECTIVES

1 Find the slope of a line given two points on the line.

2 Find the slope of a line given the equation of a line.

3 Compare the slopes of parallel and perpendicular lines.

4 Find the slopes of horizontal and vertical lines.

1 The **slope** of a line is a measure of the **steepness** of the line. Given two points on a line, slope is a ratio of the vertical change or rise between the points and the horizontal change or run between the points.

$$\text{slope} = \frac{\text{rise}}{\text{run}}$$

Consider the following line, which passes through the points (x_1, y_1) and (x_2, y_2). (The notation x_1 is read "x-sub-one.") The vertical change or rise between these points is the difference in the y-coordinates: $y_2 - y_1$. The horizontal change or run between the points is the difference of the x-coordinates: $x_2 - x_1$.

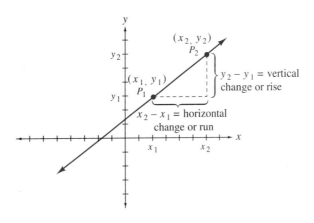

Slope of a Line

Given a line passing through points (x_1, y_1) and (x_2, y_2), the slope (denoted by m) of the line is given by

$$m = \frac{\text{rise}}{\text{run}} = \frac{y_2 - y_1}{x_2 - x_1}, \qquad \text{as long as } x_2 \neq x_1$$

EXAMPLE 1 Find the slope of the line through $(-1, 5)$ and $(2, -3)$. Graph the line.

Solution: If we let $(-1, 5)$ be (x_1, y_1), then $x_1 = -1$ and $y_1 = 5$. Also, let $(2, -3)$ be point (x_2, y_2) so that $x_2 = 2$ and $y_2 = -3$. Then, by the definition of slope,

$$m = \frac{y_2 - y_1}{x_2 - x_1}$$

$$= \frac{-3 - 5}{2 - (-1)}$$

$$= \frac{-8}{3} = -\frac{8}{3}$$

The slope of the line is $-\dfrac{8}{3}$. Its graph follows. (See next page.) ∎

In Example 1, we could just as well have identified (x_1, y_1) with $(2, -3)$ and (x_2, y_2) with $(-1, 5)$. It makes no difference which point is called (x_1, y_1) or (x_2, y_2).

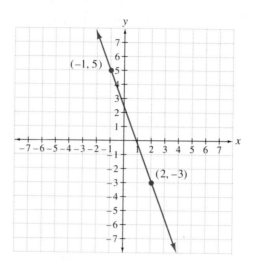

HELPFUL HINT

When finding the slope of a line through two given points, it makes no difference which given point is called (x_1, y_1) and which is called (x_2, y_2). However, once an x-coordinate is called x_1, make sure its corresponding y-coordinate is called y_1.

EXAMPLE 2   Find the slope of the line through $(-1, -2)$ and $(2, 4)$. Graph the line.

Solution: Let $(2, 4)$ be (x_1, y_1) and let $(-1, -2)$ be (x_2, y_2).

$$m = \frac{y_2 - y_1}{x_2 - x_1}$$

$$= \frac{-2 - 4}{-1 - 2}$$

$$= \frac{-6}{-3} = 2$$

The slope is 2.

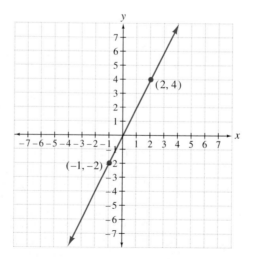

Notice that the slope of the line in Example 1 is negative, whereas the slope of the line in Example 2 is positive. Let your eye follow the line with negative slope from left to right and notice that the line "goes down." Following the line with positive slope from left to right, notice that the line "goes up." This is true in general.

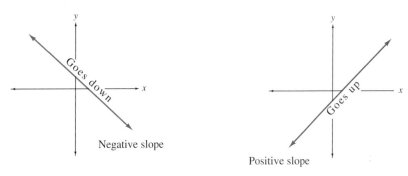

Negative slope

Positive slope

2   As we have seen, the slope of a line is defined by two points on the line. Thus, if we know the equation of a line, we can find its slope.

EXAMPLE 3   Find the slope of the line whose equation is $y = \frac{2}{3}x + 4$.

Solution: Two points are needed on the line defined by $y = \frac{2}{3}x + 4$ in order to find its slope. We will use intercepts as our two points.

If $x = 0$, the corresponding y-value is 4, the y-intercept.
If $y = 0$, the corresponding x-value is -6, the x-intercept.

Use the points $(0, 4)$ and $(-6, 0)$ to find the slope. Let $(0, 4)$ be (x_1, y_1) and $(-6, 0)$ be (x_2, y_2). Then

$$m = \frac{y_2 - y_1}{x_2 - x_1} = \frac{0 - 4}{-6 - 0} = \frac{-4}{-6} = \frac{2}{3} \quad \blacksquare$$

Analyzing the results of Example 3, you may notice a striking pattern:

The slope of $y = \frac{2}{3}x + 4$ is $\frac{2}{3}$, the same as the coefficient of x.

Also, the y-intercept is 4, the same as the constant term.

When a linear equation is written in the form $y = mx + b$, m is the slope of the line and b is its y-intercept. The form $y = mx + b$ is appropriately called the **slope–intercept form.**

> **Slope–Intercept Form**
>
> When a linear equation in two variables is written in slope–intercept form,
>
> $$y = mx + b$$
>
> then m is the slope of the line and b is the y-intercept of the line.

EXAMPLE 4   Find the slope and the y-intercept of the line whose equation is $3x - 4y = 4$.

Solution: Write the equation in slope–intercept form by solving for y.

$$3x - 4y = 4$$

$$-4y = -3x + 4 \qquad \text{Subtract } 3x \text{ from both sides.}$$

$$\frac{-4y}{-4} = \frac{-3x}{-4} + \frac{4}{-4} \qquad \text{Divide both sides by } -4.$$

$$y = \frac{3}{4}x - 1 \qquad \text{Simplify.}$$

The coefficient of x, $\dfrac{3}{4}$, is the slope, and the constant term -1 is the y-intercept.   ■

3   Slopes of lines can help us determine whether lines are parallel. Parallel lines have the same steepness, so it follows that they have the same slope.

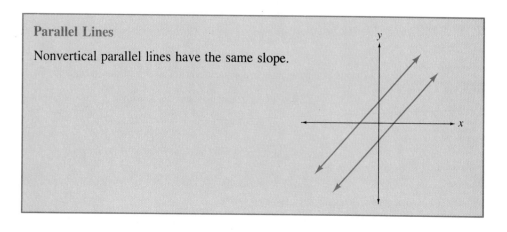

Parallel Lines

Nonvertical parallel lines have the same slope.

How do the slopes of perpendicular lines compare? Two lines that intersect at right angles are said to be **perpendicular.** The product of the slopes of two perpendicular lines is -1.

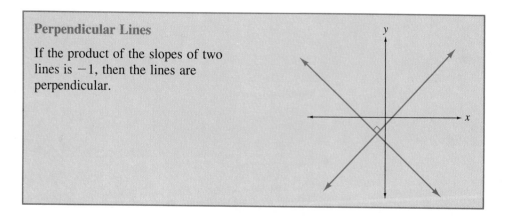

Perpendicular Lines

If the product of the slopes of two lines is -1, then the lines are perpendicular.

EXAMPLE 5   Are the following pairs of lines parallel, perpendicular, or neither?

a. $3x + 7y = 4$ **b.** $-x + 3y = 2$
   $6x + 14 = 7$ $2x + 6y = 5$

Solution: Find the slope of each line by solving each equation for y.

a. $3x + 7y = 4$ $\qquad\qquad\qquad$ $6x + 14y = 7$

$$7y = -3x + 4 \qquad\qquad 14y = -6x + 7$$

$$\frac{7y}{7} = \frac{-3x}{7} + \frac{4}{7} \qquad\qquad \frac{14y}{14} = \frac{-6x}{14} + \frac{7}{14}$$

$$y = -\frac{3}{7}x + \frac{4}{7} \qquad\qquad y = -\frac{3}{7}x + \frac{1}{2}$$

The slope of both lines is $-\dfrac{3}{7}$. The lines are parallel.

b. $-x + 3y = 2$ $\qquad\qquad\qquad$ $2x + 6y = 5$

$$3y = x + 2 \qquad\qquad\qquad 6y = -2x + 5$$

$$\frac{3y}{3} = \frac{x}{3} + \frac{2}{3} \qquad\qquad\qquad \frac{6y}{6} = \frac{-2x}{6} + \frac{5}{6}$$

$$y = \frac{1}{3}x + \frac{2}{3} \qquad\qquad\qquad y = -\frac{1}{3}x + \frac{5}{6}$$

The slope of the line $-x + 3y = 2$ is $\dfrac{1}{3}$ and the slope of the line $2x + 6y = 5$ is $-\dfrac{1}{3}$.
The slopes are not equal so the lines are not parallel. The product of the slopes is
$\dfrac{1}{3} \cdot -\dfrac{1}{3} = -\dfrac{1}{9}$, not -1, so the lines are not perpendicular. They are neither parallel
nor perpendicular.   ■

4   Next, we find the slopes of vertical and horizontal lines.

EXAMPLE 6   Find the slope of the line $x = 5$.

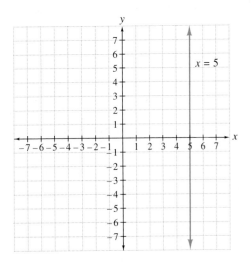

Solution: Recall that the graph of $x = 5$ is a vertical line with x-intercept 5.

To find the slope, find two ordered pair solutions of $x = 5$. Solutions of $x = 5$ must have an x-value of 5.

Let $(5, 0) = (x_1, y_1)$ and $(5, 4) = (x_2, y_2)$. Then

$$m = \frac{y_2 - y_1}{x_x - x_1} = \frac{4 - 0}{5 - 5} = \frac{4}{0}$$

Since $\frac{4}{0}$ is undefined, we say the slope of the vertical line $x = 5$ is undefined. Since all vertical lines are parallel, we can say that **vertical lines have undefined slope.** ∎

EXAMPLE 7 Find the slope of the line $y = -1$.

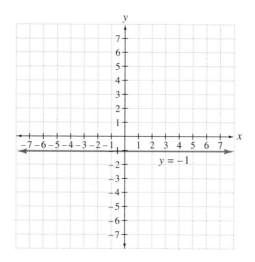

Solution: Recall that $y = -1$ is a horizontal line with y-intercept -1. To find the slope, find two ordered pair solutions of $y = -1$. Solutions of $y = -1$ must have a y-value of -1.

Let $(2, -1) = (x_1, y_1)$ and $(-3, -1) = (x_2, y_2)$. Then

$$m = \frac{y_2 - y_1}{x_2 - x_1} = \frac{-1 - (-1)}{-3 - 2} = \frac{0}{-5} = 0$$

The slope of the line $y = -1$ is 0. Since all horizontal lines are parallel, we can say that **horizontal lines have a slope of 0.** ∎

HELPFUL HINT

Slope of 0 and undefined slope are not the same. Vertical lines have undefined slope or no slope, while horizontal lines have 0 slope.

Here is a general review of slope.

Summary of Slope

$$m = \frac{y_2 - y_1}{x_2 - x_1}$$

Upward line

Positive slope: $m > 0$

Downward line

Negative slope: $m < 0$

Horizontal line
$y = c$

Zero slope: $m = 0$

Vertical line
$x = c$

No slope or undefined slope

MENTAL MATH

Decide whether a line with the given slope is upward, downward, horizontal, or vertical.

1. $m = \dfrac{7}{6}$

horizontal

2. $m = -3$

horizontal

3. $m = 0$

horizontal

4. m is undefined.

vertical

EXERCISE SET 6.3   odds

Find the slope of the line that goes through the given points. See Examples 1 and 2.

1. (0, 0) and (7, 8)

2. (−1, 5) and (0, 0)

3. (−1, 5) and (6, −2)

4. (−1, 9) and (−3, 4)

5. (1, 4) and (5, 3)

6. (3, 1) and (2, 6)

7. (−4, 3) and (−4, 5)

8. (6, −6) and (6, 2)

Find the slope and the y-intercept of each line. See Examples 3 and 4.

9. $y = 5x - 2$

10. $y = -2x + 6$

11. $2x + y = 7$

12. $-5x + y = 10$

13. $2x - 3y = 10$

14. $-3x - 4y = 6$

15. $x = 2y$

16. $x = -4y$

Determine whether the lines are parallel, perpendicular, or neither. See Example 5.

17. $y = -3x + 6$
   $y = 3x + 5$

18. $y = 5x - 6$
   $y = 5x + 2$

19. $-4x + 2y = 5$
   $2x - y = 7$

20. $2x - y = -10$
   $2x + 4y = 2$

21. $-2x + 3y = 1$
   $3x + 2y = 12$

22. $x + 4y = 7$
   $2x - 5y = 0$

Find the slope of each line. See Examples 6 and 7.

23. $x = 1$ **24.** $y = -2$ **25.** $y = -3$
26. $x = 4$ **27.** $x + 2 = 0$ **28.** $y - 7 = 0$

Find the slope of the line that goes through the given points.

29. $(-2, 8)$ and $(1, 6)$ **30.** $(4, -3)$ and $(2, 2)$ **31.** $(1, 0)$ and $(1, 1)$
32. $(0, 13)$ and $(-4, 13)$ **33.** $(5, -11)$ and $(1, -11)$ **34.** $(5, 4)$ and $(0, 5)$
35. $(0, 6)$ and $(-3, 0)$ **36.** $(5, 2)$ and $(0, 5)$ **37.** $(-1, 2)$ and $(-3, 4)$
38. $(3, -2)$ and $(-1, -6)$

Find the slope and the y-intercept of each line.

39. $x + y = 5$ **40.** $x - y = -2$ **41.** $-6x + 5y = 30$
42. $4x - 7y = 28$ **43.** $3x + 9 = y$ **44.** $2y - 7 = x$
45. $y = 4$ **46.** $x = 7$ **47.** $y = 7x$
48. $x = 7y$ **49.** $6 + y = 0$ **50.** $x - 7 = 0$
51. $2 - x = 3$ **52.** $2y + 4 = -7$

53. Find the slope of the line parallel to the line $y = -\dfrac{7}{2}x - 6$.

54. Find the slope of the line parallel to the line $y = x$.

55. Find the slope of the line perpendicular to the line $y = -\dfrac{7}{2}x - 6$.

56. Find the slope of the line perpendicular to the line $y = x$.

57. Find the slope of the line parallel to the line passing through $(-7, -5)$ and $(-2, -6)$.

58. Find the slope of the line parallel to the line passing through the origin and $(-2, 10)$.

59. Find the slope of the line perpendicular to the line passing through the origin and $(1, -3)$.

60. Find the slope of the line perpendicular to the line passing through $(-1, 2)$ and $(5, -3)$.

Writing in Mathematics

61. Explain whether two lines, both with positive slopes, can be perpendicular.

62. A horizontal line is perpendicular to a vertical line. Explain why the product of the slope of a horizontal line and a perpendicular line is not -1.

Skill Review

Solve each equation. See Section 5.6.

63. $\dfrac{3}{x} + \dfrac{2}{5} = 1$ **64.** $x - \dfrac{x}{6} = \dfrac{5}{3}$

Perform the indicated operation. See Sections 5.2 and 5.4.

65. $\dfrac{3}{x} + \dfrac{2}{5}$ **66.** $x - \dfrac{x}{6}$ **67.** $\dfrac{3}{x} \cdot \dfrac{2}{5}$ **68.** $x \div \dfrac{x}{6}$

6.4
Equations of Lines

OBJECTIVES

Tape 21

1   Use the slope–intercept form to find the equation of a line.

2   Graph a line given its slope and y-intercept.

3   Use the point–slope form to find the equation of a line.

4   Find equations of parallel lines.

1   In the last section we learned that the slope–intercept form of a linear equation is $y = mx + b$. When an equation is written in this form, the slope of the line is the same as the coefficient, m, of x. Also, the y-intercept of the line is the same as the constant term b. For example, the slope of the line defined by $y = 2x + 3$ is 2 and its y-intercept is 3.

Now we use the slope–intercept form to write the equation of a line given its slope and y-intercept.

EXAMPLE 1   Find the equation of the line with y-intercept -3 and the slope of $\frac{1}{4}$.

Solution: We are given the slope and the y-intercept. Let $m = \frac{1}{4}$ and $b = -3$, and write the equation in slope–intercept form, $y = mx + b$.

$$y = mx + b$$

$$y = \frac{1}{4}x + (-3) \qquad \text{Let } m = \frac{1}{4} \text{ and } b = -3.$$

$$y = \frac{1}{4}x - 3 \qquad \text{Simplify.} \quad \blacksquare$$

2   We can use a given slope and y-intercept of a line to graph the line as well as write its equation. Let's graph the line from Example 1. We are given that it has slope

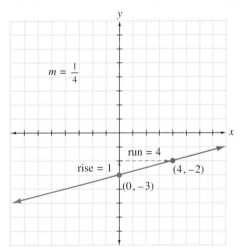

$\frac{1}{4}$ and y-intercept -3. First, plot the y-intercept point $(0, -3)$. To plot another point on the line, recall that slope is $\frac{\text{rise}}{\text{run}} = \frac{1}{4}$. Another point may then be plotted by starting at $(0, -3)$, rising 1 unit up, and then running 3 units to the right. We are now at the point $(4, -2)$. The graph of $y = \frac{1}{4}x - 3$ is the line through points $(0, -3)$ and $(4, -2)$.

EXAMPLE 2   Graph the line through $(-1, 5)$ with slope -2.

Solution: To graph the line, we need two points. One point is $(-1, 5)$, and we will use the slope -2, which can be written as $\frac{-2}{1}$, to find another point.

$$m = \frac{\text{rise}}{\text{run}} = \frac{-2}{1}$$

To find another point, start at $(-1, 5)$ and move vertically two units down, since the numerator is -2; then move horizontally 1 unit to the right. We stop at the point $(0, 3)$. The line through $(-1, 5)$ and $(0, 3)$ will have the required slope of -2.

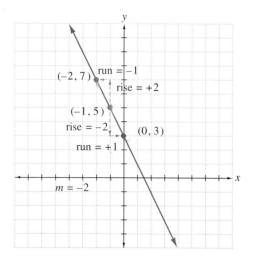

The slope -2 can also be written as $\frac{2}{-1}$, so to find another point we could start at $(-1, 5)$ and move two units up and then one unit left. We would stop at the point $(-2, 7)$. The line through $(-1, 5)$ and $(-2, 7)$ will have the required slope and will be the same line as shown previously through $(-1, 5)$ and $(0, 3)$.

3   When the slope of a line and a point on the line are known, the equation of the line can also be found. To do this, use the slope formula to write the slope of a line that passes through points (x, y) and (x_1, y_1). We have

$$m = \frac{y - y_1}{x - x_1}$$

Multiply both sides of this equation by $x - x_1$ to obtain

$$y - y_1 = m(x - x_1)$$

This form is called the **point–slope form** of the equation of a line.

Point–Slope Form of the Equation of a Line

The point–slope form of the equation of a line is $y - y_1 = m(x - x_1)$, where m is the slope of the line and (x_1, y_1) is a point on the line.

EXAMPLE 3   Find the equation of the line passing through $(-1, 5)$ with slope -2. Write the equation in standard form: $Ax + By = C$.

Solution: Since the slope and a point on the line are given, use point–slope form $y - y_1 = m(x - x_1)$ to write the equation. Let $m = -2$ and $(x_1, y_1) = (-1, 5)$.

$$y - y_1 = m(x - x_1)$$

$$y - 5 = -2[x - (-1)] \qquad \text{Let } m = -2 \text{ and } (x_1, y_1) = (-1, 5).$$

$$y - 5 = -2(x + 1) \qquad \text{Simplify.}$$

$$y - 5 = -2x - 2 \qquad \text{Use the distributive property.}$$

$$y = -2x + 3 \qquad \text{Add 5 to both sides.}$$

$$2x + y = 3 \qquad \text{Add } 2x \text{ to both sides.}$$

In standard form, the equation is $2x + y = 3$.   ∎

EXAMPLE 4   Find the equation of the line through $(2, 5)$ and $(-3, 4)$. Write the equation in standard form.

Solution: First, use the two given points to find the slope of the line. Let $(2, 5)$ be (x_1, y_1) and $(-3, 4)$ be (x_2, y_2).

$$m = \frac{y_2 - y_1}{x_2 - x_1} = \frac{4 - 5}{-3 - 2} = \frac{-1}{-5} = \frac{1}{5}$$

Next, use the slope and either one of the given points to write the equation in point–slope form. We use $(2, 5)$.

$$y - y_1 = m(x - x_1) \qquad \text{Use point–slope form.}$$

$$y - 5 = \frac{1}{5}(x - 2) \qquad \text{Let } x_1 = 2 \text{ and } y_1 = 5.$$

$$5\,(y - 5) = 5 \cdot \frac{1}{5}(x - 2) \qquad \text{Multiply both sides by 5 to clear fractions.}$$

$$5y - 25 = x - 2 \qquad \text{Use the distributive property and simplify.}$$

$$-x + 5y - 25 = -2 \qquad \text{Subtract } x \text{ from both sides.}$$

$$-x + 5y = 23 \qquad \text{Add 25 to both sides.}$$ ∎

HELPFUL HINT

Multiply both sides of the equation $-x + 5y = 23$ by -1, and it becomes $x - 5y = -23$. Both $-x + 5y = 23$ and $x - 5y = -23$ are in standard form, and they are equations of the same line.

EXAMPLE 5 Find the equation of the vertical line through $(-1, 5)$.

Solution: The equation of a vertical line can be written in the form $x = c$, so the equation for a vertical line passing through $(-1, 5)$ is $x = -1$. ■

4

EXAMPLE 6 Find the equation of the line parallel to the line $y = 7$ and passing through $(-2, -3)$.

Solution: Since the graph of $y = 7$ is a horizontal line, any line parallel to it must also be horizontal. The equation of a horizontal line can be written in the form $y = c$. An equation for the horizontal line passing through $(-2, -3)$ is $y = -3$. ■

Forms of Linear Equations

$Ax + By = C$	**Standard form** of a linear equation. A and B are not both 0.
$y = mx + b$	**Slope–intercept form** of a linear equation. The slope is m and the y-intercept is b.
$y - y_1 = m(x - x_1)$	**Point–slope form** of a linear equation. The slope is m and (x_1, y_1) is a point on the line.
$y = c$	**Horizontal line** The slope is 0 and the y-intercept is c.
$x = c$	**Vertical line** The slope is undefined and the x-intercept is c.

Parallel and Perpendicular Lines

Nonvertical parallel lines have the same slope.

The product of the slopes of two perpendicular lines is -1.

MENTAL MATH

State the slope and the y-intercept for each line for the given equation.

1. $y = -4x + 12$

2. $y = \dfrac{2}{3}x - \dfrac{7}{2}$

3. $y = 5x$

4. $y = -x$

5. $y = \dfrac{1}{2}x + 6$

6. $y = -\dfrac{2}{3}x + 5$

Decide whether the lines are parallel, perpendicular, or neither.

7. $y = 12x + 6$
$y = 12x - 2$

8. $y = -5x + 8$
$y = -5x - 8$

9. $y = -9x + 3$
$y = \dfrac{3}{2}x - 7$

10. $y = 2x - 12$
$y = \dfrac{1}{2}x - 6$

EXERCISE SET 6.4

Use the slope–intercept form of the linear equation to write the equation of each line with given slope and y-intercept. See Example 1.

1. Slope -1; y-intercept 1

2. Slope $\frac{1}{2}$; y-intercept -6

3. Slope 2; y-intercept $\frac{3}{4}$

4. Slope -3; y-intercept $-\frac{1}{5}$

5. Slope $\frac{2}{7}$; y-intercept 0

6. Slope $-\frac{4}{5}$; y-intercept 0

Graph each line passing through the given point with the given slope. See Example 2.

7. Through $(1, 3)$ with slope $\frac{3}{2}$

8. Through $(-2, -4)$ with slope $\frac{2}{5}$

9. Through $(0, 0)$ with slope 5

10. Through $(-5, 2)$ with slope 2

11. Through $(0, 7)$ with slope -1

12. Through $(3, 0)$ with slope -3

Use the point–slope form of the linear equation to find the equation of each line with the given slope and passing through the given point. Then write the equation in standard form. See Example 3.

13. Slope 6; through $(2, 2)$

14. Slope 4; through $(1, 3)$

15. Slope -8; through $(-1, -5)$

16. Slope -2; through $(-11, -12)$

17. Slope $\frac{1}{2}$; through $(5, -6)$

18. Slope $\frac{2}{3}$; through $(-8, 9)$

Find the equation of the line through the given points. Write the equation in standard form. See Example 4.

19. Through $(3, 2)$ and $(5, 6)$

20. Through $(6, 2)$ and $(8, 8)$

21. Through $(-1, 3)$ and $(-2, -5)$

22. Through $(-4, 0)$ and $(6, -1)$

23. Through $(2, 3)$ and $(-1, -1)$

24. Through $(0, 0)$ and $\left(\frac{1}{2}, \frac{1}{3}\right)$

Find the equation of each line. See Example 5.

25. Vertical line through $(0, 2)$

26. Horizontal line through $(1, 4)$

27. Horizontal line through $(-1, 3)$

28. Vertical line through $(-1, 3)$

29. Vertical line through $(-7, -2)$

30. Horizontal line through $(2, 0)$

Find the equation of each line. See Example 6.

31. Parallel to $y = 5$, through $(1, 2)$

32. Perpendicular to $y = 5$, through $(1, 2)$

33. Perpendicular to $x = -3$, through $(-2, 5)$

34. Parallel to $y = -4$, through $(0, -3)$

35. Parallel to $x = 0$, through $(6, -8)$

36. Perpendicular to $x = 7$, through $(-5, 0)$

Find the equation of each line described. Write each equation in standard form.

37. With slope $-\dfrac{1}{2}$, through $\left(0, \dfrac{5}{3}\right)$

38. With slope $\dfrac{5}{7}$, through $(0, -3)$

39. Slope 1, through $(-7, 9)$

40. Slope 5, through $(6, -8)$

41. Through $(10, 7)$ and $(7, 10)$

42. Through $(5, -6)$ and $(-6, 5)$

43. Through $(6, 7)$, parallel to the x-axis

44. Through $(0, -5)$, parallel to the y-axis

45. Slope $-\dfrac{4}{7}$, through $(-1, -2)$

46. Slope $-\dfrac{3}{5}$, through $(4, 4)$

47. Through $(-8, 1)$ and $(0, 0)$

48. Through $(2, 3)$ and $(0, 0)$

49. Through $(0, 0)$ with slope 3

50. Through $(0, -2)$ with slope -1

51. Through $(-6, -6)$ and $(0, 0)$

52. Through $(0, 0)$ and $(4, 4)$

53. Slope -5, y-intercept 7

54. Slope -2; y-intercept -4

55. Through $(-1, 5)$ and $(0, -6)$

56. Through $(4, 0)$ and $(0, -5)$

57. With undefined slope, through $\left(-\dfrac{3}{4}, 1\right)$

58. With slope 0, through $(6.7, 12.1)$

59. Through $(-2, -3)$, perpendicular to the y-axis

60. Through $(0, 12)$, perpendicular to the x-axis

61. With slope 7, through $(1, 3)$

62. With slope -10, through $(5, -1)$

63. A linear equation can be written that relates the radius of a circle to its circumference (perimeter). A circle with a radius of 10 centimeters has a circumference of approximately 63 centimeters. A circle with a radius of 15 centimeters has a circumference of approximately 94 centimeters. Use the ordered pairs $(10, 63)$ and $(15, 94)$ to find a linear equation that approximates the relationship between the radius of the circle and its circumference.

64. A rock is dropped from the top of a 300-foot building. After one second the rock is traveling 32 feet per second and after three seconds the rock is traveling 96 feet per second. Let x be the number of seconds since the rock was dropped, and let y be the speed at which it is falling after x seconds. Using the ordered pairs $(1, 32)$ and $(3, 96)$, and find a linear equation that relates time, x, to speed, y.

A Look Ahead

Find the equations of the following lines. See the following example.

EXAMPLE   Find the equation of the line through $(1, -6)$ and parallel to the line $y = 3x - 2$.

Solution: Since the line is parallel to the line $y = 3x - 2$, it has the same slope as the line $y = 3x - 2$. The slope of the line $y = 3x - 2$ is 3. Now write the equation of the line with slope 3 that passes through the point $(1, -6)$.

$$y - y_1 = m(x - x_1)$$
$$y - (-6) = 3(x - 1) \qquad \text{Let } x_1 = 1 \text{ and } y_1 = -6.$$
$$y + 6 = 3x - 3$$
$$y = 3x - 9$$

In standard form, the equation is $-3x + y = -9$.   ■

65. Through $(7, 0)$, parallel to $y = \dfrac{2}{3}x + 2$

66. Through $\left(0, \dfrac{1}{2}\right)$, parallel to $y = -2x - 10$

67. Through $(1, 7)$, parallel to $y = -x$

68. Through the origin, parallel to $y = -1.4x$

Writing in Mathematics

69. Given the equation of a nonvertical line, explain how to find the slope without finding two points on the line.

70. Given two points on a nonvertical line, explain how to use the point–slope form to find the equation of the line.

Skill Review

Solve the following equations. See Section 5.6.

71. $\dfrac{1}{x} + \dfrac{x}{3} = \dfrac{x + 5}{6}$

72. $\dfrac{2}{x} + 3 = \dfrac{3x + 2}{5}$

Find the area of each figure. See Section 5.2.

73.

$\dfrac{2 - x}{x^4}$ inches

$\dfrac{x}{x^2 - 4}$ inches

74.

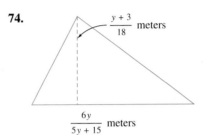

$\dfrac{y + 3}{18}$ meters

$\dfrac{6y}{5y + 15}$ meters

6.5
Graphing Linear Inequalities

OBJECTIVE	
1	Graph a linear inequality in two variables.

Tape 22

1   The linear equation $x + y = 5$ is graphed next. Recall that all points on the line correspond to ordered pairs that satisfy the equation $x + y = 5$. It can be shown that all the points above the line $x + y = 5$ have coordinates that satisfy the inequality $x + y > 5$. Similarly, all points below the line have coordinates that satisfy the inequality $x + y < 5$.

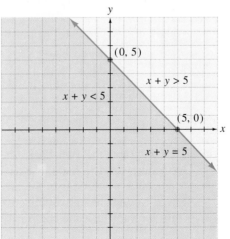

The region above the line and the region below the line are called **half-planes.** Every line divides the plane (similar to a sheet of paper extending indefinitely in all directions) into two half-planes; the line is called the **boundary.**

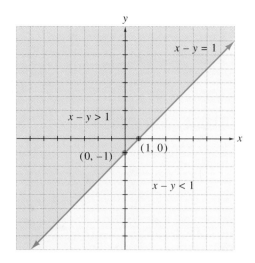

Similarly, the half-plane above the boundary $x - y = 1$ is the graph of the inequality $x - y > 1$. The half-plane below the boundary $x - y = 1$ is the graph of the inequality $x - y < 1$. Since the inequality $x - y \leq 1$ means

$$x - y = 1 \quad \text{or} \quad x - y < 1$$

the graph of $x - y \leq 1$ is the half-plane $x - y < 1$ along with the boundary line $x - y = 1$.

The steps to graph a linear inequality, are given next.

To Graph a Linear Inequality in Two Variables

Step 1 Graph the boundary line found by replacing the inequality sign with an equal sign. If the inequality sign is $>$ or $<$, graph a dashed boundary line indicating that the points on the line are not solutions of the inequality. If the inequality sign is $\geq$ or $\leq$, graph a solid boundary line indicating that the points on the line are solutions of the inequality.

Step 2 Choose a point, not on the boundary line, as a test point. Substitute the coordinates of this test point into the original inequality.

Step 3 If a true statement is obtained in step 2, shade the half-plane that contains the test point. If a false statement is obtained, shade the half-plane that does not contain the test point.

EXAMPLE 1 Graph $x + y < 7$.

Solution: First, graph the boundary line by graphing the equation $x + y = 7$. Graph this boundary as a dashed line because the inequality sign is $<$, and thus the points on the line are not solutions of the inequality $x + y < 7$. (See the next page.) Next, choose a test point, being careful not to choose a point on the boundary line. We choose $(0, 0)$. Substitute the coordinates of $(0, 0)$ into $x + y < 7$.

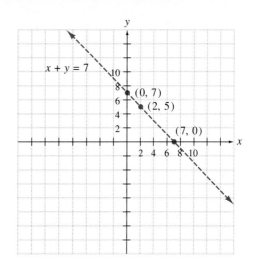

$x + y < 7$ Original inequality.

$0 + 0 < 7$ Replace x with 0 and y with 0.

$0 < 7$ True.

Since the result is a true statement, $(0, 0)$ is a solution of $x + y < 7$, and every point in the same half-plane as $(0, 0)$ is also a solution. To indicate this, shade the entire half-plane containing $(0, 0)$, as shown.

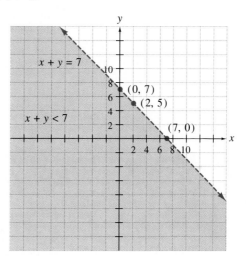

EXAMPLE 2   Graph $2x - y \geq 3$.

Solution: Graph the boundary line by graphing $2x - y = 3$. Draw this line as a solid line since the inequality sign is $\geq$, and thus the points on the line are solutions to $2x - y \geq 3$. Once again, $(0, 0)$ is a convenient test point since it is not on the boundary line.

Substitute 0 for x and 0 for y into the **original inequality.**

$$2x - y \geq 3$$

$$2(0) - 0 \geq 3 \qquad \text{Let } x = 0 \text{ and } y = 0.$$

$$0 \geq 3 \qquad \text{False.}$$

291

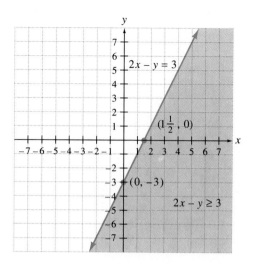

Since the statement is false, no point in the half-plane containing $(0, 0)$ is a solution. Shade the half-plane that does not contain $(0, 0)$. Every point in the shaded half-plane and every point on the boundary line satisfies $2x - y \geq 3$.

HELPFUL HINT

When graphing an inequality, make sure the test point is substituted in the **original inequality.** For example, when graphing $x + y < 3$, test $(0, 0)$ in $x + y < 3$, not $x + y = 3$. Since $(0, 0)$ is a solution of $x + y < 3$, we shade the half-plane containing $(0, 0)$ as shown.

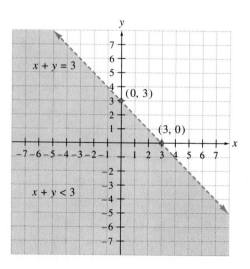

EXAMPLE 3   Graph $x > 2y$.

Solution: Find the boundary line by graphing $x = 2y$. The boundary line is a dashed line since the inequality symbol is $>$. We cannot use $(0, 0)$ as a test point because it is a point on the boundary line. Choose $(0, 2)$ as the test point.

$$x > 2y$$

$$0 > 2(2) \qquad \text{Let } x = 0 \text{ and } y = 2.$$

$$0 > 4 \qquad \text{False.}$$

Since the statement is false, shade the half-plane that does not contain the test point $(0, 2)$.

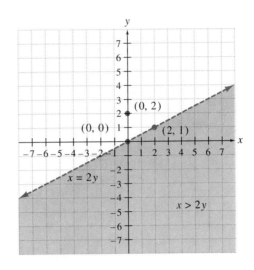

EXAMPLE 4   Graph $5x + 4y \leq 20$.

Solution: Graph the solid boundary line $5x + 4y = 20$. Choose $(0, 0)$ as the test point.

$$5x + 4y \leq 20$$

$$5(0) + 4(0) \leq 20 \qquad \text{Let } x = 0 \text{ and } y = 0.$$

$$0 \leq 20 \qquad \text{True.}$$

Shade the half-plane that contains $(0, 0)$. The graph of $5x + 4y \leq 20$ is given next.

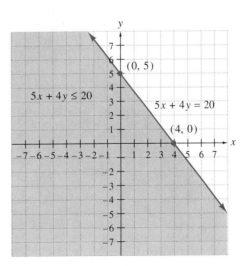

EXAMPLE 5   Graph $y > 3$.

Solution: Graph the dashed boundary line $y = 3$. Recall that the graph of $y = 3$ is a horizontal line with y-intercept 3. Choose $(0, 0)$ as the test point.

$$y > 3$$
$$0 > 3 \qquad \text{Let } y = 0.$$
$$0 > 3 \qquad \text{False.}$$

Shade the half-plane that does not contain $(0, 0)$. The graph of $y > 3$ is shown next.

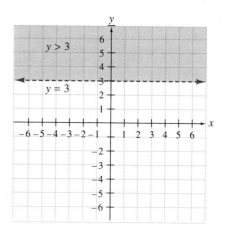

MENTAL MATH

State whether the graph of each inequality includes its corresponding boundary line.

1. $y \geq x + 4$ **2.** $x - y > -7$ **3.** $y \geq x$ **4.** $x > 0$

Decide whether $(0, 0)$ is a solution of each given inequality.

5. $x + y > -5$ **6.** $2x + 3y < 10$ **7.** $x - y \leq -1$ **8.** $\dfrac{2}{3}x + \dfrac{5}{6}y > 4$

EXERCISE SET 6.5

Graph each inequality. See Examples 1 through 5.

1. $x + y \leq 1$ **2.** $x + y \geq -2$ **3.** $2x + y > -4$ **4.** $x + 3y \leq 3$

5. $x + 6y \leq -6$

6. $7x + y > -14$

7. $2x + 5y > -10$

8. $5x + 2y \leq 10$

9. $x + 2y \leq 3$

10. $2x + 3y > -5$

11. $2x + 7y > 5$

12. $3x + 5y \leq -2$

13. $x - 2y \geq 3$

14. $4x + y \leq 2$

15. $5x + y < 3$

16. $x + 2y > -7$

17. $4x + y < 8$

18. $9x + 2y \geq -9$

19. $y \geq 2x$

20. $x < 5y$

21. $x \geq 0$

22. $y \leq 0$

23. $y \leq -3$

24. $x > -\dfrac{2}{3}$

25. $2x - 7y > 0$

26. $5x + 2y \leq 0$

27. $3x - 7y \geq 0$

28. $-2x - 9y > 0$

29. $x > y$ **30.** $x \leq -y$ **31.** $x - y \leq 6$ **32.** $x - y > 10$

33. $-\dfrac{1}{4}y + \dfrac{1}{3}x > 1$ **34.** $\dfrac{1}{2}x - \dfrac{1}{3}y \leq -1$ **35.** $-x < 0.4y$ **36.** $0.3x \geq 0.1y$

37. Write an inequality whose solutions are all pairs of numbers x and y whose sum is at least 13. Graph the inequality.

38. Write an inequality whose solutions are all the pairs of numbers x and y whose sum is at most -4. Graph the inequality.

Writing in Mathematics

39. Explain why a point on the boundary line must not be chosen as the test point.

40. Describe the graph of a linear inequality.

Skill Review

The following pairs of triangles are similar. Find the unknown lengths. See Section 5.8.

41.

6 cm

9 cm

x

12 cm

42.

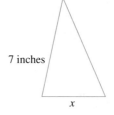

7 inches

x

5 inches

3 inches

Factor the following. See Section 4.1.

43. $8x^2 + 2x$ **44.** $7y^3 + 21y^4$ **45.** $x^2 + xy + x + y$ **46.** $z^2 + 5xz + z + 5x$

6.6
Functions

Tape 22

OBJECTIVES

1   Understand the definition of relation, domain, and range.

2   Understand the definition of function.

3   Use the vertical line test.

4   Use function notation.

1   The linear equation $y = 2x + 1$ describes a relationship between x and y. The variables x and y are related in the following way: for any given value of x, we can find the corresponding value of y by adding 1 to twice the x-value. Ordered pairs can be used to write down solutions of this equation. For example, $(1, 3)$ is a solution of $y = 2x + 1$, and this notation tells us that the x-value 1 is related to the y-value 3 for this equation.

We call a set of ordered pairs a **relation.** The **domain** of a relation is the set of all x-values, and the **range** of a relation is the set of all y-values.

We will call an equation such as $y = 2x + 1$ a relation since the equation defines a set of ordered pair solutions.

EXAMPLE 1   Find the domain and the range of the relation $\{(0, 2), (3, 3), (-1, 0), (3, -2)\}$.

Solution: The domain is the set of all x-values or $\{-1, 0, 3\}$, and the range is the set of all y-values, or $\{-2, 0, 2, 3\}$.   ■

2   Some relations are also functions.

> **Function**
>
> A function is a set of ordered pairs that assigns to each x-value exactly one y-value.

EXAMPLE 2   Which of the following relations are also functions?
a. $\{(-1, -1), (2, 3), (7, 3), (8, 6)\}$ **b.** $\{(0, -2), (1, 5), (0, 3), (7, 7)\}$

Solution: **a.** Although the ordered pairs $(2, 3)$ and $(7, 3)$ have the same y-value, each x-value is assigned to only one y-value so this set of ordered pairs is a function.

b. The x-value 0 is assigned to two y-values, -2 and 3, so this set of ordered pairs is not a function.   ■

EXAMPLE 3   Is the relation $y = 2x + 1$ a function?

Solution: The relation $y = 2x + 1$ is a function if each x-value corresponds to just one y-value. For each x-value substituted in the equation $y = 2x + 1$, the multiplication and addition performed gives a single result, so only one y-value will be associated with each x-value. Thus, $y = 2x + 1$ is a function.   ■

EXAMPLE 4   Is the equation $x = y^2$ a function?

Solution: In $x = y^2$, if $y = 3$, then $x = 9$. Also, if $y = -3$, then $x = 9$. In other words, the x-value 9 corresponds to two y-values, 3 and -3. Thus $x = y^2$ is not a function.   ■

3   As we have seen so far, not all relations are functions. Consider the graphs of $y = 2x + 1$ and $x = y^2$ shown next. (We discuss graphing quadratic equations in Chapter 9.) On the graph of $y = 2x + 1$, notice that each x-value corresponds to only one y-value. Recall from Example 3 that $y = 2x + 1$ is a function.

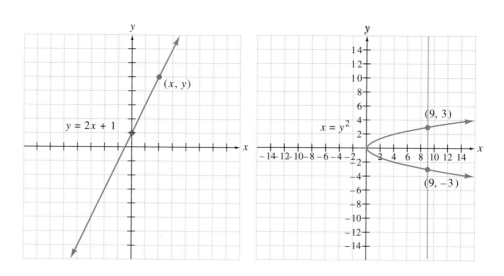

On the graph of $x = y^2$, the x-value 9, for example, corresponds to two y-values, 3 and -3, as shown by the vertical line. Recall from Example 4 that $y = x^2$ is not a function.

Graphs can be used to help determine whether a relation is also a function by the following vertical line test.

Vertical Line Test

If a vertical line can be drawn so that it intersects a graph more than once, the graph is not the graph of a function.

EXAMPLE 5   Which of the following graphs are graphs of functions?

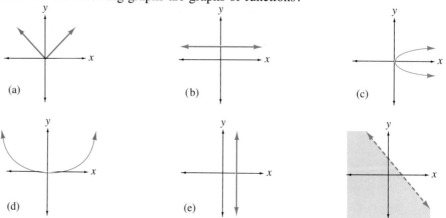

Solution: **a.** This graph is the graph of a function since no vertical line will intersect this graph more than once.

b. This graph is also the graph of a function.

c. This graph is not the graph of a function. Note that vertical lines can be drawn that intersect the graph in two points.

d. This graph is the graph of a function.

e. This graph is not the graph of a function. A vertical line can be drawn that intersects this line at every point.

f. This graph is not the graph of a function. ■

Recall that the graph of a linear equation is a line, and a line that is not vertical will pass the vertical line test. Thus, all linear equations are functions except those of the form $x = c$, which are vertical lines.

EXAMPLE 6 Which of the following are functions?

a. $y = x$

b. $y = -x + 10$

c. $y = 5$

d. $x = -1$

Solution: **(a)**, **(b)**, and **(c)** are functions because their graphs are nonvertical lines.

(d) This is not a function because its graph is a vertical line. ■

4   Earlier, we decided that $y = 2x + 1$ is a function. Many times letters such as f, g, and h are used to name functions. For example, to denote that y is a function of x in the equation $y = 2x + 1$, we can write $y = f(x)$. Then $y = 2x + 1$ can be written as $f(x) = 2x + 1$. The symbol $f(x)$ means **functions of x** and is read "f of x." This notation is called **function notation.**

The notation $f(1)$ means to replace x with 1 and find the resulting y or function value. Since

$$f(x) = 2x + 1$$

then

$$f(1) = 2(1) + 1 = 3$$

This means that, when $x = 1$, y or $f(x) = 3$. Now find $f(2)$, $f(0)$, and $f(-1)$.

$$f(x) = 2x + 1 \qquad f(x) = 2x + 1 \qquad f(x) = 2x + 1$$
$$f(2) = 2(2) + 1 \qquad f(0) = 2(0) + 1 \qquad f(-1) = 2(-1) + 1$$
$$= 4 + 1 \qquad\qquad = 0 + 1 \qquad\qquad = -2 + 1$$
$$= 5 \qquad\qquad = 1 \qquad\qquad = -1$$

HELPFUL HINT

Note that $f(x)$ is a special symbol in mathematics used to denote a function. The symbol $f(x)$ is read "f of x." It does **not** mean $f \cdot x$ (f times x).

EXAMPLE 7   Given $g(x) = x^2 - 3$, find the following.
a. $g(2)$ **b.** $g(-2)$ **c.** $g(0)$

Solution: **(a)** $g(x) = x^2 - 3$ **(b)** $g(x) = x^2 - 3$ **(c)** $g(x) = x^2 - 3$

$g(2) = 2^2 - 3$ $g(-2) = (-2)^2 - 3$ $g(0) = 0^2 - 3$

$= 4 - 3$ $= 4 - 3$ $= 0 - 3$

$= 1$ $= 1$ $= -3$ ■

We now practice finding the domain of a function. The domain of our functions will be the set of all possible real numbers that x can be replaced by.

EXAMPLE 8   Find the domain of each function.

a. $g(x) = \dfrac{1}{x}$ **b.** $f(x) = 2x + 1$

Solution: **a.** Recall that we cannot divide by 0 so that the domain of $g(x)$ is the set of all real numbers except 0.

b. In this function, x can be any real number. The domain of $f(x)$ is the set of all real numbers.   ■

EXERCISE SET 6.6

Find the domain and the range of each relation. See Example 1.

1. $\{(2, 4), (0, 0), (-7, 10), (10, -7)\}$

2. $\{(3, -6), (1, 4), (-2, -2)\}$

3. $\{(0, -2), (1, -2), (5, -2)\}$

4. $\{(5, 0), (5, -3), (5, 4) (5, 3)\}$

Decide whether each set is a function. See Example 2.

5. $\{(1, 1), (2, 2), (-3, -3), (0, 0)\}$

6. $\{(1, 2), (3, 2), (4, 2)\}$

7. $\{(-1, 0), (-1, 6), (-1, 8)\}$

8. $\{(11, 6), (-1, -2), (0, 0), (3, -2)\}$

Decide whether each is a function. See Examples 3, 4, and 6.

9. $y = x + 1$ **10.** $y = x - 1$ **11.** $x = 2y^2$ **12.** $y = x^2$

13. $y - x = 7$ **14.** $2x - 3y = 9$ **15.** $y = \dfrac{1}{x}$ **16.** $y = \dfrac{1}{x - 3}$

Use the vertical line test to determine whether each graph is the graph of a function. See Example 5.

17.

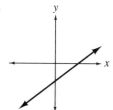

18.

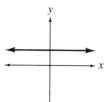

19.

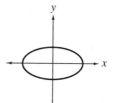

20.

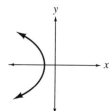

21. 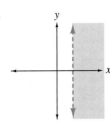   **22.** 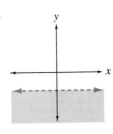   **23.** 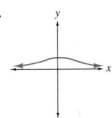   **24.**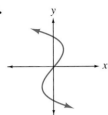

Given the following functions, find the indicated values. See Example 7.

25. Given $f(x) = 2x - 5$;   **a.** $f(-2)$ **b.** $f(0)$ **c.** $f(3)$

26. $g(x) = 3 - 7x$;   **a.** $g(-7)$ **b.** $g(7)$ **c.** $g\left(\dfrac{3}{7}\right)$

27. $h(x) = x^2 + 2$;   **a.** $h(-3)$ **b.** $h\left(\dfrac{1}{4}\right)$ **c.** $h(3)$

28. $f(x) = x^2 - 4$;   **a.** $f(7)$ **b.** $f(4)$ **c.** $f(-10)$

29. $h(x) = x^3$;   **a.** $h(-2)$ **b.** $h(-6)$ **c.** $h(0)$

30. $h(x) = -x^3$;   **a.** $h(-2)$ **b.** $h(-6)$ **c.** $h(0)$

31. $f(x) = |x|$;   **a.** $f(7)$ **b.** $f(-7)$ **c.** $f(0)$

32. $h(x) = |2 - x|$;   **a.** $h(0)$ **b.** $h(-5)$ **c.** $h(6)$

Find the domain of each function. See Example 8.

33. $f(x) = 3x - 7$ **34.** $g(x) = 5 - 2x$ **35.** $h(x) = \dfrac{1}{x + 5}$

36. $f(x) = \dfrac{1}{x - 6}$ **37.** $g(x) = |x + 1|$ **38.** $h(x) = |2x|$

Determine whether each of the following is a function.

39. $x = 5$ **40.** $y = 7$ **41.** $y = x^3$ **42.** $x = y^3$

43. $y < 2x + 1$ **44.** $y \le x - 4$ **45.** $y = x + 3$ **46.** $y = 7x + 2$

Given the following functions, find the indicated values.

47. $f(x) = 5x$;   **a.** $f(0)$ **b.** $f(2)$ **c.** $f(-2)$

48. $g(x) = -3x$;   **a.** $g(0)$ **b.** $g(-1)$ **c.** $g(3)$

49. $g(x) = 2x^2 + 3$;   **a.** $g(-11)$ **b.** $g(-1)$ **c.** $g\left(\dfrac{1}{2}\right)$

50. $h(x) = -x^2$;   **a.** $h(-5)$ **b.** $h\left(-\dfrac{1}{3}\right)$ **c.** $h\left(\dfrac{1}{3}\right)$

51. $f(x) = -x - 2x + 3$;   **a.** $f(1)$ **b.** $f(-1)$

52. $g(x) = -x^2 + 4x - 3$;   **a.** $g(-6)$ **b.** $g(2)$

53. $f(x) = 6$;   **a.** $f(2)$ **b.** $f(0)$ **c.** $f(606)$

54. $h(x) = -12$;   **a.** $h(7)$ **b.** $h(542)$ **c.** $h\left(-\dfrac{3}{4}\right)$

Find the domain of each function.

55. $g(x) = x^4$ **56.** $h(x) = x^3$

57. $f(x) = -2$ **58.** $g(x) = -4$

59. $h(x) = \dfrac{1}{(x - 2)(x + 1)}$ **60.** $f(x) = \dfrac{1}{(x - 1)(x - 3)}$

A Look Ahead

Given the following functions, find the indicated values. See the following example.

> **EXAMPLE**   If $f(x) = x^2 + 2x + 1$, find the following:
> **a.** $f(\pi)$ **b.** $f(c + d)$
>
> Solution: **a.** $f(x) = x^2 + 2x + 1$
> $f(\pi) = \pi^2 + 2\pi + 1$
>
> **b.** $f(x) = x^2 + 2x + 1$
> $f(c + d) = (c + d)^2 + 2(c + d) + 1$
> $= c^2 + 2cd + d^2 + 2c + 2d + 1$ ∎

61. $f(x) = 2x + 7$;   **a.** $f(2)$ **b.** $f(a)$ **c.** $f(a + 2)$
62. $g(x) = -3x + 12$;   **a.** $g(s)$ **b.** $g(r)$ **c.** $g(r + s)$
63. $h(x) = x^2 + 7$;   **a.** $h(3)$ **b.** $h(a)$ **c.** $h(a - 3)$
64. $f(x) = x^2 - 12$;   **a.** $f(12)$ **b.** $f(12 + a)$ **c.** $f(12 - a)$

Writing in Mathematics

65. In your own words define **(a)** function; **(b)** domain; **(c)** range.

66. Explain the vertical line test and how it is used.

67. Since $y = x + 7$ is a function, rewrite the equation using function notation.

Skill Review

Find the perimeter of the following figures. See Section 5.4.

68.

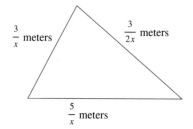

$\dfrac{3}{x}$ meters $\dfrac{3}{2x}$ meters

$\dfrac{5}{x}$ meters

69.

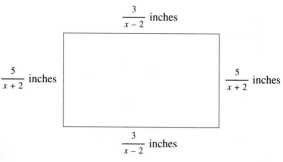

$\dfrac{3}{x - 2}$ inches

$\dfrac{5}{x + 2}$ inches $\dfrac{5}{x + 2}$ inches

$\dfrac{3}{x - 2}$ inches

Graph the following ordered pairs. Use a single coordinate system for each exercise. See Section 6.1.

70. $(0, 0)$, $(1, 1)$, $(-1, 1)$, $(2, 3)$, $(-2, 4)$

71. $(2, 0)$, $(3, 1)$, $(1, 1)$, $(4, 4)$, $(0, 4)$

CRITICAL
THINKING

In charge of arranging food for her company picnic, Karen hires a caterer. According to the caterer's rates, he charges $250 for 40 people and $1800 for 360 people. Karen expects 120 people at the picnic.

The caterer tells Karen that, since he charges $250 for 40 people, he intends to charge her $750. How do you suppose the caterer arrived at that cost?

Karen tells the caterer that, since he charges $1800 for 360 people, she should pay just $600. How do you suppose she arrived at that cost?

What charge is a reasonable compromise for Karen and the caterer?

If you plot points on a Cartesian coordinate system so that each x-coordinate is the number of people the caterer serves and the y-coordinate is the caterer's charge for that number of people, is it reasonable to assume that the points lie on a straight line?

CHAPTER 6 GLOSSARY

The **Cartesian coordinate system** or **rectangular coordinate system** consists of two number lines perpendicular to each other, intersecting at point 0 on each.

A **function** is a set of ordered pairs that assigns to each x-value exactly one y-value.

The notation $f(x)$ is called **function notation.**

The notation (x, y), where x and y represent numbers, is called an **ordered pair.**

The **slope** of a line measures its steepness.

If a graph crosses the x-axis at $(a, 0)$, the x-**intercept** of the graph is a.

If a graph crosses the y-axis at $(0, b)$, the y-**intercept** of the graph is b.

CHAPTER 6 SUMMARY

(6.1)

> **Linear Equation in Two Variables**
>
> A linear equation in two variables is of the form
>
> $$Ax + By = C$$
>
> when A, B, and C are real numbers, and A and B are not both 0.

(6.2)

The graph of a linear equation in two variables is a straight line.

(6.3)

> The slope of a line through the points (x_1, y_1) and (x_2, y_2) is
>
> $$m = \frac{y_2 - y_1}{x_2 - x_1}$$

(6.4)

Forms of Linear Equations

$Ax + By = C$	**Standard form** of a linear equation. A and B are not both 0.
$y = mx + b$	**Slope–intercept form** of a linear equation. The slope is m and the y-intercept is b.
$y - y_1 = m(x - x_1)$	**Point–slope form** of a linear equation. The slope is m and (x_1, y_1) is a point on the line.
$y = c$	**Horizontal line** The slope is 0 and the y-intercept is c.
$x = c$	**Vertical line** The slope is undefined and the x-intercept is c.

Parallel and Perpendicular Lines

Nonvertical parallel lines have the same slope.
The product of the slopes of two perpendicular lines is -1.

TO GRAPH A LINEAR INEQUALITY   (6.5)

Step 1   Graph the boundary line.
Step 2   Shade the half-plane that contains solutions to the inequality.

(6.6)

Vertical Line Test

If a vertical line can be drawn so that it intersects a graph more than once, the graph is not the graph of a function.

CHAPTER 6 REVIEW

(6.1)   *Plot the following points on a Cartesian coordinate system.*

1. $(-7, 0)$

2. $\left(0, 4\frac{4}{5}\right)$

3. $(-2, -5)$

4. $(1, -3)$

5. $(0.7, 0.7)$

6. $(-6, 4)$

Determine whether each ordered pair is a solution to the given equation.

7. $7x - 8y = 56$; $(0, 56)$, $(8, 0)$

8. $-2x + 5y = 10$; $(-5, 0)$, $(1, 1)$

9. $x = 13$; $(13, 5)$, $(13, 13)$

10. $y = 2$; $(7, 2)$, $(2, 7)$

Complete the ordered pairs so that each is a solution of the given equation.

11. $-2 + y = 6x$; $(7, \quad)$

12. $y = 3x + 5$; $(\quad, -8)$

Complete the ordered pairs so that each is a solution of the given equation; then plot the ordered pairs. Use a single coordinate system for each exercise.

13. $9 = -3x + 4y$
   a. $(\quad, 0)$
   b. $(\quad, 3)$
   c. $(9, \quad)$

14. $y = 5$
   a. $(7, \quad)$
   b. $(-7, \quad)$
   c. $(0, \quad)$

15. $x = 2y$
   a. $(\quad, 0)$
   b. $(\quad, 5)$
   c. $(\quad, -5)$

(6.2) *Graph each linear equation.*

16. $x - y = 1$

17. $x + y = 6$

18. $x - 3y = 12$

19. $5x - y = -8$

20. $x = 3y$

21. $y = -2x$

22. $x = 3$

23. $y = -2$

24. $2x - 3y = 6$

25. $4x - 3y = 12$

(6.3)   *Find the slope of the line that goes through the given points.*

26. (2, 5) and (6, 8) **27.** (4, 7) and (1, 2) **28.** (1, 3) and $(-2, -9)$ **29.** $(-4, 1)$ and $(3, -6)$

Find the slope and the y-intercept of each line.

30. $y = 3x + 7$ **31.** $x - 2y = 4$ **32.** $2x + 3y = 5$ **33.** $4x - 5y = 9$

34. Find an equation of a line perpendicular to the line in Exercise 32 and passing through $(1, -5)$.

35. Find an equation of a line parallel to the line in Exercise 33 and passing through $(-2, 3)$.

Find the equations of the following lines; then graph each line.

36. Through $(-6, 10)$ and $(-6, -1)$

37. Through (5, 5) and $(-2, 5)$

38. Through $(-4, -5)$ with no slope

39. Through $(3, -7)$ with slope 0

40. Through $(12, -1)$ with no slope

Determine whether the lines are parallel, perpendicular, or neither.

41. $x - y = -6$
$x + y = 3$

42. $3x + y = 7$
$-3x - y = 10$

(6.4)   *Find equations of the following lines. Write the answers in standard form.*

43. With slope 4, through (2, 0)

44. With slope -3, through $(0, -5)$

45. With slope $\dfrac{1}{2}$, through $\left(0, -\dfrac{7}{2}\right)$

46. With slope 0, through $(-2, -3)$

47. With 0 slope, through the origin

48. With slope -6, through $(2, -1)$

49. With slope 12, through $\left(\dfrac{1}{2}, 5\right)$

50. Through (0, 6) and (6, 0)

51. Through $(0, -4)$ and $(-8, 0)$

52. Vertical line, through (5, 7)

53. Horizontal line, through $(-6, 8)$

54. Through (6, 0), perpendicular to $y = 8$

55. Through (10, 12), perpendicular to $x = -2$

(6.5)   *Graph the following inequalities.*

56. $3x - 4y \leq 0$ **57.** $3x - 4y \geq 0$ **58.** $x + 6y < 6$ **59.** $x + y > -2$

60. $y \geq -7$ **61.** $y \leq -4$ **62.** $-x \leq y$ **63.** $x \geq -y$

(6.6)   *Which of the equations define functions?*

64. $7x - 6y = 1$ **65.** $3 + x - 7y = 0$ **66.** $y = 7$ **67.** $x = 2$

Find the domain of each function.

68. $f(x) = 2x + 7$ **69.** $g(x) = \dfrac{7}{x - 2}$

Use the vertical line test to determine whether each graph is the graph of a function.

70.

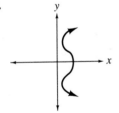

71.

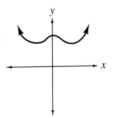

Given the following functions, find the indicated function value.

72. Given $f(x) = -2x + 6$, find   **(a)** $f(0)$ **(b)** $f(-2)$ **(c)** $f\left(\dfrac{1}{2}\right)$

73. Given $h(x) = -5 - 3x$, find   **(a)** $h(2)$ **(b)** $h(-3)$ **(c)** $h(0)$
74. Given $g(x) = x^2 + 12x$, find   **(a)** $g(3)$ **(b)** $g(-5)$ **(c)** $g(0)$
75. Given $h(x) = 6 - x^2$, find   **(a)** $h(-1)$ **(b)** $h(1)$ **(c)** $h(-4)$

CHAPTER 6 TEST

Complete the ordered pairs for the following equations.

1. $12y - 7x = 5$; $(1, \quad)$ **2.** $y = 17$; $(-4, \quad)$

Determine whether the ordered pairs are solutions to the equations.

3. $x - 2y = 3$; $(1, 1)$ **4.** $2x + 3y = 6$; $(0, -2)$

Find the slopes of the following lines.

5. $-3x + 5 = y$

6. Through $(6, -5)$ and $(-1, 2)$

Graph the following.

7. $2x + y = 8$

8. $-x + 4y = 5$

9. $x - y \geq -2$

10. $y \geq -4x$

11. $5x - 7y = 10$

12. $2x - 3y > -6$

13. $6x + y > -1$

14. $y = -1$

15. $x - 3 = 0$

16. $5x - 3y = 15$

Which of the following graphs are graphs of functions?

17.

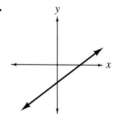

18.

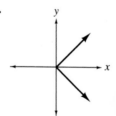

19.

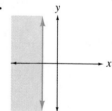

Given the following functions, find the indicated function values.

20. $f(x) = 2x - 4$ **(a)** $f(-2)$ **(b)** $f(0.2)$ **(c)** $f(0)$.

21. $h(x) = x^3 - x$ **(a)** $h(-1)$ **(b)** $h(0)$ **(c)** $h(4)$.

22. $g(x) = 6$ **(a)** $g(0)$ **(b)** $g(a)$ **(c)** $g(242)$.

Find equations of the following lines. Write the equation in standard form.

23. Through the origin and $(6, -7)$

24. Through $(2, -5)$ and $(1, 3)$

25. Through $(-5, -1)$ and parallel to $x = 7$

26. With slope $\dfrac{1}{8}$, through $(0, 12)$

27. Through $(-1, -1)$ and perpendicular to $y = -1$

29. Find the domain of each relation. Also determine whether each relation is also a function.
  a. $y = -x^2$ **b.** $x = |y|$ **c.** $y = 12$

28. Through $(9, -2)$ and parallel to the y-axis

CHAPTER 6 CUMULATIVE REVIEW

1. Translate each statement into symbols;
  a. 9 is less than or equal to 11.
  b. 8 is greater than 1.
  c. 3 is not equal to 4.
2. Find each sum;
  a. $(+3) + (-7)$ **b.** $(-2) + (10)$
  c. $2 + (-5)$
3. Simplify the following expressions;
  a. $3(2x - 5) + 1$
  b. $8 - (7x + 2) + 3x$
  c. $-2(4x + 7) - (3x - 1)$
4. If x is the first of three consecutive integers, express the sum of the three integers in terms of x. Simplify if possible;
5. Solve $y + 0.6 = -1.0$;
6. The length of a rectangular athletic field is three times its width. If 640 feet of fencing surrounds the field, find the dimensions of the field;
7. Solve $-5x + 7 < 2(x - 3)$, and graph the solution;
8. Simplify each of the following expressions;
  a. $(x^2)^5$ **b.** $(y^8)^2$ **c.** $[(-5)^3]^4$
9. Simplify the following;
  a. $\left(-\dfrac{5x^2}{y^3}\right)^2$ **b.** $\dfrac{(x^3)^4 x}{x^7}$ **c.** $\dfrac{(2x)^5}{x^3}$
  d. $\dfrac{(a^2 b)^3}{a^3 b^2}$

10. Write the following numbers in scientific notation;
  a. 367,000,000
  b. 0.000003
  c. 20,520,000,000
  d. 0.00085
11. Add $(4x^3 - 6x^2 + 2x + 7) + (5x^2 - 2x)$.
12. Multiply $(3y + 1)^2$;
13. Factor $xy + 2x + 3y + 6$ by grouping. Check by multiplying;
14. Factor $x^2 + 12x + 36$;
15. Factor $x^3 + 8$;
16. Solve $-2x^2 - 4x + 30 = 0$;
17. Simplify $\dfrac{5x - 5}{x^3 - x^2}$;
18. Subtract $\dfrac{2y}{2y - 7} - \dfrac{7}{2y - 7}$;
19. Solve $\dfrac{3a}{3a - 2} - \dfrac{5}{3a^2 + 7a - 6} = 1$;
20. Complete the following ordered pairs for the equation $3x + y = 12$;
  a. $(0, \ )$ **b.** $(\ , 6)$ **c.** $(-1, \ )$
21. Find the slope of the line through $(-1, 5)$ and $(2, -3)$. Graph the line;
22. Graph $x + y < 7$;

7.1 Solving Systems of Linear Equations by Graphing

7.2 Solving Systems of Linear Equations by Substitution

7.3 Solving Systems of Linear Equations by Addition

7.4 Applications of Systems of Linear Equations

7.5 Systems of Linear Inequalities

CHAPTER 7

Solving Systems of Linear Equations

When does an airplane in flight pass the point of no return? (See Critical Thinking, page 340.)

INTRODUCTION

In Chapter 6, we graphed equations containing two variables. Equations like these are often needed to represent relationships between two different values. For example, an economist attempts to predict what effects a price change will have on the sales prospects of calculators. There are many opportunities to compare and contrast two such equations, called a **system of equations.** This chapter presents **linear systems** and ways we solve these systems and apply them to real-life situations.

7.1
Solving Systems of Linear Equations by Graphing

OBJECTIVES

Tape 23

1 Determine if an ordered pair is a solution of a system of equations in two variables.

2 Identify a consistent system of equations, an inconsistent system of equations, and dependent equations.

3 Solve a system of linear equations by graphing.

4 Without graphing, determine the number of solutions of a system.

---

1 A **system of linear equations** consists of two or more linear equations. In this section, we focus on solving systems of linear equations containing two equations in two variables. Examples of such linear systems are

$$\begin{cases} 3x - 3y = 0 \\ x = 2y \end{cases} \qquad \begin{cases} x - y = 0 \\ 2x + y = 10 \end{cases} \qquad \begin{cases} y = 7x - 1 \\ y = 4 \end{cases}$$

A **solution** of a system of two equations in two variables is an ordered pair of numbers that is a solution of both equations in the system.

EXAMPLE 1 Which of the following ordered pairs is a solution of the given system?

$$\begin{cases} 2x - 3y = 6 & \text{First equation.} \\ x = 2y & \text{Second equation.} \end{cases}$$

a. $(12, 6)$ **b.** $(0, 2)$

Solution: If an ordered pair is a solution of both equations, it is a solution of the system.
a. Replace x with 12 and y with 6 in both equations.

$2x - 3y = 6$	First equation.		
$2(12) - 3(6) = 6$	Let $x = 12$ and $y = 6$.	$x = 2y$	Second equation.
$24 - 18 = 6$	Simplify.	$12 = 2(6)$	Let $x = 12$ and $y = 6$.
$6 = 6$	True.	$12 = 12$	True.

311

Since (12, 6) is a solution of each equation, it is a solution of the system.

b. Start by replacing x with 0 and y with 2 in the first equation.

$$2x - 3y = 6 \qquad \text{First equation.}$$
$$2(0) - 3(2) = 6 \qquad \text{Let } x = 0 \text{ and } y = 2.$$
$$0 - 6 = 6 \qquad \text{Simplify.}$$
$$-6 = 6 \qquad \text{False.}$$

The ordered pair (0, 2) is **not** a solution of $2x - 3y = 6$. Thus, there is no need to check (0, 2) in the second equation because we already know that (0, 2) is not a solution of the system.   ■

2   Since a solution of a system of two equations in two variables is an ordered pair that is a solution of both equations, it is also a point common to the graphs of the lines representing the equations in the system. Three different situations can occur when graphing the two lines associated with the equations in a linear system. These three situations are as follows:

One point of intersection : one solution

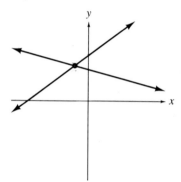

Consistent System

Parallel lines : no solution

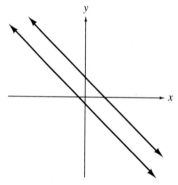

Inconsistent System

Same line : infinite number of solutions

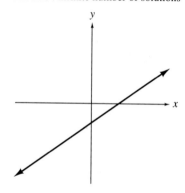

Dependent Equations

In the first graph, the lines intersect at a single point. This point represents the single solution of the linear system. A system of linear equations that has one solution is said to be a **consistent system.**

In the second graph, the two lines are parallel and so have no points in common. Thus, no ordered pair exists that makes both equations true at the same time and the system has no solution. We say that this system is an **inconsistent system.**

In the third case, the graphs of the two linear equations are identical. Each of the infinitely many solutions of one equation is a solution of the other equation, and vice versa. Thus, there are an infinite number of solutions of the system. We call these equations **dependent equations.**

3

EXAMPLE 2   Solve the system of equations by graphing.

$$\begin{cases} -x + 3y = 10 \\ x + y = 2 \end{cases}$$

Solution: On a single set of axes, graph each linear equation. The two lines appear to intersect at the point $(-1, 3)$. To verify this, replace x with -1 and y with 3 in both equations and see that true statements result. The solution is $(-1, 3)$ and the system is a consistent system.

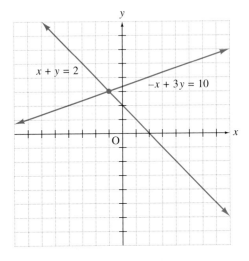

$-x + 3y = 10$

x	y
0	$\frac{10}{3}$
-4	2
2	4

$x + y = 2$

x	y
0	2
2	0
1	1

HELPFUL HINT

Neatly drawn graphs can help when "guessing" the solution of a system of linear equations by graphing.

EXAMPLE 3 Solve the following system of equations by graphing.

$$\begin{cases} 2x + y = 7 \\ 2y = -4x \end{cases}$$

Solution: Graph each of the two lines in the system.

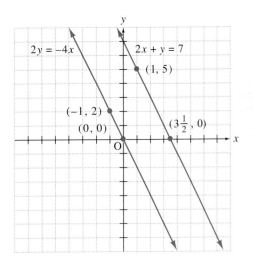

The lines **appear** to be parallel. To confirm this, write both equations in slope–intercept form by solving each for y.

$2x + y = 7$	First equation.	$2y = -4x$	Second equation.
$y = -2x + 7$	Subtract $2x$ from both sides.	$\dfrac{2y}{2} = \dfrac{-4x}{2}$	Divide both sides by 2.
		$y = -2x$	

Recall that when an equation is written in slope–intercept form, the coefficient of x is the slope. Since both equations have the same slope, -2, but different y-intercepts, the lines are parallel and have no points in common. Thus, there is no solution of the system and the system is inconsistent. ∎

EXAMPLE 4 Solve the system of equations by graphing.

$$\begin{cases} x - y = 3 \\ -x + y = -3 \end{cases}$$

Solution: Graph each line.

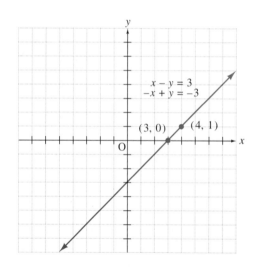

These graphs **appear** to be identical. To confirm this, write each equation in slope–intercept form.

$x - y = 3$	First equation.	$-x + y = -3$	Second equation.
$-y = -x + 3$	Subtract x from both sides.	$y = x - 3$	Add x to both sides.
$\dfrac{-y}{-1} = \dfrac{-x}{-1} + \dfrac{3}{-1}$	Divide both sides by -1.		
$y = x - 3$			

The equations are identical and so must be their graphs. The lines have an infinite number of points in common. Thus, there are an infinite number of solutions of the system. The equations are dependent equations. ∎

4 You have probably noticed by now that graphing alone is not an accurate way to solve a system of linear equations. The next two sections will present two accurate methods of solving these systems. In the meantime, we can decide how many solutions a system will have by writing each equation in the slope–intercept form.

EXAMPLE 5 Without graphing, determine the number of solutions of the system.

$$\begin{cases} \dfrac{1}{2}x - y = 2 \\ x = 2y + 5 \end{cases}$$

Solution: First write each equation in slope–intercept form.

$\dfrac{1}{2}x - y = 2$	First equation.	$x = 2y + 5$	Second equation.
$\dfrac{1}{2}x = y + 2$	Add y to both sides.	$x - 5 = 2y$	Subtract 5 from both sides.
		$\dfrac{x}{2} - \dfrac{5}{2} = \dfrac{2y}{2}$	Divide both sides by 2.
$\dfrac{1}{2}x - 2 = y$	Subtract 2 from both sides.	$\dfrac{1}{2}x - \dfrac{5}{2} = y$	Simplify.

The slope of each line is $\dfrac{1}{2}$, but they have different y-intercepts. This tells us that the lines representing these equations are parallel. Since the lines are parallel, the system has no solution and is inconsistent. ■

EXAMPLE 6 Determine the number of solutions of the system.

$$\begin{cases} 3x - y = 4 \\ x + 2y = 8 \end{cases}$$

Solution: Once again, the slope-intercept form helps determine how many solutions this system has.

$3x - y = 4$	First equation.	$x + 2y = 8$	Second equation.
$3x = y + 4$	Add y to both sides.	$x = -2y + 8$	Add $-2y$ to both sides.
$3x - 4 = y$	Subtract 4 from both sides.	$x - 8 = -2y$	Subtract 8 from both sides.
		$\dfrac{x}{-2} - \dfrac{8}{-2} = \dfrac{-2y}{-2}$	Divide both sides by -2.
		$-\dfrac{1}{2}x + 4 = y$	Simplify.

The slope of the second line is $-\dfrac{1}{2}$, whereas the first line has a slope of 3. Since the slopes are not equal, the two lines are neither parallel nor identical and must intersect. Therefore, this system has one solution and is consistent. ■

MENTAL MATH

Identify each system of linear equations as consistent or inconsistent. Also state whether the equations of the lines are dependent.

1.

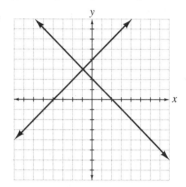

2.

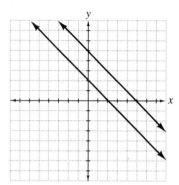

3.

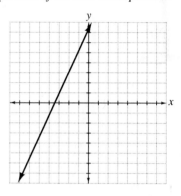

4.

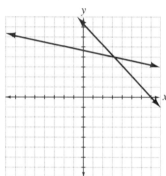

5.

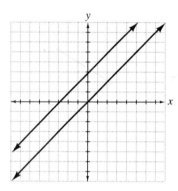

6.

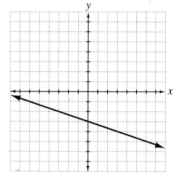

7.

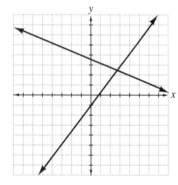

8.

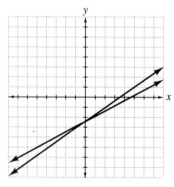

EXERCISE SET 7.1

Determine whether any of the following ordered pairs satisfy the system of linear equations. See Example 1.

1. $\begin{cases} x + y = 8 \\ 3x + 2y = 21 \end{cases}$

   a. $(2, 4)$
   b. $(5, 3)$
   c. $(1, 9)$

2. $\begin{cases} 2x + y = 5 \\ x + 3y = 5 \end{cases}$

   a. $(5, 0)$
   b. $(1, 2)$
   c. $(2, 1)$

3. $\begin{cases} 3x - y = 5 \\ x + 2y = 11 \end{cases}$

   a. $(2, -1)$
   b. $(3, 4)$
   c. $(0, -5)$

4. $\begin{cases} 2x - 3y = 8 \\ x - 2y = 6 \end{cases}$

   a. $(4, 0)$
   b. $(-2, -4)$
   c. $(7, 2)$

5. $\begin{cases} 2y = 4x \\ 2x - y = 0 \end{cases}$

   a. $(-3, -6)$
   b. $(0, 0)$
   c. $(1, 2)$

6. $\begin{cases} 4x = 1 - y \\ x - 3y = -8 \end{cases}$

   a. $(0, 1)$
   b. $(1, -3)$
   c. $(-2, 2)$

Solve each system of equations by graphing the equations on the same set of axes. See Examples 2 through 4.

7. $\begin{cases} y = x + 1 \\ y = 2x - 1 \end{cases}$

8. $\begin{cases} y = 3x - 4 \\ y = x + 2 \end{cases}$

9. $\begin{cases} 2x + y = 0 \\ 3x + y = 1 \end{cases}$

10. $\begin{cases} 2x + y = 1 \\ 3x + y = 0 \end{cases}$

11. $\begin{cases} x + y = -1 \\ -2x + y = 5 \end{cases}$

12. $\begin{cases} y - x = -1 \\ 3x + y = -5 \end{cases}$

13. $\begin{cases} 2x - y = 6 \\ y = 2 \end{cases}$

14. $\begin{cases} x + y = 5 \\ x = 4 \end{cases}$

Without graphing, decide:

  a. Are the graphs of the equations identical lines, parallel lines, or lines intersecting at a single point? **b.** How many solutions does the system have? See Examples 5 and 6.

15. $\begin{cases} 4x + y = 24 \\ x + 2y = 2 \end{cases}$

16. $\begin{cases} 3x + y = 1 \\ 3x + 2y = 6 \end{cases}$

17. $\begin{cases} 2x + y = 0 \\ 2y = 6 - 4x \end{cases}$ graph

18. $\begin{cases} 3x + y = 0 \\ 2y = -6x \end{cases}$ graph

19. $\begin{cases} 6x - y = 4 \\ \dfrac{1}{2}y = -2 + 3x \end{cases}$ graph

20. $\begin{cases} 3x - y = 2 \\ \dfrac{1}{3}y = -2 + 3x \end{cases}$

Solve each system of equations by the graphing method.

21. $\begin{cases} y = x - 2 \\ y = 2x + 3 \end{cases}$

22. $\begin{cases} y = x + 5 \\ y = -2x - 4 \end{cases}$

23. $\begin{cases} x + y = 7 \\ x - y = 3 \end{cases}$

24. $\begin{cases} x + y = -4 \\ x - y = 2 \end{cases}$

25. $\begin{cases} x + y = 5 \\ x + y = 6 \end{cases}$

26. $\begin{cases} 2x + y = 4 \\ x + y = 2 \end{cases}$

27. $\begin{cases} y - 3x = -2 \\ 6x - 2y = 4 \end{cases}$

28. $\begin{cases} y + 2x = 3 \\ 4x = 2 - 2y \end{cases}$

29. $\begin{cases} x - 2y = 2 \\ 3x + 2y = -2 \end{cases}$

30. $\begin{cases} x + 3y = 7 \\ 2x - 3y = -4 \end{cases}$

31. $\begin{cases} \frac{1}{2}x + y = -1 \\ x = 4 \end{cases}$

32. $\begin{cases} x + \frac{3}{4}y = 2 \\ x = -1 \end{cases}$

Without graphing, decide:

a. Are the graphs of the equations identical lines, parallel lines, or lines intersecting at a single point?

b. How many solutions does the system have?

33. $\begin{cases} 3y - 2x = 3 \\ x + 2y = 9 \end{cases}$

34. $\begin{cases} 2y = x + 2 \\ y - 2x = 3 \end{cases}$

35. $\begin{cases} 6y + 4x = 6 \\ 3y - 3 = -2x \end{cases}$

36. $\begin{cases} 8y + 6x = 4 \\ 4y - 2 = 3x \end{cases}$

37. $\begin{cases} x + y = 4 \\ x + y = 3 \end{cases}$

38. $\begin{cases} 2x + y = 0 \\ y = -2x + 1 \end{cases}$

Writing in Mathematics

39. Explain how to use a graph to determine the number of solutions a system has.

40. Explain how writing each equation in a linear system in the point–slope form helps determine the number of solutions a system has.

Skill Review

Solve each linear equation. See Section 2.4.

41. $2(x + 7) = 13$

42. $3(x + 4) = 16$

43. $3 - 4x = -5$

44. $8 - 3x = -7$

45. $4x - 8 = 9x + 2$

46. $2x + 6 = 5x + 9$

7.2
Solving Systems of Linear Equations by Substitution

OBJECTIVE   **1**   Use substitution to solve a system of linear equations.

Tape 23

1   As we mentioned in the previous section, graphing is not an accurate method for solving systems of equations. For example, a solution of $\left(\dfrac{1}{2}, \dfrac{2}{9}\right)$ is unlikely to be read correctly from a graph. In this section, we discuss a second, more accurate method for solving systems of equations. This method is called the **substitution method** and is introduced in the next example.

EXAMPLE 1   Solve the system

$$\begin{cases} 2x + y = 10 & \text{First equation.} \\ x = y + 2 & \text{Second equation.} \end{cases}$$

Solution: Notice that the second equation is solved for x. Substituting $y + 2$ for x in the first equation results in an equation containing just one variable, y, which we can then solve.

$$2\,x + y = 10 \qquad \text{First equation.}$$
$$2\,(y + 2) + y = 10 \qquad \text{Substitute } y + 2 \text{ for } x \text{ since } x = y + 2.$$
$$2y + 4 + y = 10 \qquad \text{Use the distributive property.}$$
$$3y + 4 = 10$$
$$3y = 6$$
$$y = 2$$

The y-value of the ordered pair solution of the system is 2. To find the corresponding x-value, let $y = 2$ in the equation $x = y + 2$ and solve for x.

$$x = y + 2$$
$$x = 2 + 2 \qquad \text{Let } y = 2.$$
$$x = 4$$

The solution of the system is the ordered pair $(4, 2)$. Since an ordered pair solution must satisfy both equations in the system, we could have chosen the equation $2x + y = 10$ to find the corresponding x-value. The resulting x-value will be the same.

To check, see that $(4, 2)$ satisfies both equations.

First equation	Second equation	
$2x + y = 10$	$x = y + 2$	
$2(4) + 2 = 10$	$4 = 2 + 2$	Let $x = 4$ and $y = 2$.
$10 = 10$ True.	$4 = 4$	True.

The solution of the system is (4, 2).

A graph of the two equations shows two lines intersecting at the point (4, 2).

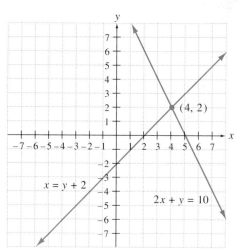

To solve a system of equations by substitution, we need an equation solved for one of the variables.

EXAMPLE 2   Solve the system $\begin{cases} x + 2y = 7 \\ 2x + 2y = 13 \end{cases}$.

Solution: Choose an equation and solve for x or y. We will solve the first equation for x by subtracting $2y$ from both sides.

$$x + 2y = 7 \qquad \text{First equation.}$$

$$x = 7 - 2y$$

Since $x = 7 - 2y$, substitute $7 - 2y$ for x in the second equation and solve for y.

$$2x + 2y = 13 \qquad \text{Second equation}$$

$$2(7 - 2y) + 2y = 13 \qquad \text{Let } x = 7 - 2y.$$

$$14 - 4y + 2y = 13 \qquad \text{Distributive property.}$$

$$14 - 2y = 13 \qquad \text{Simplify.}$$

$$-2y = -1 \qquad \text{Subtract 14 from both sides.}$$

$$y = \frac{1}{2} \qquad \text{Divide both sides by } -2.$$

To find x, let $y = \frac{1}{2}$ in the equation $x = 7 - 2y$.

$$x = 7 - 2y$$

$$x = 7 - 2\left(\frac{1}{2}\right) \qquad \text{Let } y = \frac{1}{2}.$$

$$x = 7 - 1$$

$$x = 6$$

The solution is $\left(6, \dfrac{1}{2}\right)$. Check the solution in both of the original linear equations.   ∎

The following steps may be used to solve a system of equations by the substitution method.

> **To Solve a System of Equations by the Substitution Method**
>
> *Step 1* Solve one of the equations for one of its variables.
> *Step 2* Substitute the expression for the variable found in step 1 into the other equation.
> *Step 3* Find the value of one variable by solving the equation from step 2.
> *Step 4* Find the value of the other variable by substituting the value found in step 3 in any equation containing both variables.
> *Step 5* Check the proposed solution in the original system.

EXAMPLE 3 Solve the system $\begin{cases} 7x - 3y = -14 \\ -3x + y = 6 \end{cases}$.

Solution: To avoid introducing fractions, we will solve the second equation for y.

$$-3x + y = 6 \qquad \text{Second equation.}$$

$$y = 3x + 6$$

Next, substitute $3x + 6$ for y in the first equation.

$$7x - 3y = -14 \qquad \text{First equation.}$$

$$7x - 3(3x + 6) = -14$$

$$7x - 9x - 18 = -14$$

$$-2x - 18 = -14$$

$$-2x = 4$$

$$\frac{-2x}{-2} = \frac{4}{-2}$$

$$x = -2$$

To find the corresponding y-value, substitute -2 for x in the equation $y = 3x + 6$. Then $y = 3(-2) + 6$ or $y = 0$. The solution of the system is $(-2, 0)$. Check this solution in both equations of the system.   ∎

HELPFUL HINT

When solving a system of equations by the substitution method, begin by solving an equation for one of its variables. If possible, solve for a variable that has a coefficient of 1. This way, we avoid working with time-consuming fractions.

EXAMPLE 4   Solve the system $\begin{cases} \frac{1}{2}x - y = 3 \\ x = 6 + 2y \end{cases}$ by substitution.

Solution: The second equation is already solved for x in terms of y. Thus. we substitute $6 + 2y$ for x in the first equation and solve for y.

$$\frac{1}{2}x - y = 3 \qquad \text{First equation.}$$

$$\frac{1}{2}(6 + 2y) - y = 3 \qquad \text{Let } x = 6 + 2y.$$

$$3 + y - y = 3$$

$$3 = 3$$

Arriving at a true statement such as $3 = 3$ indicates that the two equations in the system are equivalent. This means that their graphs are identical and there are an infinite number of solutions to the system. Any solution of one equation is also a solution of the other. The equations are dependent.   ■

EXAMPLE 5   Use substitution to solve the system.

$$\begin{cases} 6x + 12y = 5 \\ -4x - 8y = 0 \end{cases}$$

Solution: Choose the second equation and solve for y.

$$-4x - 8y = 0 \qquad \text{Second equation.}$$

$$-8y = 4x \qquad \text{Add } 4x \text{ to both sides.}$$

$$\frac{-8y}{-8} = \frac{4x}{-8} \qquad \text{Divide both sides by } -8.$$

$$y = -\frac{1}{2}x \qquad \text{Simplify.}$$

Now replace y with $-\frac{1}{2}x$ in the first equation.

$$6x + 12y = 5 \qquad \text{First equation.}$$

$$6x + 12\left(-\frac{1}{2}x\right) = 5 \qquad \text{Let } y = -\frac{1}{2}x.$$

$$6x + (-6x) = 5 \qquad \text{Simplify.}$$

$$0 = 5 \qquad \text{Combine like terms.}$$

The false statement $0 = 5$ indicates that this system has no solution and is inconsistent. The graph of the equations in the system is a pair of parallel lines.   ■

EXERCISE SET 7.2

Solve each system of equations by the substitution method. See Example 1.

1. $\begin{cases} x + y = 3 \\ x = 2y \end{cases}$

2. $\begin{cases} x + y = 20 \\ x = 3y \end{cases}$

3. $\begin{cases} x + y = 6 \\ y = -3x \end{cases}$

4. $\begin{cases} x + y = 6 \\ y = -4x \end{cases}$

5. $\begin{cases} 3x + 2y = 16 \\ x = 3y - 2 \end{cases}$

6. $\begin{cases} 2x + 3y = 18 \\ x = 2y - 5 \end{cases}$

Solve each system of equations by the substitution method. See Examples 2 and 3.

7. $\begin{cases} x + 2y = 6 \\ 2x + 3y = 8 \end{cases}$

8. $\begin{cases} x + 3y = -5 \\ 2x + 2y = 6 \end{cases}$

9. $\begin{cases} 2x - 5y = 1 \\ 3x + y = -7 \end{cases}$

10. $\begin{cases} 4x + 2y = 5 \\ 2x + y = -4 \end{cases}$

Solve each system of equations by the substitution method. See Examples 4 and 5.

11. $\begin{cases} 2y = x + 2 \\ 6x - 12y = 0 \end{cases}$

12. $\begin{cases} 3y = x + 6 \\ 4x + 12y = 0 \end{cases}$

13. $\begin{cases} \dfrac{1}{3}x - y = 2 \\ x - 3y = 6 \end{cases}$

14. $\begin{cases} \dfrac{1}{4}x - 2y = 1 \\ x - 8y = 4 \end{cases}$

Solve each system of equations by the substitution method.

15. $\begin{cases} 3x - 4y = 10 \\ x = 2y \end{cases}$

16. $\begin{cases} 3x - 4y = 10 \\ y = 2x \end{cases}$

17. $\begin{cases} y = 3x + 1 \\ 4y - 8x = 12 \end{cases}$

18. $\begin{cases} y = 2x + 3 \\ 5y - 7x = 18 \end{cases}$

19. $\begin{cases} 4x + y = 11 \\ 2x + 5y = 1 \end{cases}$

20. $\begin{cases} 3x + y = -14 \\ 4x + 3y = -22 \end{cases}$

21. $\begin{cases} 2x - 3y = -9 \\ 3x = y + 4 \end{cases}$

22. $\begin{cases} 8x - 3y = -4 \\ 7x = y + 3 \end{cases}$

23. $\begin{cases} 6x - 3y = 5 \\ x + 2y = 0 \end{cases}$

24. $\begin{cases} 10x - 5y = -21 \\ x + 3y = 0 \end{cases}$

25. $\begin{cases} 3x - y = 1 \\ 2x - 3y = 10 \end{cases}$

26. $\begin{cases} 2x - y = -7 \\ 4x - 3y = -11 \end{cases}$

27. $\begin{cases} -x + 2y = 10 \\ -2x + 3y = 18 \end{cases}$

28. $\begin{cases} -x + 3y = 18 \\ -3x + 2y = 19 \end{cases}$

29. $\begin{cases} 5x + 10y = 20 \\ 2x + 6y = 10 \end{cases}$

30. $\begin{cases} 2x + 4y = 6 \\ 5x + 10y = 15 \end{cases}$

31. $\begin{cases} 3x + 6y = 9 \\ 4x + 8y = 16 \end{cases}$

32. $\begin{cases} 6x + 3y = 12 \\ 9x + 6y = 15 \end{cases}$

33. $\begin{cases} y = 2x + 9 \\ y = 7x + 10 \end{cases}$

34. $\begin{cases} y = 5x - 3 \\ y = 8x + 4 \end{cases}$

Writing in Mathematics

35. Explain how to identify an inconsistent system when using the substitution method.

36. Occasionally, when using the substitution method, the equation $0 = 0$ is obtained. Explain how this result indicates that the linear system is dependent.

Skill Review

Perform the indicated operation. See Section 5.3.

37. $\dfrac{5}{x + 2} - \dfrac{x}{x + 2}$

38. $\dfrac{y}{y - 1} + \dfrac{3}{y - 1}$

39. $\dfrac{3x}{x + 5} + \dfrac{15}{x + 5}$

40. $\dfrac{x}{x + 1} - \dfrac{2x + 1}{x + 1}$

Solve the following proportions. See Section 5.7.

41. $\dfrac{2x}{5} = \dfrac{x + 2}{3}$

42. $\dfrac{a + 1}{7} = \dfrac{a}{2}$

7.3
Solving Systems of Linear Equations by Addition

OBJECTIVE   **1**   Use addition to solve a system of linear equations.

Tape 24

1   We have seen that substitution is an accurate way to solve a linear system. Another method for solving a system of equations accurately is the **addition** or **elimination method.** The addition method is based on the fact that adding equal quantities to both sides of an equation does not change the solution of the equation. In symbols,

$$\text{if } A = B \text{ and } C = D, \text{ then } A + C = B + C.$$

EXAMPLE 1   Solve the system $\begin{cases} x + y = 7 \\ x - y = 5 \end{cases}$ by the addition method.

Solution: Since the left side of each equation is equal to the right side, we add equal quantities by adding the left sides of the equations together and the right sides of the equations together. This sum gives us an equation in one variable, x, which we can solve for x.

$$
\begin{aligned}
x + y &= 7 && \text{First equation.} \\
\underline{x - y} &= \underline{5} && \text{Second equation.} \\
2x &= 12 && \text{Add the equations.} \\
x &= 6 && \text{Divide both sides by 2.}
\end{aligned}
$$

The x-value of the solution is 6. To find the corresponding y-value, let $x = 6$ in either equation of the system. We will use the first equation.

$$
\begin{aligned}
x + y &= 7 && \text{First equation.} \\
6 + y &= 7 && \text{Let } x = 6. \\
y &= 7 - 6 && \text{Solve for } y. \\
y &= 1 && \text{Simplify.}
\end{aligned}
$$

The solution is (6, 1). Check this in both equations.
Check:

First equation	Second equation	
$x + y = 7$	$x - y = 5$	
$6 + 1 = 7$	$6 - 1 = 5$	Let $x = 6$ and $y = 1$.
$7 = 7$ True.	$5 = 5$ True.	

Thus, the graphs of the two equations intersect at the point (6, 1) as shown.

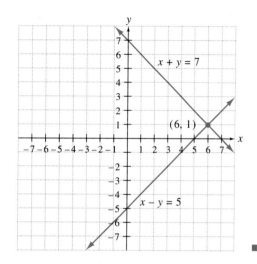

EXAMPLE 2   Use the addition method to solve $\begin{cases} -2x + y = 2 \\ -x + 3y = -4 \end{cases}$.

Solution: If we simply add the two equations, the result will still be an equation in two variables. However, our goal is to eliminate one of the variables. Notice what happens if we multiply **both sides** of the first equation by -3. The system

$$\begin{cases} -3(-2x + y) = -3(2) \\ -x + 3y = -4 \end{cases} \quad \text{simplifies to} \quad \begin{cases} 6x - 3y = -6 \\ -x + 3y = -4 \end{cases}$$

Now add the resulting equations and the y variable is eliminated.

$$\begin{array}{rl} 6x - 3y &= -6 \\ \underline{-x + 3y} &= \underline{-4} \\ 5x\phantom{{}- 3y} &= -10 \qquad \text{Add.} \\ x &= -2 \qquad \text{Divide both sides by 5.} \end{array}$$

To find the corresponding y-value, let $x = -2$ in any of the preceding equations containing both variables. We use the first equation of the original system.

$$\begin{array}{rl} -2x + y &= 2 \qquad \text{First equation.} \\ -2(-2) + y &= 2 \qquad \text{Let } x = -2. \\ 4 + y &= 2 \\ y &= -2 \end{array}$$

The solution is $(-2, -2)$. Check this ordered pair in both equations of the original system. ■

In Example 2, the decision to multiply the first equation by -3 was no accident. To eliminate a variable when adding two equations, the coefficient of the variable in one equation must be the opposite of its coefficient in the other equation.

HELPFUL HINT

Be sure to multiply **both sides** of an equation by a chosen number when solving by the addition method. A common mistake is to only multiply the side containing the variables.

EXAMPLE 3   Use the addition method to solve $\begin{cases} 2x - y = 7 \\ 8x - 4y = 1 \end{cases}$.

Solution: Multiply both sides of the first equation by -4 and the coefficient of x will be -8, the opposite of 8, the coefficient of x in the second equation. The system

$$\begin{cases} -4(2x - y) = -4(7) \\ 8x - 4y = 1 \end{cases} \quad \text{simplifies to} \quad \begin{cases} -8x + 4y = -28 \\ 8x - 4y = 1 \end{cases}$$

Now add the resulting equations.

$$-8x + 4y = -28$$
$$\underline{8x - 4y = 1}$$
$$0 = -27 \qquad \text{False.}$$

Since $0 = -27$ is a false statement, the system has no solution. This system is inconsistent. The equations, if graphed, are parallel lines.   ■

As we discovered with the substitution method, if our attempts to solve a system yield a true statement like $0 = 0$, the system has an infinite number of solutions. The equations, if graphed, are identical lines and dependent.

EXAMPLE 4   Solve the system $\begin{cases} 3x + 4y = 13 \\ 5x - 9y = 6 \end{cases}$ by the addition method.

Solution: We can eliminate the variable y by multiplying the first equation by 9 and the second equation by 4.

$$\begin{cases} 9(3x + 4y) = 9(13) \\ 4(5x - 9y) = 4(6) \end{cases} \quad \text{simplifies to} \quad \begin{cases} 27x + 36y = 117 \\ 20x - 36y = 24 \end{cases}$$
$$47x = 141 \qquad \text{Add the equations.}$$
$$x = 3$$

To find the corresponding y-value, let $x = 3$ in any equation containing 2 variables. Doing so in any of these equations will give $y = 1$. The solution to this system is $(3, 1)$. Check to see that $(3, 1)$ satisfies each equation in the original system.   ■

If we had decided to eliminate x's instead of y's in Example 4, the first equation could have been multiplied by 5 and the second by -3. Try solving the original system this way to see that the solution is still $(3, 1)$.

To Solve a System of Two Linear Equations by the Addition Method

Step 1   Rewrite each equation in standard form $Ax + By = C$.

> *Step 2*   If necessary, multiply one or both equations by a nonzero number so that the coefficient of one variable in one equation is the opposite of its coefficient in the other equation.
>
> *Step 3*   Add the equations.
>
> *Step 4*   Find the value of one variable by solving the resulting equation from step 3.
>
> *Step 5*   Find the value of the second variable by substituting the value found in step 4 into either of the original equations.
>
> *Step 6*   Check the proposed solution in the original system.

EXAMPLE 5   Use the addition method to solve $\begin{cases} -x - \dfrac{y}{2} = \dfrac{5}{2} \\ -\dfrac{x}{2} + \dfrac{y}{4} = 0 \end{cases}$.

Solution: Begin by clearing each equation of fractions. Multiply both sides of the first equation by the LCD 2 and multiply both sides of the second equation by the LCD 4. Then

$$\begin{cases} 2\left(-x - \dfrac{y}{2}\right) = 2\left(\dfrac{5}{2}\right) \\ 4\left(-\dfrac{x}{2} + \dfrac{y}{4}\right) = 4(0) \end{cases} \quad \text{simplifies to} \quad \begin{cases} -2x - y = 5 \\ -2x + y = 0 \end{cases}$$

Now add the resulting equations in the simplified system.

$$\begin{aligned} -2x - y &= 5 \\ \underline{-2x + y} &= \underline{0} \\ -4x \quad\;\; &= 5 \qquad \text{Add.} \end{aligned}$$

$$x = -\frac{5}{4}$$

To find y, we could replace x with $-\dfrac{5}{4}$ in an equation with two variables. Instead, let's go back to the simplified system and multiply by appropriate factors to eliminate the variable x and solve for y. To do this, multiply the first equation in the simplified system by -1. Then

$$\begin{cases} -1(-2x - y) = -1(5) \\ -2x + y = 0 \end{cases} \quad \text{simplifies to} \quad \begin{cases} 2x + y = -5 \\ \underline{-2x + y = 0} \\ \qquad\quad 2y = -5 \qquad \text{Add.} \end{cases}$$

$$y = -\frac{5}{2}$$

The solution is $\left(-\dfrac{5}{4}, -\dfrac{5}{2}\right)$.   ∎

EXERCISE SET 7.3

Solve each system of equations by the addition method. See Example 1.

1. $\begin{cases} 3x + y = 5 \\ 6x - y = 4 \end{cases}$

2. $\begin{cases} 4x + y = 13 \\ 2x - y = 5 \end{cases}$

3. $\begin{cases} x - 2y = 8 \\ -x + 5y = -17 \end{cases}$

4. $\begin{cases} x - 2y = -11 \\ -x + 5y = 23 \end{cases}$

5. $\begin{cases} 3x + 2y = 11 \\ 5x - 2y = 29 \end{cases}$

6. $\begin{cases} 4x + 2y = 2 \\ 3x - 2y = 12 \end{cases}$

Solve each system of equations by the addition method. See Examples 2 through 4.

7. $\begin{cases} 3x + y = -11 \\ 6x - 2y = -2 \end{cases}$

8. $\begin{cases} 4x + y = -13 \\ 6x - 3y = -15 \end{cases}$

9. $\begin{cases} x + 5y = 18 \\ 3x + 2y = -11 \end{cases}$

10. $\begin{cases} x + 4y = 14 \\ 5x + 3y = 2 \end{cases}$

11. $\begin{cases} 2x - 5y = 4 \\ 3x - 2y = 4 \end{cases}$

12. $\begin{cases} 6x - 5y = 7 \\ 4x - 6y = 7 \end{cases}$

13. $\begin{cases} 2x + 3y = 0 \\ 4x + 6y = 3 \end{cases}$

14. $\begin{cases} -x + 5y = -1 \\ 3x - 15y = 3 \end{cases}$

Solve each system of equations by the addition method. See Example 5.

15. $\begin{cases} \dfrac{x}{3} + \dfrac{y}{6} = 1 \\ \dfrac{x}{2} - \dfrac{y}{4} = 0 \end{cases}$

16. $\begin{cases} \dfrac{x}{2} + \dfrac{y}{8} = 3 \\ x - \dfrac{y}{4} = 0 \end{cases}$

17. $\begin{cases} x - \dfrac{y}{3} = -1 \\ -\dfrac{x}{2} + \dfrac{y}{8} = \dfrac{1}{4} \end{cases}$

18. $\begin{cases} 2x - \dfrac{3y}{4} = -3 \\ x + \dfrac{y}{9} = \dfrac{13}{3} \end{cases}$

Solve each system by the addition method.

19. $\begin{cases} x + y = 6 \\ x - y = 6 \end{cases}$

20. $\begin{cases} x - y = 1 \\ -x + 2y = 0 \end{cases}$

21. $\begin{cases} 3x + y = 4 \\ 9x + 3y = 6 \end{cases}$

22. $\begin{cases} 2x + y = 6 \\ 4x + 2y = 12 \end{cases}$

23. $\begin{cases} 3x - 2y = 7 \\ 5x + 4y = 8 \end{cases}$

24. $\begin{cases} 6x - 5y = 25 \\ 4x + 15y = 13 \end{cases}$

25. $\begin{cases} \dfrac{2}{3}x + 4y = -4 \\ 5x + 6y = 18 \end{cases}$

26. $\begin{cases} \dfrac{3}{2}x + 4y = 1 \\ 9x + 24y = 5 \end{cases}$

27. $\begin{cases} 4x - 6y = 8 \\ 6x - 9y = 12 \end{cases}$

28. $\begin{cases} 9x - 3y = 12 \\ 12x - 4y = 18 \end{cases}$

29. $\begin{cases} \dfrac{x}{3} - y = 2 \\ -\dfrac{x}{2} + \dfrac{3y}{2} = -3 \end{cases}$

30. $\begin{cases} \dfrac{x}{2} + \dfrac{y}{4} = 1 \\ -\dfrac{x}{4} - \dfrac{y}{8} = 1 \end{cases}$

31. $\begin{cases} 8x + 11y = -16 \\ 2x + 3y = -4 \end{cases}$

32. $\begin{cases} 10x + 3y = -12 \\ 5x + 4y = -16 \end{cases}$

Solve each system by either the addition method or the substitution method.

33. $\begin{cases} 2x - 3y = -11 \\ y = 4x - 3 \end{cases}$

34. $\begin{cases} 4x - 5y = 6 \\ y = 3x - 10 \end{cases}$

35. $\begin{cases} x + 2y = 1 \\ 3x + 4y = -1 \end{cases}$

36. $\begin{cases} x + 3y = 5 \\ 5x + 6y = -2 \end{cases}$

37. $\begin{cases} 2y = x + 6 \\ 3x - 2y = -6 \end{cases}$

38. $\begin{cases} 3y = x + 14 \\ 2x - 3y = -16 \end{cases}$

39. $\begin{cases} y = 2x - 3 \\ y = 5x - 18 \end{cases}$

40. $\begin{cases} y = 6x - 5 \\ y = 4x - 11 \end{cases}$

41. $\begin{cases} x + \dfrac{1}{6}y = \dfrac{1}{2} \\ 3x + 2y = 3 \end{cases}$

42. $\begin{cases} x + \dfrac{1}{3}y = \dfrac{5}{12} \\ 8x + 3y = 4 \end{cases}$

43. $\begin{cases} \dfrac{x + 2}{2} = \dfrac{y + 11}{3} \\ \dfrac{x}{2} = \dfrac{2y + 16}{6} \end{cases}$

44. $\begin{cases} \dfrac{x + 5}{2} = \dfrac{y + 14}{4} \\ \dfrac{x}{3} = \dfrac{2y + 2}{6} \end{cases}$

Writing in Mathematics

45. To solve the system $\begin{cases} 2x - 3y = 5 \\ 5x + 2y = 6 \end{cases}$, explain why we might prefer the addition method rather than the substitution method.

Skill Review

Rewrite the following sentences using mathematical symbols. Do not solve the equations. See Section 2.4.

46. Twice a number added to 6 is 3 less than the number.

47. The sum of three consecutive integers is 66.

48. Three times a number subtracted from 20 is 2.

49. The product of 4 and the sum of a number and 6 is twice a number.

50. The quotient of twice a number and 7 is subtracted from the reciprocal of the number.

7.4
Applications of Systems of Linear Equations

OBJECTIVE

Tape 24

1   Use a system of equations to solve word problems.

1   Many of the word problems solved earlier using one-variable equations can also be solved using two equations in two variables. Use the following steps for solving applications using two variables in this section. Notice how similar these steps are to those for solving a word problem or application in a single variable given earlier.

> **To Solve a Word Problem Using Linear Systems in Two Variables**
>
> *Step 1* Read and reread the application. Determine what two quantities are unknown and choose **two** variables to represent the two unknown quantities.
> *Step 2* If possible, draw a diagram to visualize the known facts.
> *Step 3* Translate the problem into **two** equations using both variables.
> *Step 4* Solve the system of equations.
> *Step 5* State the solution and determine if the solution is reasonable.
> *Step 6* Check the solution in the originally stated problem.

EXAMPLE 1 Find two numbers whose sum is 37 and whose difference is 21.

Solution: *Step 1* Choose two variables to represent two unknown quantities.
We are asked to find 2 different numbers. Let $x =$ first number and $y =$ second number.

Step 3 Translate the problem into two equations using both variables.
Since the sum of the numbers is 37 and their difference is 21, we have the system

$$\begin{cases} x + y = 37 \\ x - y = 21 \end{cases}$$

Step 4 Solve the system.

$$x + y = 37$$
$$\underline{x - y = 21}$$
$$2x \quad\quad = 58 \qquad\qquad \text{Add the equations.}$$

$$x = \frac{58}{2} = 29 \qquad \text{Divide both sides by 2.}$$

Let $x = 29$ in the first equation to find y.

$$x + y = 37 \qquad\qquad \text{First equation.}$$
$$29 + y = 37$$
$$y = 37 - 29 = 8$$

Step 5 State the solution.
The first number is 29 and the second number is 8.

Step 6 Check.
The sum of 29 and 8 is 37 and their difference is 21. Since these two numbers satisfy the original stated conditions, 29 and 8 are the solutions. ∎

EXAMPLE 2 The Barnum and Bailey Circus is in town. Admission for 4 adults and 2 children is $22, while admission for 2 adults and 3 children is $16.

a. What is the price of an adult's ticket?

b. What is the price of a child's ticket?

c. A special rate of $60 is charged for advance sales to groups of 20 persons. Should a group of 4 adults and 16 children use the group rate? Why or why not?

Solution: *Step 1* Choose two variables to represent two unknown quantities.
The unknown quantities are the price of an adult's ticket and the price of a child's ticket. Let A = price of an adult's ticket and C = price of a child's ticket.

Step 3 Translate the problem into two equations using both variables.
Since admission for 1 adult is A, admission for 4 adults is $4A$. Also, admission for 2 children is $2C$ and admission for 3 children is $3C$.

admission for 4 adults	plus	admission for 2 children	is	$22	or
$4A$	$+$	$2C$	$=$	22	

admission for 2 adults	plus	admission for 3 children	is	$16	or
$2A$	$+$	$3C$	$=$	16	

Step 4 Solve the system.
Multiply the second equation by -2 to eliminate the variable A. Then

$$\begin{cases} 4A + 2C = 22 \\ -2(2A + 3C) = -2(16) \end{cases} \quad \text{simplifies to} \quad \begin{cases} 4A + 2C = 22 \\ -4A - 6C = -32 \end{cases}$$
$$\overline{-4C = -10} \quad \text{Add the equations.}$$

$$C = \frac{5}{2} = 2.5 \text{ or } \$2.50$$

To find A, replace C with 2.5 in the first equation.

$$4A + 2C = 22 \qquad \text{First equation.}$$

$$4A + 2(2.5) = 22 \qquad \text{Let } C = 2.5.$$

$$4A + 5 = 22$$

$$4A = 17$$

$$A = \frac{17}{4} = 4.25 \text{ or } \$4.25$$

Step 5 State the solution and answer the three questions.

a. Since $A = 4.25$, the price of an adult's ticket is $4.25.

b. Since $C = 2.5$, the price of a child's ticket is $2.50.

c. The regular admission price for 4 adults and 16 children is

$$4(\$4.25) + 16(\$2.50) = \$17.00 + \$40.00$$

$$= \$57.00$$

This is $3 less than the special group rate of $60, so they should **not** request the group rate.

Step 6 Check.

To check these solutions, notice that 4 adults and 2 children will pay $4(\$4.25) + 2(\$2.50) = \$17 + \$5 = \$22$. Also, the price for 2 adults and 3 children is $2(\$4.25) + 3(\$2.50) = \$8.50 + \$7.50 = \$16$, which agrees with the original information.   ■

EXAMPLE 3   Albert and Louis are 15 miles away from each other when they start walking toward one another. After 2 hours they meet. If Louis walks one mile per hour faster than Albert, find both walking speeds.

Solution: *Step 1* Choose two variables to represent two unknown quantities.

$$\text{Let } x = \text{Albert's rate in miles per hour}$$

$$y = \text{Louis's rate in miles per hour}$$

Step 2 Draw a diagram.

Use the facts stated in the problem and the formula $d = rt$ to fill in the following chart.

	r	$\cdot$ t	$= d$
Albert	x	2	$2x$
Louis	y	2	$2y$

Translate the problem into two equations using both variables. To write one equation, use the fact that

Albert's distance	+	Louis's distance	=	total distance

or

$$2x \quad + \quad 2y \quad = \quad 15$$

Also, since Louis walks one mile per hour faster than Albert, we know that

$$y = x + 1$$

Step 4 Solve the system of equations.

The system of equations we are solving is

$$\begin{cases} y = x + 1 \\ 2x + 2y = 15 \end{cases}$$

Use substitution to solve the system since the first equation is solved for y.

$2x + 2y = 15$	Second equation.
$2x + 2(x + 1) = 15$	Replace y with $x + 1$.
$2x + 2x + 2 = 15$	
$4x = 13$	
$x = \dfrac{13}{4} = 3.25$	

Step 5 State the solution.

Albert walks at a rate of 3.25 miles per hour. Since $y = x + 1$, we have $y = 3.25 + 1 = 4.25$. Louis walks at a rate of 4.25 mph.

Step 6 Check.

Use the formula $d = rt$, and find that in 2 hours Albert's distance is $(3.25)(2)$ miles or 6.5 miles. In 2 hours, Louis's distance is $(4.25)(2)$ miles or 8.5 miles. The total distance walked is 6.5 miles + 8.5 miles or 15 miles, the given distance. ■

EXAMPLE 4 Guy, a chemistry teaching assistant, needs 10 liters of a 20% saline solution (salt water) for his 2 P.M. laboratory class. Unfortunately, the only mixtures on hand are a 5% saline solution and a 25% saline solution. How much of each solution should he mix to produce the 20% solution?

Solution: Let x = necessary liters of the 5% solution

y = necessary liters of the 25% solution

Since a total of 10 liters is needed, one equation is $x + y = 10$. Now consider the mixture being made.

	5% solution		25% solution		mixture
Liters of solution	x		y		10
Strength of solution (percent salt)	0.05	+	0.25	=	0.20
Liters of salt	$0.05x$	+	$0.25y$	=	2

Since the liters of salt in the first solution plus the liters of salt in the second solution should equal the liters of salt in the mixture, the second equation in the system is

$$\mathbf{0.05x + 0.25y = 2}$$

The system to solve is

$$\begin{cases} x + y = 10 \\ 0.05x + 0.25y = 2 \end{cases}$$

To solve, multiply the first equation by -25 and the second equation by 100. Then

$$\begin{cases} -25(x + y) = -25(10) \\ 100(0.05x + 0.25y) = 100(2) \end{cases} \quad \text{simplifies to} \quad \begin{cases} -25x - 25y = -250 \\ 5x + 25y = 200 \end{cases}$$

$$\begin{aligned} -20x &= -50 \quad \text{Add.} \\ x &= 2.5 \end{aligned}$$

To find y, let $x = 2.5$ in the first equation of the original system.

$$x + y = 10$$

$$2.5 + y = 10 \qquad \text{Let } x = 2.5.$$

$$y = 7.5$$

Thus, Guy needs to mix 2.5 liters of 5% saline solution with 7.5 liters of 25% saline solution. This checks since $2.5 + 7.5 = 10$, the required number of liters. Also, the sum of the liters of salt in the two solutions equals the liters of salt in the required mixture:

$$0.05(2.5) + 0.25(7.5) = 0.20(10)$$

$$0.125 + 1.875 = 2 \quad \blacksquare$$

EXERCISE SET 7.4

Write a system of equations describing the known facts. Do not solve the system.

1. Two numbers add up to 15 and have a difference of 7.

2. The total of two numbers is 16. The first number plus 2 more than 3 times the second equals 18.

3. Keiko has a total of $6500, which she has invested in two accounts. The larger account is $800 greater than the smaller account.

4. Fran has four times as much invested in his savings account as in his checking account. The total amount invested is $2300.

Assuming the facts given, choose the correct solution from among the options given.

5. The length of a rectangle is 3 feet longer than the width. The perimeter is 30 feet. Find the dimensions of the rectangle.
  a. Length = 8 feet; width = 5 feet
  b. Length = 8 feet; width = 7 feet
  c. Length = 9 feet; width = 6 feet

6. An isosceles triangle, a triangle with two sides of equal length, has a perimeter of 20 inches. Each of the equal sides is one inch longer than the unequal side. Find the lengths of the three sides.
  a. 6 inches, 6 inches, and 7 inches
  b. 7 inches, 7 inches, and 6 inches
  c. 6 inches, 7 inches, and 8 inches

7. A small plane takes 4 hours to travel 400 miles when flying against the wind. Only 2 hours are needed to complete the trip when the plane flies with the wind. Find the wind speed and the speed of the plane in still air.
  a. Wind = 50 mph; plane = 150 mph
  b. Wind = 150 mph; plane = 50 mph
  c. Wind = 100 mph; plane = 100 mph

8. Two CDs and 4 cassette tapes cost a total of $40. However, 3 CDs and 5 cassette tapes cost $55. Find the price of each type of recording.
  a. CDs = $12; cassettes = $4
  b. CDs = $15; cassettes = $2
  c. CDs = $10; cassettes = $5

9. Lynn has a total of 100 coins, all of which are either dimes or quarters. The total value of the coins is $13.00. Find the number of each type of coin.
  a. 80 dimes; 20 quarters
  b. 20 dimes; 44 quarters
  c. 60 dimes; 40 quarters

10. Charlie has 28 gallons of saline solution available in the pharmacy in two large containers. One container holds three times as much as the other container. Find the capacity of each container.
  a. 15 gallons; 5 gallons
  b. 20 gallons; 8 gallons
  c. 21 gallons; 7 gallons

Solve. See Example 1.

11. Two numbers total 83 and have a difference of 17. Find the two numbers.

12. The sum of two numbers is 76 and their difference is 52. Find the two numbers.

Solve. See Example 2.

13. Jane has been pricing Amtrak train fares for a group trip to New York. Three adults and 4 children must pay $159. Two adults and 3 children must pay $112. Find the price of an adult's ticket, and find the price of a child's ticket.

14. Last month Kathryn purchased 5 cassettes and 2 compact disks at Wall-to-Wall Sound for $65. This month she bought 3 cassettes and 4 compact disks for $81. Find the price for each cassette, and find the price of each compact disk.

15. Taylor has 80 coins in a jar, all of which are either quarters or nickels. The total value is $14.60. How many of each type of coin does he have?

16. Rita purchased 40 stamps, a mixture of 25¢ and 15¢ stamps. Find the number of each type of stamp if she spent $8.80.

Solve. See Example 3.

17. Alan can row 18 miles down the Delaware River in 2 hours, but the return trip took him $4\frac{1}{2}$ hours. Find the rate Alan could row in still water, and find the rate of the current.

18. The Schultz family took a canoe 10 miles down the Allegheny River in 1 hour and 15 minutes. After lunch it took them 4 hours to return. Find the rate of the current.

19. Rich and Pat fly from Philadelphia to Chicago, a distance of 780 miles. On the trip they fly into the wind, and the flight takes 2 hours. The return trip, with the wind behind them, only takes $1\frac{1}{2}$ hours. Find the speed of the wind and find the speed of the plane in still air.

20. With a strong wind behind it, a United Airlines jet flies 2400 miles from Los Angeles to Orlando in 4 hours and 45 minutes. The return trip takes 6 hours, as the plane flies into the wind. Find the speed of the plane in still air, and find the wind speed to the nearest tenth.

Solve. See Example 4.

21. Kay needs to prepare 12 ounces of a 9% hydrochloric acid solution. To get this solution, find the amount of 4% and the amount of 12% solution she should mix.

22. Hope is preparing 15 liters of a 25% saline solution. Hope has two other saline solutions with strengths of 40% and 10%. Find the amount of 40% solution and the amount of 10% solution she should mix to get 15 liters of a 25% solution.

23. Guillermo blends coffee for Maxwell House. He needs to prepare 200 pounds of blended coffee beans selling for $3.95 per pound. He intends to do this by blending together a high-quality bean costing $4.95 per pound and a cheaper bean costing $2.65 per pound. To the nearest pound, find how much high-quality coffee and how much cheaper coffee he should blend.

24. Macadamia nuts cost an astounding $16.50 per pound, but research by Planter's Peanuts says that mixed nuts sell better if macadamias are included. The standard mix costs $9.25 per pound. Find how many pounds of macadamias and how many pounds of the standard mix should be combined to produce 40 pounds that will cost $10 per pound. Find the amounts to the nearest tenth of a pound.

Solve.

25. Ray is thinking of two numbers. The first number plus twice the second number is 8. Twice the first number plus the second totals 25. Find the numbers.

26. One number is 4 more than twice the second number. Their total is 25. Find the numbers.

27. Carrie and Raymond had a pottery stand at the annual Skippack Craft Fair. They sold some of their pottery at the original price of $9.50 each, but later dropped the price to $7.50 each. If they sold all 90 pieces and took in $721, find how many they sold at the original price.

28. Trinity Church held its annual supper and fed a total of 387 people. They charged $6.75 for adults and $3.50 for children. If they took in $2433.50, find how many children attended the dinner.

29. Joe has decided to fence off a garden plot behind his house, using his house as the "fence" along one side of the garden. The length (which runs parallel to the house) is 3 feet less than twice the width. Find the dimensions if 33 feet of fencing is used along the three sides requiring it.

30. Anne plans to erect 152 feet of fencing around her rectangular horse pasture. A river bank serves as one side of the rectangle. If each width is 4 feet longer than half the length, find the dimensions.

31. Jim began a 186-mile bicycle trip to build up stamina for a triathlete competition. Unfortunately, his bicycle chain broke, so he finished the trip walking. The whole trip took 6 hours. If Jim walks at a rate of 4 miles per hour and rides at 40 mph, find the amount of time he spent on the bicycle.

32. In Canada, eastbound and westbound trains travel along the same track, with sidings to pull onto to avoid accidents. Two trains are now 150 miles apart, with the westbound train traveling twice as fast as the eastbound train. A warning must be issued to pull one train onto a siding or else the trains will crash in $1\frac{1}{4}$ hours. Find the speed of the eastbound train and the speed of the westbound train.

33. Joan rented a car from Hertz, which rents its car for a daily fee plus an additional charge per mile driven. Joan recalls that a car rented for 5 days and driven for 300 miles cost her $178, while a car rented for 4 days and driven for 500 miles costs $197. Find the daily fee, and find the mileage charge.

34. Cyril and Anoa operate a small construction and supply company. In July they charged the Shaffers $1702.50 for 65 hours of labor and 3 tons of material. In August the Shaffers paid $1349 for 49 hours of labor and $2\frac{1}{2}$ tons of material. Find the cost per hour of labor and the cost per ton of material.

Skill Review

Graph each linear inequality. See Section 6.5.

35. $y < 3 - x$

36. $y \geq 4 - 2x$

Simplify the following. See Section 3.2.

37. 4^{-2}

38. 2^{-3}

7.5
Systems of Linear Inequalities

	OBJECTIVE		
		1	Solve a system of a linear inequalities.

Tape 25

1   Earlier we solved linear inequalities in two variables. Just as two linear equations comprise a system of linear equations, two linear inequalities comprise a **system of linear inequalities.** Systems of inequalities are very important in a process called linear programming. Many businesses use linear programming to find the most profitable way to use limited resources such as employees, machines, or buildings.

A **solution of a system of linear inequalities** is an ordered pair that satisfies each inequality in the system. By graphing each inequality in a system, we can discover the solutions of the system.

EXAMPLE 1   Find the graph of the solution of the system $\begin{cases} 3x \geq y \\ x + 2y \leq 8 \end{cases}$.

Solution: We begin by graphing each inequality on the same set of axes. The graph of the solution of the system is the region contained in the graphs of both inequalities. It is their intersection.

First, graph $3x \geq y$. The boundary line is the graph of $3x = y$. Sketch a solid boundary line since the inequality $3x \geq y$ means $3x > y$ or $3x = y$. The test point $(1, 0)$ satisfies the inequality, so shade the half-plane that includes $(1, 0)$.

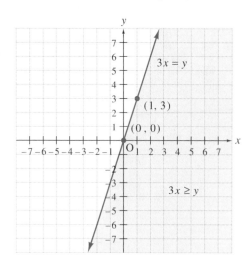 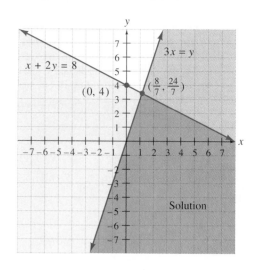

Next, sketch a solid boundary line $x + 2y = 8$ on the same set of axes. The test point $(0, 0)$ satisfies the inequality $x + 2y \leq 8$, so shade the half-plane that includes $(0, 0)$.

An ordered pair solution of the system must satisfy both inequalities. These solutions are points that lie in both shaded regions. The solution of the system is the darkest shaded region. This solution includes parts of both boundary lines.   ■

It is sometimes necessary to find the coordinates of the **corner point:** the point at which the two boundary lines intersect. To find the point of intersection, solve the related linear system

$$\begin{cases} 3x = y \\ x + 2y = 8 \end{cases}$$

by the substitution method or the addition method. The lines intersect at $\left(\dfrac{8}{7}, \dfrac{24}{7}\right)$, the corner point of the graph.

To Graph the Solutions of a System of Linear Inequalities
Step 1   Graph each inequality in the system on the same set of axes.
Step 2   The solutions of the system are the points common to the graphs of all the inequalities in the system.

EXAMPLE 2   Graph the solutions of the system $\begin{cases} x - y < 2 \\ x + 2y > -1 \end{cases}$.

Solution: Graph both inequalities on the same set of axes. Both boundary lines are dashed lines since the inequality symbols are $<$ and $>$. The solution of the system is the region shown by the darkest shading. In this example, the boundary lines are not a part of the solution.

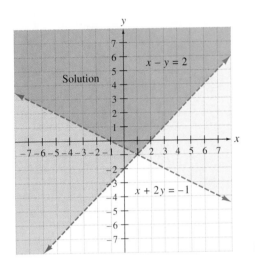

EXAMPLE 3   Graph the solution of the system $\begin{cases} -3x + 4y < 12 \\ x \geq 2 \end{cases}$.

Solution: Graph both inequalities on the same set of axes.

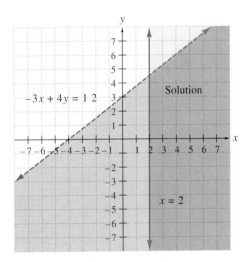

The solution of the system is the darkest shaded region.   ■

EXERCISE SET 7.5

Graph the solutions of each system of linear inequalities. See Examples 1 through 3.

1. $\begin{cases} y \geq x + 1 \\ y \geq 3 - x \end{cases}$

2. $\begin{cases} y \geq x - 3 \\ y \geq -1 - x \end{cases}$

3. $\begin{cases} y < 3x - 4 \\ y \leq x + 2 \end{cases}$

4. $\begin{cases} y \leq 2x + 1 \\ y > x + 2 \end{cases}$

5. $\begin{cases} y \leq -2x - 2 \\ y \geq x + 4 \end{cases}$

6. $\begin{cases} y \leq 2x + 4 \\ y \geq -x - 5 \end{cases}$

7. $\begin{cases} y \geq -x + 2 \\ y \leq 2x + 5 \end{cases}$

8. $\begin{cases} y \geq x - 5 \\ y \leq -3x + 3 \end{cases}$

9. $\begin{cases} x \geq 3y \\ x + 3y \leq 6 \end{cases}$

10. $\begin{cases} -2x < y \\ x + 2y < 3 \end{cases}$

11. $\begin{cases} y + 2x \geq 0 \\ 5x - 3y \leq 12 \end{cases}$

12. $\begin{cases} y + 2x \leq 0 \\ 5x + 3y \geq -2 \end{cases}$

13. $\begin{cases} 3x - 4y \geq -6 \\ 2x + y \leq 7 \end{cases}$

14. $\begin{cases} 4x - y \geq -2 \\ 2x + 3y \leq -8 \end{cases}$

15. $\begin{cases} x \leq 2 \\ y \geq -3 \end{cases}$

16. $\begin{cases} x \geq -3 \\ y \geq -2 \end{cases}$

17. $\begin{cases} y \geq 1 \\ x < -3 \end{cases}$

18. $\begin{cases} y > 2 \\ x \geq -1 \end{cases}$

19. $\begin{cases} 2x + 3y < -8 \\ x \geq -4 \end{cases}$

20. $\begin{cases} 3x + 2y \leq 6 \\ x < 2 \end{cases}$

21. $\begin{cases} 2x - 5y \le 9 \\ y \le -3 \end{cases}$

22. $\begin{cases} 2x + 5y \le -10 \\ y \ge 1 \end{cases}$

23. $\begin{cases} y \ge \dfrac{1}{2}x + 2 \\ y \le \dfrac{1}{2}x - 3 \end{cases}$

24. $\begin{cases} y \ge \dfrac{-3}{2}x + 3 \\ y < \dfrac{-3}{2}x + 6 \end{cases}$

25. Tony budgets his time at work today. Part of the day he can write bills; the rest of the day he can use to write purchase orders. The total time available is at most 8 hours. Less than 3 hours is to be spent writing bills.
   a. Write a system of inequalities to describe the situation. (Let x = hours available for writing bills and y = hours available for writing purchase orders.)
   b. Graph the solutions of the system.

26. Marty plans to invest money in two different accounts: a standard savings account and a riskier money market fund. The total of the two investments can be no more than $1000. At least $300 is to be put into the money market fund.
   a. Write a system of inequalities to describe the situation. Let x = amount invested in the savings account and y = amount invested in the money market fund.
   b. Graph the solution to the system in part **a.**

Writing in Mathematics

27. Explain how to decide which region to shade to show the solution of the following system. $\begin{cases} x \ge 3 \\ y \ge -2 \end{cases}$

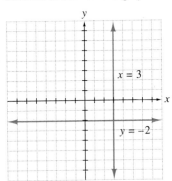

28. Describe the region of solutions of the system $\begin{cases} x > 0 \\ y > 0 \end{cases}$.

Skill Review

Evaluate each expression. See Section 3.1.

29. $(-3)^2$

30. $(-5)^3$

31. $\left(\dfrac{2}{3}\right)^2$

32. $\left(\dfrac{3}{4}\right)^3$

Perform the following operations.

33. $-2 - (-3)$

34. $5 - 11$

35. $8 + (-13)$

36. $-12 + (-1)$

CRITICAL THINKING

An airplane that encounters mechanical or fuel problems may need to make an emergency landing at the nearest available airport. For a plane flying over the Pacific Ocean from Los Angeles to Honolulu, the only reasonable airports to land at are Los Angeles' International Airport and Honolulu's International Airport. Between these two airports is a flying distance of 2560 miles. It is reasonable to think that exactly halfway along the route (1280 miles into the flight) the airplane passes the "point of no return": From that halfway point, the airplane would take as long to fly back to Los Angeles to make an emergency landing as it would to fly the rest of the way to Honolulu.

The halfway point, though, is only the point of no return when the speed of the plane is exactly the same flying out from Los Angeles as it is flying back to Los Angeles. In fact, flying out from Los Angeles, the plane flies against strong winds; flying back to Los Angeles, the plane flies with the strong winds. At the halfway point, the plane actually takes longer to continue flying to Honolulu than it does to fly back to Los Angeles, since the winds decrease the speed of the plane.

Given this information, is the point of no return closer to the Los Angeles airport or the Honolulu airport? Can you predict the point of no return for a plane flying from Los Angeles to Honolulu? What facts must you know or assume to know to make such a prediction?

CHAPTER 7 GLOSSARY

A linear system is **consistent** if it has a solution.

Equations that have identical graphs are **dependent** equations.

A linear system is **inconsistent** if it has no solution.

A **solution of a system of equations** is an ordered pair of numbers that is a solution of all the equations of the system.

A **system of linear equations** consists of two or more linear equations.

A **system of linear inequalities** consists of two or more linear inequalities.

CHAPTER 7 SUMMARY

(7.2)

To Solve a System of Equations by the Substitution Method

Step 1   Solve one equation for one of its variables.

Step 2   Substitute the expression for the variable found in step 1 into the other equation.

Step 3   Find the value of one variable by solving the equation from step 2.

Step 4   Find the value of the other variable by substituting the value found in step 3 in any equation containing both variables.

Step 5   Check the proposed solution in the original system.

(7.3)

> ### To Solve a System of Two Linear Equations by the Addition Method
>
> *Step 1* Rewrite each equation in standard form, $Ax + By = C$.
>
> *Step 2* If necessary, multiply one or both equations by a nonzero number so that the coefficient of one variable in one equation is the opposite of its coefficient in the other equation.
>
> *Step 3* Add the equations.
>
> *Step 4* Find the value of one variable by solving the resulting equation from step 3.
>
> *Step 5* Find the value of the second variable by substituting the value found in step 4 into either of the original equations.
>
> *Step 6* Check the proposed solution in the original system.

CHAPTER 7 REVIEW

(7.1) *Determine whether any of the following ordered pairs satisfy the system of linear equations.*

1. $\begin{cases} 2x - 3y = 12 \\ 3x + 4y = 1 \end{cases}$
  a. $(12, 4)$
  b. $(3, -2)$
  c. $(-3, 6)$

2. $\begin{cases} 4x + y = 0 \\ -8x - 5y = 9 \end{cases}$
  a. $\left(\dfrac{3}{4}, -3\right)$
  b. $(-2, 8)$
  c. $\left(\dfrac{1}{2}, -2\right)$

3. $\begin{cases} 5x - 6y = 18 \\ 2y - x = -4 \end{cases}$
  a. $(-6, -8)$
  b. $\left(3, \dfrac{5}{2}\right)$
  c. $\left(3, -\dfrac{1}{2}\right)$

4. $\begin{cases} 2x + 3y = 1 \\ 3y - x = 4 \end{cases}$
  a. $(2, 2)$
  b. $(-1, 1)$
  c. $(2, -1)$

Solve each system of equations by graphing.

5. $\begin{cases} 2x + y = 5 \\ 3y = -x \end{cases}$

6. $\begin{cases} 3x + y = -2 \\ 2x - y = -3 \end{cases}$

7. $\begin{cases} y - 2x = 4 \\ x + y = -5 \end{cases}$

8. $\begin{cases} y - 3x = 0 \\ 2y - 3 = 6x \end{cases}$

9. $\begin{cases} 3x + y = 2 \\ 3x - 6 = -9y \end{cases}$

10. $\begin{cases} 2y + x = 2 \\ x - y = 5 \end{cases}$

Without graphing, decide whether the system is consistent or inconsistent and whether the equations are dependent.

11. $\begin{cases} 2x + y = 4 \\ 3x + 2y = 8 \end{cases}$

12. $\begin{cases} 6x + 2y = 4 \\ 3x + y = 2 \end{cases}$

13. $\begin{cases} x + 2y = 4 \\ y = -\dfrac{1}{2}x + 3 \end{cases}$

14. $\begin{cases} \dfrac{2}{3}x + 3y = 1 \\ 2x = 6 - 9y \end{cases}$

Without graphing, decide whether the graphs of the system are identical, parallel lines, or lines intersecting at a single point.

15. $\begin{cases} 2x - y = 3 \\ y = 3x + 1 \end{cases}$

16. $\begin{cases} 3x + y = 4 \\ y = -3x + 1 \end{cases}$

17. $\begin{cases} \dfrac{2}{3}x + \dfrac{1}{6}y = 0 \\ \quad\quad y = -4x \end{cases}$

18. $\begin{cases} \dfrac{1}{4}x + \dfrac{1}{8}y = 0 \\ \quad\quad y = -6x \end{cases}$

(7.2)   *Solve the following systems of equations by the substitution method. If there is a single solution, give the ordered pair. State whether the system is inconsistent or whether the equations are dependent.*

19. $\begin{cases} y = 2x + 6 \\ 3x - 2y = -11 \end{cases}$

20. $\begin{cases} y = 3x - 7 \\ 2x - 3y = 7 \end{cases}$

21. $\begin{cases} x + 3y = -3 \\ 2x + y = 4 \end{cases}$

22. $\begin{cases} 3x + y = 11 \\ x + 2y = 12 \end{cases}$

23. $\begin{cases} 4y = 2x - 3 \\ x - 2y = 4 \end{cases}$

24. $\begin{cases} 2x = 3y - 18 \\ x + 4y = 2 \end{cases}$

25. $\begin{cases} 2(3x - y) = 7x - 5 \\ 3(x - y) = 4x - 6 \end{cases}$

26. $\begin{cases} 4(x - 3y) = 3x - 1 \\ 3(4y - 3x) = 1 - 8x \end{cases}$

27. $\begin{cases} \dfrac{3}{4}x + \dfrac{2}{3}y = 2 \\ \quad\quad 3x + y = 18 \end{cases}$

28. $\begin{cases} \dfrac{2}{5}x + \dfrac{3}{4}y = 1 \\ \quad\quad x + 3y = -2 \end{cases}$

(7.3)   *Solve the following systems of equations by the addition method.*

29. $\begin{cases} 2x + 3y = -6 \\ x - 3y = -12 \end{cases}$

30. $\begin{cases} 4x + y = 15 \\ -4x + 3y = -19 \end{cases}$

31. $\begin{cases} 2x - 3y = -15 \\ x + 4y = 31 \end{cases}$

32. $\begin{cases} x - 5y = -22 \\ 4x + 3y = 4 \end{cases}$

33. $\begin{cases} 2x = 6y - 1 \\ \dfrac{1}{3}x - y = \dfrac{-1}{6} \end{cases}$

34. $\begin{cases} 8x = 3y - 2 \\ \dfrac{4}{7}x - y = \dfrac{-5}{2} \end{cases}$

35. $\begin{cases} 5x = 6y + 25 \\ -2y = 7x - 9 \end{cases}$

36. $\begin{cases} -4x = 8 + 6y \\ -3y = 2x - 3 \end{cases}$

37. $\begin{cases} 3(x - 4) = -2y \\ 2x = 3(y - 19) \end{cases}$

38. $\begin{cases} 4(x + 5) = -3y \\ 3x = 2(y + 18) \end{cases}$

39. $\begin{cases} \dfrac{2x + 9}{3} = \dfrac{y + 1}{2} \\ \dfrac{x}{3} = \dfrac{y - 7}{6} \end{cases}$

40. $\begin{cases} \dfrac{2 - 5x}{4} = \dfrac{2y - 4}{2} \\ \dfrac{x + 5}{3} = \dfrac{y}{5} \end{cases}$

(7.4)   *Solve by writing and solving a system of linear equations.*

41. The sum of two numbers is 16. Three times the larger number decreased by the smaller number is 72. Find the two numbers.

42. The Forrest Theater can seat a total of 360 people. They take in $15,150 when every seat is sold. If orchestra section tickets cost $45 and balcony tickets cost $35, find the number of people that can be seated in the orchestra section.

43. A passenger ship can head 340 miles upriver in 19 hours, but the return trip takes only 14 hours. Find the current of the river and find the speed of the ship in still water to the nearest tenth of a mile.

44. Sam invested $9000 one year ago. Part of the money was invested at 6%, the rest at 10%. If the total interest earned in one year was $652.80, find how much was invested at each rate.

45. The most pleasing dimensions for a picture are those where the length is approximately 1.6 times longer than the width. This ratio is known as the Golden Ratio. If Bob has 6 feet of framing material, find the dimensions of the largest frame he can make that satisfies the Golden Ratio. Find the dimensions to the nearest hundredth of a foot.

46. Find the amount of 6% acid solution and the amount of 14% acid solution Pat should combine to prepare 50 cc's (cubic centimeters) of a 12% solution.

47. The Deli charges $3.80 for a breakfast of 3 eggs and 4 strips of bacon. The charge is $2.75 for 2 eggs and 3 strips of bacon. Find the cost of each egg and the cost of each strip of bacon.

48. Chris alternates between jogging and walking. He traveled 15 miles during the past 3 hours. He jogs at a rate of 7.5 miles per hour and walks at a rate of 4 miles per hour. Find how much time, to the nearest hundredth of an hour, he actually spent jogging.

(7.5)   *Graph the solutions of the following systems of linear inequalities.*

49. $\begin{cases} y \geq 2x - 3 \\ y \leq -2x + 1 \end{cases}$

50. $\begin{cases} y \leq -3x - 3 \\ y \leq 2x + 7 \end{cases}$

51. $\begin{cases} x + 2y > 0 \\ x - y \leq 6 \end{cases}$

52. $\begin{cases} x - 2y \geq 7 \\ x + y \leq -5 \end{cases}$

53. $\begin{cases} 3x - 2y \leq 4 \\ 2x + y \geq 5 \end{cases}$

54. $\begin{cases} 4x - y \leq 0 \\ 3x - 2y \geq -5 \end{cases}$

55. $\begin{cases} -3x + 2y > -1 \\ y < -2 \end{cases}$

56. $\begin{cases} -2x + 3y > -7 \\ x \geq -2 \end{cases}$

CHAPTER 7 TEST

Is the ordered pair a solution of the given linear system?

1. $\begin{cases} 2x - 3y = 5 \\ 6x + y = 1 \end{cases}$; $(1, -1)$

2. $\begin{cases} 4x - 3y = 24 \\ 4x + 5y = -8 \end{cases}$; $(3, -4)$

3. Use graphing to find the solutions of the system $\begin{cases} y - x = 6 \\ y + 2x = -6 \end{cases}$.

4. Use the substitution method to solve the system $\begin{cases} 3x - 2y = -14 \\ x + 3y = -1 \end{cases}$.

5. Use the substitution method to solve the system
$$\begin{cases} \frac{1}{2}x + 2y = -\frac{15}{4} \\ 4x = -y \end{cases}.$$

6. Use the addition method to solve the system $\begin{cases} 3x + 5y = 2 \\ 2x - 3y = 14 \end{cases}$.

7. Use the addition method to solve the system $\begin{cases} 5x - 6y = 7 \\ 7x - 4y = 12 \end{cases}$.

Solve each system using the substitution method or the addition method.

8. $\begin{cases} 3x + y = 7 \\ 4x + 3y = 1 \end{cases}$

9. $\begin{cases} 3(2x + y) = 4x + 20 \\ x - 2y = 3 \end{cases}$

10. $\begin{cases} \dfrac{x - 3}{2} = \dfrac{2 - y}{4} \\ \dfrac{7 - 2x}{3} = \dfrac{y}{2} \end{cases}$

11. Lisa has a bundle of money consisting of $1 bills and $5 bills. There are 62 bills in the bundle. The total value of the bundle is $230. Find the number of $1 bills and the number of $5 bills.

12. Don has invested $4000, part at 5% interest and the rest at 9% interest. Find how much he invested at each rate if the total interest after 1 year is $311.

Graph the solutions of the following systems of linear inequalities.

13. $\begin{cases} y + 2x \le 4 \\ y \ge 2 \end{cases}$

14. $\begin{cases} 2y - x \ge 1 \\ x + y \ge -4 \end{cases}$

CHAPTER 7 CUMULATIVE REVIEW

1. If $x = 2$ and $y = -5$, find the value of the following expressions.

   a. $\dfrac{x - y}{12 + x}$ **b.** $x^2 - y$

2. Solve for x: $\dfrac{5}{2}x = 15$.

3. A 10-foot piece of wire is to be cut into two pieces so that the longer piece is 4 times the shorter. If x represents the length of the shorter piece, find the length of each piece.

4. Solve $V = lwh$ for l.

5. Simplify each expression. Write answers with positive exponents:

   a. $\left(\dfrac{2}{3}\right)^{-4}$ **b.** $2^{-1} + 4^{-1}$ **c.** $(-2)^{-4}$

6. Find $\left(x^2 - \dfrac{1}{3}y\right)\left(x^2 + \dfrac{1}{3}y\right)$.

7. Factor $25x^2 + 25xy + 4y^2$.

8. Factor $2x^3 + 3x^2 - 2x - 3$.

9. Simplify $\dfrac{2x^2 - 2xy + 3x - 3y}{2x + 3}$.

10. Divide $\dfrac{2x^2 - 11x + 5}{5x - 25} \div \dfrac{4x - 2}{10}$.

11. Add $\dfrac{2}{3t} + \dfrac{5}{t + 1}$.

12. Solve the equation $x + \dfrac{14}{x - 2} = \dfrac{7x}{x - 2} + 1$.

13. Graph $y = 4x - 8$.

14. Graph $x = 2$.

15. Find the equation of a vertical line through the point $(-1, 5)$.

16. Graph $5x + 4y \leq 20$.

17. Graph $y > 3$.

18. Solve the system $\begin{cases} x + 2y = 7 \\ 2x + 2y = 13 \end{cases}$.

19. Use the addition method to solve $\begin{cases} 2x - y = 7 \\ 8x - 4y = 1 \end{cases}$.

20. Graph the solutions of the system $\begin{cases} x - y < 2 \\ x + 2y > -1 \end{cases}$.

8.1 Introduction to Radicals

8.2 Simplifying Radicals

8.3 Adding and Subtracting Radicals

8.4 Multiplying and Dividing Radicals

8.5 Solving Equations Containing Radicals

8.6 Applications of Equations Containing Radicals

8.7 Rational Exponents

CHAPTER 8

Roots and Radicals

Joe Gerken of Gerken's Nursery expands his business to include growing and selling Scotch pine Christmas trees. The efficient use of his land depends on an optimal pattern for planting the trees. (See Critical Thinking, page 382.)

INTRODUCTION

Having spent the last two chapters studying equations, we return now to algebraic expressions. We expand on your skills of operating on expressions—adding, subtracting, multiplying, dividing, and raising to powers—to include finding roots. Just as subtraction is defined by addition and division by multiplication, finding roots is defined by raising out powers. This chapter also includes work with equations that contain roots and with word problems that can be solved by such equations.

8.1
Introduction to Radicals

OBJECTIVES

Tape 26

1 Find square roots of perfect squares.

2 Find cube roots of perfect cubes.

3 Find nth roots.

4 Identify rational and irrational numbers.

1   In this section, we define finding the **root** of a number by its reverse operation, raising a number to a power. We begin with squares and square roots.

The square of 5 is $5^2 = 25$.
The square of -5 is $(-5)^2 = 25$.
The square of $\dfrac{1}{2}$ is $\left(\dfrac{1}{2}\right)^2 = \dfrac{1}{4}$.

The reverse operation of squaring a number is finding the **square root** of a number. For example:

A square root of 25 is 5, because $5^2 = 25$.
A square root of 25 is also -5, because $(-5)^2 = 25$.
A square root of $\dfrac{1}{4}$ is $\dfrac{1}{2}$, because $\left(\dfrac{1}{2}\right)^2 = \dfrac{1}{4}$.

Notice that both 5 and -5 are square roots of 25. We will use the symbol $\sqrt{}$ to denote the **positive** or **principal** square root of a number. For example,

$$\sqrt{25} = 5 \text{ since } 5^2 = 25$$

The symbol $-\sqrt{}$ will be used to denote the negative square root. For example,

$$-\sqrt{25} = -5$$

Square Root

The positive or principal square root of a positive number a is written as $\sqrt{a}$. The negative square root of a is written as $-\sqrt{a}$.

$$\sqrt{a} = b, \quad \text{if } b^2 = a$$

Also, the square root of 0 is 0, written as $\sqrt{0} = 0$.

The symbol $\sqrt{}$ is called a **radical** or **radical sign.** The expression within or under a radical sign is called the **radicand.** An expression containing a radical is called a **radical expression.**

$$\sqrt{a} \quad \begin{matrix} \text{radical sign} \\ \text{radicand} \end{matrix}$$

EXAMPLE 1 Find each square root.

  a. $\sqrt{36}$ **b.** $\sqrt{64}$ **c.** $-\sqrt{25}$ **d.** $\sqrt{\dfrac{9}{100}}$ **e.** $\sqrt{0}$

Solution: **a.** $\sqrt{36} = 6$ because $6^2 = 36$ and 6 is positive.

  b. $\sqrt{64} = 8$ because $8^2 = 64$ and 8 is positive.

  c. $-\sqrt{25} = -5$. The negative sign in front of the radical indicates the negative square root of 25.

  d. $\sqrt{\dfrac{9}{100}} = \dfrac{3}{10}$ because $\left(\dfrac{3}{10}\right)^2 = \dfrac{9}{100}$ and $\dfrac{3}{10}$ is positive.

  e. $\sqrt{0} = 0$ because $0^2 = 0$. ∎

Can we find the square root of a negative number, say $\sqrt{-4}$? That is, can we find a real number whose square is -4? No. There is no real number whose square is -4, and we say that $\sqrt{-4}$ is not a real number. In general:

The square root of a negative number is not a real number.

2 Finding roots can be extended to other roots such as cube roots. For example, since $2^3 = 8$, we call 2 the **cube root** of 8. In symbols, this is

$$\sqrt[3]{8} = 2$$

Cube Root

A **cube root** of a real number a is written as $\sqrt[3]{a}$, and

$$\sqrt[3]{a} = b, \quad \text{if } b^3 = a$$

From the above definition, we have

$$\sqrt[3]{27} = 3, \quad \text{since } 3^3 = 27$$

$$\sqrt[3]{-64} = -4, \quad \text{since } (-4)^3 = -64$$

Notice that unlike square roots, it is possible to have a negative radicand when finding a cube root. This is so because the cube of a negative number is a negative number. Therefore, the cube root of a negative number is a negative number.

EXAMPLE 2   Find the cube roots.

   a. $\sqrt[3]{1}$ **b.** $\sqrt[3]{-27}$ **c.** $\sqrt[3]{\dfrac{1}{25}}$

Solution: **a.** $\sqrt[3]{1} = 1$ because $1^3 = 1$.

   b. $\sqrt[3]{-27} = -3$ because $(-3)^3 = -27$

   c. $\sqrt[3]{\dfrac{1}{125}} = \dfrac{1}{5}$ because $\left(\dfrac{1}{5}\right)^3 = \dfrac{1}{125}$.   ∎

   3   Just as we can raise a real number to powers other than 2 or 3, we can find roots other than square roots and cube roots. In fact, we can take the nth root of a number where n is any natural number. In symbols, the nth root of a is written as $\sqrt[n]{a}$, where n is called the **index.** The index 2 is usually omitted for square roots.

HELPFUL HINT

If the index is even, such as $\sqrt{}$, $\sqrt[4]{}$, $\sqrt[6]{}$, and so on, the radicand must be nonnegative for the root to be a real number. For example,

$$\sqrt[4]{-16} \quad \text{is not a real number}$$

$$\sqrt[6]{-64} \quad \text{is not a real number}$$

EXAMPLE 3   Evaluate the following expressions.

   a. $\sqrt[4]{16}$ **b.** $\sqrt[5]{-32}$ **c.** $-\sqrt[3]{8}$ **d.** $\sqrt[4]{-81}$

Solution: **a.** $\sqrt[4]{16} = 2$ because $2^4 = 16$.

   b. $\sqrt[5]{-32} = -2$ because $(-2)^5 = -32$.

   c. $-\sqrt[3]{8} = -2$ since $\sqrt[3]{8} = 2$.

   d. $\sqrt[4]{-81}$ is not a real number.   ∎

   4   Recall that numbers such as 1, 4, 9, and 25 are called **perfect squares,** since $1^2 = \boxed{1}$, $2^2 = \boxed{4}$, $3^2 = \boxed{9}$, and $5^2 = \boxed{25}$. Square roots of perfect square radicands simplify to rational numbers. What happens when we try to simplify a root such as $\sqrt{3}$? Since 3 is not a perfect square, $\sqrt{3}$ is not a rational number. It is called an **irrational number** and we can find a decimal **approximation** of it. To find decimal approximations, use the table in Appendix D or a calculator. (For calculator help, see the box at the end of this section.) For example, an approximation for $\sqrt{3}$ is

$$\sqrt{3} \approx 1.732$$

$$\uparrow$$

approximation symbol

   Radicands can also contain variables. Since the square root of a negative number is not a real number, we want to make sure variables in the radicand do not have replacement values that would make the radicand negative. To avoid negative radicands, assume for the rest of this chapter that **if a variable appears in the radicand of a radical expression, it will represent positive numbers only.** Then

$$\sqrt{y^2} = y \quad \text{because } y \text{ times itself equals } y^2$$

Also,

$$\sqrt{x^8} = x^4 \quad \text{because } (x^4)^2 = x^8$$

Also,

$$\sqrt{9x^2} = 3x \quad \text{because } (3x)^2 = 9x^2$$

EXAMPLE 4   Simplify the following expressions. Assume that each variable represents a positive number.

a. $\sqrt{x^2}$ **b.** $\sqrt{x^6}$ **c.** $\sqrt[3]{27y^6}$ **d.** $\sqrt{16x^{16}}$

Solution: **a.** $\sqrt{x^2} = x$ because x times itself equals x^2.

b. $\sqrt{x^6} = x^3$ because $(x^3)^2 = x^6$.

c. $\sqrt[3]{27y^6} = 3y^2$ because $(3y^2)^3 = 27y^6$.

d. $\sqrt{16x^{16}} = 4x^8$ because $(4x^8)^2 = 16x^{16}$.   ■

CALCULATOR BOX

To simplify or approximate square roots using a calculator, locate the key marked $\boxed{\sqrt{}}$.

To simplify $\sqrt{25}$, press $\boxed{25}\ \boxed{\sqrt{}}$. The display should read $\boxed{5}$.

To approximate $\sqrt{30}$, press $\boxed{30}\ \boxed{\sqrt{}}$. The display should read $\boxed{5.4772256}$. This is an approximation for $\sqrt{30}$. Then a three-decimal-place approximation is

$$\sqrt{30} \approx 5.477$$

Is this answer reasonable? Since 30 is between perfect squares 25 and 36, $\sqrt{30}$ is between $\sqrt{25} = 5$ and $\sqrt{36} = 6$. Our answer is then reasonable since 5.4772256 is between 5 and 6.

Use a calculator to find a three-decimal-place approximation for the following.

1. $\sqrt{7}$ **2.** $\sqrt{14}$ **3.** $\sqrt{10}$

4. $\sqrt{200}$ **5.** $\sqrt{82}$ **6.** $\sqrt{46}$

EXERCISE SET 8.1

Evaluate each square root. See Example 1.

1. $\sqrt{16}$ **2.** $\sqrt{9}$ **3.** $\sqrt{81}$ **4.** $\sqrt{49}$

5. $\sqrt{\dfrac{1}{25}}$ **6.** $\sqrt{\dfrac{1}{64}}$ **7.** $-\sqrt{100}$ **8.** $-\sqrt{36}$

Find each cube root. See Example 2.

9. $\sqrt[3]{64}$ **10.** $\sqrt[3]{-1}$ **11.** $-\sqrt[3]{27}$ **12.** $-\sqrt[3]{8}$

13. $\sqrt[3]{\dfrac{1}{8}}$ **14.** $\sqrt[3]{\dfrac{1}{64}}$ **15.** $\sqrt[3]{-125}$ **16.** $\sqrt[3]{-27}$

Find each root that is a real number. See Example 3.

17. $\sqrt[5]{32}$ **18.** $\sqrt[4]{-1}$ **19.** $\sqrt[4]{81}$ **20.** $\sqrt{121}$

21. $\sqrt{-4}$ **22.** $\sqrt[5]{\dfrac{1}{32}}$ **23.** $\sqrt[3]{\dfrac{1}{27}}$ **24.** $\sqrt[4]{256}$

25. $\sqrt{\dfrac{9}{25}}$ **26.** $\sqrt[3]{\dfrac{8}{27}}$ **27.** $-\sqrt{49}$ **28.** $-\sqrt[4]{625}$

Find each root. Assume that each variable represents a nonnegative real number. See Example 4.

29. $\sqrt{z^2}$ **30.** $\sqrt{y^{10}}$ **31.** $\sqrt{x^4}$ **32.** $\sqrt{z^6}$

33. $\sqrt{9x^8}$ **34.** $\sqrt{36x^{12}}$ **35.** $\sqrt{x^2y^6}$ **36.** $\sqrt{y^4z^{18}}$

Find each root that is a real number.

37. $\sqrt{0}$ **38.** $\sqrt[3]{0}$ **39.** $-\sqrt[5]{\dfrac{1}{32}}$ **40.** $-\sqrt[3]{\dfrac{27}{125}}$

41. $\sqrt{-64}$ **42.** $\sqrt[3]{-64}$ **43.** $-\sqrt{64}$ **44.** $\sqrt[6]{64}$

45. $-\sqrt{169}$ **46.** $\sqrt[4]{-16}$ **47.** $\sqrt{1}$ **48.** $\sqrt[3]{1}$

49. $\sqrt{\dfrac{25}{64}}$ **50.** $\sqrt{\dfrac{1}{100}}$ **51.** $-\sqrt[3]{-8}$ **52.** $-\sqrt[3]{-27}$

*Determine whether each square root is rational or irrational. If it is rational, find its **exact value.** If it is irrational, use a calculator or the table in the appendix to write a three-decimal-place **approximation.***

53. $\sqrt{9}$ **54.** $\sqrt{8}$ **55.** $\sqrt{37}$ **56.** $\sqrt{36}$

57. $\sqrt{169}$ **58.** $\sqrt{160}$ **59.** $\sqrt{4}$ **60.** $\sqrt{27}$

Find each root. Assume that each variable represents a nonnegative real number.

61. $\sqrt[3]{x^{15}}$ **62.** $\sqrt[3]{y^{12}}$ **63.** $\sqrt{x^{12}}$ **64.** $\sqrt{z^{16}}$

65. $\sqrt{81x^2}$ **66.** $\sqrt{100z^4}$ **67.** $-\sqrt{144y^{14}}$ **68.** $-\sqrt{121z^{22}}$

69. $\sqrt{x^2y^2}$ **70.** $\sqrt{y^{20}z^{30}}$ **71.** $\sqrt{16x^{16}}$ **72.** $\sqrt{36y^{36}}$

73. A fence is to be erected around a square garden with an area of 324 square feet. Each side of this garden has a length of $\sqrt{324}$ feet. Write a one-decimal-point approximation of this length.

74. A standard baseball diamond is a square with 90 foot sides connecting the bases. The distance from first base to third base is $90 \cdot \sqrt{2}$ feet. Approximate $\sqrt{2}$ accurate to two decimal places and use it to approximate the distance $90 \cdot \sqrt{2}$ feet.

75. The roof of the warehouse shown needs to be shingled. The total area of the roof is exactly $240\sqrt{41}$ square feet. Approximate this area to the nearest whole number.

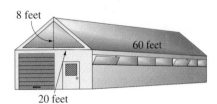

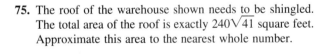

8 feet

60 feet

20 feet

Writing in Mathematics

76. Explain why the square root of a negative number is not a real number.

Skill Review

If $f(x) = 3x - 5$ and $g(x) = x^2 + 2x + 1$, find the following function values. See Section 6.6.

77. $f(2)$ **78.** $f(0)$ **79.** $g(0)$

80. $g(-3)$ **81.** $g(-1)$ **82.** $f(100)$

Find the equation of the line. See Section 6.4

83. Through the point $(2, 4)$ with slope -5.   **84.** Through the point $(0, 3)$ with slope $\dfrac{1}{2}$.

8.2
Simplifying Radicals

OBJECTIVES

Tape 26

1   Use the product rule to simplify radicals.

2   Use the quotient rule to simplify radicals.

1   Much of our work with expressions in this book has involved finding ways to write expressions in their simplest form. Writing radicals in simplest form requires recognizing several patterns, or rules, which we present here. Notice that

$$\sqrt{9 \cdot 16} = \sqrt{144} = 12$$

Also,

$$\sqrt{9 \cdot 16} = 3 \cdot 4 = 12$$

Since both expressions simplify to 12, we can say that

$$\sqrt{9 \cdot 16} = \sqrt{9} \cdot \sqrt{16}$$

This suggests the following product rule for square roots.

> **Product Rule for Square Roots**
>
> If a and b are nonnegative numbers, then
> $$\sqrt{a \cdot b} = \sqrt{a} \cdot \sqrt{b}$$

The product rule states that the square root of a product is equal to the product of the square roots. We use this rule to simplify radicals such as $\sqrt{20}$. A radical is **simplified** if the radicand has no perfect square factors other than 1. To simplify $\sqrt{20}$, factor 20 so that one of its factors is a perfect square factor.

$$\sqrt{20} = \sqrt{4 \cdot 5} \qquad \text{Factor 20.}$$
$$= \sqrt{4} \cdot \sqrt{5} \qquad \text{Use the product rule.}$$
$$= 2\sqrt{5} \qquad \text{Write } \sqrt{4} \text{ as 2.}$$

The notation $2\sqrt{5}$ means $2 \cdot \sqrt{5}$. Since the radicand 5 has no perfect square factor other than 1, $2\sqrt{5}$ is in **simplified form.**

When factoring a radicand, look for at least one factor that is a perfect square. Review the table of perfect squares in Appendix D to help locate perfect square factors more quickly.

When simplifying a radical, realize that it does not mean getting a decimal approximation. The simplified form of a radical is an exact form and may still contain a radical.

HELPFUL HINT

When simplifying a radical, use **factors** of the radicand; **do not** write the radicand as a sum. For example, **do not** rewrite $\sqrt{20}$ as $\sqrt{4 + 16}$ because $\sqrt{4 + 16} \neq \sqrt{4} + \sqrt{16}$. Correctly simplified, $\sqrt{20} = \sqrt{4 \cdot 5} = \sqrt{4} \cdot \sqrt{5} = 2\sqrt{5}$.

EXAMPLE 1   Simplify each expression.
   a. $\sqrt{54}$ **b.** $\sqrt{12}$ **c.** $\sqrt{200}$ **d.** $\sqrt{35}$

Solution: **a.** Try to factor 54 so that at least one of the factors is a perfect square. Since $54 = 9 \cdot 6$,

$$\sqrt{54} = \sqrt{9 \cdot 6} \qquad \text{Factor.}$$
$$= \sqrt{9} \cdot \sqrt{6} \qquad \text{Use the product rule.}$$
$$= 3\sqrt{6} \qquad \text{Write } \sqrt{9} \text{ as 3.}$$

b.
$$\sqrt{12} = \sqrt{4 \cdot 3} \qquad \text{Factor 12.}$$
$$= \sqrt{4} \cdot \sqrt{3} \qquad \text{Use the product rule.}$$
$$= 2\sqrt{3} \qquad \text{Write } \sqrt{4} \text{ as 2.}$$

c. The largest perfect square factor of 200 is 100.

$$\sqrt{200} = \sqrt{100 \cdot 2} \qquad \text{Factor 200.}$$
$$= \sqrt{100} \cdot \sqrt{2} \qquad \text{Use the product rule.}$$
$$= 10\sqrt{2} \qquad \text{Write } \sqrt{100} \text{ as 10.}$$

d. The radicand 35 contains no perfect square factors other than 1. Thus $\sqrt{35}$ is in simplified form.   ∎

2   Next, let's examine the square root of a quotient.

$$\sqrt{\frac{16}{4}} = \sqrt{4} = 2$$

Also,

$$\frac{\sqrt{16}}{\sqrt{4}} = \frac{4}{2} = 2$$

Since both expressions equal 2, we have that

$$\sqrt{\frac{16}{4}} = \frac{\sqrt{16}}{\sqrt{4}}$$

This suggests the following quotient rule.

> **Quotient Rule for Square Roots**
>
> If a and b are nonnegative numbers and $b \neq 0$, then
>
> $$\sqrt{\frac{a}{b}} = \frac{\sqrt{a}}{\sqrt{b}}$$

The quotient rule states that the square root of a quotient is equal to the quotient of the square roots.

EXAMPLE 2 Simplify the following.

a. $\sqrt{\dfrac{25}{36}}$ **b.** $\sqrt{\dfrac{3}{64}}$ **c.** $\sqrt{\dfrac{40}{81}}$

Solution: Use the quotient rule.

a. $\sqrt{\dfrac{25}{36}} = \dfrac{\sqrt{25}}{\sqrt{36}} = \dfrac{5}{6}$ **b.** $\sqrt{\dfrac{3}{64}} = \dfrac{\sqrt{3}}{\sqrt{64}} = \dfrac{\sqrt{3}}{8}$

c. $\sqrt{\dfrac{40}{81}} = \dfrac{\sqrt{40}}{\sqrt{81}}$ Use the quotient rule.

$$= \frac{\sqrt{4} \cdot \sqrt{10}}{9}$$ Write $\sqrt{81}$ as 9 and use the product rule.

$$= \frac{2\sqrt{10}}{9}$$ Write $\sqrt{4}$ as 2.   ∎

EXAMPLE 3 Simplify each expression. Assume that variables represent positive numbers only.

a. $\sqrt{x^5}$ **b.** $\sqrt{8y^2}$ **c.** $\sqrt{\dfrac{45}{x^6}}$

Solution: **a.** $\sqrt{x^5} = \sqrt{x^4 \cdot x} = \sqrt{x^4} \cdot \sqrt{x} = x^2\sqrt{x}$

b. $\sqrt{8y^2} = \sqrt{4 \cdot 2 \cdot y^2} = \sqrt{4y^2 \cdot 2} = \sqrt{4y^2} \cdot \sqrt{2} = 2y\sqrt{2}$

c. $\sqrt{\dfrac{45}{x^6}} = \dfrac{\sqrt{45}}{\sqrt{x^6}} = \dfrac{\sqrt{9 \cdot 5}}{x^3} = \dfrac{\sqrt{9} \cdot \sqrt{5}}{x^3} = \dfrac{3\sqrt{5}}{x^3}$ ∎

The product and quotient rules also apply to roots other than square roots. In general, we have the following product and quotient rules for radicals:

> **Product Rule for Radicals**
>
> If $\sqrt[n]{a}$ and $\sqrt[n]{b}$ exist, then
>
> $$\sqrt[n]{a} \cdot \sqrt[n]{b} = \sqrt[n]{a \cdot b}$$
>
> **Quotient Rule for Radicals**
>
> If $\sqrt[n]{a}$ and $\sqrt[n]{b}$ exist, then
>
> $$\frac{\sqrt[n]{a}}{\sqrt[n]{b}} = \sqrt[n]{\frac{a}{b}}, \text{ providing } b \neq 0$$

For example, to simplify cube roots, look for perfect cube factors of the radicand. For example, 8 is a perfect cube, since $2^3 = 8$.

To simplify $\sqrt[3]{48}$, factor 48 as $8 \cdot 6$.

$$\sqrt[3]{48} = \sqrt[3]{8 \cdot 6} \qquad \text{Factor 48.}$$

$$= \sqrt[3]{8} \cdot \sqrt[3]{6} \qquad \text{Use the product rule.}$$

$$= 2\sqrt[3]{6} \qquad \text{Write } \sqrt[3]{8} \text{ as 2.}$$

$2\sqrt[3]{6}$ is in simplest form since the radicand 6 contains no perfect cube factors other than 1.

EXAMPLE 4 Simplify each expression.

    a. $\sqrt[3]{54}$ **b.** $\sqrt[3]{18}$ **c.** $\sqrt[3]{\dfrac{7}{8}}$ **d.** $\sqrt[3]{\dfrac{40}{27}}$

Solution: **a.** $\sqrt[3]{54} = \sqrt[3]{27 \cdot 2} = \sqrt[3]{27} \cdot \sqrt[3]{2} = 3\sqrt[3]{2}$

    b. The number 18 contains no perfect cube factors, so $\sqrt[3]{18}$ cannot be simplified further.

    c. $\sqrt[3]{\dfrac{7}{8}} = \dfrac{\sqrt[3]{7}}{\sqrt[3]{8}} = \dfrac{\sqrt[3]{7}}{2}$

    d. $\sqrt[3]{\dfrac{40}{27}} = \dfrac{\sqrt[3]{40}}{\sqrt[3]{27}} = \dfrac{\sqrt[3]{8 \cdot 5}}{3} = \dfrac{\sqrt[3]{8} \cdot \sqrt[3]{5}}{3} = \dfrac{2\sqrt[3]{5}}{3}$ ∎

EXAMPLE 5 Simplify each expression. Assume variables represent positive numbers only.

    a. $\sqrt[3]{x^5}$ **b.** $\sqrt[3]{40y^7}$ **c.** $\sqrt[3]{\dfrac{16}{x^6}}$

Solution: **a.** $\sqrt[3]{x^5} = \sqrt[3]{x^3 \cdot x^2} = \sqrt[3]{x^3} \cdot \sqrt[3]{x^2} = x\sqrt[3]{x^2}$

    b. $\sqrt[3]{40y^7} = \sqrt[3]{8 \cdot 5 \cdot y^6 \cdot y} = \sqrt[3]{8y^6 \cdot 5y} = \sqrt[3]{8y^6} \cdot \sqrt[3]{5y} = 2y^2\sqrt[3]{5y}$

    c. $\sqrt[3]{\dfrac{16}{x^6}} = \dfrac{\sqrt[3]{16}}{\sqrt[3]{x^6}} = \dfrac{\sqrt[3]{8 \cdot 2}}{x^2} = \dfrac{\sqrt[3]{8} \cdot \sqrt[3]{2}}{x^2} = \dfrac{2\sqrt[3]{2}}{x^2}$ ∎

MENTAL MATH

Simplify each expression. Assume that all variables represent nonnegative real numbers.

1. $\sqrt{9}$ **2.** $\sqrt{16}$ **3.** $\sqrt{x^2}$ **4.** $\sqrt{y^4}$

5. $\sqrt{0}$ **6.** $\sqrt{1}$ **7.** $\sqrt{25x^4}$ **8.** $\sqrt{49x^2}$

EXERCISE SET 8.2

Simplify each expression. See Example 1.

1. $\sqrt{20}$ **2.** $\sqrt{44}$ **3.** $\sqrt{18}$ **4.** $\sqrt{45}$

5. $\sqrt{50}$ **6.** $\sqrt{28}$ **7.** $\sqrt{33}$ **8.** $\sqrt{98}$

Simplify each expression. See Example 2.

9. $\sqrt{\dfrac{8}{25}}$ **10.** $\sqrt{\dfrac{63}{16}}$ **11.** $\sqrt{\dfrac{27}{121}}$ **12.** $\sqrt{\dfrac{24}{169}}$

13. $\sqrt{\dfrac{9}{4}}$ **14.** $\sqrt{\dfrac{100}{49}}$ **15.** $\sqrt{\dfrac{125}{9}}$ **16.** $\sqrt{\dfrac{27}{100}}$

Simplify each expression. Assume that all variables represent positive numbers only. See Example 3.

17. $\sqrt{x^7}$ **18.** $\sqrt{y^3}$ **19.** $\sqrt{\dfrac{88}{x^4}}$ **20.** $\sqrt{\dfrac{x^{11}}{81}}$

Simplify each expression. See Example 4.

21. $\sqrt[3]{24}$ **22.** $\sqrt[3]{81}$ **23.** $\sqrt[3]{\dfrac{5}{64}}$ **24.** $\sqrt[3]{\dfrac{32}{125}}$

Simplify each expression. Assume that all variables represent positive numbers only. See Example 5.

25. $\sqrt[3]{x^{16}}$ **26.** $\sqrt[3]{y^{20}}$ **27.** $\sqrt[3]{\dfrac{2}{x^9}}$ **28.** $\sqrt[3]{\dfrac{48}{x^{12}}}$

Simplify each expression. Assume that all variables represent positive numbers only.

29. $\sqrt{60}$ **30.** $\sqrt{90}$ **31.** $\sqrt{180}$ **32.** $\sqrt{150}$

33. $\sqrt{52}$ **34.** $\sqrt{75}$ **35.** $\sqrt{\dfrac{11}{36}}$ **36.** $\sqrt{\dfrac{30}{49}}$

37. $-\sqrt{\dfrac{27}{144}}$ **38.** $-\sqrt{\dfrac{84}{121}}$ **39.** $\sqrt[3]{\dfrac{15}{64}}$ **40.** $\sqrt[3]{\dfrac{4}{27}}$

41. $\sqrt[3]{80}$ **42.** $\sqrt[3]{108}$ **43.** $\sqrt[4]{48}$ **44.** $\sqrt[4]{162}$

45. $\sqrt{x^{13}}$ **46.** $\sqrt{y^{17}}$ **47.** $\sqrt{75x^2}$ **48.** $\sqrt{72y^2}$

49. $\sqrt{96x^4y^2}$ **50.** $\sqrt{40x^8y^{10}}$ **51.** $\sqrt{\dfrac{12}{y^2}}$ **52.** $\sqrt{\dfrac{63}{x^4}}$

53. $\sqrt{\dfrac{9x}{y^2}}$ **54.** $\sqrt{\dfrac{6y^2}{x^4}}$ **55.** $\sqrt[3]{-8x^6}$ **56.** $\sqrt[3]{-54y^6}$

57. If a cube is to have a volume of 80 cubic inches, then each side must be $\sqrt[3]{80}$ inches long. Simplify the radical representing the side length.

58. Jeannie is swimming across a 40-foot-wide river, trying to head straight across to the opposite shore. However, the current is strong enough to move her downstream 100 feet by the time she reaches land. (See the figure.) <u>Because of the current, the actual distance she swam is $\sqrt{11,600}$ feet.</u> Simplify this radical.

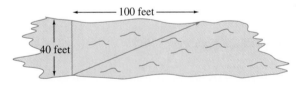

59. By using replacement values for a and b, show that $\sqrt{a^2 + b^2}$ does not equal $a + b$.

Skill Review

Perform the following operations. See Sections 3.2, 3.3, and 3.4.

60. $6x + 8x$ **61.** $(6x)(8x)$ **62.** $(2x + 3)(x - 5)$

63. $(2x + 3) + (x - 5)$ **64.** $9y^2 - 9y^2$ **65.** $(9y^2)(-8y^2)$

8.3
Adding and Subtracting Radicals

OBJECTIVES

1 Add or subtract like radicals.

2 Simplify radical expressions, and then add or subtract any like radicals.

Tape 27

1 To combine like terms, we use the distributive property.

$$5x + 3x = (5 + 3)x = 8x$$

The distributive property can also be applied to expressions containing radicals. For example,

$$5\sqrt{2} + 3\sqrt{2} = (5 + 3)\sqrt{2} = 8\sqrt{2}$$

Also,

$$9\sqrt{5} - 6\sqrt{5} = (9 - 6)\sqrt{5} = 3\sqrt{5}$$

Radical terms $5\sqrt{2}$ and $3\sqrt{2}$ are **like radicals** as are $9\sqrt{5}$ and $6\sqrt{5}$.

> **Like Radicals**
>
> **Like radicals** are radicals that have the same index and the same radicand.

From these examples, we can say that **only like radicals can be combined.** For example, the terms in the expression $2\sqrt{3} + 3\sqrt{2}$ cannot be added since the radicals are not like radicals. Also, the terms in the expression $4\sqrt{7} + 4\sqrt[3]{7}$ cannot be added because the radicals are not like radicals since the indices are different.

EXAMPLE 1 Add or subtract by combining like radical terms.
a. $4\sqrt{5} + 3\sqrt{5}$ **b.** $\sqrt{10} - 6\sqrt{10}$ **c.** $2\sqrt[3]{7} - 5\sqrt[3]{7} - 3\sqrt[3]{7}$ **d.** $2\sqrt{6} + 2\sqrt[3]{6}$

Solution: **a.** $4\sqrt{5} + 3\sqrt{5} = (4 + 3)\sqrt{5} = 7\sqrt{5}$

b. $\sqrt{10} - 6\sqrt{10} = 1\sqrt{10} - 6\sqrt{10} = (1 - 6)\sqrt{10} = -5\sqrt{10}$

c. $2\sqrt[3]{7} - 5\sqrt[3]{7} - 3\sqrt[3]{7} = (2 - 5 - 3)\sqrt[3]{7} = -6\sqrt[3]{7}$

d. $2\sqrt{6} + 2\sqrt[3]{6}$ cannot be simplified further since the indices are not the same. ∎

2 At first glance, it appears that the expression $\sqrt{50} + \sqrt{8}$ cannot be simplified further because the radicands are different. However, the product rule can be used to simplify each radical, and then further simplification might be possible.

EXAMPLE 2 Add or subtract by first simplifying each radical.
a. $\sqrt{50} + \sqrt{8}$
b. $7\sqrt{12} - \sqrt{75}$
c. $\sqrt{25} - \sqrt{27} - 2\sqrt{18} - \sqrt{9}$
d. $2\sqrt[3]{27} - \sqrt[3]{54}$

a. First simplify each radical.

$$\sqrt{50} + \sqrt{8} = \sqrt{25 \cdot 2} + \sqrt{4 \cdot 2} \qquad \text{Factor radicands.}$$
$$= \sqrt{25} \cdot \sqrt{2} + \sqrt{4} \cdot \sqrt{2} \qquad \text{Use the product rule.}$$
$$= 5\sqrt{2} + 2\sqrt{2} \qquad \text{Simplify.}$$
$$= 7\sqrt{2} \qquad \text{Add like radical terms.}$$

b. $7\sqrt{12} - \sqrt{75} = 7\sqrt{4 \cdot 3} - \sqrt{25 \cdot 3} \qquad \text{Factor radicands.}$
$$= 7\sqrt{4} \cdot \sqrt{3} - \sqrt{25} \cdot \sqrt{3} \qquad \text{Use the product rule.}$$
$$= 7 \cdot 2\sqrt{3} - 5\sqrt{3} \qquad \text{Simplify.}$$
$$= 14\sqrt{3} - 5\sqrt{3} \qquad \text{Multiply.}$$
$$= 9\sqrt{3} \qquad \text{Subtract like radical terms.}$$

c. $\sqrt{25} - \sqrt{27} - 2\sqrt{18} - \sqrt{9}$

$$= 5 - \sqrt{9 \cdot 3} - 2\sqrt{9 \cdot 2} - 3 \qquad \text{Factor radicands.}$$
$$= 5 - \sqrt{9} \cdot \sqrt{3} - 2\sqrt{9} \cdot \sqrt{2} - 3 \qquad \text{Use the product rule.}$$
$$= 5 - 3\sqrt{3} - 2 \cdot 3\sqrt{2} - 3 \qquad \text{Simplify.}$$
$$= 2 - 3\sqrt{3} - 6\sqrt{2} \qquad \text{Write } 5 - 3 \text{ as 2 and } 2 \cdot 3 \text{ as 6.}$$

d. $2\sqrt[3]{27} - \sqrt[3]{54} = 2 \cdot 3 - \sqrt[3]{27 \cdot 2}$
$$= 6 - 3\sqrt[3]{2}$$

No further simplification is possible. ∎

If radical expressions contain variables, we proceed in a similar way. Simplify radicals using the product and quotient rules. Then add or subtract any like radical terms.

EXAMPLE 3 Simplify each radical expression.

a. $2\sqrt{x^2} - \sqrt{25x} + \sqrt{x}$ **b.** $3\sqrt[3]{54x^4} + 5x\sqrt[3]{16x}$

Solution: **a.** $2\sqrt{x^2} - \sqrt{25x} + \sqrt{x}$

$$= 2x - \sqrt{25} \cdot \sqrt{x} + \sqrt{x} \qquad \text{Use the product rule.}$$
$$= 2x - 5\sqrt{x} + 1\sqrt{x} \qquad \text{Simplify.}$$
$$= 2x - 4\sqrt{x} \qquad \text{Add like terms.}$$

b. $3\sqrt[3]{54x^4} + 5x\sqrt[3]{16x}$

$$= 3\sqrt[3]{27x^3 \cdot 2x} + 5x\sqrt[3]{8 \cdot 2x} \qquad \text{Factor.}$$
$$= 3\sqrt[3]{27x^3} \cdot \sqrt[3]{2x} + 5x\sqrt[3]{8} \cdot \sqrt[3]{2x} \qquad \text{Use the product rule.}$$
$$= 3 \cdot 3x \cdot \sqrt[3]{2x} + 5x \cdot 2 \cdot \sqrt[3]{2x} \qquad \text{Simplify.}$$
$$= 9x\sqrt[3]{2x} + 10x\sqrt[3]{2x} \qquad \text{Multiply.}$$
$$= 19x\sqrt[3]{2x} \qquad \text{Add like terms.} \quad ∎$$

MENTAL MATH

Simplify each expression by combining like radicals.

1. $3\sqrt{2} + 5\sqrt{2}$

2. $2\sqrt{3} + 7\sqrt{3}$

3. $5\sqrt{x} + 2\sqrt{x}$

4. $8\sqrt{x} + 3\sqrt{x}$

5. $5\sqrt{7} - 2\sqrt{7}$

6. $8\sqrt{6} - 5\sqrt{6}$

EXERCISE SET 8.3

Simplify each expression by combining like radicals where possible. See Example 1.

1. $4\sqrt{3} - 8\sqrt{3}$

2. $\sqrt{5} - 9\sqrt{5}$

3. $3\sqrt{6} + 8\sqrt{6} - 2\sqrt{6}$

4. $12\sqrt{2} - 3\sqrt{2} + 8\sqrt{2}$

5. $6\sqrt{5} - 5\sqrt{5} + \sqrt{2}$

6. $4\sqrt{3} + \sqrt{5} - 3\sqrt{3}$

7. $2\sqrt[3]{3} + 5\sqrt[3]{3}$

8. $8\sqrt[3]{4} + 2\sqrt[3]{4}$

9. $2\sqrt[3]{2} - 7\sqrt[3]{2}$

Add or subtract by first simplifying each radical and then combining any like radical terms. See Example 2.

10. $\sqrt{12} + \sqrt{27}$

11. $\sqrt{50} + \sqrt{18}$

12. $\sqrt{45} + 3\sqrt{20}$

13. $2\sqrt{54} - \sqrt{20} + \sqrt{45} - \sqrt{24}$

14. $2\sqrt{8} - \sqrt{128} + \sqrt{48} + \sqrt{18}$

15. $\sqrt[3]{81} + \sqrt[3]{24}$

16. $\sqrt[3]{32} - \sqrt[3]{4}$

17. $4\sqrt[3]{9} - \sqrt[3]{243}$

Simplify the following. Assume that all variables represent nonnegative real numbers. See Example 3.

18. $4x - 3\sqrt{x^2} + \sqrt{x}$

19. $x - 6\sqrt{x^2} + 2\sqrt{x}$

20. $\sqrt{25x} + \sqrt{36x} - 11\sqrt{x}$

21. $3\sqrt{x^3} - x\sqrt{4x}$

22. $\sqrt{16x} - \sqrt{x^3}$

23. $\sqrt[3]{8x^3} + x\sqrt[3]{27}$

Simplify the following. Assume that all variables represent nonnegative real numbers.

24. $12\sqrt{5} - \sqrt{5} - 4\sqrt{5}$

25. $\sqrt{5} + \sqrt[3]{5}$

26. $\sqrt{5} + \sqrt{5}$

27. $4 + 8\sqrt{2} - 9$

28. $6 - 2\sqrt{3} - \sqrt{3}$

29. $8 - \sqrt{2} - 5\sqrt{2}$

30. $\sqrt{75} + \sqrt{48}$

31. $5\sqrt{32} - \sqrt{72}$

32. $2\sqrt{80} - \sqrt{45}$

33. $\sqrt{8} + \sqrt{9} + \sqrt{18} + \sqrt{81}$

34. $\sqrt{6} + \sqrt{16} + \sqrt{24} + \sqrt{25}$

35. $\sqrt{\dfrac{5}{9}} + \sqrt{\dfrac{5}{81}}$

36. $\sqrt{\dfrac{3}{64}} + \sqrt{\dfrac{3}{16}}$

37. $\sqrt{\dfrac{3}{4}} - \sqrt{\dfrac{3}{64}}$

38. $2\sqrt[3]{8} + 2\sqrt[3]{16}$

39. $3\sqrt[3]{27} + 3\sqrt[3]{81}$

40. $\sqrt[3]{8} + \sqrt[3]{54} - 5$

41. $2\sqrt{45} - 2\sqrt{20}$

42. $5\sqrt{18} + 2\sqrt{32}$

43. $\sqrt{35} - \sqrt{140}$

44. $5\sqrt{2xz^2} + z\sqrt{98x}$

45. $3\sqrt{9x} + 2\sqrt{x}$

46. $5\sqrt{x} + 4\sqrt{4x}$

47. $\sqrt{9x} + \sqrt{81x} - 11\sqrt{x}$

48. $\sqrt{3x^3} + 3x\sqrt{x}$

49. $x\sqrt{4x} + \sqrt{9x^3}$

50. $\sqrt{32x^2} + \sqrt[3]{32x^2} + \sqrt{4x^2}$

51. $\sqrt{18x^2} + \sqrt[3]{24x^3} + \sqrt{2x^2}$

52. $\sqrt{40x} + \sqrt[3]{40x^4} - 2\sqrt{10x} - \sqrt[3]{5x^4}$

53. $\sqrt{72x^2} + \sqrt[3]{54x} - x\sqrt{50} - 3\sqrt[3]{2x}$

54. Find the perimeter of the rectangular picture frame.

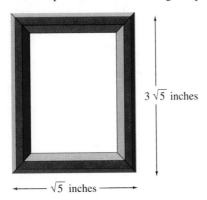

3 $\sqrt{5}$ inches

$\sqrt{5}$ inches

55. Find the perimeter of the plot of land.

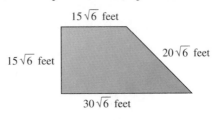

15 $\sqrt{6}$ feet

15 $\sqrt{6}$ feet

20 $\sqrt{6}$ feet

30 $\sqrt{6}$ feet

56. A water trough is to be made of wood. Each of the two triangular end pieces has an area of $\dfrac{3\sqrt{27}}{4}$ square feet. The two side panels are both rectangular. In simplest radical form, find the total area of the wood needed.

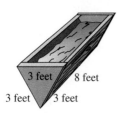

3 feet   8 feet

3 feet   3 feet

57. Eight wooden braces are to be attached along the diagonals of the vertical sides of a storage bin. Each of four of these diagonals has a length of $\sqrt{52}$ feet, while each of the other four has a length of $\sqrt{80}$ feet. In simplest radical form, find the total length of the wood needed for these braces.

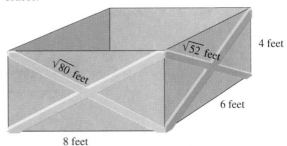

$\sqrt{52}$ feet   4 feet

$\sqrt{80}$ feet

6 feet

8 feet

Writing in Mathematics

58. In your own words, describe like radicals.

Skill Review

Square each binomial. See Section 3.4.

59. $(x + 6)^2$

60. $(3x + 2)^2$

61. $(2x - 1)^2$

62. $(x - 5)^2$

Solve each system of linear equations. See Section 7.2.

63. $\begin{cases} x = 2y \\ x + 5y = 14 \end{cases}$

64. $\begin{cases} y = -5x \\ x + y = 16 \end{cases}$

8.4
Multiplying and Dividing Radicals

OBJECTIVES

Tape 27

1   Multiply radicals.

2   Divide radicals.

3   Rationalize denominators containing square roots.

1   In Section 8.2 we used the product rule for radicals to help us simplify radicals. In this section, we use this rule to multiply radicals.

Product Rule for Radicals

If $\sqrt[n]{a}$ and $\sqrt[n]{b}$ exist, then

$$\sqrt[n]{a} \cdot \sqrt[n]{b} = \sqrt[n]{a \cdot b}$$

This property says that the product of two nth roots is the nth root of their product. For example,

$$\sqrt{3} \cdot \sqrt{2} = \sqrt{3 \cdot 2} = \sqrt{6}$$

Also,

$$\sqrt[3]{5} \cdot \sqrt[3]{7} = \sqrt[3]{5 \cdot 7} = \sqrt[3]{35}$$

EXAMPLE 1   Find the following products.

a. $\sqrt{7} \cdot \sqrt{3}$

b. $\sqrt{3} \cdot \sqrt{15}$

c. $\sqrt[3]{4} \cdot \sqrt[3]{18}$

d. $2\sqrt{6} \cdot 5\sqrt{2}$

e. $(3\sqrt{2})^2$

Solution: **a.** $\sqrt{7} \cdot \sqrt{3} = \sqrt{7 \cdot 3} = \sqrt{21}$

b. $\sqrt{3} \cdot \sqrt{15} = \sqrt{45}$. Next, simplify $\sqrt{45}$.

$$\sqrt{45} = \sqrt{9 \cdot 5} = \sqrt{9} \cdot \sqrt{5} = 3\sqrt{5}$$

c. $\sqrt[3]{4} \cdot \sqrt[3]{18} = \sqrt[3]{4 \cdot 18} = \sqrt[3]{4 \cdot 2 \cdot 9} = \sqrt[3]{8 \cdot 9} = \sqrt[3]{8} \cdot \sqrt[3]{9} = 2\sqrt[3]{9}$

d. $2\sqrt{6} \cdot 5\sqrt{2} = 2 \cdot 5\sqrt{6 \cdot 2} = 10\sqrt{12}$. Next, simplify $\sqrt{12}$.

$$10\sqrt{12} = 10\sqrt{4 \cdot 3} = 10\sqrt{4} \cdot \sqrt{3} = 10 \cdot 2 \cdot \sqrt{3} = 20\sqrt{3}$$

e. $(3\sqrt{2})^2 = 3^2 \cdot (\sqrt{2})^2 = 9 \cdot 2 = 18$ ■

When multiplying radical expressions containing more than one term, use the same techniques we use to multiply other algebraic expressions with more than one term.

EXAMPLE 2 Find the product and simplify.

a. $\sqrt{5}(\sqrt{5} - \sqrt{2})$ **b.** $(\sqrt{x} + \sqrt{2})(\sqrt{3} - \sqrt{2})$

Solution: **a.** Using the distributive property, we have

$$\sqrt{5}(\sqrt{5} - \sqrt{2}) = \sqrt{5} \cdot \sqrt{5} - \sqrt{5} \cdot \sqrt{2}$$

$$= 5 - \sqrt{10}$$

b. Use the FOIL method of multiplication.

$$(\sqrt{x} + \sqrt{2})(\sqrt{3} - \sqrt{2}) = \overset{F}{\sqrt{x} \cdot \sqrt{3}} - \overset{O}{\sqrt{x} \cdot \sqrt{2}} + \overset{I}{\sqrt{2} \cdot \sqrt{3}} - \overset{L}{\sqrt{2} \cdot \sqrt{2}}$$

$$= \sqrt{3x} - \sqrt{2x} + \sqrt{6} - \sqrt{4}$$

$$= \sqrt{3x} - \sqrt{2x} + \sqrt{6} - 2 \qquad \text{Simplify.} \quad \blacksquare$$

Special products can be used to multiply expressions containing radicals.

EXAMPLE 3 Find the product and simplify.

a. $(\sqrt{5} - 7)(\sqrt{5} + 7)$ **b.** $(\sqrt{7x} + 2)^2$

Solution: **a.** Recall from Chapter 3 that $(a - b)(a + b) = a^2 - b^2$. Then

$$(\sqrt{5} - 7)(\sqrt{5} + 7) = (\sqrt{5})^2 - 7^2$$

$$= 5 - 49$$

$$= -44$$

b. Recall that $(a + b)^2 = a^2 + 2ab + b^2$. Then

$$(\sqrt{7x} + 2)^2 = (\sqrt{7x})^2 + 2(\sqrt{7x})(2) + (2)^2$$

$$= 7x + 4\sqrt{7x} + 4 \quad \blacksquare$$

1 To divide rational expressions, we use the quotient rule.

> **Quotient Rule for Radicals**
>
> If $\sqrt[n]{a}$ and $\sqrt[n]{b}$ exist, then
>
> $$\frac{\sqrt[n]{a}}{\sqrt[n]{b}} = \sqrt[n]{\frac{a}{b}}, \text{ providing } b \neq 0$$

EXAMPLE 4 Find the quotient and simplify.

a. $\dfrac{\sqrt{14}}{\sqrt{2}}$ **b.** $\dfrac{\sqrt{100}}{\sqrt{5}}$ **c.** $\dfrac{\sqrt[3]{32}}{\sqrt[3]{4}}$ **d.** $\dfrac{\sqrt{12x^3}}{\sqrt{3x}}$

Solution: Use the quotient rule and then simplify the resulting radicand.

a. $\dfrac{\sqrt{14}}{\sqrt{2}} = \sqrt{\dfrac{14}{2}} = \sqrt{7}$

b. $\dfrac{\sqrt{100}}{\sqrt{5}} = \sqrt{\dfrac{100}{5}} = \sqrt{20} = \sqrt{4} \cdot \sqrt{5} = 2\sqrt{5}$

c. $\dfrac{\sqrt[3]{32}}{\sqrt[3]{4}} = \sqrt[3]{\dfrac{32}{4}} = \sqrt[3]{8} = 2$

d. $\dfrac{\sqrt{12x^3}}{\sqrt{3x}} = \sqrt{\dfrac{12x^3}{3x}} = \sqrt{4x^2} = 2x$ ∎

3 It is sometimes easier to work with radical expressions if the denominator does not contain a radical. To eliminate the radical in the denominator of a radical expression, we use the fact that we can multiply the numerator and the denominator of a fraction by the same nonzero number. This is equivalent to multiplying by 1. To eliminate the radical in the denominator of $\dfrac{\sqrt{5}}{\sqrt{2}}$, multiply the numerator and the denominator by $\sqrt{2}$. Then

$$\frac{\sqrt{5}}{\sqrt{2}} = \frac{\sqrt{5} \cdot \sqrt{2}}{\sqrt{2} \cdot \sqrt{2}} = \frac{\sqrt{10}}{2}$$

This process is called **rationalizing** the denominator.

EXAMPLE 5 Rationalize each denominator.

    a. $\dfrac{2}{\sqrt{7}}$ **b.** $\sqrt{\dfrac{5}{12}}$ **c.** $\sqrt{\dfrac{1}{18x}}$

Solution: **a.** To eliminate the radical in the denominator of $\dfrac{2}{\sqrt{7}}$, multiply the numerator and the denominator by $\sqrt{7}$.

$$\frac{2}{\sqrt{7}} = \frac{2 \cdot \sqrt{7}}{\sqrt{7} \cdot \sqrt{7}} = \frac{2\sqrt{7}}{7}$$

b. We can multiply the numerator and denominator by $\sqrt{12}$, but see what happens if we simplify first.

$$\sqrt{\frac{5}{12}} = \frac{\sqrt{5}}{\sqrt{12}} = \frac{\sqrt{5}}{\sqrt{4 \cdot 3}} = \frac{\sqrt{5}}{2\sqrt{3}}$$

To rationalize the denominator now, multiply the numerator and the denominator by $\sqrt{3}$.

$$\frac{\sqrt{5} \cdot \sqrt{3}}{2\sqrt{3} \cdot \sqrt{3}} = \frac{\sqrt{15}}{2\sqrt{9}} = \frac{\sqrt{15}}{2 \cdot 3} = \frac{\sqrt{15}}{6}$$

c. $\sqrt{\dfrac{1}{18x}} = \dfrac{1}{\sqrt{18x}} = \dfrac{1}{\sqrt{9} \cdot \sqrt{2x}} = \dfrac{1}{3\sqrt{2x}}$

To rationalize the denominator, multiply the numerator and denominator by $\sqrt{2x}$.

$$\frac{1 \cdot \sqrt{2x}}{3\sqrt{2x} \cdot \sqrt{2x}} = \frac{\sqrt{2x}}{3 \cdot 2x} = \frac{\sqrt{2x}}{6x}$$ ∎

As a general rule, simplify a radical expression first and then rationalize the denominator.

To rationalize a denominator such as the one found in

$$\frac{2}{4 + \sqrt{3}}$$

we multiply the numerator and the denominator by $4 - \sqrt{3}$. The expressions $4 + \sqrt{3}$ and $4 - \sqrt{3}$ are called **conjugates.** When a radical expression such as $4 + \sqrt{3}$ is multiplied by its conjugate $4 - \sqrt{3}$, the product simplifies to an expression that contains no radicals.

$$(4 + \sqrt{3})(4 - \sqrt{3}) = 4^2 - (\sqrt{3})^2 = 16 - 3 = 13$$

Then

$$\frac{2}{4 + \sqrt{3}} = \frac{2(4 - \sqrt{3})}{(4 + \sqrt{3})(4 - \sqrt{3})} = \frac{2(4 - \sqrt{3})}{13}$$

EXAMPLE 6 Rationalize each denominator and simplify.

a. $\dfrac{2}{1 + \sqrt{3}}$ **b.** $\dfrac{\sqrt{5} + 4}{\sqrt{5} - 1}$

Solution: **a.** Multiply the numerator and the denominator of this fraction by the conjugate of $1 + \sqrt{3}$, that is, by $1 - \sqrt{3}$.

$$\frac{2}{1 + \sqrt{3}} = \frac{2\,(1 - \sqrt{3})}{(1 + \sqrt{3})\,(1 - \sqrt{3})}$$

$$= \frac{2(1 - \sqrt{3})}{1 - 3}$$

$$= \frac{2(1 - \sqrt{3})}{-2}$$

$$= \frac{2\,(1 - \sqrt{3})}{-1 \cdot 2}$$

$$= -1(1 - \sqrt{3}) \qquad \text{Simplify.}$$

$$= -1 + \sqrt{3}$$

b. $\dfrac{\sqrt{5} + 4}{\sqrt{5} - 1} = \dfrac{(\sqrt{5} + 4)\,(\sqrt{5} + 1)}{(\sqrt{5} - 1)\,(\sqrt{5} + 1)}$ Multiply the numerator and denominator by $\sqrt{5} + 1$, the conjugate of $\sqrt{5} - 1$.

$$= \frac{5 + \sqrt{5} + 4\sqrt{5} + 4}{5 - 1} \qquad \text{Multiply.}$$

$$= \frac{9 + 5\sqrt{5}}{4} \qquad \text{Simplify.} \qquad \blacksquare$$

EXAMPLE 7 Simplify $\dfrac{12 - \sqrt{18}}{9}$.

Solution: First simplify $\sqrt{18}$.

$$\frac{12 - \sqrt{18}}{9} = \frac{12 - \sqrt{9 \cdot 2}}{9} = \frac{12 - 3\sqrt{2}}{9}$$

Next, factor out a common factor of 3 from the terms in the numerator and simplify.

$$\frac{12 - 3\sqrt{2}}{9} = \frac{3\,(4 - \sqrt{2})}{3 \cdot 3} = \frac{4 - \sqrt{2}}{3} \qquad \blacksquare$$

MENTAL MATH

Find each product. Assume that variables represent nonnegative real numbers.

1. $\sqrt{2} \cdot \sqrt{3}$ **2.** $\sqrt{5} \cdot \sqrt{7}$ **3.** $\sqrt{1} \cdot \sqrt{6}$

4. $\sqrt{7} \cdot \sqrt{x}$ **5.** $\sqrt{10} \cdot \sqrt{y}$ **6.** $\sqrt{x} \cdot \sqrt{y}$

EXERCISE SET 8.4

Find each product and simplify. See Example 1.

1. $\sqrt{8} \cdot \sqrt{2}$ **2.** $\sqrt{3} \cdot \sqrt{12}$ **3.** $\sqrt{10} \cdot \sqrt{5}$

4. $3\sqrt{2} \cdot 5\sqrt{14}$ **5.** $\sqrt[3]{12} \cdot \sqrt[3]{4}$ **6.** $\sqrt[3]{9} \cdot \sqrt[3]{6}$

Find each product and simplify. See Example 2.

7. $\sqrt{10}(\sqrt{2} + \sqrt{5})$ **8.** $\sqrt{6}(\sqrt{3} + \sqrt{2})$

9. $(3\sqrt{5} - \sqrt{10})(\sqrt{5} - 4\sqrt{3})$ **10.** $(2\sqrt{3} - 6)(\sqrt{3} - 4\sqrt{2})$

Find each product and simplify. See Example 3.

11. $(\sqrt{x} + 6)(\sqrt{x} - 6)$ **12.** $(2\sqrt{5} + 1)(2\sqrt{5} - 1)$

13. $(\sqrt{3} + 8)^2$ **14.** $(\sqrt{x} - 7)^2$

Find each quotient and simplify. See Example 4.

15. $\dfrac{\sqrt{32}}{\sqrt{2}}$ **16.** $\dfrac{\sqrt{40}}{\sqrt{10}}$ **17.** $\dfrac{\sqrt{90}}{\sqrt{5}}$

18. $\dfrac{\sqrt{96}}{\sqrt{8}}$ **19.** $\dfrac{\sqrt{75y^5}}{\sqrt{3}}$ **20.** $\dfrac{\sqrt{24x^7}}{\sqrt{6}}$

Rationalize each denominator and simplify. See Example 5.

21. $\sqrt{\dfrac{3}{5}}$ **22.** $\sqrt{\dfrac{2}{3}}$ **23.** $\dfrac{1}{\sqrt{6}}$

24. $\dfrac{1}{\sqrt{10}}$ **25.** $\sqrt{\dfrac{5}{18}}$ **26.** $\sqrt{\dfrac{7}{12}}$

Rationalize each denominator and simplify. See Example 6.

27. $\dfrac{3}{\sqrt{2} + 1}$ **28.** $\dfrac{6}{\sqrt{5} + 2}$ **29.** $\dfrac{2}{\sqrt{10} - 3}$

30. $\dfrac{4}{2 - \sqrt{3}}$ **31.** $\dfrac{\sqrt{5} + 1}{\sqrt{6} - \sqrt{5}}$ **32.** $\dfrac{\sqrt{3} + 1}{\sqrt{3} - \sqrt{2}}$

Simplify the following. See Example 7.

33. $\dfrac{6 + 2\sqrt{3}}{2}$ **34.** $\dfrac{9 + 6\sqrt{2}}{3}$ **35.** $\dfrac{18 - 12\sqrt{5}}{6}$

36. $\dfrac{8 - 20\sqrt{3}}{4}$ **37.** $\dfrac{15\sqrt{3} + 5}{5}$ **38.** $\dfrac{8 + 16\sqrt{2}}{8}$

Multiply or divide as indicated and simplify.

39. $2\sqrt{3} \cdot 4\sqrt{15}$

40. $3\sqrt{14} \cdot 4\sqrt{2}$

41. $(\sqrt{11})^2$

42. $(\sqrt{7})^2$

43. $(2\sqrt{5})^2$

44. $(3\sqrt{10})^2$

45. $(6\sqrt{x})^2$

46. $(8\sqrt{y})^2$

47. $\sqrt{6}(\sqrt{5} + \sqrt{7})$

48. $\sqrt{10}(\sqrt{3} - \sqrt{7})$

49. $4\sqrt{5x}(\sqrt{x} - 3\sqrt{5})$

50. $3\sqrt{7y}(\sqrt{y} - 2\sqrt{7})$

51. $(\sqrt{3} + \sqrt{5})(\sqrt{2} - \sqrt{5})$

52. $(\sqrt{6} + \sqrt{3})(\sqrt{6} - \sqrt{3})$

53. $(\sqrt{7} - 2\sqrt{3})(\sqrt{7} + 2\sqrt{3})$

54. $(\sqrt{2} - 4\sqrt{5})(\sqrt{2} + 4\sqrt{5})$

55. $(\sqrt{x} - 3)(\sqrt{x} + 3)$

56. $(2\sqrt{y} + 5)(2\sqrt{y} - 5)$

57. $(\sqrt{6} + 3)^2$

58. $(2 + \sqrt{7})^2$

59. $(3\sqrt{x} - 5)^2$

60. $(2\sqrt{x} - 7)^2$

61. $\dfrac{\sqrt{150}}{\sqrt{2}}$

62. $\dfrac{\sqrt{120}}{\sqrt{3}}$

63. $\dfrac{\sqrt{72y^5}}{\sqrt{3y^3}}$

64. $\dfrac{\sqrt{54x^3}}{\sqrt{2x}}$

65. $\dfrac{\sqrt{48x^3}}{\sqrt{6x}}$

66. $\dfrac{\sqrt{90x^7}}{\sqrt{2x^3}}$

67. $\dfrac{\sqrt{24x^3y^4}}{\sqrt{2xy}}$

68. $\dfrac{\sqrt{96x^5y^3}}{\sqrt{3x^2y}}$

69. $2\sqrt[3]{5} \cdot 6\sqrt[3]{2}$

70. $8\sqrt[3]{4} \cdot 7\sqrt[3]{7}$

71. $\sqrt[3]{15} \cdot \sqrt[3]{25}$

72. $\sqrt[3]{4} \cdot \sqrt[3]{4}$

73. $\dfrac{\sqrt[3]{54}}{\sqrt[3]{2}}$

74. $\dfrac{\sqrt[3]{80}}{\sqrt[3]{10}}$

75. $\dfrac{\sqrt[3]{600}}{\sqrt[3]{5}}$

76. $\dfrac{\sqrt[3]{96}}{\sqrt[3]{6}}$

Rationalize each denominator and simplify.

77. $\sqrt{\dfrac{2}{15}}$

78. $\sqrt{\dfrac{11}{14}}$

79. $\sqrt{\dfrac{3}{20}}$

80. $\sqrt{\dfrac{3}{50}}$

81. $\sqrt{\dfrac{5}{72}}$

82. $\sqrt{\dfrac{11}{48}}$

83. $\dfrac{3x}{\sqrt{2x}}$

84. $\dfrac{5y}{\sqrt{3y}}$

85. $\dfrac{4}{2 - \sqrt{5}}$

86. $\dfrac{2}{1 - \sqrt{2}}$

87. $\dfrac{5}{3 + \sqrt{10}}$

88. $\dfrac{5}{\sqrt{6} + 2}$

89. $\dfrac{8}{\sqrt{3} - 1}$

90. $\dfrac{6}{\sqrt{7} - 2}$

91. $\dfrac{2\sqrt{3}}{\sqrt{15} + 2}$

92. $\dfrac{3\sqrt{2}}{\sqrt{10} + 2}$

93. $\dfrac{\sqrt{3} + 1}{\sqrt{2} - 1}$

94. $\dfrac{\sqrt{2} - 2}{2 - \sqrt{3}}$

95. $\dfrac{\sqrt{2} + 3}{\sqrt{6} - \sqrt{2}}$

96. $\dfrac{\sqrt{3} + 2}{\sqrt{6} - \sqrt{3}}$

Simplify the following.

97. $\dfrac{16 + 8\sqrt{3}}{4}$

98. $\dfrac{12 + 24\sqrt{3}}{6}$

99. $\dfrac{18 + 24\sqrt{2}}{12}$

100. $\dfrac{10 + 20\sqrt{3}}{15}$

101. Find the area of a rectangular room whose length is $13\sqrt{2}$ meters and width is $5\sqrt{6}$ meters.

102. Find the volume of a microwave oven whose length is $\sqrt{3}$ feet, width is $\sqrt{2}$ feet, and height is $\sqrt{2}$ feet.

103. If a round ball has volume V, then the formula for the radius r of the ball is

$$r = \sqrt[3]{\dfrac{3V}{4\pi}}$$

Simplify this expression by rationalizing the denominator.

Writing in Mathematics

104. When rationalizing the denominator of $\dfrac{\sqrt{2}}{\sqrt{3}}$, explain why both the numerator and the denominator must be multiplied by $\sqrt{3}$.

Skill Review

Simplify the following expressions. See Section 5.1.

105. $\dfrac{3x + 12}{3}$

106. $\dfrac{12x + 8}{4}$

107. $\dfrac{6x^2 - 3x}{3x}$

108. $\dfrac{8y^2 - 2y}{2y}$

8.5
Solving Equations Containing Radicals

OBJECTIVE

1   Solve equations containing square roots of variables.

Tape 28

1   In this section, we will solve **radical equations** such as

$$\sqrt{x + 3} = 5$$

To do so, we rely on the following squaring property.

Squaring Property of Equality

$$\text{If } a = b, \quad \text{then } a^2 = b^2$$

Unfortunately, this squaring property does not guarantee that all solutions of the new equation will be solutions of the original equation. For example, if we square both sides of the equation

$$x = 2$$

we have

$$x^2 = 4$$

This new equation has two solutions, 2 and -2, while the original equation $x = 2$ has only one solution. For this reason:

Always check proposed solutions in the original equation.

EXAMPLE 1 Solve for x: $\sqrt{x + 3} = 5$.

Solution: To solve for x, use the squaring property of equality and square both sides of the equation.

$$\sqrt{x + 3} = 5$$
$$(\sqrt{x + 3})^2 = 5^2 \qquad \text{Square both sides.}$$
$$x + 3 = 25 \qquad \text{Simplify.}$$
$$x = 22 \qquad \text{Subtract 3 from both sides.}$$

To check this proposed solution, replace x with 22 in the original equation.

$$\sqrt{x + 3} = 5 \qquad \text{Original equation.}$$
$$\sqrt{22 + 3} = 5 \qquad \text{Let } x = 22.$$
$$\sqrt{25} = 5$$
$$5 = 5 \qquad \text{True.}$$

Since a true statement results, 22 is the solution. ∎

EXAMPLE 2 Solve the equation $\sqrt{y + 2} + 9 = 16$.

Solution: First, write the equation so that the radical is by itself on one side of the equation. Then square both sides.

$$\sqrt{y + 2} + 9 = 16$$
$$\sqrt{y + 2} = 7 \qquad \text{Subtract 9 from both sides.}$$
$$(\sqrt{y + 2})^2 = 7^2 \qquad \text{Square both sides.}$$
$$y + 2 = 49 \qquad \text{Simplify.}$$
$$y = 47 \qquad \text{Subtract 2 from both sides.}$$

Check this solution by replacing y with 47 in the original equation.

$$\sqrt{y + 2} + 9 = 16$$
$$\sqrt{47 + 2} + 9 = 16 \qquad \text{Replace } y \text{ with 47.}$$
$$\sqrt{49} + 9 = 16$$
$$7 + 9 = 16$$
$$16 = 16 \qquad \text{True.}$$

The solution is 47. ∎

EXAMPLE 3 Solve for x: $\sqrt{x} + 6 = 4$.

Solution: First isolate the radical. Then square both sides.

$$\sqrt{x} + 6 = 4$$
$$\sqrt{x} = -2 \qquad \text{Subtract 6 from both sides to isolate the radical.}$$
$$(\sqrt{x})^2 = (-2)^2 \qquad \text{Square both sides.}$$
$$x = 4 \qquad \text{Simplify.}$$

To check, replace x with 4 in the original equation.

$$\sqrt{x} + 6 = 4 \qquad \text{Original equation.}$$

$$\sqrt{4} + 6 = 4 \qquad \text{Let } x = 4.$$

$$2 + 6 = 4 \qquad \textbf{False.}$$

Since 4 **does not** satisfy the original equation, there is no solution.   ■

Example 3 makes it very clear that we **must** check proposed solutions in the original equation to determine if they are truly solutions. If a proposed solution does not work, we say that the value is an **extraneous solution.**

The following set of guidelines will help you solve radical equations containing square roots.

To Solve a Radical Equation Containing Square Roots

Step 1   Arrange terms so that one radical is by itself on one side of the equation. That is, isolate a radical.

Step 2   Square both sides of the equation.

Step 3   Simplify both sides of the equation.

Step 4   If the equation still contains a radical term, repeat steps 1 through 3.

Step 5   Solve the equation.

Step 6   Check all solutions in the original equation for extraneous solutions.

EXAMPLE 4   Solve the equation $\sqrt{x} = \sqrt{5x - 2}$.

Solution: Each of the radicals is already isolated, since each is by itself on one side of the equation. Begin solving by squaring both sides.

$$\sqrt{x} = \sqrt{5x - 2} \qquad \text{Original equation.}$$

$$(\sqrt{x})^2 = (\sqrt{5x - 2})^2 \qquad \text{Square both sides.}$$

$$x = 5x - 2 \qquad \text{Simplify.}$$

$$-4x = -2 \qquad \text{Subtract } 5x \text{ from both sides.}$$

$$x = \frac{-2}{-4} = \frac{1}{2} \qquad \text{Divide both sides by } -4 \text{ and simplify.}$$

The proposed solution is $\frac{1}{2}$. To check, replace x with $\frac{1}{2}$ in the original equation.

$$\sqrt{x} = \sqrt{5x - 2} \qquad \text{Original equation.}$$

$$\sqrt{\frac{1}{2}} = \sqrt{5 \cdot \frac{1}{2} - 2} \qquad \text{Let } x = \frac{1}{2}.$$

$$\sqrt{\frac{1}{2}} = \sqrt{\frac{5}{2} - 2} \qquad \text{Multiply.}$$

$$\sqrt{\frac{1}{2}} = \sqrt{\frac{5}{2} - \frac{4}{2}} \qquad \text{Write 2 as } \frac{4}{2}.$$

$$\sqrt{\frac{1}{2}} = \sqrt{\frac{1}{2}} \qquad \text{True.}$$

This statement is true, so the solution is $\frac{1}{2}$.   ■

It is helpful to recall the following fact before proceeding to the next example.

HELPFUL HINT

$$(a + b)^2 = a^2 + 2ab + b^2$$

For example,

$$(x - 3)^2 = x^2 - 6x + 9$$

EXAMPLE 5   Solve the equation $\sqrt{x + 3} - x = -3$.

Solution: First, isolate the radical by adding x to both sides. Then square both sides.

$$\sqrt{x + 3} - x = -3$$

$$\sqrt{x + 3} = x - 3 \qquad \text{Add } x \text{ to both sides.}$$

$$(\sqrt{x + 3})^2 = (x - 3)^2 \qquad \text{Square both sides.}$$

$$x + 3 = x^2 - 6x + 9 \qquad \text{Write } (x - 3)^2 \text{ as } x^2 - 6x + 9.$$

To solve this resulting quadratic equation, write the equation in standard form by subtracting x and 3 from both sides.

$$3 = x^2 - 7x + 9 \qquad \text{Subtract } x \text{ from both sides.}$$

$$0 = x^2 - 7x + 6 \qquad \text{Subtract 3 from both sides.}$$

$$0 = (x - 6)(x - 1) \qquad \text{Factor.}$$

$$0 = x - 6 \quad \text{or} \quad 0 = x - 1 \qquad \text{Set each factor equal to zero.}$$

$$6 = x \qquad \text{or} \quad 1 = x \qquad \text{Solve for } x.$$

Both of these proposed solutions must be checked.

$$\text{Let } x = 6 \qquad\qquad\qquad \text{Let } x = 1$$

$$\sqrt{x + 3} - x = -3 \qquad\qquad \sqrt{x + 3} - x = -3$$

$$\sqrt{6 + 3} - 6 = -3 \qquad\qquad \sqrt{1 + 3} - 1 = -3$$

$$\sqrt{9} - 6 = -3 \qquad\qquad\qquad \sqrt{4} - 1 = -3$$

$$3 - 6 = -3 \qquad\qquad\qquad 2 - 1 = -3$$

$$-3 = -3 \quad \text{True.} \qquad\qquad 1 = -3 \quad \text{False.}$$

Since replacing x with 1 resulted in a false statement, 1 is not a solution. The only solution is 6.   ■

EXAMPLE 6   Solve for y: $\sqrt{4y^2 + 5y - 15} = 2y$.

Solution: The radical is already isolated, so start by squaring both sides.

$$\sqrt{4y^2 + 5y - 15} = 2y$$

$$(\sqrt{4y^2 + 5y - 15})^2 = (2y)^2$$

$$4y^2 + 5y - 15 = 4y^2 \qquad \text{Simplify.}$$

$$5y - 15 = 0 \qquad \text{Subtract } 4y^2 \text{ from both sides.}$$

$$5y = 15 \qquad \text{Add 15 to both sides.}$$
$$y = 3 \qquad \text{Divide both sides by 5.}$$

Replace y with 3 in the original equation to check this proposed solution.

$$\sqrt{4y^2 + 5y - 15} = 2y \qquad \text{Original equation.}$$
$$\sqrt{4 \cdot 3^2 + 5 \cdot 3 - 15} = 2 \cdot 3 \qquad \text{Let } y = 3.$$
$$\sqrt{4 \cdot 9 + 15 - 15} = 6 \qquad \text{Simplify.}$$
$$\sqrt{36} = 6$$
$$6 = 6 \qquad \text{True.}$$

This statement is true, so the solution is 3.   ■

EXERCISE SET 8.5

Solve each equation. See Examples 1 through 3.

1. $\sqrt{x} = 9$

2. $\sqrt{x} = 4$

3. $\sqrt{x + 5} = 2$

4. $\sqrt{x + 12} = 3$

5. $\sqrt{x} + 3 = 7$

6. $\sqrt{x} + 5 = 10$

Solve each equation. Be sure to check for extraneous solutions. See Examples 4 through 6.

7. $\sqrt{4x - 3} = \sqrt{x + 3}$

8. $\sqrt{5x - 4} = \sqrt{x + 8}$

9. $\sqrt{x + 7} = x + 5$

10. $\sqrt{x + 5} = x - 1$

11. $\sqrt{9x^2 + 2x - 4} = 3x$

12. $\sqrt{4x^2 + 3x - 9} = 2x$

13. $\sqrt{x + 6} + 5 = 3$

14. $\sqrt{2x - 1} + 7 = 1$

Solve each equation. Be sure to check for extraneous solutions.

15. $\sqrt{2x + 6} = 4$

16. $\sqrt{3x + 7} = 5$

17. $\sqrt{x} - 2 = 5$

18. $4\sqrt{x} - 7 = 5$

19. $3\sqrt{x} + 5 = 2$

20. $3\sqrt{x} + 5 = 8$

21. $\sqrt{x + 6} + 1 = 3$

22. $\sqrt{x + 5} + 2 = 5$

23. $\sqrt{2x + 1} + 3 = 5$

24. $\sqrt{3x - 1} + 4 = 1$

25. $\sqrt{x} = \sqrt{3x - 8}$

26. $\sqrt{x} = \sqrt{4x - 3}$

27. $\sqrt{4x} - \sqrt{2x + 6} = 0$

28. $\sqrt{5x + 6} - \sqrt{8x} = 0$

29. $\sqrt{x} = x - 6$

30. $\sqrt{x} = x + 6$

31. $\sqrt{2x + 1} = x - 7$

32. $\sqrt{2x + 5} = x - 5$

33. $x = \sqrt{2x - 2} + 1$

34. $\sqrt{1 - 8x} + 2 = x$

35. $\sqrt{1 - 8x} - x = 4$

36. $\sqrt{3x + 7} - x = 3$

37. $\sqrt{2x + 5} - 1 = x$

38. $x = \sqrt{4x - 7} + 1$

39. $\sqrt{16x^2 - 3x + 6} = 4x$

40. $\sqrt{9x^2 - 2x + 8} = 3x$

41. $\sqrt{16x^2 + 2x + 2} = 4x$

42. $\sqrt{4x^2 + 3x - 2} = 2x$

43. $\sqrt{2x^2 + 6x + 9} = 3$

44. $\sqrt{3x^2 + 6x + 4} = 2$

45. A number is 6 more than its principal square root. Find the number.

46. The formula $b = \sqrt{\dfrac{3V}{h}}$ can be used in connection with square-based pyramids, where b represents the length of one side of the base, V represents the volume, and h represents the height. A pyramid is 6 feet high with base length of 3 feet. Using these values in the formula gives us $3 = \sqrt{\dfrac{3V}{6}}$. Solve this equation for V.

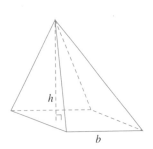

47. The maximum distance d in kilometers that we can see from a given height h above the ground is given by the formula $d = 111.7\sqrt{h}$. Find the height that would allow us to see 80 kilometers by solving the equation $80 = 111.7\sqrt{h}$ to the nearest hundredth of a kilometer.

Writing in Mathematics

48. Explain why apparent solutions of radical equations must be checked.

Skill Review

Translate the sentences into equations, and then solve the equations. See Section 2.6.

49. If 8 is subtracted from the product of 3 and x, the result is 19. Find x.

50. If 3 more than x is subtracted from twice x, the result is 11. Find x.

51. The length of a rectangle is twice the width. The perimeter is 24 inches. Find the length.

52. The length of a rectangle is 2 inches longer than the width. The perimeter is 24 inches. Find the length.

Translate the sentences into two equations with two unknowns and solve the resulting systems of equations. See Section 7.4.

53. The sum of two numbers is 20 and their differences is 8. Find the numbers.

54. The difference of two numbers is 58. If one number is three times the other number, find the numbers.

8.6
Applications of Equations Containing Radicals

<table>
<tr><td>OBJECTIVES</td><td>1</td><td>Use the Pythagorean formula to solve applications.</td></tr>
<tr><td></td><td>2</td><td>Use the distance formula to solve applications.</td></tr>
<tr><td>Tape 28</td><td>3</td><td>Solve applications using a formula containing radicals.</td></tr>
</table>

1　Applications of radicals can be found in geometry, finance, science, and other areas of technology. Our first application involves the Pythagorean theorem, giving a formula that relates the lengths of the three sides of a right triangle. We first studied the Pythagorean theorem in Chapter 4 and we review it here.

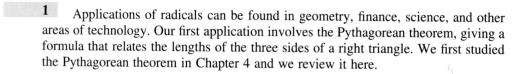

The Pythagorean Theorem

If a and b are the lengths of the legs of a right triangle and c is the length of the hypotenuse, then $a^2 + b^2 = c^2$.

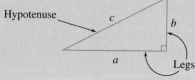

This property, along with the square root property, allows us to find unknown sides of right triangles.

Square Root Property

If a and b are **positive numbers** and if

$$a = b, \qquad \text{then } \sqrt{a} = \sqrt{b}$$

The following examples demonstrate how to use this property.

EXAMPLE 1   Find the length of the hypotenuse of a right triangle whose legs are 6 inches and 8 inches long.

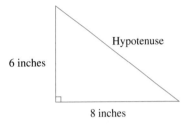

Solution: Because this is a right triangle, we use the Pythagorean theorem. The legs are 6 inches and 8 inches long, so let $a = 6$ and $b = 8$ in the formula.

$$a^2 + b^2 = c^2$$

$$6^2 + 8^2 = c^2 \qquad \text{Substitute the lengths of the legs.}$$

$$36 + 64 = c^2 \qquad \text{Simplify.}$$

$$100 = c^2$$

Next, use the square root property. Since c represents a length, we assume that c is positive.

$$100 = c^2$$

$$\sqrt{100} = \sqrt{c^2} \qquad \text{Apply the square root property.}$$

$$10 = c \qquad \text{Simplify.}$$

The hypothenuse has a length of 10 inches.   ■

EXAMPLE 2   A surveyor must determine the distance across a lake. He first inserts poles at points P and Q on opposite sides of the lake, as pictured. Perpendicular to line PQ, he finds another point R. If the length of PR is 320 feet and the length of QR is 240 feet, what is the distance across the lake?

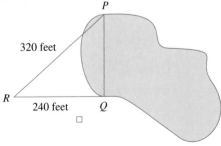

Solution: By creating a line perpendicular to line PQ, the surveyor deliberately constructed a right triangle. The hypotenuse, PR, has a length of 320 feet, so we let $c = 320$ in the Pythagorean formula. The side QR is one of the legs, so let $a = 240$.

$$a^2 + b^2 = c^2 \qquad \text{Use the Pythagorean theorem.}$$

$$240^2 + b^2 = 320^2 \qquad \text{Let } c = 320 \text{ and } a = 240.$$

$$57{,}600 + b^2 = 102{,}400$$

$$b^2 = 44{,}800 \qquad \text{Subtract } 57{,}600 \text{ from both sides.}$$

$$\sqrt{b^2} = \sqrt{44{,}800} \qquad \text{Apply the square root property.}$$

The distance across the lake is $\sqrt{44{,}800}$ feet. The surveyor can now use a calculator to find that $\sqrt{44{,}800}$ feet is approximately 211.6601 feet, so the distance across the lake is roughly 212 feet. ■

2 A second place that radicals are commonly used is in finding the distance between two points in the plane. By using the Pythagorean theorem, the following formula can be derived.

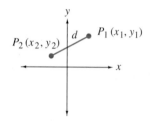

> **Distance Formula**
>
> The distance d between two points $P_1 = (x_1, y_1)$ and $P_2 = (x_2, y_2)$ is given by
> $$d = \sqrt{(x_2 - x_1)^2 + (y_2 - y_1)^2}$$

EXAMPLE 3 Find the distance between $(-1, 9)$ and $(-3, -5)$.

Solution: Use the distance formula with $(x_1, y_1) = (-1, 9)$ and $(x_2, y_2) = (-3, -5)$.

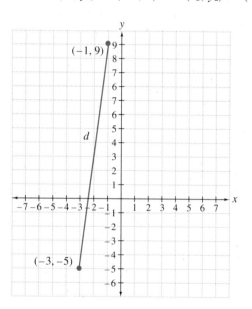

Watch your signs.

$$d = \sqrt{(x_2 - x_1)^2 + (y_2 - y_1)^2} \qquad \text{The distance formula.}$$

$$= \sqrt{[-3 - (-1)]^2 + (-5 - 9)^2} \qquad \text{Substitute known values.}$$

$$= \sqrt{(-2)^2 + (-14)^2} \qquad \text{Simplify.}$$

$$= \sqrt{4 + 196}$$

$$= \sqrt{200} = 10\sqrt{2} \qquad \text{Simplify the radical.}$$

The distance is **exactly** $10\sqrt{2}$ units or **approximately** 14.1 units.   ∎

3   The Pythagorean theorem and the distance formula are both extremely important results in mathematics and should be memorized. But there are other applications involving formulas containing radicals that are not quite as well known. For example:

EXAMPLE 4   A formula used to determine the velocity of an object (neglecting air resistance) after it has fallen a certain height is $v = \sqrt{2gh}$, where g is the acceleration due to gravity, and h is the height the object has fallen. On Earth, the acceleration g due to gravity is approximately 32 feet per second squared. Find the velocity of a watermelon after it has fallen 5 feet.

Solution: We are told that $g = 32$ feet per second squared. To find the velocity v when $h = 5$ feet, use the velocity formula.

$$v = \sqrt{2gh} \qquad \text{The velocity formula.}$$

$$= \sqrt{2 \cdot 32 \cdot 5} \qquad \text{Substitute known values.}$$

$$= \sqrt{320} \quad \text{or} \quad 8\sqrt{5} \text{ feet per second}$$

The velocity of the watermelon after 5 feet is **exactly** $8\sqrt{5}$ feet per second, or **approximately** 17.9 feet per second.   ∎

EXERCISE SET 8.6

Use the Pythagorean theorem to find the unknown side of each right triangle. See Example 1.

1.

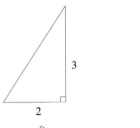

2.

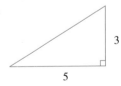

3.

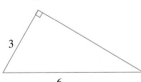

4.

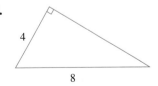

Find the length of the unknown side of each right triangle with sides a, b, and c, where a and b are legs and c is the hypotenuse. See Example 1.

5. $a = 9, b = 12$ **6.** $a = 12, b = 16$ **7.** $a = 8, c = 9$ **8.** $a = 6, c = 7$

Solve the following. See Example 2.

9. A wire is used to anchor a 20-foot-high pole. One end of the wire is attached to the top of the pole. The other end is fastened to a stake five feet away from the bottom of the pole. Find the length of the wire, to the nearest tenth of a foot.

10. David and Stephanie leave the seashore at the same time. David drives northward at a rate of 30 miles per hour, while Stephanie drives west at 60 mph. Find how far apart they are after 3 hours to the nearest mile.

Use the distance formula to find the distance between the points given. See Example 3.

11. $P_1(8, 3), P_2(11, 7)$

12. $P_1(2, 1), P_2(14, 6)$

13. $P_1(-1, 3), P_2(2, 8)$

14. $P_1(3, -2), P_2(5, 7)$

15. $P_1(-2, -3), P_2(-1, 4)$

16. $P_1(-4, -1), P_2(-6, 3)$

Solve the following. See Example 4.

17. For a square-based pyramid, the formula $b = \sqrt{\dfrac{3V}{H}}$ describes the relationship between the length b of one side of the base, the volume V, and the height H. Find the volume if each edge of the base is 6 feet long, and the pyramid is 2 feet high.

18. The formula $d = 16t^2$ relates the distance d, in feet, that an object falls in t seconds, assuming that air resistance does not slow down the object. Find how long, to the nearest hundredth of a second, it takes an object to reach the ground from the top of the Sears Tower in Chicago, a distance of 1454 feet.

Find the unknown side of each right triangle.

19.
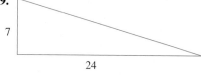

20.
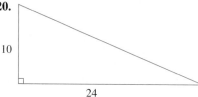

21.

22.
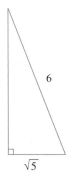

23.
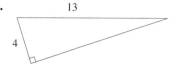

24.
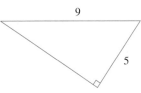

Find the length of the unknown side of each right triangle with sides a, b, and c, where c is the hypotenuse.

25. $a = 4, b = 5$

26. $a = 2, b = 7$

27. $b = 2, c = 6$

28. $b = 1, c = 5$

29. $a = \sqrt{10}, c = 10$

30. $a = \sqrt{7}, c = \sqrt{35}$

Use the distance formula to find the distance between the points given.

31. $P_1(3, 6), P_2(5, 11)$

32. $P_1(2, 3), P_2(9, 7)$

33. $P_1(-3, 1), P_2(5, -2)$

34. $P_1(-2, 6), P_2(3, -2)$

35. $P_1(3, -2), P_2(1, -8)$

36. $P_1(-5, 8), P_2(-2, 2)$

37. $P_1\left(\dfrac{1}{2}, 2\right), P_2(2, -1)$

38. $P_1\left(\dfrac{1}{3}, 1\right), P_2(1, -1)$

39. Beverly wants to determine the distance across a pond on her property. She is able to measure the distances shown on the following diagram. Find how wide the lake is to the nearest tenth of a foot.

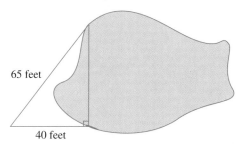

65 feet

40 feet

40. Carl needs to connect two underground pipelines, which are offset by 3 feet, as pictured in the diagram. Neglecting the joints needed to join the pipes, find the length of the shortest possible connecting pipe rounded to the nearest hundredth of a foot.

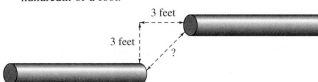

3 feet

3 feet

?

41. Bruce needs to attach a diagonal brace to a rectangular frame in order to make it structurally sound. If the framework is 6 feet by 10 feet, find how long the brace needs to be to the nearest tenth of a foot.

42. Connie is flying a kite. She let out 80 feet of string and attached the string to a stake in the ground. The kite is now directly above her brother Mike, who is 32 feet away from Connie. Find how high the kite is to the nearest foot.

43. Using the formula $b = \sqrt{\dfrac{3V}{H}}$, find the exact length of the base b, if the volume V is 18 cubic inches and the height H is 1 foot.

44. Police use the formula $s = \sqrt{30fd}$ to estimate the speed s of a car in miles per hour. In this formula, d represents the distance the car skidded in feet and f represents the coefficient of friction. The value of f depends on the type of road surface, and for wet concrete f is 0.35. Find how

fast a car was moving if it skidded 280 feet on wet concrete, to the nearest mile per hour.

45. The coefficient of friction of a certain dry road is 0.95. Use the formula in Exercise 44 to find how far a car will skid on this dry road if it is traveling at a rate of 60 mph. Round the length to the nearest foot.

46. The formula $v = \sqrt{2.5r}$ can be used to estimate the maximum safe velocity, v, in miles per hour, at which a car can travel if it is driven along a curved road with a **radius of curvature,** r, in feet. To the nearest whole number, find the maximum safe speed if a cloverleaf exit on an expressway has a radius of curvature of 300 feet.

47. Use the formula from Exercise 4.6 to find the radius of curvature if the safe velocity is 30 mph.

48. The maximum distance d in kilometers that you can see from the top of a tall building is given by $d = 111.7\sqrt{h}$. In this formula, h represents the height of the building in kilometers. Knowing that the Sears Tower is approximately 0.443 kilometers tall, find how far you can see on a clear day.

49. Use the formula from Exercise 4.8 to determine how high above the ground you need to be to see 40 kilometers.

50. Railroad tracks are invariably made up of relatively short sections of rail connected by expansion joints. To see why this construction is necessary, consider a single rail 100 feet long (or 1200 inches). On an extremely hot day, suppose it expands 1 inch in the hot sun to a new length of 1201 inches. Theoretically, the track would bow upward as pictured.

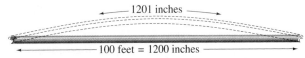

1201 inches

100 feet = 1200 inches

Let us approximate the bulge in the railroad this way.

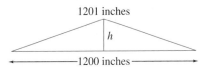

1201 inches

h

1200 inches

Calculate the height h of the bulge to the nearest tenth of an inch.

Writing in Mathematics

51. Based on the results of Exercise 50, explain why railroads need to use short sections of rail connected by expansion joints.

Skill Review

Simplify using rules for exponents. See Sections 3.1 and 3.2.

52. $x^2 \cdot x^3$ **53.** $x^4 \cdot x^2$ **54.** $y^3 \cdot y$ **55.** $x \cdot x^7$

Solve each system of linear equations by the graphing method. See Section 7.1.

56. $\begin{cases} x = 3y \\ 2x + y = 7 \end{cases}$

57. $\begin{cases} x + y = 3 \\ -3x + y = 1 \end{cases}$

8.7
Rational Exponents

OBJECTIVES

Tape 29

1 Evaluate exponential expressions of the form $a^{1/n}$.

2 Evaluate exponential expressions of the form $a^{m/n}$.

3 Evaluate exponential expressions of the form $a^{-m/n}$.

4 Use rules for exponents to simplifying expressions containing fractional exponents.

---

1 Thus far, we have simplified exponential expressions containing integer exponents. In this section, we explore the meaning of expressions containing rational exponents such as:

$$3^{1/2}, \quad 2^{-3/4}, \quad \text{and} \quad y^{5/6}$$

In simplifying fractional expressions, keep in mind that we want the rules for operating with rational exponents to be the same as the rules for operating with integer exponents. These rules are repeated here for review.

Summary of Exponent Rules

If m and n are integers and a, b, and c are real numbers, then

Product Rule for Exponents: $\quad a^m \cdot a^n = a^{m+n}$

Power Rule for Exponents: $\quad (a^m)^n = a^{m \cdot n}$

Power Rules for Products and Quotients: $\quad (ab)^n = a^n b^n$ and

$$\left(\frac{a}{c}\right)^n = \frac{a^n}{c^n}, \quad c \neq 0$$

Quotient Rule for Exponents: $\quad \dfrac{a^m}{a^n} = a^{m-n}, \quad a \neq 0$

Zero Exponent: $\quad a^0 = 1, \quad a \neq 0$

Negative Exponent: $\quad a^{-n} = \dfrac{1}{a^n}, \quad a \neq 0$

If the rule $(a^m)^n = a^{m \cdot n}$ is to hold for fractional exponents, it should be true that

$$(3^{1/2})^2 = 3^{1/2 \cdot 2} = 3^1 = 3$$

Also, we know that

$$(\sqrt{3})^2 = 3$$

Since both expressions simplify to 3, it would be reasonable to define

$$3^{1/2} \quad \text{as} \quad \sqrt{3}$$

In general, we have the following:

> **Definition of $a^{1/n}$**
>
> If n is a positive integer and $\sqrt[n]{a}$ is a real number, then
> $$a^{1/n} = \sqrt[n]{a}$$

Notice that the denominator of the rational exponent corresponds to the index of the radical.

EXAMPLE 1   Simplify each expression.
   a. $25^{1/2}$ **b.** $8^{1/3}$ **c.** $-16^{1/4}$ **d.** $(-27)^{1/3}$ **e.** $\left(\dfrac{1}{9}\right)^{1/2}$

Solution: **a.** $25^{1/2} = \sqrt{25} = 5$.
   b. $8^{1/3} = \sqrt[3]{8} = 2$.
   c. In $-16^{1/4}$, the base of the exponent is 16. Thus **the negative sign is not affected by the exponent;** so $-16^{1/4} = -\sqrt[4]{16} = -2$.
   d. $(-27)^{1/3} = \sqrt[3]{-27} = -3$.
   e. $\left(\dfrac{1}{9}\right)^{1/2} = \sqrt{\dfrac{1}{9}} = \dfrac{1}{3}$.   ■

2   In Example 1, each rational exponent has a numerator of 1. What happens if the numerator is some other positive integer? Consider $8^{2/3}$. Since $\dfrac{2}{3}$ is the same $\dfrac{1}{3} \cdot 2$, we reason that

$$8^{2/3} = 8^{1/3 \cdot 2} = (8^{1/3})^2 = (\sqrt[3]{8})^2 = 2^2 = 4$$

The denominator 3 of the rational exponent corresponds to the index of the radical. The numerator 2 of the fractional exponent indicates that the base is to be squared.

> **Definition of $a^{m/n}$**
>
> If m and n are integers with $n > 0$ and a is positive, then
> $$a^{m/n} = a^{(1/n) \cdot m} = \sqrt[n]{a^m} = (\sqrt[n]{a})^m$$

EXAMPLE 2 Simplify each expression.

a. $4^{3/2}$ **b.** $(-27)^{2/3}$ **c.** $-16^{3/4}$

Solution: **a.** $4^{3/2} = (4^{1/2})^3 = (\sqrt{4})^3 = 2^3 = 8$.

b. $(-27)^{2/3} = (-27^{1/3})^2 = (\sqrt[3]{-27})^2 = (-3)^2 = 9$.

c. The negative sign is **not** affected by the exponent since the base of the exponent is 16. $-16^{3/4} = -(16^{1/4})^3 = -(\sqrt[4]{16})^3 = -2^3 = -8$.   ∎

HELPFUL HINT

Recall that

$$-3^2 = -(3 \cdot 3) = -9$$

and

$$(-3)^2 = (-3)(-3) = 9$$

In other words, without parentheses the exponent 2 applies to the base of 3, **not** -3. The same is true of rational exponents. For example,

$$-16^{1/2} = -\sqrt{16} = -4$$

and

$$(-27)^{1/3} = \sqrt[3]{-27} = -3.$$

---

3   If the exponent is a negative rational number, use the following definition.

Definition of $a^{-m/n}$

If $a^{m/n}$ is a nonzero real number, then

$$a^{-m/n} = \frac{1}{a^{m/n}}$$

EXAMPLE 3 Simplify each expression.

a. $36^{-1/2}$ **b.** $16^{-3/4}$ **c.** $-9^{1/2}$ **d.** $32^{-4/5}$

Solution: **a.** $36^{-1/2} = \dfrac{1}{36^{1/2}} = \dfrac{1}{\sqrt{36}} = \dfrac{1}{6}$.

b. $16^{-3/4} = \dfrac{1}{16^{3/4}} = \dfrac{1}{(\sqrt[4]{16})^3} = \dfrac{1}{2^3} = \dfrac{1}{8}$.

c. $-9^{1/2} = -\sqrt{9} = -3$.

d. $32^{-4/5} = \dfrac{1}{32^{4/5}} = \dfrac{1}{(\sqrt[5]{32})^4} = \dfrac{1}{2^4} = \dfrac{1}{16}$.   ∎

4   It can be shown that the properties of integer exponents hold for rational exponents. By using these properties and definitions, we can now simplify expressions containing rational exponents.

EXAMPLE 4   Simplify each expression. Write answers with positive exponents only. Assume that all variables represent positive numbers.

a. $3^{1/2} \cdot 3^{3/2}$ **b.** $\dfrac{5^{1/3}}{5^{2/3}}$ **c.** $(x^{1/4})^{12}$ **d.** $\dfrac{x^{1/5}}{x^{-4/5}}$ **e.** $\left(\dfrac{y^{3/5}}{z^{1/4}}\right)^2$

Solution: **a.** $3^{1/2} \cdot 3^{3/2} = 3^{1/2 + 3/2} = 3^{4/2} = 3^2 = 9.$

b. $\dfrac{5^{1/3}}{5^{2/3}} = 5^{(1/3)-(2/3)} = 5^{-1/3} = \dfrac{1}{5^{1/3}}.$

c. $(x^{1/4})^{12} = x^{1/4 \cdot 12} = x^3.$

d. $\dfrac{x^{1/5}}{x^{-4/5}} = x^{1/5-(-4/5)} = x^{5/5} = x^1$ or $x.$

e. $\left(\dfrac{y^{3/5}}{z^{1/4}}\right)^2 = \dfrac{y^{3/5 \cdot 2}}{z^{1/4 \cdot 2}} = \dfrac{y^{6/5}}{z^{1/2}}.$ ∎

EXERCISE SET 8.7

Simplify each expression. See Examples 1 and 2.

1. $8^{1/3}$ **2.** $16^{1/4}$ **3.** $9^{1/2}$ **4.** $16^{1/2}$

5. $16^{3/4}$ **6.** $27^{2/3}$ **7.** $32^{2/5}$ **8.** $64^{5/6}$

Simplify each expression. See Example 3.

9. $-16^{-1/4}$ **10.** $-8^{-1/3}$ **11.** $16^{-3/2}$ **12.** $27^{-4/3}$

13. $81^{-3/2}$ **14.** $32^{-2/5}$ **15.** $\left(\dfrac{4}{25}\right)^{-1/2}$ **16.** $\left(\dfrac{8}{27}\right)^{-1/3}$

Simplify each expression. Write each answer with positive exponents. Assume that all variables represent positive numbers. See Example 4.

17. $2^{1/3} \cdot 2^{2/3}$ **18.** $4^{2/5} \cdot 4^{3/5}$ **19.** $\dfrac{4^{3/4}}{4^{1/4}}$ **20.** $\dfrac{9^{7/2}}{9^{3/2}}$

21. $\dfrac{x^{1/6}}{x^{5/6}}$ **22.** $\dfrac{x^{1/4}}{x^{3/4}}$ **23.** $(x^{1/2})^6$ **24.** $(x^{1/3})^6$

Simplify each expression.

25. $81^{1/2}$ **26.** $(-27)^{1/3}$ **27.** $(-8)^{1/3}$ **28.** $36^{1/2}$

29. $-81^{1/4}$ **30.** $-64^{1/3}$ **31.** $\left(\dfrac{1}{81}\right)^{1/2}$ **32.** $\left(\dfrac{9}{16}\right)^{1/2}$

33. $\left(\dfrac{27}{64}\right)^{1/3}$ **34.** $\left(\dfrac{16}{81}\right)^{1/4}$ **35.** $9^{3/2}$ **36.** $16^{3/2}$

37. $64^{3/2}$ **38.** $64^{2/3}$ **39.** $-8^{2/3}$ **40.** $(-8)^{2/3}$

41. $4^{5/2}$ **42.** $9^{4/2}$ **43.** $\left(\dfrac{4}{9}\right)^{3/2}$ **44.** $\left(\dfrac{8}{27}\right)^{2/3}$

45. $\left(\dfrac{1}{81}\right)^{3/4}$ **46.** $\left(\dfrac{1}{32}\right)^{3/5}$ **47.** $4^{-1/2}$ **48.** $9^{-1/2}$

49. $125^{-1/3}$ **50.** $216^{-1/3}$ **51.** $625^{-3/4}$ **52.** $256^{-5/8}$

Simplify each expression. Write each answer with positive exponents. Assume that all variables represent positive numbers.

53. $3^{4/3} \cdot 3^{2/3}$

54. $2^{5/4} \cdot 2^{3/4}$

55. $\dfrac{6^{2/3}}{6^{1/3}}$

56. $\dfrac{3^{3/5}}{3^{1/5}}$

57. $(x^{2/3})^9$

58. $(x^6)^{3/4}$

59. $\dfrac{6^{1/3}}{6^{-5/3}}$

60. $\dfrac{2^{-3/4}}{2^{5/4}}$

61. $\dfrac{3^{-3/5}}{3^{2/5}}$

62. $\dfrac{5^{1/4}}{5^{-3/4}}$

63. $\left(\dfrac{x^{1/3}}{y^{3/4}}\right)^2$

64. $\left(\dfrac{x^{1/2}}{y^{2/3}}\right)^6$

65. $\left(\dfrac{x^{2/5}}{y^{3/4}}\right)^8$

66. $\left(\dfrac{x^{3/4}}{y^{1/6}}\right)^3$

67. If a population grows at a rate of 8% annually, the formula $P = P_O(1.08)^N$ can be used to estimate the total population P after N years have passed, assuming the original population is P_O. Find the population after $1\frac{1}{2}$ years if the original population of 10,000 people is growing at a rate of 8% annually.

Writing in Mathematics

68. Explain what each of the numbers 2, 3, and 4 signifies in the expression $4^{3/2}$.

69. Explain why $-4^{1/2}$ is a real number but $(-4)^{1/2}$ is not.

Skill Review

Solve each system of linear inequalities by graphing on a single coordinate system. See Section 7.5.

70. $\begin{cases} x + y < 6 \\ \quad y \geq 2x \end{cases}$

71. $\begin{cases} 2x - y \geq 3 \\ \quad\; x < 5 \end{cases}$

Solve each quadratic equation. See Section 4.6.

72. $x^2 - 4 = 3x$

73. $x^2 + 2x = 8$

74. $2x^2 - 5x - 3 = 0$

75. $3x^2 + x - 2 = 0$

CRITICAL THINKING   The owner of Gerken's Nursery has decided to plant Scotch pines on a portion of his land, thinking he can sell the trees as Christmas trees in future years. Originally, he planned to plant the seedlings in a boxlike pattern, so that four neighboring trees are located at the corners of a square. He begins to wonder, however, if another pattern is more efficient.

What pattern yields a maximum number of trees per acre? What additional facts do you need in order to determine how many trees in this pattern can be planted? If you assume these facts, how many trees can be planted?

CHAPTER 8 GLOSSARY

The **conjugate** of $a + b$ is $a - b$.

A number b is the **cube root** of a if $b^3 = a$. That is, $\sqrt[3]{a} = b$ if $b^3 = a$.

Radicals are **like radicals** if they have the same radicand and the same index.

The **principal square root** of a positive real number a, written as $\sqrt{a}$, is the positive number whose square root equals a. That is, $\sqrt{a} = b$ if $a = b^2$.

The **radical sign** is $\sqrt{}$.

The **radicand** is the expression under or within the radical sign.

Rationalizing the denominator is the process of removing the radical from the denominator.

CHAPTER 8 SUMMARY

(8.2)

> **Product Rule for Radicals**
>
> If $\sqrt[n]{a}$ and $\sqrt[n]{b}$ exist, then
> $$\sqrt[n]{a} \cdot \sqrt[n]{b} = \sqrt[n]{a \cdot b}$$
>
> **Quotient Rule for Radicals**
>
> If $\sqrt[n]{a}$ and $\sqrt[n]{b}$ exist, with $b \neq 0$, then
> $$\frac{\sqrt[n]{a}}{\sqrt[n]{b}} = \sqrt[n]{\frac{a}{b}}$$

THE PYTHAGOREAN THEOREM   (8.6)

If a and b are the lengths of the legs of a right triangle and c is the length of the hypotenuse, then
$$a^2 + b^2 = c^2$$

DISTANCE FORMULA   (8.6)

The distance between points $P_1 = (x_1, y_1)$ and $P_2 = (x_2, y_2)$ is given by the formula
$$d = \sqrt{(x_2 - x_1)^2 + (y_2 - y_1)^2}$$

DEFINITION OF $a^{1/n}$ (8.7)

$a^{1/n} = \sqrt[n]{a}$, where n is a positive integer and $\sqrt[n]{a}$ exists.

DEFINITION OF $a^{m/n}$ (8.7)

$a^{m/n} = \sqrt[n]{a^m} = (\sqrt[n]{a})^m$, where m and n are integers, $n > 0$, and $\sqrt[n]{a}$ and $\sqrt[n]{a^m}$ exist.

DEFINITION OF $a^{-m/n}$ (8.7)

$a^{-m/n} = \dfrac{1}{a^{m/n}}$, if $a^{m/n}$ is a nonzero real number.

CHAPTER 8 REVIEW

(8.1) *Find the root.*

1. $\sqrt{81}$

2. $-\sqrt{49}$

3. $\sqrt[3]{27}$

4. $\sqrt[4]{16}$

5. $-\sqrt{\dfrac{9}{64}}$

6. $\sqrt{\dfrac{36}{81}}$

7. $\sqrt[4]{-\dfrac{16}{81}}$

8. $\sqrt[3]{-\dfrac{27}{64}}$

Determine whether each of the following is rational or irrational. If rational, find the exact value. If irrational, use a caiculator or the appendix to find an approximation accurate to three decimal places.

9. $\sqrt{76}$

10. $\sqrt{576}$

Find the following roots. Assume that variables represent positive numbers only.

11. $\sqrt{x^{12}}$

12. $\sqrt{x^8}$

13. $\sqrt{9x^6y^2}$

14. $\sqrt{25x^4y^{10}}$

15. $-\sqrt[3]{8x^6}$

16. $-\sqrt[4]{16x^8}$

(8.2) *Simplify each expression using the product rule.*

17. $\sqrt{54}$

18. $\sqrt{88}$

19. $\sqrt{150x^3y^6}$

20. $\sqrt{92x^8y^5}$

21. $\sqrt[3]{54}$

22. $\sqrt[3]{88}$

23. $\sqrt[4]{48x^3y^6}$

24. $\sqrt[4]{162x^8y^5}$

Simplify each expression using the quotient rule.

25. $\sqrt{\dfrac{18}{25}}$

26. $\sqrt{\dfrac{75}{64}}$

27. $\sqrt{\dfrac{45x^2y^2}{4x^6}}$

28. $\sqrt{\dfrac{20x^5}{9x^2}}$

29. $\sqrt[4]{\dfrac{9}{16}}$

30. $\sqrt[3]{\dfrac{40}{27}}$

31. $\sqrt[3]{\dfrac{3y^6}{8x^3}}$

32. $\sqrt[4]{\dfrac{5x^6}{81x^8}}$

(8.3) *Add or subtract by combining like radicals.*

33. $3\sqrt[3]{2} + 2\sqrt[3]{3} - 4\sqrt[3]{2}$

34. $5\sqrt{2} + 2\sqrt[3]{2} - 8\sqrt{2}$

35. $\sqrt{6} + 2\sqrt[3]{6} - 4\sqrt[3]{6} + 5\sqrt{6}$

36. $3\sqrt{5} - \sqrt[3]{5} - 2\sqrt{5} + 3\sqrt[3]{5}$

Add or subtract by simplifying each radical and then combining like terms.

37. $\sqrt{28} + \sqrt{63} + \sqrt[3]{56}$

38. $\sqrt{75} + \sqrt{48} - \sqrt[4]{16}$

39. $\sqrt{\dfrac{5}{9}} - \sqrt{\dfrac{5}{36}}$

40. $\sqrt{\dfrac{11}{25}} + \sqrt{\dfrac{11}{16}}$

41. $2\sqrt[3]{125x^3} - 5x\sqrt[3]{8}$

42. $3\sqrt[3]{16x^4} - 2x\sqrt[3]{2x}$

(8.4) *Find the product and simplify if possible.*

43. $3\sqrt{10} \cdot 2\sqrt{5}$

44. $2\sqrt[3]{4} \cdot 5\sqrt[3]{6}$

45. $\sqrt{3}(2\sqrt{6} - 3\sqrt{12})$

46. $4\sqrt{5}(2\sqrt{10} - 5\sqrt{5})$

47. $(\sqrt{3} + 2)(\sqrt{6} - 5)$

48. $(2\sqrt{5} + 1)(4\sqrt{5} - 3)$

Find the quotient and simplify if possible.

49. $\dfrac{\sqrt{96}}{\sqrt{3}}$

50. $\dfrac{\sqrt{160}}{\sqrt{8}}$

51. $\dfrac{\sqrt{15x^6y}}{\sqrt{12x^3y^9}}$

52. $\dfrac{\sqrt{50xy^8}}{\sqrt{72x^7y^3}}$

Rationalize each denominator and simplify.

53. $\sqrt{\dfrac{5}{6}}$

54. $\sqrt{\dfrac{7}{10}}$

55. $\sqrt{\dfrac{3}{2x}}$

56. $\sqrt{\dfrac{6}{5y}}$

57. $\sqrt{\dfrac{7}{20y^2}}$ **58.** $\sqrt{\dfrac{5z}{12x^2}}$ **59.** $\sqrt[3]{\dfrac{7}{9}}$ **60.** $\sqrt[3]{\dfrac{3}{4}}$

61. $\sqrt[3]{\dfrac{3}{2x^2}}$ **62.** $\sqrt[3]{\dfrac{5x}{4y}}$ **63.** $\dfrac{3}{\sqrt{5}-2}$ **64.** $\dfrac{8}{\sqrt{10}-3}$

65. $\dfrac{8}{\sqrt{6}+2}$ **66.** $\dfrac{12}{\sqrt{15}-3}$ **67.** $\dfrac{\sqrt{2}}{4+\sqrt{2}}$ **68.** $\dfrac{\sqrt{3}}{5+\sqrt{3}}$

69. $\dfrac{2\sqrt{3}}{\sqrt{3}-5}$ **70.** $\dfrac{7\sqrt{2}}{\sqrt{2}-4}$

(8.5) *Solve the following radical equations.*

71. $\sqrt{2x}=6$ **72.** $\sqrt{x+3}=4$ **73.** $\sqrt{x}+3=8$
74. $\sqrt{x+8}=3$ **75.** $\sqrt{2x+1}=x-7$ **76.** $\sqrt{3x+1}=x-1$
77. $\sqrt{x+3}+x=9$ **78.** $\sqrt{2x}+x=4$

(8.6) *Use the Pythagorean theorem to find the length of the unknown side.*

79.

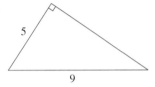

80.

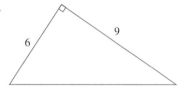

81. Romeo is standing 20 feet away from the wall below Juliet's balcony. Juliet is on the balcony, 12 feet above the ground. Find how far apart Romeo and Juliet are.

82. The diagonal of a rectangle is 10 inches long. If the width of the rectangle is 5 inches, find the length of the rectangle.

Use the distance formula to find the distance between the points.

83. $(6, -2)$ and $(-3, 5)$ **84.** $(2, 8)$ and $(-6, 10)$

Use the formula $r = \sqrt{\dfrac{A}{4\pi}}$, where r = the radius of a sphere, A = the surface area of the sphere, and 3.14 is the approximation to be used for π.

85. Find the radius of a sphere if the area is 72 square inches.

86. To the nearest square inch, find the surface area of a sphere if its radius is 6 inches.

(8.7) *Simplify each of the following expressions.*

87. $16^{1/2}$ **88.** $36^{1/2}$ **89.** $(-8)^{1/3}$ **90.** $(-32)^{1/5}$

91. $-64^{3/2}$ **92.** $-8^{2/3}$ **93.** $\left(\dfrac{16}{81}\right)^{3/4}$ **94.** $\left(\dfrac{9}{25}\right)^{3/2}$

95. $25^{-1/2}$ **96.** $64^{-2/3}$

Write each of the following with fractional exponents and simplify if possible.

97. $\sqrt{a^5}$ **98.** $\sqrt[5]{a^3}$ **99.** $\sqrt[6]{x^{15}}$ **100.** $\sqrt[4]{x^{12}}$

Simplify each expression using positive exponents only.

101. $8^{1/3} \cdot 8^{4/3}$ **102.** $4^{3/2} \cdot 4^{1/2}$ **103.** $\dfrac{3^{1/6}}{3^{5/6}}$

104. $\dfrac{2^{1/4}}{2^{-3/5}}$ **105.** $(x^{-1/3})^6$ **106.** $\left(\dfrac{x^{1/2}}{y^{1/3}}\right)^2$

CHAPTER 8 TEST

Simplify the following. Indicate if the expression is not a real number.

1. $\sqrt{16}$

2. $\sqrt[3]{125}$

3. $16^{3/4}$

4. $\left(\dfrac{9}{16}\right)^{1/2}$

5. $\sqrt[4]{-81}$

6. $27^{-2/3}$

Simplify each radical expression. Assume that variables represent positive numbers only.

7. $\sqrt{54}$

8. $\sqrt{92}$

9. $\sqrt{3x^6}$

10. $\sqrt{8x^4y^7}$

11. $\sqrt{9x^9}$

12. $\sqrt[3]{40}$

13. $\sqrt[3]{8x^6y^{10}}$

14. $\sqrt{12} - 2\sqrt{75}$

15. $\sqrt{2x^2} + \sqrt[3]{54} - x\sqrt{18}$

16. $\sqrt{\dfrac{5}{16}}$

17. $\sqrt[3]{\dfrac{2x^3}{27}}$

18. $\sqrt{\dfrac{2}{3}}$

19. $3\sqrt{8x}$

20. $\sqrt[3]{\dfrac{5}{9}}$

21. $\sqrt[3]{\dfrac{3}{4x^2}}$

22. $\dfrac{8}{\sqrt{6} + 2}$

23. $\dfrac{2\sqrt{3}}{\sqrt{3} - 3}$

Solve each of the following radical equations.

24. $\sqrt{x} + 8 = 11$

25. $\sqrt{3x - 6} = \sqrt{x + 4}$

26. $\sqrt{2x - 2} = x - 5$

27. Find the length of the unknown leg of a right triangle if the other leg is 8 inches long and the hypothenuse is 12 inches long.

28. Find the distance between $(-3, 6)$ and $(-2, 8)$.

Simplify each expression using positive exponents only.

29. $16^{-3/4} \cdot 16^{-1/4}$

30. $(x^{2/3})^5$

CHAPTER 8 CUMULATIVE REVIEW

1. Write a mathematical expression that represents each of the following phrases. Let x represent the unknown number:
   a. The sum of a number and 3
   b. The product of 3 and a number
   c. Twice a number
   d. 10 decreased by a number
   e. 7 more than 5 times a number

2. Simplify each expression:
   a. $\dfrac{(-12)(-3) + 4}{-7 - (-2)}$ **b.** $\dfrac{2(-3)^2 - 20}{-5 + 4}$

3. Solve $-3x = 33$ for x.

4. Solve $3(x - 4) = 3x - 12$.

5. When Hiroto empties the coke machine he finds 130 coins with a total value of $25.75. If the machine accepts only quarters and dimes, find the number of quarters and dimes Hiroto emptied out of the machine.

6. Solve $2(x - 3) - 5 \le 3(x + 2) - 18$ and graph the solution.

7. Simplify $(2x^2)(-3x^5)$.

8. Combine like terms.
   a. $-3x + 7x$ **b.** $11x^2 + 5 + 2x^2 - 7$

9. Multiply $(5x - 7)(x - 2)$.

10. Factor $8x^2 - 22x + 5$.

11. Factor $4m^2 - 4m + 1$.

12. Solve $(5x - 1)(2x^2 + 15x + 18) = 0$.

13. Simplify $\dfrac{x^2 + 4x + 4}{2x + 4}$.

14. Divide $\dfrac{(x - 1)(x + 2)}{10} \div \dfrac{(2x + 4)}{5}$.

15. Add $1 + \dfrac{m}{m + 1}$.

16. a. Solve for x: $\dfrac{x}{4} + 2x = 9$.

   b. Add $\dfrac{x}{4} + 2x$.

17. Graph $4x = 3y - 9$.

18. Solve the system $\begin{cases} 7x - 3y = -14 \\ -3x + y = 6 \end{cases}$

19. Evaluate the following expressions.

   a. $\sqrt[4]{16}$ **b.** $\sqrt[5]{-32}$ **c.** $-\sqrt[3]{8}$ **d.** $\sqrt[4]{-81}$

20. Simplify each expression. Assume that variables represent positive numbers only.

   a. $\sqrt{x^5}$ **b.** $\sqrt{8y^2}$ **c.** $\sqrt{\dfrac{45}{x^6}}$

21. Add or subtract by first simplifying each radical and then combining like radicals.

   a. $\sqrt{50} + \sqrt{8}$

   b. $7\sqrt{12} - \sqrt{75}$

   c. $\sqrt{25} - \sqrt{27} - 2\sqrt{18} - \sqrt{9}$

   d. $2\sqrt[3]{27} - \sqrt[3]{54}$

22. Simplify $\dfrac{12 - \sqrt{18}}{9}$.

23. Solve the equation $\sqrt{x} = \sqrt{5x - 2}$.

9.1 Solving Quadratic Equations by the Square Root Method

9.2 Solving Quadratic Equations by Completing the Square

9.3 Solving Quadratic Equations by the Quadratic Formula

9.4 Summary of Methods for Solving Quadratic Equations

9.5 Complex Solutions of Quadratic Equations

9.6 Graphing Quadratic Equations

CHAPTER 9

Solving Quadratic Equations

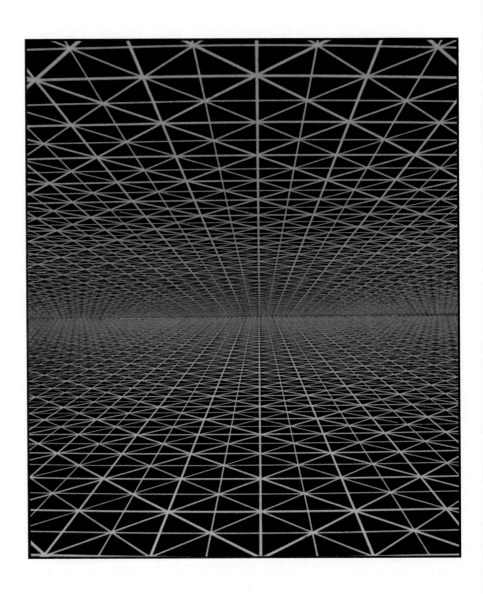

Amidst the disorder of these right triangles is a remarkable pattern, bridging geometry and the algebra of quadratic equations. (See Critical Thinking, page 419.)

INTRODUCTION

An important part of the study of algebra is learning to use methods for solving equations. In Chapter 2, we presented techniques for solving linear equations in one variable. In Chapter 4, we solved quadratic equations in one variable by factoring the quadratic expressions. We now present other methods for solving quadratic equations in one variable.

9.1
Solving Quadratic Equations by the Square Root Method

OBJECTIVES

Tape 30

1   Review factoring to solve quadratic equations.

2   Use the square root property to solve quadratic equations.

1   Recall that a quadratic equation is an equation that can be written in the form

$$ax^2 + bx + c = 0$$

where a, b, and c are real numbers and $a \neq 0$.

To solve quadratic equations by factoring, use the **zero factor theorem:** If the product of two numbers is zero, then at least one of the two numbers is zero. Example 1 reviews the process of solving quadratic equations by factoring.

EXAMPLE 1   Solve $x^2 - 4 = 0$.

Solution:

$$x^2 - 4 = 0$$

$$(x + 2)(x - 2) = 0 \qquad \text{Factor.}$$

$$x + 2 = 0 \quad \text{or} \quad x - 2 = 0 \qquad \text{Apply the zero factor theorem.}$$

$$x = -2 \quad \text{or} \quad x = 2 \qquad \text{Solve each equation.} \quad \blacksquare$$

2   Consider solving $x^2 - 4 = 0$ another way. First, add 4 to both sides of the equation.

$$x^2 - 4 = 0$$

$$x^2 = 4 \qquad \text{Add 4 to both sides.}$$

The variable x must be a number whose square is 4. Therefore, $x = 2$ or $x = -2$, written compactly as $x = \pm 2$. This reasoning is an example of the square root property.

> **Square Root Property**
>
> If $X^2 = a$ for $a \geq 0$, then $X = \pm\sqrt{a}$.

389

EXAMPLE 2 Use the square root property to solve $x^2 - 9 = 0$.

Solution:
$$x^2 - 9 = 0$$
$$x^2 = 9 \qquad \text{Add 9 to both sides.}$$

Next, apply the square root property.
$$x = \pm\sqrt{9} = \pm 3$$

Check to verify these solutions in the original equation. ∎

EXAMPLE 3 Use the square root property to solve $2x^2 = 7$.

Solution: Before we use the square root property, divide both sides by 2.
$$2x^2 = 7$$
$$x^2 = \frac{7}{2} \qquad \text{Divide both sides by 2.}$$
$$x = \pm\sqrt{\frac{7}{2}} \qquad \text{Use the square root property.}$$

Next, we rationalize the denominators of our solutions by multiplying numerator and denominator by $\sqrt{2}$.
$$x = \pm\sqrt{\frac{7}{2}}$$
$$x = \pm\frac{\sqrt{7}}{\sqrt{2}} \qquad\qquad \text{Apply the quotient rule.}$$
$$x = \pm\frac{\sqrt{7} \cdot \sqrt{2}}{\sqrt{2} \cdot \sqrt{2}} = \pm\frac{\sqrt{14}}{2}$$

Remember to check both solutions. ∎

EXAMPLE 4 Use the square root property to solve $(x - 3)^2 = 16$.

Solution: To use the square root property, consider X to be $x - 3$ and a to be 16.
$$(x - 3)^2 = 16$$
$$(x - 3) = \pm\sqrt{16} \qquad \text{Use the square root property.}$$
$$x - 3 = \pm 4 \qquad\qquad \text{Simplify.}$$

Now solve for x by adding 3 to both sides.
$$x = 3 \pm 4$$
$$x = 3 + 4 \quad \text{or} \quad x = 3 - 4$$
$$x = 7 \quad \text{or} \quad x = -1 \qquad ∎$$

EXAMPLE 5 Use the square root property to solve $(x + 1)^2 = 8$.

Solution:
$$(x + 1)^2 = 8$$
$$x + 1 = \pm\sqrt{8} \qquad \text{Apply the square root property.}$$

$$x + 1 = \pm 2\sqrt{2} \qquad \text{Simplify the radical.}$$
$$x = -1 \pm 2\sqrt{2} \qquad \text{Solve for } x.$$

The solutions are both $-1 + 2\sqrt{2}$ and $-1 - 2\sqrt{2}$.   ■

EXAMPLE 6   Use the square root property to solve $(x - 1)^2 = -2$.

Solution: This equation has no real solution because there is no real number whose square is -2.

■

EXAMPLE 7   Use the square root property to solve $(5x - 2)^2 = 10$.

Solution: $(5x - 2)^2 = 10$

$$5x - 2 = \pm\sqrt{10} \qquad\qquad \text{Apply the square root property.}$$
$$5x = 2 \pm \sqrt{10} \qquad\qquad \text{Add 2.}$$
$$x = \frac{2 \pm \sqrt{10}}{5} \qquad\qquad \text{Divide by 5.}$$
$$x = \frac{2 + \sqrt{10}}{5} \quad \text{or} \quad x = \frac{2 - \sqrt{10}}{5} \quad ■$$

EXERCISE SET 9.1

Solve each equation by factoring. See Example 1.

1. $k^2 - 9 = 0$ **2.** $k^2 - 49 = 0$ **3.** $m^2 + 2m = 15$ **4.** $m^2 + 6m = 7$

5. $2x^2 - 81 = 0$ **6.** $3p^4 - 9p^3 = 0$ **7.** $4a^2 - 36 = 0$ **8.** $7a^2 - 175 = 0$

Use the square root property to solve each quadratic equation. See Examples 2 and 3.

9. $x^2 = 64$ **10.** $x^2 = 121$ **11.** $p^2 = \dfrac{1}{49}$ **12.** $p^2 = \dfrac{1}{16}$

13. $y^2 = -36$ **14.** $y^2 = -25$ **15.** $2x^2 = 50$ **16.** $5x^2 = 20$

17. $3x^2 = 4$ **18.** $5x^2 = 9$

Use the square root property to solve each quadratic equation. See Examples 4 and 5.

19. $(x - 5)^2 = 49$ **20.** $(x + 2)^2 = 25$ **21.** $(x + 2)^2 = 7$ **22.** $(x - 7)^2 = 2$

23. $\left(m - \dfrac{1}{2}\right)^2 = \dfrac{1}{4}$ **24.** $\left(m + \dfrac{1}{3}\right)^2 = \dfrac{1}{9}$ **25.** $(p + 2)^2 = 121$ **26.** $(p - 7)^2 = 16$

Use the square root property to solve each quadratic equation. See Examples 6 and 7.

27. $(3y + 2)^2 = 100$ **28.** $(4y - 3)^2 = 81$ **29.** $(z - 4)^2 = -9$

30. $(z + 7)^2 = -20$ **31.** $(2x - 11)^2 = 50$ **32.** $(3x - 17)^2 = 28$

Solve each quadratic equation by using the square root property.

33. $q^2 = 100$ **34.** $q^2 = 25$ **35.** $(x - 13)^2 = 16$

36. $(x + 10)^2 = 25$ **37.** $z^2 = 12$ **38.** $z^2 = 18$

39. $(x + 5)^2 = 10$

40. $(x - 2)^2 = 13$

41. $m^2 = -10$

42. $m^2 = -25$

43. $2y^2 = 11$

44. $3y^2 = 13$

45. $(2p - 5)^2 = 121$

46. $(3p - 1)^2 = 4$

47. $(3x - 1)^2 = 7$

48. $(5x + 3)^2 = 3$

49. $(3x - 7)^2 = 32$

50. $(5x - 11)^2 = 54$

51. The area of a square room is 225 square feet. Find the dimensions of the room.

52. The area of a circle is 36π square inches. Find the radius of the circle.

53. An isosceles right triangle has legs of equal length. If the hypotenuse is 20 centimeters long, find the length of the legs.

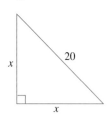

54. A 27-inch-square TV is advertised in the local paper. If 27 inches is the measure of the diagonal of the picture tube, find the measure of the side of the picture tube.

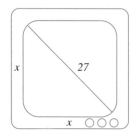

Writing in Mathematics

55. Explain why there is no real solution of the equation $x^2 = -9$.

Skill Review

Find the slope of the line through the following pairs of points. See Section 6.3.

56. $(2, 1), (0, -3)$

57. $(5, 8), (-2, -4)$

58. $(-7, 1), (-6, 2)$

59. $(4, 0), (-6, -6)$

Solve the following systems of equations by addition. See Section 7.3.

60. $x + y = 40$
$x - y = 20$

61. $x + y = 7$
$-x + y = 7$

9.2
Solving Quadratic Equations by Completing the Square

Tape 30

OBJECTIVES

1 Find perfect square trinomials.

2 Solve quadratic equations by completing the square.

1   In the last section, we solved equations such as

$$(x + 1)^2 = 8$$

and

$$(5x - 2)^2 = 10$$

Thus, if we can write a quadratic equation in a form similar to these examples, we can then solve using the square root property. Notice that the left side of each equation is a quantity squared and each right side is a constant. For example, to solve $x^2 + 2x - 4 = 0$, add 4 to both sides and get

$$x^2 + 2x = 4$$

Next, notice that if we add 1 to both sides of the equation the left side is a perfect square trinomial that can be factored.

$$x^2 + 2x + 1 = 4 + 1$$
$$(x + 1)^2 = 5$$

Then solve this equation as we did in the previous section by using the square root property.

$$(x + 1)^2 = 5$$
$$x + 1 = \pm\sqrt{5}$$
$$x = -1 \pm \sqrt{5}$$

The process of adding a number to $x^2 + 2x$ to form a perfect square trinomial is called **completing the square.** In general, if $x^2 + bx$ is a binomial, then adding $\left(\dfrac{b}{2}\right)^2$ gives a perfect square trinomial.

To complete the square of $x^2 + bx$, **add the square of half of the coefficient of x.**

Here are more examples of completing the square.

EXAMPLE 1   Complete the square for each expression and then factor the resulting perfect square trinomial.
a. $x^2 + 10x$ **b.** $m^2 - 6m$ **c.** $x^2 + x$

Solution: **a.** The coefficient of the x-term is 10. Half of 10 is 5, and $5^2 = 25$. Add 25.

$$x^2 + 10x + 25 = (x + 5)^2$$

b. Half the coefficient of m is -3, and $(-3)^2$ is 9. Add 9.

$$m^2 - 6m + 9 = (m - 3)^2$$

c. Half the coefficient of x is $\dfrac{1}{2}$ and $\left(\dfrac{1}{2}\right)^2 = \dfrac{1}{4}$. Add $\dfrac{1}{4}$.

$$x^2 + x + \frac{1}{4} = \left(x + \frac{1}{2}\right)^2 \qquad \blacksquare$$

2 By completing the square, a quadratic equation can be solved using the square root property.

EXAMPLE 2 Solve the quadratic equation $x^2 + 6x + 3 = 0$ by completing the square.

Solution: Isolate the variable terms by subtracting 3 from both sides of the equation.

$$x^2 + 6x + 3 = 0$$

$$x^2 + 6x = -3$$

Next, add the square of half the coefficient of the x-term to **both sides** of the equation so that the left side becomes a perfect square trinomial. The coefficient of x is 6, and half of 6 is 3. Add 3^2 or 9 to both sides.

$$x^2 + 6x + 9 = -3 + 9 \qquad \text{Complete the square.}$$

$$(x + 3)^2 = 6 \qquad \text{Factor the trinomial } x^2 + 6x + 9.$$

$$x + 3 = \pm\sqrt{6} \qquad \text{Use the square root property.}$$

$$x = -3 \pm \sqrt{6} \qquad \text{Subtract 3 from both sides.} \quad \blacksquare$$

HELPFUL HINT

Remember, when solving a quadratic equation by completing the square, to add the number that completes the square to **both sides of the equation.**

EXAMPLE 3 Solve the quadratic equation $y^2 - 10y = -14$ by completing the square.

Solution: The variable terms are already isolated on one side of the equation. The coefficient of y is -10. Half of -10 is -5, and $(-5)^2 = 25$. Add 25 to both sides.

$$y^2 - 10y = -14$$

$$y^2 - 10y + 25 = -14 + 25$$

$$(y - 5)^2 = 11 \qquad \text{Factor the trinomial and simplify} \\ -14 + 25.$$

$$y - 5 = \pm\sqrt{11} \qquad \text{Apply the square root property.}$$

$$y = 5 \pm \sqrt{11} \qquad \text{Add 5 to both sides.} \quad \blacksquare$$

When the coefficient of the squared variable is not 1, first divide both sides of the equation by the coefficient of the squared variable so that the coefficient is 1. Then complete the square.

EXAMPLE 4 Solve the quadratic equation $4x^2 - 8x - 5 = 0$ by completing the square.

Solution: $$4x^2 - 8x - 5 = 0$$

$$x^2 - 2x - \frac{5}{4} = 0 \qquad \text{Divide by 4.}$$

$$x^2 - 2x = \frac{5}{4} \qquad \text{Isolate variable terms.}$$

The coefficient of x is -2. Half of -2 is -1, and $(-1)^2 = 1$. Add 1 to both sides.

$$x^2 - 2x \boxed{+ 1} = \frac{5}{4} \boxed{+ 1}$$

$$(x - 1)^2 = \frac{9}{4} \qquad \text{Factor } x^2 - 2x + 1 \text{ and simplify } \frac{5}{4} + 1.$$

$$x - 1 = \pm\sqrt{\frac{9}{4}} \qquad \text{Use the square root property.}$$

$$x = 1 \pm \frac{3}{2} \qquad \text{Add 1 to both sides and simplify the radical.}$$

$$x = 1 + \frac{3}{2} = \frac{5}{2} \text{ or } x = 1 - \frac{3}{2} = -\frac{1}{2}. \qquad \blacksquare$$

The following are steps that may be used to solve a quadratic equation in x by completing the square.

To Solve a Quadratic Equation in x by Completing the Square

Step 1   If the coefficient of x^2 is 1, go to step 2. If not, divide both sides of the equation by the coefficient of x^2.

Step 2   Isolate all terms with variables on one side of the equation.

Step 3   Complete the square for the resulting binomial expression by adding the square of half of the coefficient of x to both sides of the equation.

Step 4   Factor the resulting perfect square trinomial.

Step 5   Use the square root property to solve the equation.

EXAMPLE 5   Solve the quadratic equation $2y^2 + 6y = -7$ by completing the square.

Solution: The coefficient of y^2 is not 1. Divide both sides by 2, the coefficient of y^2.

$$2y^2 + 6y = -7$$

$$y^2 + 3y = -\frac{7}{2} \qquad \text{Divide by 2.}$$

$$y^2 + 3y \boxed{+ \frac{9}{4}} = -\frac{7}{2} \boxed{+ \frac{9}{4}} \qquad \text{Add } \left(\frac{3}{2}\right)^2 \text{ or } \frac{9}{4} \text{ to both sides.}$$

$$\left(y + \frac{3}{2}\right)^2 = -\frac{5}{4} \qquad \text{Factor the left side and simplify the right.}$$

There is no real solution to this equation, since the square of a real number cannot be negative.   ∎

EXAMPLE 6   Solve the quadratic equation $2x^2 = 10x + 1$ by completing the square.

Solution: First divide both sides of the equation by 2, the coefficient of x^2.

$$2x^2 = 10x + 1$$

$$x^2 = 5x + \frac{1}{2} \qquad \text{Divide by 2.}$$

Next, isolate the variable terms by subtracting $5x$ from both sides.

$$x^2 - 5x = \frac{1}{2}$$

$$x^2 - 5x \boxed{+ \frac{25}{4}} = \frac{1}{2} \boxed{+ \frac{25}{4}} \qquad \text{Add } \left(\frac{-5}{2}\right)^2 \text{ or } \frac{25}{4} \text{ to both sides.}$$

$$\left(x - \frac{5}{2}\right)^2 = \frac{27}{4} \qquad \text{Factor the left side and simplify the right.}$$

$$x - \frac{5}{2} = \pm\sqrt{\frac{27}{4}} \qquad \text{Apply the square root property.}$$

$$x - \frac{5}{2} = \pm\frac{3\sqrt{3}}{2} \qquad \text{Simplify.}$$

$$x = \frac{5}{2} \pm \frac{3\sqrt{3}}{2} = \frac{5 \pm 3\sqrt{3}}{2} \qquad \blacksquare$$

MENTAL MATH

Determine the number to add to make each expression a perfect square trinomial. See Example 1.

1. $p^2 + 8p$ **2.** $p^2 + 6p$ **3.** $x^2 + 20x$

4. $x^2 + 18x$ **5.** $y^2 + 14y$ **6.** $y^2 + 2y$

EXERCISE SET 9.2

Complete the square for each expression and then factor the resulting perfect square trinomial. See Example 1.

1. $x^2 - 4x$ **2.** $x^2 - 6x$ **3.** $k^2 - 12k$ **4.** $k^2 - 16k$

5. $x^2 - 3x$ **6.** $x^2 - 5x$ **7.** $m^2 - m$ **8.** $y^2 + y$

Solve each quadratic equation by completing the square. See Examples 2 and 3.

9. $x^2 - 6x = 0$ **10.** $y^2 + 4y = 0$ **11.** $z^2 + 8z = -12$

12. $x^2 - 10x = -24$ **13.** $x^2 + 2x - 5 = 0$ **14.** $z^2 + 6z - 9 = 0$

Solve each quadratic equation by completing the square. See Examples 4 through 6.

15. $4x^2 - 24x = 13$ **16.** $2x^2 + 8x = 10$ **17.** $5x^2 + 10x + 6 = 0$

18. $3x^2 - 12x + 14 = 0$ **19.** $2x^2 - 6x - 5 = 0$ **20.** $4x^2 + 20x - 3 = 0$

Solve each quadratic equation by completing the square.

21. $x^2 + 6x - 25 = 0$ **22.** $x^2 - 6x + 7 = 0$ **23.** $z^2 + 5z = 7$

24. $z^2 - 7z = 5$

25. $x^2 - 2x - 1 = 0$

26. $x^2 - 4x + 2 = 0$

27. $y^2 + 5y + 4 = 0$

28. $y^2 - 5y + 6 = 0$

29. $3x^2 - 6x = 24$

30. $2x^2 + 18x = -40$

31. $2y^2 + 8y + 5 = 0$

32. $3z^2 + 6z + 4 = 0$

33. $2y^2 - 3y + 1 = 0$

34. $2y^2 - y - 1 = 0$

35. $3y^2 - 2y - 4 = 0$

36. $4y^2 - 2y - 3 = 0$

37. $y^2 = 5y + 14$

38. $y^2 = 3y + 10$

39. $m(m + 3) = 18$

40. $m(m - 3) = 18$

Writing in Mathematics

41. Describe how to find the number to add to $x^2 - 7x$ to make a perfect square trinomial.

Skill Review

Simplify each expression. See Section 8.3.

42. $\dfrac{3}{4} - \sqrt{\dfrac{25}{16}}$

43. $\dfrac{3}{5} + \sqrt{\dfrac{16}{25}}$

44. $\dfrac{1}{2} - \sqrt{\dfrac{9}{4}}$

45. $\dfrac{9}{10} - \sqrt{\dfrac{49}{100}}$

Simplify each expression. See Section 8.5.

46. $\dfrac{6 + 4\sqrt{5}}{2}$

47. $\dfrac{10 - 20\sqrt{3}}{2}$

48. $\dfrac{3 - 9\sqrt{2}}{6}$

49. $\dfrac{12 - 8\sqrt{7}}{16}$

9.3
Solving Quadratic Equations by the Quadratic Formula

OBJECTIVE

1   Use the quadratic formula to solve quadratic equations.

Tape 31

1   We can use the completing the square method to develop a formula to find solutions of any quadratic equation. We develop and use the **quadratic formula** in this section.

Recall that a quadratic equation in **standard form** is

$$ax^2 + bx + c = 0, \text{ providing } a \neq 0$$

To develop and use the quadratic formula, we need to practice identifying the values of a, b, and c in a quadratic equation.

Quadratic Equations in Standard Form

$5x^2 - 6x + 2 = 0$	$a = 5, b = -6, c = 2$
$4y^2 - 9 = 0$	$a = 4, b = 0, c = -9$
$x^2 + x = 0$	$a = 1, b = 1, c = 0$
$\sqrt{2}x^2 + \sqrt{5}x + \sqrt{3} = 0$	$a = \sqrt{2}, b = \sqrt{5}, c = \sqrt{3}$

To derive the quadratic formula, we complete the square for the quadratic equation

$$ax^2 + bx + c = 0$$

First, divide both sides of the equation by the coefficient of x^2 and then isolate the variable terms.

$$x^2 + \frac{b}{a}x + \frac{c}{a} = 0 \qquad \text{Divide by } a; \text{ recall that } a \text{ cannot be 0.}$$

$$x^2 + \frac{b}{a}x = -\frac{c}{a} \qquad \text{Isolate the variable terms.}$$

The coefficient of x is $\frac{b}{a}$. Half of $\frac{b}{a}$ is $\frac{b}{2a}$ and $\left(\frac{b}{2a}\right)^2 = \frac{b^2}{4a^2}$. Add $\frac{b^2}{4a^2}$ to both sides of the equation.

$$x^2 + \frac{b}{a}x + \boxed{\frac{b^2}{4a^2}} = -\frac{c}{a} + \boxed{\frac{b^2}{4a^2}} \qquad \text{Add } \frac{b^2}{4a^2} \text{ to both sides.}$$

$$\left(x + \frac{b}{2a}\right)^2 = -\frac{c}{a} + \frac{b^2}{4a^2} \qquad \text{Factor the left side.}$$

$$\left(x + \frac{b}{2a}\right)^2 = -\frac{4ac}{4a^2} + \frac{b^2}{4a^2} \qquad \begin{array}{l}\text{Multiply } -\frac{c}{a} \text{ by } \frac{4a}{4a} \text{ so that both} \\ \text{terms on the right side will have} \\ \text{a common denominator.}\end{array}$$

$$\left(x + \frac{b}{2a}\right)^2 = \frac{b^2 - 4ac}{4a^2} \qquad \text{Simplify the right side.}$$

Now use the square root property.

$$x + \frac{b}{2a} = \pm\sqrt{\frac{b^2 - 4ac}{4a^2}} \qquad \text{Apply the square root property.}$$

$$x + \frac{b}{2a} = \frac{\pm\sqrt{b^2 - 4ac}}{2a} \qquad \text{Simplify the radical.}$$

$$x = -\frac{b}{2a} \pm \frac{\sqrt{b^2 - 4ac}}{2a} \qquad \text{Subtract } \frac{b}{2a} \text{ from both sides.}$$

$$= \frac{-b \pm \sqrt{b^2 - 4ac}}{2a} \qquad \text{Simplify.}$$

This final equation is called the **quadratic formula** and gives the solutions of any quadratic equation.

Quadratic Formula

If a, b, and c are real numbers and $a \neq 0$, a quadratic equation written in the form $ax^2 + bx + c = 0$ has solutions

$$x = \frac{-b \pm \sqrt{b^2 - 4ac}}{2a}$$

EXAMPLE 1 Use the quadratic formula to solve $3x^2 + x - 3 = 0$.

Solution: This equation is in standard form with $a = 3$, $b = 1$, and $c = -3$. By the quadratic formula,

$$x = \frac{-1 \pm \sqrt{1^2 - 4 \cdot 3 \cdot (-3)}}{2 \cdot 3} \qquad \text{Let } a = 3, b = 1, \text{ and } c = -3.$$

$$= \frac{-1 \pm \sqrt{1 + 36}}{6} \qquad \text{Simplify.}$$

$$= \frac{-1 \pm \sqrt{37}}{6}$$

Remember that the $\pm$ is shorthand for

$$x = \frac{-1 + \sqrt{37}}{6} \quad \text{or} \quad x = \frac{-1 - \sqrt{37}}{6} \qquad \blacksquare$$

EXAMPLE 2 Use the quadratic formula to solve $2x^2 - 9x = 5$.

Solution: First, write the equation in standard form by subtracting 5 from both sides.

$$2x^2 - 9x = 5$$
$$2x^2 - 9x - 5 = 0$$

Next, $a = 2$, $b = -9$, and $c = -5$. Substitute these values into the quadratic formula.

$$x = \frac{-(-9) \pm \sqrt{(-9)^2 - 4 \cdot 2 \cdot (-5)}}{2 \cdot 2} \qquad \text{Substitute in the formula.}$$

$$= \frac{9 \pm \sqrt{81 + 40}}{4} \qquad \text{Simplify.}$$

$$= \frac{9 \pm \sqrt{121}}{4} = \frac{9 \pm 11}{4}$$

Then, $x = \dfrac{9 - 11}{4} = -\dfrac{1}{2}$ or $x = \dfrac{9 + 11}{4} = 5$ $\quad \blacksquare$

Check by substituting $-\dfrac{1}{2}$ and 5 into the original equation. In this example, the radicand of $\sqrt{121}$ is a perfect square, so the square root is rational.

To Solve a Quadratic Equation Using the Quadratic Formula

Step 1 Write the quadratic equation in standard form: $ax^2 + bx + c = 0$. Clear the equation of fractions to simplify calculations.

Step 2 Identify a, b, and c.

Step 3 Replace a, b, and c in the quadratic formula by known values, and simplify.

EXAMPLE 3   Use the quadratic formula to solve $7x^2 = 1$.

Solution: Write the equation in standard form by subtracting 1 from both sides.

$$7x^2 = 1$$
$$7x^2 - 1 = 0$$

Next, replace a, b, and c with values: $a = 7$, $b = 0$, $c = -1$.

$$x = \frac{0 \pm \sqrt{0^2 - 4 \cdot 7 \cdot (-1)}}{2 \cdot 7} \qquad \text{Substitute in the formula.}$$

$$= \frac{\pm\sqrt{28}}{14} \qquad \text{Simplify.}$$

$$= \frac{\pm 2\sqrt{7}}{14}$$

$$= \pm\frac{\sqrt{7}}{7}$$

The solutions are $\pm\dfrac{\sqrt{7}}{7}$.   ■

EXAMPLE 4   Use the quadratic formula to solve $x^2 = -x - 1$.

Solution: First, write the equation in standard form.

$$x^2 + x + 1 = 0$$

Next, replace a, b, and c in the quadratic formula by $a = 1$, $b = 1$, and $c = 1$.

$$x = \frac{-1 \pm \sqrt{1^2 - 4 \cdot 1 \cdot 1}}{2 \cdot 1} \qquad \text{Substitute in the formula.}$$

$$= \frac{-1 \pm \sqrt{-3}}{2} \qquad \text{Simplify.}$$

There is no real number solution, because the radicand is negative.   ■

EXAMPLE 5   Use the quadratic formula to solve $\dfrac{1}{2}x^2 - x = 2$.

Solution: The calculations are simpler if we clear the equation of fractions first by multiplying both sides by the LCD 2.

$$\frac{1}{2}x^2 - x = 2$$

$$x^2 - 2x = 4 \qquad \text{Multiply both sides by 2.}$$

Write the equation in standard form by subtracting 4 from both sides.

$$x^2 - 2x - 4 = 0 \qquad \text{Write in standard form.}$$

Here, $a = 1$, $b = -2$, and $c = -4$. Substitute these values into the quadratic formula.

$$x = \frac{-(-2) \pm \sqrt{(-2)^2 - 4 \cdot 1 \cdot (-4)}}{2 \cdot 1}$$

$$= \frac{2 \pm \sqrt{20}}{2} = \frac{2 \pm 2\sqrt{5}}{2} \qquad \text{Simplify.}$$

$$= \frac{2\,(1 \pm \sqrt{5})}{2} = 1 \pm \sqrt{5} \qquad \text{Factor and simplify.} \qquad \blacksquare$$

HELPFUL HINT

When simplifying expressions such as

$$\frac{3 \pm 6\sqrt{2}}{6}$$

first factor out a common factor from the terms of the numerator and then simplify.

$$\frac{3 \pm 6\sqrt{2}}{6} = \frac{3\,(1 \pm 2\sqrt{2})}{2 \cdot 3} = \frac{1 \pm 2\sqrt{2}}{2}$$

MENTAL MATH

Identify the value of a, b, and c in each quadratic equation.

1. $2x^2 + 5x + 3 = 0$

2. $5x^2 - 7x + 1 = 0$

3. $10x^2 - 13x - 2 = 0$

4. $x^2 + 3x - 7 = 0$

5. $x^2 - 6 = 0$

6. $9x^2 - 4 = 0$

EXERCISE SET 9.3

Simplify the following.

1. $\dfrac{-1 \pm \sqrt{1^2 - 4(1)(-2)}}{2(1)}$

2. $\dfrac{-(-5) \pm \sqrt{(-5)^2 - 4(2)(3)}}{2(2)}$

3. $\dfrac{-5 \pm \sqrt{5^2 - 4(1)(2)}}{2(1)}$

4. $\dfrac{-7 \pm \sqrt{7^2 - 4(2)(1)}}{2(2)}$

5. $\dfrac{-(-4) \pm \sqrt{(-4)^2 - 4(2)(1)}}{2(2)}$

6. $\dfrac{-6 \pm \sqrt{6^2 - 4(3)(1)}}{2(3)}$

Use the quadratic formula to solve each quadratic equation. See Examples 1 and 2.

7. $x^2 - 3x + 2 = 0$

8. $x^2 - 5x - 6 = 0$

9. $3k^2 + 7k + 1 = 0$

10. $7k^2 + 3k - 1 = 0$

11. $49x^2 - 4 = 0$

12. $25x^2 - 15 = 0$

13. $5z^2 - 4z + 3 = 0$

14. $3z^2 + 2x + 1 = 0$

Use the quadratic formula to solve each quadratic equation. See Examples 3 and 4.

15. $y^2 = 7y + 30$

16. $y^2 = 5y + 36$

17. $2x^2 = 10$

18. $5x^2 = 15$

19. $m^2 - 12 = m$

20. $m^2 - 14 = 5m$

21. $3 - x^2 = 4x$

22. $10 - x^2 = 2x$

Use the quadratic formula to solve each quadratic equation. See Example 5.

23. $3p^2 - \dfrac{2}{3}p + 1 = 0$

24. $\dfrac{5}{2}p^2 - p + \dfrac{1}{2} = 0$

25. $\dfrac{m^2}{2} = m + \dfrac{1}{2}$

26. $\dfrac{m^2}{2} = 3m - 1$

27. $4p^2 + \dfrac{3}{2} = -5p$

28. $4p^2 + \dfrac{3}{2} = 5p$

Use the quadratic formula to solve each quadratic equation.

29. $2a^2 - 7a + 3 = 0$

30. $3a^2 - 7a + 2 = 0$

31. $x^2 - 5x - 2 = 0$

32. $x^2 - 2x - 5 = 0$

33. $3x^2 - x - 14 = 0$

34. $5x^2 - 13x - 6 = 0$

35. $6x^2 + 9x = 2$

36. $3x^2 - 9x = 8$

37. $7p^2 + 2 = 8p$

38. $11p^2 + 2 = 10p$

39. $a^2 - 6a + 2 = 0$

40. $a^2 - 10a + 19 = 0$

41. $2x^2 - 6x + 3 = 0$

42. $5x^2 - 8x + 2 = 0$

43. $3x^2 = 1 - 2x$

44. $5y^2 = 4 - x$

45. $20y^2 = 3 - 11y$

46. $2z^2 = z + 3$

47. $x^2 + x + 1 = 0$

48. $k^2 + 2k + 5 = 0$

49. $4y^2 = 6y + 1$

50. $6z^2 + 3z + 2 = 0$

51. $5x^2 = \dfrac{7}{2}x + 1$

52. $2x^2 = \dfrac{5}{2}x + \dfrac{7}{2}$

53. $28x^2 + 5x + \dfrac{11}{4} = 0$

54. $\dfrac{2}{3}x^2 - 2x - \dfrac{2}{3} = 0$

55. $5z^2 - 2z = \dfrac{1}{5}$

56. $9z^2 + 12z = -1$

Writing in Mathematics

57. Explain how the quadratic formula is derived and why it
it useful.

Skill Review

Solve the following linear equations. See Section 2.4.

58. $\dfrac{7x}{2} = 3$

59. $\dfrac{5x}{3} = 1$

60. $\dfrac{5}{7}x - \dfrac{2}{3} = 0$

61. $\dfrac{6}{11}x + \dfrac{1}{5} = 0$

62. $\dfrac{3}{4}z + 3 = 0$

63. $\dfrac{5}{2}z + 10 = 0$

9.4
Summary of Methods for Solving Quadratic Equations

OBJECTIVE **1**   Review methods for solving quadratic equations.

Tape 31

1   An important skill in mathematics is learning when to use one technique in favor of another. We now practice this skill by deciding which method to use when solving quadratic equations. Although both the quadratic formula and completing the square can be used to solve any quadratic equation, the quadratic formula is usually less tedious and thus preferred. The following steps may be used to solve a quadratic equation.

> **To Solve a Quadratic Equation**
>
> *Step 1*   If the equation is in the form $(ax + b)^2 = c$, use the square root property and solve. If not, go to step 2.
>
> *Step 2*   Write the equation in standard form.
>
> *Step 3*   Try to solve the equation by the factoring method. If not, go to step 4.
>
> *Step 4*   Solve the equation by the quadratic formula.

EXAMPLE 1   Solve $m^2 - 2m - 7 = 0$.

Solution: The expression $m^2 - 2m - 7$ is not factorable, so use the quadratic formula with $a = 1$, $b = -2$, and $c = -7$.

$$m^2 - 2m - 7 = 0$$

$$m = \frac{-(-2) \pm \sqrt{(-2)^2 - 4 \cdot 1 \cdot (-7)}}{2 \cdot 1} = \frac{2 \pm \sqrt{32}}{2}$$

$$m = \frac{2 \pm 4\sqrt{2}}{2} = \frac{2\,(1 \pm 2\sqrt{2})}{2} = 1 \pm 2\sqrt{2} \quad \blacksquare$$

EXAMPLE 2   Solve $(3x + 1)^2 = 20$.

Solution: This equation is in a form that makes the square root property easy to apply.

$$(3x + 1)^2 = 20$$

$$3x + 1 = \pm\sqrt{20} \qquad \text{Apply the square root property.}$$

$$3x + 1 = \pm 2\sqrt{5} \qquad \text{Simplify } \sqrt{20}.$$

$$3x = -1 \pm 2\sqrt{5}$$

$$x = \frac{-1 \pm 2\sqrt{5}}{3} \quad \blacksquare$$

EXAMPLE 3   Solve $x^2 - \dfrac{11}{2}x = -\dfrac{5}{2}$.

Solution: The fractions make factoring more difficult and also complicate the calculations for using the quadratic formula. Clear the equation of fractions by multiplying both sides of the equation by the LCD 2.

$$x^2 - \frac{11}{2}x = -\frac{5}{2}$$

$$2x^2 - 11x = -5 \qquad \text{Multiply both sides by 2.}$$

$$2x^2 - 11x + 5 = 0 \qquad \text{Write in standard form.}$$

$$(2x - 1)(x - 5) = 0 \qquad \text{Factor}$$

$$2x - 1 = 0 \quad \text{or} \quad x - 5 = 0 \qquad \text{Apply the zero factor theorem.}$$

$$2x = 1 \quad \text{or} \quad x = 5$$

$$x = \frac{1}{2} \quad \text{or} \quad x = 5 \quad \blacksquare$$

EXERCISE SET 9.4

Choose and use a method to solve each equation.

1. $5x^2 - 11x + 2 = 0$

2. $5x^2 + 13x - 6 = 0$

3. $x^2 - 1 = 2x$

4. $x^2 + 7 = 6x$

5. $a^2 = 20$

6. $a^2 = 72$

7. $x^2 - x + 4 = 0$

8. $x^2 - 2x + 7 = 0$

9. $3x^2 - 12x + 12 = 0$

10. $5x^2 - 30x + 45 = 0$

11. $9 - 6p + p^2 = 0$

12. $49 - 28p + 4p^2 = 0$

13. $4y^2 - 16 = 0$

14. $3y^2 - 27 = 0$

15. $x^4 - 3x^3 + 2x^2 = 0$

16. $x^3 + 7x^2 + 12x = 0$

17. $(2z + 5)^2 = 25$

18. $(3z - 4)^2 = 16$

19. $30x = 25x^2 + 2$

20. $12x = 4x^2 + 4$

21. $\frac{2}{3}m^2 - \frac{1}{3}m - 1 = 0$

22. $\frac{5}{8}m^2 + m - \frac{1}{2} = 0$

23. $x^2 - \frac{1}{2}x - \frac{1}{5} = 0$

24. $x^2 + \frac{1}{2}x - \frac{1}{8} = 0$

25. $4x^2 - 27x + 35 = 0$

26. $9x^2 - 16x + 7 = 0$

27. $(7 - 5x)^2 = 18$

28. $(5 - 4x)^2 = 75$

29. $3z^2 - 7z = 12$

30. $6z^2 + 7z = 6$

31. $x = x^2 - 110$

32. $x = 56 - x^2$

33. $\frac{3}{4}x^2 - \frac{5}{2}x - 2 = 0$

34. $x^2 - \frac{6}{5}x - \frac{8}{5} = 0$

35. $x^2 - 0.6x + 0.05 = 0$

36. $x^2 - 0.1x - 0.06 = 0$

37. $10x^2 - 11x + 2 = 0$

38. $20x^2 - 11x + 1 = 0$

39. $\frac{1}{2}z^2 - 2z + \frac{3}{4} = 0$

40. $\frac{1}{5}z^2 - \frac{1}{2}z - 2 = 0$

Writing in Mathematics

41. Explain the advantage of using the quadratic formula to solve quadratic equations.

Skill Review

Simplify each expression. See Section 8.2.

42. $\sqrt{48}$

43. $\sqrt{104}$

44. $\sqrt{50}$

45. $\sqrt{80}$

Solve the following. See Section 2.6.

46. The height of a triangle is 4 times the length of the base. The area of the triangle is 18 square feet. Find the height and base of the triangle.

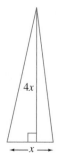

47. The height of a rectangle is 6 inches more than its width. The area of the rectangle is 391 square inches. Find the dimensions of the rectangle.

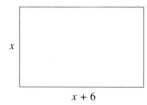

9.5
Complex Solutions of Quadratic Equations

OBJECTIVES

1 Write complex numbers using i notation.

2 Add and subtract complex numbers.

Tape 32

3 Multiply complex numbers.

4 Divide complex numbers.

5 Solve quadratic equations that have complex solutions.

In this chapter, we have seen many quadratic equations that have no real solutions. For example, the equation $x^2 = -4$. If we apply the square root property, we have

$$x^2 = -4$$
$$x = \pm\sqrt{-4}$$

Since there is no real number whose square is -4, we say there is no real number solution. However, our real number system can be extended to include numbers like $\sqrt{-4}$. This extended number system is called the **complex number** system. The complex number system includes the **imaginary unit i,** which is defined next.

Imaginary Unit i

The imaginary unit, written i, is the number whose square is -1. That is,

$$i^2 = -1 \quad \text{and} \quad i = \sqrt{-1}$$

---

1   We use i to write numbers like $\sqrt{-6}$ as the product of a real number and i. Since $i = \sqrt{-1}$, we have

$$\sqrt{-6} = \sqrt{-1 \cdot 6} = \sqrt{-1} \cdot \sqrt{6} = i\sqrt{6}$$

EXAMPLE 1   Write each radical as the product of a real number and i.
a. $\sqrt{-4}$ **b.** $\sqrt{-11}$ **c.** $\sqrt{-20}$

Solution: Write each negative radicand as a product of a positive number and -1. Then write $\sqrt{-1}$ as i.
a. $\sqrt{-4} = \sqrt{-1 \cdot 4} = \sqrt{-1} \cdot \sqrt{4} = i \cdot 2 = 2i$
b. $\sqrt{-11} = \sqrt{-1 \cdot 11} = \sqrt{-1} \cdot \sqrt{11} = i\sqrt{11}$
c. $\sqrt{-20} = \sqrt{-1 \cdot 20} = \sqrt{-1} \cdot \sqrt{20} = i \cdot 2\sqrt{5} = 2i\sqrt{5}$ ∎

The numbers $2i$, $i\sqrt{11}$, and $2i\sqrt{5}$ are called **imaginary numbers.** Both real numbers and imaginary numbers are complex numbers.

> **Complex Number and Imaginary Number**
>
> A complex number is a number that can be written in the form
>
> $$a + bi$$
>
> where a and b are real numbers. A complex number that can be written in the form $0 + bi$, $b \neq 0$, is also called an **imaginary number.**

A complex number written in the form $a + bi$ is in **standard form.**

Examples of Complex Numbers

7 is a complex number since $7 = 7 + 0i$
0 is a complex number since $0 = 0 + 0i$
$\sqrt{20}$ is a complex number since $\sqrt{20} = 2\sqrt{5} = 2\sqrt{5} + 0i$
$\sqrt{-27}$ is a complex number since $\sqrt{-27} = i \cdot 3\sqrt{3} = 0 + 3i\sqrt{3}$
$2 + \sqrt{-4}$ is a complex number since $2 + \sqrt{-4} = 2 + 2i$

2 We now present arithmetic operations—addition, subtraction, multiplication, and division—for the complex number system. Complex numbers are added and subtracted in the same way as we add and subtract polynomials.

EXAMPLE 2 Find the sum or difference. Write the answer in standard form.
a. $(2 + 3i) + (-6 - i)$ **b.** $-i + (3 + 7i)$ **c.** $(5 - i) - 4$

Solution: **a.** $(2 + 3i) + (-6 - i) = 2 + (-6) + (3i - i) = -4 + 2i$
b. $-i + (3 + 7i) = 3 + (-i + 7i) = 3 + 6i$
c. $(5 - i) - 4 = (5 - 4) - i = 1 - i$ ∎

EXAMPLE 3 Subtract $(11 - i)$ from $(1 + i)$.

Solution: $(1 + i) - (11 - i) = 1 + i - 11 + i = (1 - 11) + (i + i) = -10 + 2i$ ∎

3 Use the distributive property and the FOIL method to multiply complex numbers.

EXAMPLE 4 Find the following products and write in standard form.
a. $5i(2 - i)$ **b.** $(7 - 3i)(4 + 2i)$ **c.** $(2 + 3i)(2 - 3i)$

Solution: **a.** By the distributive property, we have

$$5i(2 - i) = 5i \cdot 2 - 5i \cdot i$$

$$= 10i - 5i^2 \qquad \text{Apply the distributive property.}$$

$$= 10i - 5(-1) \qquad \text{Write } i^2 \text{ as } -1.$$

$$= 10i + 5$$

$$= 5 + 10i \qquad \text{Write in standard form.}$$

b.
$$\overset{\text{F} \quad \text{O} \quad \text{I} \quad \text{L}}{(7 - 3i)(4 + 2i) = 28 + 14i - 12i - 6i^2}$$

$$= 28 + 2i - 6(-1) \qquad \text{Write } i^2 \text{ as } -1.$$

$$= 28 + 2i + 6$$

$$= 34 + 2i$$

c. $(2 + 3i)(2 - 3i) = 4 - 6i + 6i - 9i^2$

$$= 4 - 9(-1)$$

$$= 13 \qquad \blacksquare$$

The product in part (c) is the real number 13. When the complex numbers are related as these two are, their product is a real number. Notice that one is the sum of 2 and $3i$, and one is the difference of 2 and $3i$. In general,

$$(a + bi)(a - bi) = a^2 + b^2, \text{ a real number}$$

The complex numbers $a + bi$ and $a - bi$ are called **complex conjugates** of each other. For example, $2 - 3i$ is the conjugate of $2 + 3i$, and $2 + 3i$ is the conjugate of $2 - 3i$. Also,

$3 + 10i$ is the conjugate of $3 - 10i$.

5 is the conjugate of 5. (Note that $5 = 5 + 0i$ and its conjugate is $5 - 0i = 5$.)

$-4i$ is the conjugate of $4i$. ($0 - 4i$ is the conjugate of $0 + 4i$.)

4   The fact that the product of a complex number and its conjugate is a real number provides a method for dividing by a complex number and for simplifying fractions whose denominators are complex numbers.

EXAMPLE 5   Write $\dfrac{4 + i}{3 - 4i}$ in standard form.

Solution: To write this quotient as a complex number in the standard form $a + bi$, we need to find an equivalent fraction whose denominator is a real number. By multiplying both numerator and denominator by the denominator's conjugate, the new fraction is an equivalent fraction with a real number denominator.

$$\frac{4 + i}{3 - 4i} = \frac{(4 + i)}{(3 - 4i)} \cdot \frac{(3 + 4i)}{(3 + 4i)} \qquad \text{Multiply numerator and denominator by } 3 + 4i.$$

$$= \frac{12 + 16i + 3i + 4i^2}{9 - 16i^2}$$

$$= \frac{12 + 19i + 4(-1)}{9 - 16(-1)}$$

$$= \frac{12 + 19i - 4}{9 + 16} = \frac{8 + 19i}{25}$$

$$= \frac{8}{25} + \frac{19}{25}i \qquad \text{Write in standard form.}$$

Note that our last step was to write $\dfrac{4 + i}{3 - 4i}$ in standard form $a + bi$, where a and b are real numbers. $\blacksquare$

5   Next, we solve quadratic equations with complex solutions.

EXAMPLE 6   Solve $(x + 2)^2 = -25$ for x.

Solution: Begin by applying the square root property.

$$(x + 2)^2 = -25$$
$$x + 2 = \pm\sqrt{-25}$$
$$x + 2 = \pm 5i$$
$$x = -2 \pm 5i$$
$$x = -2 + 5i \quad \text{or} \quad x = -2 - 5i \quad \blacksquare$$

EXAMPLE 7   Solve $m^2 = 4m - 5$.

Solution: Write the equation in standard form and use the quadratic formula to solve.

$$m^2 = 4m - 5$$
$$m^2 - 4m + 5 = 0 \qquad\qquad \text{Write the equation in standard form.}$$

Apply the quadratic formula with $a = 1$, $b = -4$, and $c = 5$.

$$m = \frac{4 \pm \sqrt{16 - 4 \cdot 1 \cdot 5}}{2 \cdot 1}$$
$$= \frac{4 \pm \sqrt{-4}}{2}$$
$$= \frac{4 \pm 2i}{2}$$
$$= \frac{2\,(2 \pm i)}{2} = 2 \pm i \quad \blacksquare$$

EXAMPLE 8   Solve $x^2 + x = -1$.

Solution:
$$x^2 + x = -1$$
$$x^2 + x + 1 = 0 \qquad\qquad \text{Write in standard form.}$$
$$x = \frac{-1 \pm \sqrt{1 - 4 \cdot 1 \cdot 1}}{2 \cdot 1} \qquad \text{Apply the quadratic formula with } a = 1, b = 1, \text{ and } c = 1.$$
$$= \frac{-1 \pm \sqrt{-3}}{2}$$
$$= \frac{-1 \pm i\sqrt{3}}{2} \quad \blacksquare$$

EXERCISE SET 9.5

Write each expression in i notation. See Example 1.

1. $\sqrt{-9}$ **2.** $\sqrt{-64}$ **3.** $\sqrt{-100}$ **4.** $\sqrt{-16}$

5. $\sqrt{-50}$ **6.** $\sqrt{-98}$ **7.** $\sqrt{-63}$ **8.** $\sqrt{-44}$

Add or subtract as indicated. See Examples 2 and 3.

9. $(2 - i) + (-5 + 10i)$ **10.** $(-7 + 2i) + (5 - 3i)$ **11.** $(3 - 4i) - (2 - i)$ **12.** $(-6 + i) - (3 + i)$

Multiply. See Example 4.

13. $4i(3 - 2i)$ **14.** $-2i(5 + 4i)$ **15.** $(6 - 2i)(4 + i)$ **16.** $(6 + 2i)(4 - i)$

Divide. Write each of the following in standard form. See Example 5.

17. $\dfrac{8 - 12i}{4}$ **18.** $\dfrac{14 + 28i}{-7}$ **19.** $\dfrac{7 - i}{4 - 3i}$ **20.** $\dfrac{4 - 3i}{7 - i}$

Solve the following quadratic equations for complex solutions. See Example 6.

21. $(x + 1)^2 = -9$ **22.** $(y - 2)^2 = -25$ **23.** $(2z - 3)^2 = -12$ **24.** $(3p + 5)^2 = -18$

Solve the following quadratic equations for complex solutions. See Examples 7 and 8.

25. $y^2 + 6y + 13 = 0$ **26.** $y^2 - 2y + 5 = 0$ **27.** $4x^2 + 7x + 4 = 0$

28. $8x^2 - 7x + 2 = 0$ **29.** $2m^2 - 4m + 5 = 0$ **30.** $5m^2 - 6m + 7 = 0$

Perform the indicated operations. Write results in standard form.

31. $3 + (12 - 7i)$ **32.** $(-14 + 5i) + 3i$ **33.** $-9i(5i - 7)$

34. $10i(4i - 1)$ **35.** $(2 - i) - (3 - 4i)$ **36.** $(3 + i) - (-6 + i)$

37. $\dfrac{15 + 10i}{5i}$ **38.** $\dfrac{-18 + 12i}{-6i}$ **39.** Subtract $2 + 3i$ from $-5 + i$.

40. Subtract $-8 - i$ from $7 - 4i$.   **41.** $(4 - 3i)(4 + 3i)$ **42.** $(12 - 5i)(12 + 5i)$

43. $\dfrac{4 - i}{1 + 2i}$ **44.** $\dfrac{9 - 2i}{-3 + i}$ **45.** $(5 + 2i)^2$

46. $(9 - 7i)^2$

Solve the following quadratic equations for complex solutions.

47. $(y - 4)^2 = -64$ **48.** $(x + 7)^2 = -1$ **49.** $4x^2 = -100$

50. $7x^2 = -28$ **51.** $z^2 + 6z + 10 = 0$ **52.** $z^2 + 4z + 13 = 0$

53. $2a^2 - 5a + 9 = 0$ **54.** $4a^2 + 3a + 2 = 0$ **55.** $(2x + 8)^2 = -20$

56. $(6z - 4)^2 = -24$ **57.** $3m^2 + 108 = 0$ **58.** $5m^2 + 80 = 0$

59. $x^2 + 14x + 50 = 0$ **60.** $x^2 + 8x + 25 = 0$

Writing in Mathematics

61. In an earlier section, we learned that $\sqrt{-4}$ is not a real number. Explain what this means and explain what type of number $\sqrt{-4}$ is.

Skill Review

Graph the following linear equations in two variables. See Section 6.2.

62. $y = -3$ **63.** $x = 4$ **64.** $y = 3x - 2$ **65.** $y = 2x + 3$

Find the length of the unknown side of each triangle.

66.

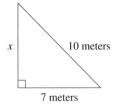

67.

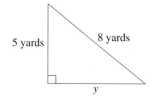

9.6
Graphing Quadratic Equations

OBJECTIVES

Tape 32

1	Identify the graph of a quadratic equation as a parabola.
2	Graph quadratic equations of the form $y = ax^2 + bx + c$.
3	Find the intercept points of a parabola.
4	Determine the vertex of a parabola.

---

1   Recall that the graph of a linear equation in two variables $Ax + By = C$ is a straight line. A **quadratic equation in two variables** is an equation that can be written in the form

$$y = ax^2 + bx + c$$

where a, b, and c are real numbers and a is not 0. Its graph is not a straight line since it is not a linear equation. In this section we learn to graph such equations. For example, $y = x^2$ is a quadratic equation in two variables. The graph of $y = x^2$ is the graph of its solutions. That is, the graph contains each point (x, y) whose coordinates make the equation true. To graph the equation, select a few values for x and find the corresponding y-values. Make a table of values to keep track. Then plot the points corresponding to these solutions.

        If $x = 0$, then $y = 0^2 = 0$.
        If $x = -2$, then $y = (-2)^2 = 4$. And so on.

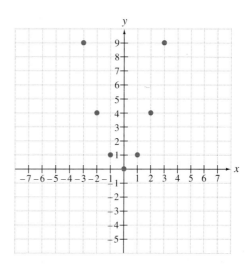

$y = x^2$	
x	y
0	0
1	1
2	4
3	9
−1	1
−2	4
−3	9

Clearly, these points are not on one straight line. The graph of $y = x^2$ is a smooth curve through the plotted points. This curve is called a **parabola.** The lowest point on a parabola opening upward is called the **vertex.** The vertex is (0, 0) for the parabola $y = x^2$. If we fold the graph paper along the y-axis, the two pieces of the parabola match perfectly. For this reason, we say the graph is **symmetric about the y-axis,** and we call the y-axis the **axis of symmetry.**

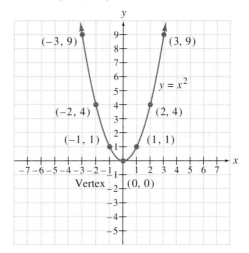

2   Notice that the parabola that corresponds to the equation $y = x^2$ opens upward. This happens when the coefficient of x^2 is positive. In $y = x^2$, the coefficient of x^2 is 1. Example 1 shows the appearance of the graph when the coefficient of x^2 is negative.

EXAMPLE 1   Graph $y = -2x^2$.

Solution: Select x-values and calculate the corresponding y-values. Plot the ordered pairs found. Then draw a smooth curve through those points. When the coefficient of x^2 is negative, the corresponding parabola opens downward. When a parabola opens downward, the vertex is the highest point of the parabola. The vertex of this parabola is (0, 0) and the axis of symmetry is again the y-axis. (See the next page.)

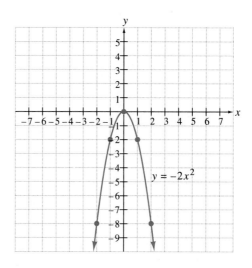

$y = -2x^2$

x	y
0	0
1	-2
2	-8
3	-18
-1	-2
-2	-8
-3	-18

EXAMPLE 2   Graph $y = x^2 - 4$.

Solution: Select x-values, find y-values, plot the points, and draw a smooth curve through the points. The vertex is $(0, -4)$, and the axis of symmetry is the y-axis. This graph has the same shape as the graph of $y = x^2$. It is different from the graph of $y = x^2$ because the vertex is 4 units lower on the y-axis.

$y = x^2 - 4$

x	y
0	-4
1	-3
2	0
3	5
-1	-3
-2	0
-3	5

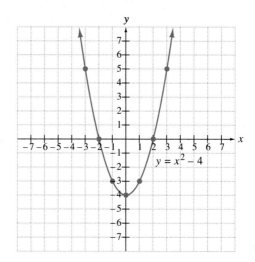

3   Just as for linear equations, we can use x- and y-intercepts to help graph quadratic equations. Recall from Chapter 6 that an x-intercept is the x-coordinate of the point where the graph intersects the x-axis. A y-intercept is the y-coordinate of the point where the graph intersects the y-axis.

HELPFUL HINT

Recall that:

To find x-intercepts, let $y = 0$ and solve for x.

To find y-intercepts, let $x = 0$ and solve for y.

EXAMPLE 3 Graph $y = (x + 2)^2$.

Solution: Find the intercepts. To find x-intercepts, let $y = 0$.

$$0 = (x + 2)^2, \quad \text{so } x = -2$$

The x-intercept point is $(-2, 0)$.
    To find any y-intercepts, let $x = 0$.

$$y = (0 + 2)^2 = 4$$

The y-intercept point is $(0, 4)$.
    Plot the points $(-2, 0)$ and $(0, 4)$ and then select other values for x to obtain more ordered pairs.

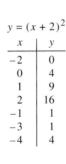

$y = (x + 2)^2$

x	y
-2	0
0	4
1	9
2	16
-1	1
-3	1
-4	4

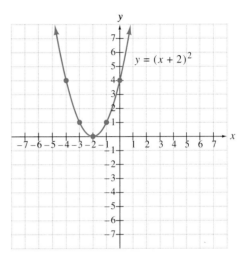

Notice that the graph of $y = (x + 2)^2$ is the same as the graph of $y = x^2$, except the vertex is shifted 2 units to the left, to $(-2, 0)$. The axis of symmetry is not the y-axis, but the vertical line through the vertex $(-2, 0)$. The axis of symmetry is the line $x = -2$. ∎

4   So far we have located the vertex by graphing the equation. However, by writing the equation in a particular form, we can determine the vertex, the axis of symmetry, and whether the parabola opens upward or downward. This can be stated as follows.

The Graph of a Quadratic Equation

The graph of the quadratic equation $y = a(x - h)^2 + k, a \neq 0$, is a parabola whose:

1. Vertex is (h, k).
2. Axis of symmetry is the line $x = h$.
3. Direction of opening is upward if $a > 0$ and downward if $a < 0$.

EXAMPLE 4 Graph $y = 3(x - 5)^2 + 2$.

Solution: The equation is written in the form $y = a(x - h)^2 + k$, with $a = 3$, $h = 5$, and $k = 2$. The vertex is then $(5, 2)$, the axis of symmetry is the line $x = 5$, and the graph opens upward since $a = 3$. By letting $x = 0$, we find that the y-intercept is 77, but our grid does not show coordinates as large as 77. Instead, take a point or two on each side of the vertex to determine how "fat" the parabola is. The x-coordinate of the vertex is 5, so we select $x = 4$ and $x = 6$, for example, to find two other points, as shown in the table.

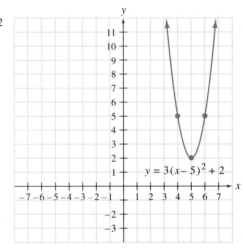

$y = 3(x - 5)^2 + 2$

x	y
5	2
4	5
6	5

As this parabola demonstrates, some parabolas do not have an x-intercept. In this parabola, the vertex is above the x-axis and the graph opens upward, so clearly the graph cannot intersect the x-axis.

If we tried to find the x-intercept by replacing y with 0 in the equation, we would solve $3(x - 5)^2 = -2$ for x. But this equation has no real number solution. Since the x and y-axes are real number lines, there is no x-intercept.

EXAMPLE 5 Graph $y = -(x - 2)^2 + 1$.

Solution: The vertex is $(2, 1)$, the axis of symmetry is the line $x = 2$, and the parabola opens downward since $a = -1$. Again, selecting an x-value on each side of the vertex helps to determine the shape of the parabola. Finding the intercepts often yields a more accurate graph.

To find x-intercepts, let $y = 0$.

$$0 = -(x - 2)^2 + 1$$
$$(x - 2)^2 = 1$$
$$x - 2 = \pm 1$$
$$x = 2 \pm 1$$
$$x = 3 \quad \text{or} \quad x = 1$$

There are two x-intercept points: $(3, 0)$ and $(1, 0)$. Plot $(3, 0)$ and $(1, 0)$.

To find y-intercepts, let $x = 0$.

$$y = -(0 - 2)^2 + 1 = -3$$

Plot the y-intercept point, $(0, -3)$

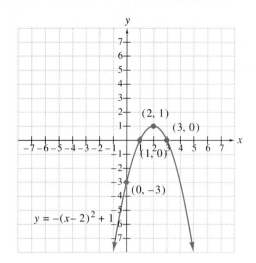

$$y = -(x-2)^2 + 1$$

If the equation is not in the form $y = a(x - h)^2 + k$, then we cannot determine the vertex immediately. By completing the square, however, the equation can be written in this form.

EXAMPLE 6   Graph $y = x^2 - 6x + 8$.

Solution: Since the equation is not in a form that lets us easily determine the vertex and axis of symmetry, complete the square. In this example, half the coefficient of x is -3, and $(-3)^2 = 9$. Here we **add and subtract 9 on one side.** Since we are adding and subtracting the same number, we are not changing the solutions of the equations.

$$y = (x^2 - 6x \;\; + 9 \;) \;\; - 9 \;\; + 8 \qquad \text{Add and subtract 9.}$$

$$y = (x - 3)^2 - 1 \qquad\qquad\qquad \text{Factor } x^2 - 6x + 9.$$

Now determine from the equation that the vertex is $(3, -1)$, the axis of symmetry is the line $x = 3$, and the parabola opens upward since $a = 1$. We also plot intercepts.

To find x-intercepts, let $y = 0$.

$$0 = x^2 - 6x + 8$$

Factor the expression $x^2 - 6x + 8$ as $(x - 4)(x - 2) = 0$. The x-intercepts are 4 and 2.

If we let $x = 0$ in the original equation, then $y = 8$, the y-intercept. Plot the vertex $(3, -1)$ and the intercept points $(4, 0)$, $(2, 0)$, and $(0, 8)$. Then sketch the parabola.

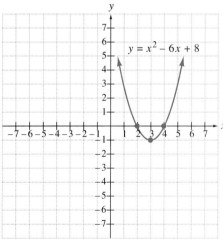

$$y = x^2 - 6x + 8$$

415

EXAMPLE 7 Graph $y = x^2 + 2x + 5$.

Solution: First, complete the square to find the vertex. Half the coefficient of x is 1 and $1^2 = 1$. Add and subtract 1.

$$y = x^2 + 2x + 5$$
$$= (x^2 + 2x \boxed{+ 1}) \boxed{- 1} + 5 \qquad \text{Add and subtract 1.}$$
$$= (x + 1)^2 + 4 \qquad \text{Factor and simplify.}$$
$$= [x - (-1)]^2 + 4 \qquad \text{Write in the form } y = a(x - h)^2 + k.$$

Vertex: $(-1, 4)$

Axis of symmetry: $x = -1$

Opens: Upward

To find x-intercepts, let $y = 0$.

$$0 = (x + 1)^2 + 4$$
$$-4 = (x + 1)^2$$

The equation has no real number solution. Therefore, the graph has no x-intercepts. To find y-intercepts, let $x = 0$ in the original equation and find that $y = 5$, and the resulting y-intercept is $(0, 5)$.

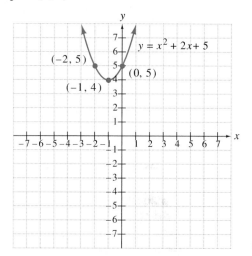

EXERCISE SET 9.6

Graph each quadratic equation. Find the vertex and intercepts. See Examples 1 through 5.

1. $y = 2x^2$

2. $y = -2x^2$

3. $y = (x - 1)^2$

4. $y = (x + 2)^2$

5. $y = -x^2 + 4$

6. $y = x^2 - 4$

7. $y = \dfrac{1}{3}x^2$

8. $y = -\dfrac{1}{2}x^2$

9. $y = (x - 2)^2 + 1$

10. $y = -(x - 2)^2 - 1$

11. $y = -(x + 1)^2 + 4$

12. $y = (x - 1)^2 - 4$

13. $y = -4x^2 + 1$

14. $y = 4x^2 - 1$

Write the letter of the graph corresponding to each equation. See Examples 1 through 5.

15. $y = -x^2$

16. $y = x^2$

17. $y = (x - 2)^2$

18. $y = -(x - 1)^2$

19. $y = (x + 3)^2 - 1$

20. $y = -(x - 2)^2 + 3$

21. $y = 2(x + 3)^2$

22. $y = -3(x + 1)^2$

23. $y = -\dfrac{1}{2}x^2 + 1$

a.

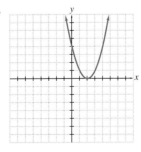

b.

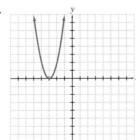

c.

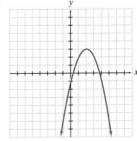

d.

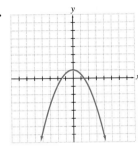

e.

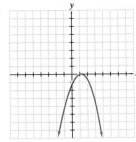

f.

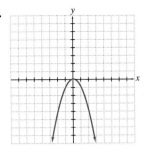

g.

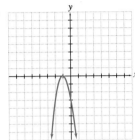

h.

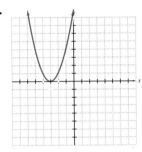

i.

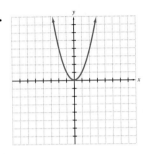

Write each quadratic equation in standard form to identify the vertex. Find the intercepts and graph the equation. See Examples 6 and 7.

24. $y = x^2 + 6x$

25. $y = x^2 - 4x$

26. $y = x^2 + 2x - 8$

27. $y = x^2 - 2x - 3$

28. $y = x^2 - x - 2$

29. $y = x^2 + 2x + 1$

30. $y = x^2 + 5x + 4$

31. $y = x^2 + 7x + 10$

32. $y = x^2 - 4x + 3$

33. $y = x^2 - 6x + 8$

Writing in Mathematics

34. Describe the key features and general shape of a parabola whose equation is $y = a(x - h)^2 + k$, if a, h, and k are positive real numbers.

Skill Review

Simplify the following complex fractions. See Section 5.5.

35. $\dfrac{\dfrac{1}{7}}{\dfrac{2}{5}}$

36. $\dfrac{\dfrac{3}{8}}{\dfrac{1}{7}}$

37. $\dfrac{\dfrac{1}{x}}{\dfrac{2}{x^2}}$

38. $\dfrac{\dfrac{x}{5}}{\dfrac{2}{x}}$

39. $\dfrac{2x}{1 - \dfrac{1}{x}}$

40. $\dfrac{x}{x - \dfrac{1}{x}}$

41. $\dfrac{\dfrac{a - b}{2b}}{\dfrac{b - a}{8b^2}}$

42. $\dfrac{\dfrac{2a^2}{a - 3}}{\dfrac{a}{3 - a}}$

CRITICAL THINKING   Out of sturdy paper, cut at least 10 different triangles so that each triangle is a right triangle whose hypotenuse is 2 inches long. For each triangle, choose a vertex at one end of the hypotenuse and label the vertex O. Label the right angle vertex X and the remaining vertex Y. Draw a Cartesian coordinate system, using 1 inch as the unit distance for each axis. Arrange the triangles so that all the vertices labeled O coincide at the origin and all the vertices labeled X lie somewhere on the x-axis.

What pattern do you see taking form? Using the Pythagorean theorem, can you express this pattern in the form of an equation?

For each vertex labeled Y, mark on the paper a point coinciding with the vertex. Draw a smooth curve connecting the points. Can you write the equation of this graph?

What conclusions do you make about the Pythagorean theorem and the equation of this graph?

CHAPTER 9 GLOSSARY

The **axis of symmetry** is a line about which a parabola is symmetric.

The process of adding a constant to a binomial to form a perfect square trinomial is called **completing the square**.

A **complex number** is a number that can be written in the form $a + bi$.

The **imaginary unit** i is the number whose square is -1.

The graph of a quadratic equation is a **parabola**.

A **perfect square trinomial** is a trinomial that can be written as the square of a binomial.

The **vertex** of a parabola is the highest point if the parabola opens downward or the lowest point if the parabola opens upward.

CHAPTER 9 SUMMARY

SQUARE ROOT PROPERTY (9.1)

If $X^2 = a$, then $X = \pm\sqrt{a}$ for $a \geq 0$.

(9.3)

Quadratic Formula

A quadratic equation of the form $ax^2 + bx + c = 0$, $a \neq 0$, has solutions

$$x = \frac{-b \pm \sqrt{b^2 - 4ac}}{2a}$$

(9.6)

The Graph of a Quadratic Equation

The graph of the quadratic equation $y = a(x - h)^2 + k$, $a \neq 0$, is a parabola whose:

1. Vertex is (h, k).
2. Axis of symmetry is the line $x = h$.
3. Direction of opening is upward if $a > 0$ and downward if $a < 0$.

CHAPTER 9 REVIEW

(9.1) *Solve each quadratic equation by factoring or using the square root property.*

1. $(x - 4)(5x + 3) = 0$
2. $(x + 7)(3x + 4) = 0$
3. $3m^2 - 5m = 2$
4. $7m^2 + 2m = 5$
5. $k^2 = 50$
6. $k^2 = 45$
7. $(x - 5)(x - 1) = 12$
8. $(x - 3)(x + 2) = 6$
9. $(x - 11)^2 = 49$
10. $(x + 3)^2 = 100$
11. $6x^3 - 54x = 0$
12. $2x^2 - 8 = 0$
13. $(4p + 2)^2 = 100$
14. $(3p + 6)^2 = 81$

(9.2) *Complete the square for the following expressions and then factor the resulting perfect square trinomial.*

15. $x^2 - 10x$
16. $x^2 + 16x$
17. $a^2 + 4a$
18. $a^2 - 12a$
19. $m^2 - 3m$
20. $m^2 + 5m$

Solve each quadratic equation by completing the square.

21. $x^2 - 6x + 7 = 0$ **22.** $x^2 + 6x + 7 = 0$ **23.** $2y^2 + y - 1 = 0$ **24.** $y^2 + 3y - 1 = 0$

(9.3) *Solve each quadratic equation by using the quadratic formula.*

25. $x^2 - 10x + 7 = 0$ **26.** $x^2 + 4x - 7 = 0$ **27.** $2x^2 + x - 1 = 0$

28. $x^2 + 3x - 1 = 0$ **29.** $9x^2 + 30x + 25 = 0$ **30.** $16x^2 - 72x + 81 = 0$

31. $15x^2 + 2 = 11x$ **32.** $15x^2 + 2 = 13x$ **33.** $2x^2 + x + 5 = 0$

34. $7x^2 - 3x + 1 = 0$

(9.4) *Solve the following equations by using the most appropriate method.*

35. $5z^2 + z - 1 = 0$ **36.** $4z^2 + 7z - 1 = 0$ **37.** $4x^4 = x^2$

38. $9x^3 = x$ **39.** $2x^2 - 15x + 7 = 0$ **40.** $x^2 - 6x - 7 = 0$

41. $(3x - 1)^2 = 0$ **42.** $(2x - 3)^2 = 0$ **43.** $x^2 = 6x - 9$

44. $x^2 = 10x - 25$ **45.** $\left(\dfrac{1}{2}x - 3\right)^2 = 64$ **46.** $\left(\dfrac{1}{3}x + 1\right)^2 = 49$

47. $x^2 - 0.3x + 0.01 = 0$ **48.** $x^2 + 0.6x - 0.16 = 0$ **49.** $\dfrac{1}{10}x^2 + x - \dfrac{1}{2} = 0$

50. $\dfrac{1}{12}x^2 - \dfrac{1}{2}x + \dfrac{1}{3} = 0$

(9.5) *Perform the indicated operations. Write the resulting complex number in standard form.*

51. $\sqrt{-144}$ **52.** $\sqrt{-36}$ **53.** $\sqrt{-108}$

54. $\sqrt{-500}$ **55.** $(7 - i) + (14 - 9i)$ **56.** $(10 - 4i) + (9 - 21i)$

57. $3 - (11 + 2i)$ **58.** $(-4 - 3i) + 5i$ **59.** $(2 - 3i)(3 - 2i)$

60. $(2 + 5i)(5 - i)$ **61.** $(3 - 4i)(3 + 4i)$ **62.** $(7 - 2i)(7 - 2i)$

63. $\dfrac{2 - 6i}{4i}$ **64.** $\dfrac{5 - i}{2i}$ **65.** $\dfrac{4 - i}{1 + 2i}$

66. $\dfrac{1 + 3i}{2 - 7i}$

Solve each quadratic equation.

67. $3x^2 = -48$ **68.** $5x^2 = -125$ **69.** $x^2 - 4x + 13 = 0$ **70.** $x^2 + 4x + 11 = 0$

(9.6) *Identify the vertex, axis of symmetry, and whether the parabola opens upward or downward for the given quadratic equation.*

71. $y = -3x^2$ **72.** $y = -\dfrac{1}{2}x^2$ **73.** $y = (x - 3)^2$ **74.** $y = (x - 5)^2$

75. $y = 3x^2 - 7$ **76.** $y = -2x^2 + 25$ **77.** $y = -5(x - 72)^2 + 14$ **78.** $y = 2(x - 35)^2 - 21$

Graph the following quadratic equations. Label the vertex and the intercept points with their coordinates.

79. $y = -(x + 1)^2$ **80.** $y = -(x - 2)^2$ **81.** $y = (x - 2)^2$ **82.** $y = (x + 3)^2$

83. $y = 2x^2 - 3$ **84.** $y = 3x^2 - 5$ **85.** $y = \dfrac{1}{3}x^2 + 3$ **86.** $y = \dfrac{1}{2}x^2 - 2$

CHAPTER 9 TEST

Solve by factoring.

1. $2x^2 - 11x = 21$

2. $x^4 + x^3 - 2x^2 = 0$

Solve using the square root property.

3. $5k^2 = 80$

4. $(3m - 5)^2 = 8$

Solve by completing the square.

5. $x^2 - 26x + 160 = 0$

6. $5x^2 + 9x = 2$

Solve using the quadratic formula.

7. $x^2 - 3x - 10 = 0$

8. $p^2 - \dfrac{5}{3}p - \dfrac{1}{3} = 0$

Solve by the most appropriate method.

9. $(3x - 5)(x + 2) = -6$ **10.** $(3x - 1)^2 = 16$

11. $3x^2 - 7x - 2 = 0$ **12.** $x^2 - 4x + 5 = 0$

13. $3x^2 - 7x + 2 = 0$ **14.** $2x^2 - 6x + 1 = 0$

15. $2x^5 + 5x^4 - 3x^3 = 0$ **16.** $9x^3 = x$

Perform the indicated operations. Write the resulting complex number in standard form.

17. $\sqrt{-25}$ **18.** $\sqrt{-200}$ **19.** $(3 + 2i) + (5 - i)$

20. $(4 - i) - (-3 + 5i)$ **21.** $(3 + 2i) - (3 - 2i)$ **22.** $(3 + 2i) + (3 - 2i)$

23. $(3 + 2i)(3 - 2i)$ **24.** $\dfrac{3 - i}{1 + 2i}$

Graph the quadratic equations. Label the vertex and the intercept points with their coordinates.

25. $y = x^2$ **26.** $y = (x - 3)^2$ **27.** $y = -x^2 + 2$

28. $y = (x + 1)^2 - 2$ **29.** $y = x^2 - 4x + 7$ **30.** $y = x^2 + 2x + 3$

CHAPTER 9 CUMULATIVE REVIEW

1. Solve for x: $5x - 2 = 18$.

2. Solve $P = 2L + 2W$ for W.

3. Solve $-1 \le 2x - 3 < 5$, and then graph the solution.

4. Simplify the following expressions.
   a. $(st)^4$ **b.** $\left(\dfrac{m}{n}\right)^7$ **c.** $(2a)^3$
   d. $(-5x^2y^3z)^2$ **e.** $\left(\dfrac{2x^4}{3y^5}\right)^4$

5. Subtract $(2x^3 + 8x^2 - 6x) - (2x^3 - x^2 + 1)$.

6. Use the distributive property to find each product.
   a. $5x(2x^3 + 6)$
   b. $-3x^2(5x^2 + 6x - 1)$
   c. $(3n^2 - 5n + 4)(2n)$

7. Find $\dfrac{4x^2 + 7 + 8x^3}{2x + 3}$.

8. Factor $10x^2 - 13xy - 3y^2$.

9. Factor $9x^2 - 36$.

10. Solve $2x^3 - 4x^2 - 30x = 0$.

11. Simplify $\dfrac{4 - x^2}{3x^2 - 5x - 2}$.

12. Divide $\dfrac{6x + 2}{x^2 - 1} \div \dfrac{3x^2 + x}{x - 1}$.

13. Subtract $\dfrac{6x}{x^2 - 4} - \dfrac{3}{x + 2}$.

14. Simplify $\dfrac{\dfrac{1}{z} - \dfrac{1}{2}}{\dfrac{1}{3} - \dfrac{z}{6}}$.

15. Solve for x: $\dfrac{2x}{a} - 5 = \dfrac{3x}{b} + a$.

16. Graph $x - 3y = 6$ by plotting intercept points.

17. Find the slope of the line through $(-1, -2)$ and $(2, 4)$. Graph the line.

18. Find the equation of the vertical line through $(-1, 5)$.

19. Find each square root.
   a. $\sqrt{36}$ **b.** $\sqrt{64}$ **c.** $-\sqrt{25}$
   d. $\sqrt{\dfrac{9}{100}}$ **e.** $\sqrt{0}$

20. Find the following products.
   a. $\sqrt{7} \cdot \sqrt{3}$ **b.** $\sqrt{3} \cdot \sqrt{15}$ **c.** $\sqrt[3]{4} \cdot \sqrt[3]{18}$
   d. $2\sqrt{6} \cdot 5\sqrt{2}$ **e.** $(3\sqrt{2})^2$

21. Find the product and simplify.
   a. $(\sqrt{5} - 7)(\sqrt{5} + 7)$ **b.** $(\sqrt{7x} + 2)^2$

22. Solve for x: $\sqrt{x} + 6 = 4$.

23. Use the square root property to solve $(x - 3)^2 = 16$.

24. Use the quadratic formula to solve $3x^2 + x - 3 = 0$.

Operations
on Decimals

To **add** or **subtract** decimals, write the numbers vertically with decimal points lined up. Add or subtract as with whole numbers and place the decimal point in the answer directly below the decimal points in the problem.

EXAMPLE 1   Add 5.87 + 23.279 + 0.003.

   Solution:

$$\begin{array}{r} 5.87 \\ 23.279 \\ +\ 0.003 \\ \hline 29.152 \end{array}$$ ∎

EXAMPLE 2   Subtract 32.15 − 11.237.

   Solution:

$$\begin{array}{cccccc} & 1 & 11 & 4 & 10 & \\ 3 & 2 & . & 1 & 5 & 0 \\ -\ 1 & 1 & . & 2 & 3 & 7 \\ \hline 2 & 0 & . & 9 & 1 & 3 \end{array}$$ ∎

To **multiply** decimals, multiply the numbers as if they were whole numbers. The decimal point in the product is placed so that the number of decimal places in the product is the same as the sum of the number of decimal places in the factors.

EXAMPLE 3   Multiply 0.072 × 3.5.

   Solution:

$$\begin{array}{rl} 0.072 & \text{3 decimal places} \\ \times\ \ \ 3.5 & \text{1 decimal place} \\ \hline 360 & \\ 216\ \ \ & \\ \hline 0.2520 & \text{4 decimal places} \end{array}$$ ∎

To **divide** decimals, move the decimal point in the divisor to the right of the last digit. Move the decimal point in the dividend the same number of places that the decimal point in the divisor was moved. The decimal point in the quotient lies directly above the decimal point in the dividend.

EXAMPLE 4   Divide $9.46 \div 0.04$.

Solution:

$$
\begin{array}{r}
236.5 \\
04.\overline{)946.0} \\
-8 \\
\hline
14 \\
-12 \\
\hline
26 \\
-24 \\
\hline
20 \\
-20 \\
\hline
\end{array}
$$ ■

APPENDIX A EXERCISE SET

Perform the indicated operations.

1. $9.076 + 8.004$

2. $\begin{array}{r} 6.3 \\ \times 0.05 \\ \hline \end{array}$

3. $\begin{array}{r} 27.006 \\ -14.2 \\ \hline \end{array}$

4. $\begin{array}{r} 0.0036 \\ 7.12 \\ 32.502 \\ +\ 0.05 \\ \hline \end{array}$

5. $\begin{array}{r} 107.92 \\ +\ 3.04 \\ \hline \end{array}$

6. $7.2 \div 4$

7. $10 - 7.6$

8. $40 \div 0.25$

9. $126.32 - 97.89$

10. $\begin{array}{r} 3.62 \\ 7.11 \\ 12.36 \\ 4.15 \\ +\ 2.29 \\ \hline \end{array}$

11. $\begin{array}{r} 3.25 \\ \times\ 70 \\ \hline \end{array}$

12. $\begin{array}{r} 26.014 \\ -\ 7.8 \\ \hline \end{array}$

13. $8.1 \div 3$

14. $\begin{array}{r} 1.2366 \\ 0.005 \\ 15.17 \\ +\ 0.97 \\ \hline \end{array}$

15. $55.405 - 6.1711$

16. $8.09 + 0.22$

17. $60 \div 0.75$

18. $20 - 12.29$

19. $7.612 \div 100$

20. $\begin{array}{r} 8.72 \\ 1.12 \\ 14.86 \\ 3.98 \\ +\ 1.99 \\ \hline \end{array}$

21. $12.312 \div 2.7$

22. $0.443 \div 100$

23. $\begin{array}{r} 569.2 \\ 71.25 \\ +\ 8.01 \\ \hline \end{array}$

24. $3.706 - 2.91$

25. 768 − 0.17

26. 63 ÷ 0.28

27. 12 + 0.062

28. 0.42 + 18

29. 76 − 14.52

30. 1.1092 ÷ 0.47

31. 3.311 ÷ 0.43

32. 7.61 + 0.0004

33.   762.12
        89.7
     + 11.55

34. 444 ÷ 0.6

35. 23.4 − 0.821

36. 3.7 + 5.6

37. 476.12 − 112.97

38. 19.872 ÷ 0.54

39. 0.007 + 7

40.   51.77
       + 3.6

Review of Angles, Lines, and Special Triangles

The word **geometry** is formed from the Greek words, **geo,** meaning earth, and **metron,** meaning measure. Geometry literally means to measure the earth.

This section contains a review of some basic geometric ideas. It will be assumed that fundamental ideas of geometry such as point, line, ray, and angle are known. In this appendix, the notation $\angle 1$ is read "angle 1" and the notation $m\angle 1$ is read "the measure of angle 1."

We first review types of angles.

Angles

A **right angle** is an angle whose measure is 90°. A right angle can be indicated by a square drawn at the vertex of the angle, as shown below. An angle whose measure is more than 0° but less than 90° is called an **acute angle.**

An angle whose measure is greater than 90° but less than 180° is called an **obtuse angle.**

An angle whose measure is 180° is called a **straight angle.**

Two angles are said to be **complementary** if the sum of their measures is 90°. Each angle is called the **complement** of the other.

Two angles are said to be **supplementary** if the sum of their measures is 180°. Each angle is called the **supplement** of the other.

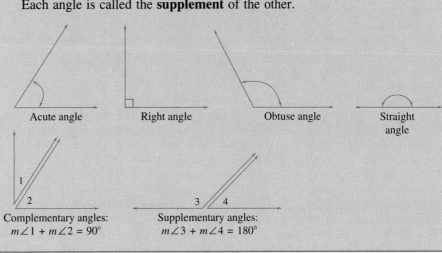

Acute angle   Right angle   Obtuse angle   Straight angle

Complementary angles:
$m\angle 1 + m\angle 2 = 90°$

Supplementary angles:
$m\angle 3 + m\angle 4 = 180°$

EXAMPLE 1   If an angle measures 28°, find its complement.

Solution: Two angles are complementary if the sum of their measures is 90°. The complement of a 28° angle is an angle whose measure is 90° − 28° = 62°. To check, notice that 28° + 62° = 90°.   ∎

Plane is an undefined term that we will describe. A plane can be thought of as a flat surface with infinite length and width, but no thickness. A plane is two dimensional. The arrows in the following diagram indicate that a plane extends indefinitely and has no boundaries.

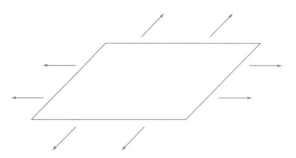

Figures that lie on a plane are called **plane figures.** (See the description of common plane figures in Appendix C.) Lines that lie in the same plane are called **coplanar.**

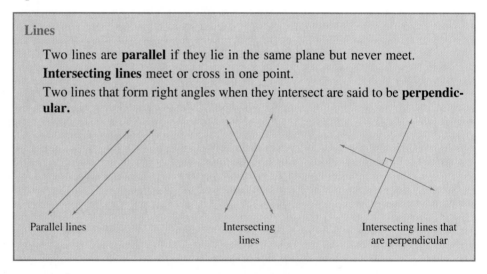

Lines

Two lines are **parallel** if they lie in the same plane but never meet.
Intersecting lines meet or cross in one point.
Two lines that form right angles when they intersect are said to be **perpendicular.**

Parallel lines   Intersecting lines   Intersecting lines that are perpendicular

Two intersecting lines form **vertical angles.** Angles 1 and 3 are vertical angles. Also angles 2 and 4 are vertical angles. It can be shown that **vertical angles have equal measures.**

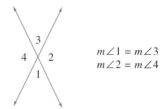

$m\angle 1 = m\angle 3$
$m\angle 2 = m\angle 4$

Adjacent angles have the same vertex and share a side. Angles 1 and 2 are adjacent angles. Other pairs of adjacent angles are angles 2 and 4, angles 3 and 4, and angles 3 and 1.

A **transversal** is a line that intersects two or more lines in the same plane. Line l is a transversal that intersects lines m and n. The eight angles formed are numbered and certain pairs of these angles are given special names.

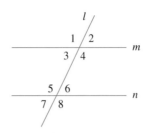

Corresponding angles: $\angle 1$ and $\angle 5$, $\angle 3$ and $\angle 7$, $\angle 2$ and $\angle 6$, and $\angle 4$ and $\angle 8$.

Exterior angles: $\angle 1$, $\angle 2$, $\angle 7$, and $\angle 8$.

Interior angles: $\angle 3$, $\angle 4$, $\angle 5$, and $\angle 6$.

Alternate interior angles: $\angle 3$ and $\angle 6$, $\angle 4$ and $\angle 5$.

These angles and parallel lines are related in the following manner.

Parallel Lines Cut by a Transversal

1. If two parallel lines are cut by a transversal, then
   a. **corresponding angles are equal** and
   b. **alternate interior angles are equal.**
2. If corresponding angles formed by two lines and a transversal are equal, then the lines are parallel.
3. If alternate interior angles formed by two lines and a transversal are equal, then the lines are parallel.

EXAMPLE 2   Given that lines m and n are parallel and that the measure of angle 1 is $100°$, find the measures of angles 2, 3, and 4.

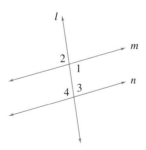

Solution: $m\angle 2 = 100°$,   since angles 1 and 2 are vertical angles.

$m\angle 4 = 100°$,   since angles 1 and 4 are alternate interior angles.

$m\angle 3 = 180° - 100° = 80°$,   since angles 4 and 3 are supplementary angles.   ∎

A **polygon** is the union of three or more coplanar line segments that intersect each other only at each end point, with each end point shared by exactly two segments.

A **triangle** is a polygon with three sides. The sum of the measures of the three angles of a triangle is 180°. In the following figure, $m\angle 1 + m\angle 2 + m\angle 3 = 180°$.

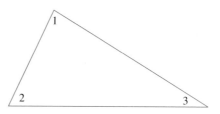

EXAMPLE 3 Find the measure of the third angle of the triangle shown.

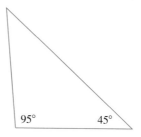

Solution: The sum of the measures of the angles of a triangle is 180°. Since one angle measures 45° and the other angle measures 95°, the third angle measures $180° - 45° - 95° = 40°$. ■

Two triangles are **congruent** if they have the same size and the same shape. In congruent triangles, the measures of corresponding angles are equal and the lengths of corresponding sides are equal. The following triangles are congruent.

Corresponding angles are equal: m∠1 = m∠4, m∠2 = m∠5, and m∠3 = m∠6. Also, lengths of corresponding sides are equal: $a = x$, $b = y$, and $c = z$.

Any one of the following may be used to determine whether two triangles are congruent.

Congruent Triangles

1. If the measures of two angles of a triangle equal the measures of two angles of another triangle and the lengths of the sides between each pair of angles are equal, the triangles are congruent.

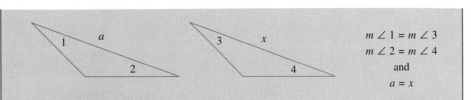

$$m \angle 1 = m \angle 3$$
$$m \angle 2 = m \angle 4$$
and
$$a = x$$

2. If the lengths of the three sides of a triangle equal the lengths of corresponding sides of another triangle, the triangles are congruent.

$$a = x$$
$$b = y$$
and
$$c = z$$

3. If the lengths of two sides of a triangle equal the lengths of corresponding sides of another triangle, and the measures of the angles between each pair of sides are equal, the triangles are congruent.

$$a = x$$
$$b = y$$
and
$$m \angle 1 = m \angle 2$$

Two triangles are similar if they have the same shape. In similar triangles, the measures of corresponding angles are equal and corresponding sides are in proportion. The following triangles are similar. (All similar triangles drawn in this appendix will be oriented the same.)

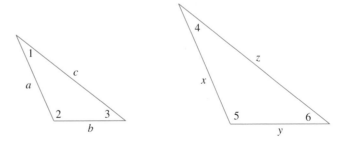

Corresponding angles are equal: $m \angle 1 = m \angle 4$, $m \angle 2 = m \angle 5$, and $m \angle 3 = m \angle 6$. Also, corresponding sides are proportional: $\dfrac{a}{x} = \dfrac{b}{y} = \dfrac{c}{z}$.

Any one of the following may be used to determine whether two triangles are similar.

Similar Triangles

1. If the measures of two angles of a triangle equal the measures of two angles of another triangle, the triangles are similar.

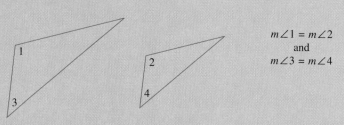

$m\angle 1 = m\angle 2$
and
$m\angle 3 = m\angle 4$

2. If three sides of one triangle are proportional to three sides of another triangle, the triangles are similar.

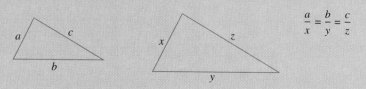

$\dfrac{a}{x} = \dfrac{b}{y} = \dfrac{c}{z}$

3. If two sides of a triangle are proportional to two sides of another triangle and the measures of the included angles are equal, the triangles are similar.

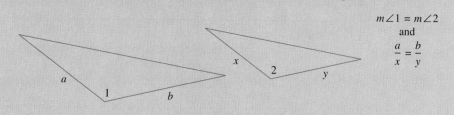

$m\angle 1 = m\angle 2$
and
$\dfrac{a}{x} = \dfrac{b}{y}$

EXAMPLE 4 Given that the following triangles are similar, find the missing length x.

Solution: Since the triangles are similar, corresponding sides are in proportion. Thus, $\dfrac{2}{3} = \dfrac{10}{x}$. To solve this equation for x, we multiply both sides by the LCD $3x$.

$$3x\left(\frac{2}{3}\right) = 3x\left(\frac{10}{x}\right)$$
$$2x = 30$$
$$x = 15$$

The missing length is 15 units. ∎

A **right triangle** contains a right angle. The side opposite the right angle is called the **hypotenuse,** and the other two sides are called the **legs.** The **Pythagorean theorem** gives a formula that relates the lengths of the three sides of a right triangle.

The Pythagorean Theorem

If a and b are the lengths of the legs of a right triangle, and c is the length of the hypotenuse, then $a^2 + b^2 = c^2$.

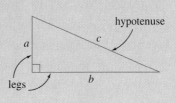

EXAMPLE 5   Find the length of the hypotenuse of a right triangle whose legs have lengths of 3 centimeters and 4 centimeters.

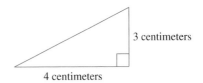

Solution: Because we have a right triangle, we use the Pythagorean theorem. The legs are 3 centimeters and 4 centimeters, so let $a = 3$ and $b = 4$ in the formula.

$$a^2 + b^2 = c^2$$
$$3^2 + 4^2 = c^2$$
$$9 + 16 = c^2$$
$$25 = c^2$$

Since c represents a length, we assume that c is positive. Thus, if c^2 is 25, c must be 5. The hypotenuse has a length of 5 centimeters.   ■

APPENDIX B EXERCISE SET

Find the complement of each angle. See Example 1.

1. $19°$

2. $65°$

3. $70.8°$

4. $45\frac{2}{3}°$

5. $11\frac{1}{4}°$

6. $19.6°$

Find the supplement of each angle.

7. $150°$

8. $90°$

9. $30.2°$

10. $81.9°$

11. $79\frac{1}{2}°$

12. $165\frac{8}{9}°$

13. If lines *m* and *n* are parallel, find the measures of angles 1 through 7. See Example 2.

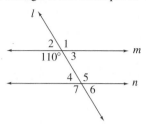

14. If lines *m* and *n* are parallel, find the measures of angles 1 through 5. See Example 2.

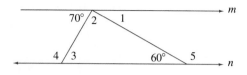

In each of the following, the measures of two angles of a triangle are given. Find the measure of the third angle. See Example 3.

15. 11°, 79°

16. 8°, 102°

17. 25°, 65°

18. 44°, 19°

19. 30°, 60°

20. 67°, 23°

In each of the following, the measure of one angle of a right triangle is given. Find the measures of the other two angles.

21. 45°

22. 60°

23. 17°

24. 30°

25. $39\frac{3}{4}°$

26. 72.6°

Given that each of the following pairs of triangles is similar, find the missing lengths. See Example 4.

27.

28.

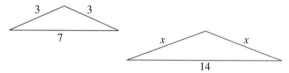

29.

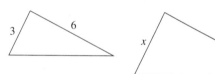

30.

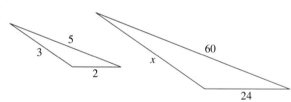

Use the Pythagorean Theorem to find the missing lengths in the right triangles. See Example 5.

31.

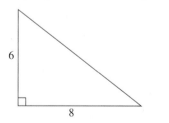

32.

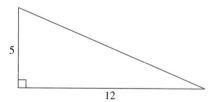

33.

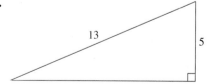

34.

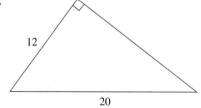

Review of Geometric Figures

Plane figures have length and width but no thickness or depth.		
Name	**Description**	**Figure**
Polygon	Union of three or more coplanar line segments that intersect each other only at each end point, with each end point shared by two segments.	
Triangle	Polygon with three sides (sum of measures of three angles is 180°).	
Scalene triangle	Triangle with no sides of equal length.	
Isosceles triangle	Triangle with two sides of equal length.	
Equilateral triangle	Triangle with all sides of equal length.	
Right triangle	Triangle that contains a right angle.	leg hypotenuse leg
Quadrilateral	Polygon with four sides (sum of measures of four angles is 360°).	

Plane figures have length and width but no thickness or depth.

Name	Description	Figure
Trapezoid	Quadrilateral with exactly one pair of opposite sides parallel.	
Isosceles trapezoid	Trapezoid with legs of equal length.	
Parallelogram	Quadrilateral with both pair of opposite sides parallel and equal in length.	
Rhombus	Parallelogram with all sides of equal length.	
Rectangle	Parallelogram with four right angles.	
Square	Rectangle with all sides of equal length.	
Circle	All points in a plane the same distance from a fixed point called the **center**.	

Solid figures have length, width, and height or depth.		
Name	**Description**	**Figure**
Rectangular solid	A solid with six sides, all of which are rectangles.	
Cube	A rectangular solid whose six sides are squares.	
Sphere	All points the same distance from a fixed point, called the **center**.	radius / center
Right circular cylinder	A cylinder consisting of two circular bases that are perpendicular to its altitude.	
Right circular cone	A cone with a circular base that is perpendicular to its altitude.	

APPENDIX D

Table of Squares and Square Roots

n	n^2	$\sqrt{n}$	n	n^2	$\sqrt{n}$
1	1	1.000	51	2,601	7.141
2	4	1.414	52	2,704	7.211
3	9	1.732	53	2,809	7.280
4	16	2.000	54	2,916	7.348
5	25	2.236	55	3,025	7.416
6	36	2.449	56	3,136	7.483
7	49	2.646	57	3,249	7.550
8	64	2.828	58	3,364	7.616
9	81	3.000	59	3,481	7.681
10	100	3.162	60	3,600	7.746
11	121	3.317	61	3,721	7.810
12	144	3.464	62	3,844	7.874
13	169	3.606	63	3,969	7.937
14	196	3.742	64	4,096	8.000
15	225	3.873	65	4,225	8.062
16	256	4.000	66	4,356	8.124
17	289	4.123	67	4,489	8.185
18	324	4.243	68	4,624	8.246
19	361	4.359	69	4,761	8.307
20	400	4.472	70	4,900	8.367
21	441	4.583	71	5,041	8.426
22	484	4.690	72	5,184	8.485
23	529	4.796	73	5,329	8.544
24	576	4.899	74	5,476	8.602
25	625	5.000	75	5,625	8.660
26	676	5.099	76	5,776	8.718
27	729	5.196	77	5,929	8.775
28	784	5.292	78	6,084	8.832
29	841	5.385	79	6,241	8.888
30	900	5.477	80	6,400	8.944
31	961	5.568	81	6,561	9.000
32	1,024	5.657	82	6,724	9.055
33	1,089	5.745	83	6,889	9.110
34	1,156	5.831	84	7,056	9.165
35	1,225	5.916	85	7,225	9.220
36	1,296	6.000	86	7,396	9.274
37	1,369	6.083	87	7,569	9.327
38	1,444	6.164	88	7,744	9.381
39	1,521	6.245	89	7,921	9.434
40	1,600	6.325	90	8,100	9.487
41	1,681	6.403	91	8,281	9.539
42	1,764	6.481	92	8,464	9.592
43	1,849	6.557	93	8,649	9.644
44	1,936	6.633	94	8,836	9.695
45	2,025	6.785	95	9,025	9.747
46	2,116	6.782	96	9,216	9.798
47	2,209	6.856	97	9,409	9.849
48	2,304	6.928	98	9,604	9.899
49	2,401	7.000	99	9,801	9.950
50	2,500	7.071	100	10,000	10.000

Answers to Selected Exercises

CHAPTER 1
Real Numbers and Their Properties

Exercise Set 1.1
1. $<$ **3.** $>$ **5.** $=$ **7.** $<$ **9.** True **11.** False **13.** False **15.** True **17.** $8 < 12$ **19.** $5 \geq 4$
21. $2 + 3 < 6$ **23.** $\dfrac{10}{2} = 5$ **25.** $4 \leq (3)(5)$ **27.** $<$ **29.** $>$ **31.** $<$ **33.** $=$ **35.** $>$ **37.** True
39. False **41.** True **43.** $25 \leq 25$ **45.** $6 > 0$ **47.** $b \geq a$ **49.** $y < x$ **51.** $4 > 2.5$ **53.** $8 \leq 12$
55. $5 \leq 6$ **57.** $5 + 6 > 10$ **59.** $(3)(5) > 12$ **61.** $\dfrac{12}{6} > 1$ **63.** $3 > 2$ **65.** $4 = 4 + 0$ **67.** $a \leq 5$
69. $c < d$

Exercise Set 1.2
1. whole, integers, rational, real **3.** integers, rational, real **5.** natural, whole, integers, rational, real **7.** rational, real
9. $<$ **11.** $>$ **13.** $=$ **15.** $<$ **17.** $=$ **19.** $>$ **21.** $<$ **23.** integers, rational, real **25.** irrational, real
27. natural, whole, integers, rational, real **29.** True **31.** False **33.** True **35.** True **37.** False
39. True **41.** False **43.** $>$ **45.** $>$ **47.** $<$ **49.** $<$ **51.** $>$ **53.** $=$ **55.** False **57.** True
59. False **61.** False **63.** the distance between a given number and 0 on the number line

Exercise Set 1.3
1. $2 \cdot 2 \cdot 5$ **3.** $3 \cdot 5 \cdot 5$ **5.** $3 \cdot 3 \cdot 5$ **7.** $\dfrac{1}{2}$ **9.** $\dfrac{2}{3}$ **11.** $\dfrac{3}{7}$ **13.** $\dfrac{3}{5}$ **15.** $\dfrac{3}{8}$ **17.** $\dfrac{1}{2}$ **19.** $\dfrac{6}{7}$ **21.** 15
23. $\dfrac{1}{6}$ **25.** $\dfrac{25}{27}$ **27.** $\dfrac{3}{5}$ **29.** 1 **31.** $\dfrac{1}{3}$ **33.** $\dfrac{9}{35}$ **35.** $\dfrac{21}{30}$ **37.** $\dfrac{4}{18}$ **39.** $\dfrac{16}{20}$ **41.** $\dfrac{23}{21}$ **43.** $\dfrac{2}{3}$ **45.** $\dfrac{5}{66}$
47. $\dfrac{7}{5}$ **49.** $\dfrac{5}{7}$ **51.** $\dfrac{65}{21}$ **53.** $\dfrac{2}{5}$ **55.** $\dfrac{9}{7}$ **57.** $\dfrac{3}{4}$ **59.** $\dfrac{17}{3}$ **61.** $\dfrac{7}{26}$ **63.** 1 **65.** $\dfrac{1}{5}$ **67.** $\dfrac{31}{6}$ **69.** $\dfrac{17}{18}$
71. $12\dfrac{3}{4}$ lbs. **73.** $\dfrac{11}{36}$ yd.

Calculator Box 1.4
1. 125 **3.** $59{,}049$ **5.** 30 **7.** 9857 **9.** 2376

Exercise Set 1.4
1. 243 **3.** 27 **5.** 1 **7.** 5 **9.** $\dfrac{1}{125}$ **11.** $\dfrac{16}{81}$ **13.** 17 **15.** 20 **17.** 10 **19.** 21 **21.** 50 **23.** $\dfrac{25}{24}$

25. $\dfrac{7}{18}$ **27.** $\dfrac{21}{8}$ **29.** $\dfrac{7}{5}$ **31.** 88   **33.** 49   **35.** 32   **37.** 0   **39.** 16   **41.** 1.44   **43.** =   **45.** >

47. >   **49.** <   **51.** 21   **53.** 16   **55.** $\dfrac{3}{4}$ **57.** 30   **59.** 225   **61.** 18   **63.** 3   **65.** 9   **67.** $\dfrac{17}{24}$ **69.** $\dfrac{7}{6}$

71. 6   **73.** $\dfrac{7}{5}$ **75.** $\dfrac{27}{55}$ **77.** 2.86   **79.** 3.752   **81.** to eliminate confusion

83. yes, otherwise the product of 3 and 5 would be taken first.

Exercise Set 1.5

1. $x + 15$ **3.** $x - 5$ **5.** $3x + 22$ **7.** 1   **9.** 11   **11.** 8   **13.** 10   **15.** 15   **17.** $2x = 17$

19. $x - 7 = 0$ **21.** $8 + 2x = 42$ **23.** 0   **25.** $\dfrac{9}{5}$ **27.** 5   **29.** 42   **31.** 15   **33.** 98   **35.** 120   **37.** 36

39. 1   **41.** $\dfrac{19}{42}$ **43.** $7(x + 19)$ **45.** $4x - x = 75.6$ **47.** $12 - x$ **49.** $\dfrac{3}{4}(x + 1) = 9$

51. $1\dfrac{11}{12} + 3x = x + 2$ **53.** $\dfrac{11}{14}$ yd.   **55.** 55 sq. in.   **57.** 51 mph

Exercise Set 1.6

1. 9   **3.** -14 **5.** -15 **7.** -16 **9.** 11   **11.** $2\dfrac{5}{8}$ **13.** $-6°$ **15.** -6 **17.** 2   **19.** 0   **21.** -6

23. -2 **25.** 0   **27.** $-\dfrac{2}{3}$ **29.** -8 **31.** -12 **33.** 6   **35.** -4 **37.** 7   **39.** 12   **41.** -8 **43.** -59

45. -8 **47.** $-\dfrac{3}{16}$ **49.** $-\dfrac{13}{10}$ **51.** -19 **53.** 31   **55.** 59   **57.** -2.1 **59.** 38   **61.** -300 **63.** -9

65. -13.1 **67.** 5   **69.** -24 **71.** 19   **73.** $\dfrac{1}{14}$ **75.** 146 ft.   **77.** $-\dfrac{3}{8}$ **79.** answers vary
81. answers vary

Exercise Set 1.7

1. -10 **3.** -5 **5.** 19   **7.** -15 **9.** -34 **11.** -5 **13.** 3   **15.** -45 **17.** -4 **19.** 9   **21.** -3

23. -16 **25.** 2   **27.** -10 **29.** $\dfrac{11}{6}$ **31.** $100°$ **33.** 23 yd. loss   **35.** -11 **37.** 11   **39.** 5   **41.** $5\dfrac{7}{10}$

43. $-\dfrac{31}{36}$ **45.** 4.1   **47.** 1   **49.** -7 **51.** 37   **53.** -17 **55.** $4\dfrac{1}{4}$ **57.** $\dfrac{31}{40}$ **59.** -6.4 **61.** 5.17

63. -36 **65.** -73 **67.** -25 **69.** -223 **71.** -13 **73.** 25   **75.** -18 **77.** 9   **79.** 23   **81.** $\dfrac{10}{3}$

83. 105   **85.** $\dfrac{5}{12}$ **87.** 63 B.C.   **89.** -308 ft.

Calculator Box 1.8

1. 8   **3.** -441 **5.** 2   **7.** 274   **9.** 9

Exercise Set 1.8

1. -12 **3.** 42   **5.** 0   **7.** -18 **9.** $-\dfrac{2}{3}$ **11.** 2   **13.** -30 **15.** 90   **17.** 16   **19.** -9 **21.** 3

23. 5   **25.** 0   **27.** undefined   **29.** 16   **31.** -3 **33.** $-\dfrac{16}{7}$ **35.** -21 **37.** 41   **39.** -134 **41.** 3

43. -1 **45.** 12   **47.** -14 **49.** -6 **51.** undefined   **53.** 5   **55.** 0   **57.** -36 **59.** -125 **61.** 16

63. -16 **65.** 2   **67.** $\dfrac{6}{5}$ **69.** -5 **71.** 18   **73.** -30 **75.** -24 **77.** $-\dfrac{1}{2}$ **79.** $-4\dfrac{1}{5}$ **81.** -8.372

83. 14   **85.** 22   **87.** 2   **89.** 3   **91.** -3 **93.** $\dfrac{2}{3}$ **95.** \$420   **97.** answers vary

Exercise 1.9

1. commutative property of multiplication   **3.** associative property of addition   **5.** distributive property   **7.** $18 + 3x$

9. $-2y + 2z$ **11.** $-21y + 35$ **13.** -16 **15.** 8   **17.** -9 **19.** $\dfrac{3}{2}$ **21.** $-\dfrac{6}{5}$ **23.** $\dfrac{1}{6}$ **25.** $-\dfrac{1}{2}$

27. $\dfrac{2}{3} \cdot \dfrac{3}{2} = 1$ **29.** $(-3)(-4)$ **31.** $(3 + 8) + 9$ **33.** associative property of multiplication

35. identity property of addition   **37.** distributive property   **39.** associative property of multiplication

41. $5x + 20m + 10$ **43.** $-4 + 8m - 4n$ **45.** $-5x - 2$ **47.** $-r + 3 + 7p$ **49.** 0   **51.** -2 **53.** -8

55. -3 **57.** 2   **59.** 5   **61.** $\dfrac{9}{3} = 3$ **63.** -1 **65.** undefined; there is none.   **67.** $-\dfrac{5}{3}$ **69.** $\dfrac{6}{23}$

71. $y + 0 = y$ **73.** $xa + xb$ **75.** $(b + c)a$ **77.** answers vary   **79.** answers vary

Chapter 1 Review

1. $<$ **3.** $=$ **5.** $4 \geq 3$ **7.** $(8 + 4) \leq 12$ **9.** $7 = 3 + 4$

11. (a) $\{1, 3\}$ **(b)** $\{0, 1, 3\}$ **(c)** $\{-6, 0, 1, 3\}$ **(d)** $\left\{-6, 0, 1, 1\dfrac{1}{2}, 3, 9.62\right\}$ **(e)** $\{\pi\}$ **(f)** $\left\{-6, 0, 1, 1\dfrac{1}{2}, 3, 9.62, \pi\right\}$

13. $>$ **15.** $<$ **17.** $2 \cdot 2 \cdot 3 \cdot 3$ **19.** $\dfrac{12}{25}$ **21.** $\dfrac{13}{10}$ **23.** $9\dfrac{3}{8}$ **25.** 70   **27.** 37   **29.** $\dfrac{18}{7}$ **31.** 18

33. 5   **35.** 9   **37.** -11 **39.** 12   **41.** -13.9 **43.** -14 **45.** 5   **47.** -19 **49.** 15   **51.** -48

53. 3   **55.** -36 **57.** undefined   **59.** 6   **61.** $-\dfrac{1}{6}$ **63.** commutative property of addition

65. distributive property   **67.** associative property of addition   **69.** distributive property

71. multiplicative inverse   **73.** commutative property of addition

Chapter 1 Test

1. $|-7| > 5$ **2.** $(9 + 5) \geq 4$ **3.** -5 **4.** -11 **5.** -14 **6.** -39 **7.** 12   **8.** -2 **9.** undefined

10. 4   **11.** $-\dfrac{1}{3}$ **12.** $4\dfrac{5}{8}$ **13.** $\dfrac{51}{40}$ **14.** -32 **15.** -48 **16.** 3   **17.** 0   **18.** $>$ **19.** $>$ **20.** $>$

21. $=$ **22. (a)** $\{1, 7\}$ **(b)** $\{0, 1, 7\}$ **(c)** $\{-5, -1, 0, 1, 7\}$ **(d)** $\left\{-5, -1, \dfrac{1}{4}, 0, 1, 7, 11.6\right\}$ **(e)** $\{\sqrt{7}, 3\pi\}$

(f) $\left\{-5, -1, \dfrac{1}{4}, 0, 1, 7, 11.6, \sqrt{7}, 3\pi\right\}$ **23.** 40   **24.** 12   **25.** 22   **26.** -1 **27.** associative property of addition

28. commutative property of multiplication   **29.** distributive property   **30.** multiplicative inverse   **31.** 9   **32.** -3

CHAPTER 2
Solving Linear Equations and Inequalities

Mental Math, Sec. 2.1

1. -7 **3.** 1   **5.** 17   **7.** like   **9.** unlike   **11.** like

Exercise Set 2.1

1. $15y$ **3.** $13w$ **5.** $-7b - 9$ **7.** $-m - 6$ **9.** $5y - 20$ **11.** $7d - 11$ **13.** $-3x + 2y - 1$

15. $2x + 14$ **17.** $10x - 3$ **19.** $-4x - 9$ **21.** $2x - 4$ **23.** $\dfrac{3}{4}x + 12$ **25.** $12 - z$ **27.** $(n + 284)$ votes

29. $x + 2$ **31.** $2x + 2$ **33.** $5x^2$ **35.** $4x - 3$ **37.** $8x - 53$ **39.** -8 **41.** $7.2x - 5.2$ **43.** $k - 6$

45. $0.9m + 1$ **47.** $-12y + 16$ **49.** $x + 5$ **51.** -11 **53.** $1.3x + 3.5$ **55.** $x + 2$ **57.** $-4m - 3$

59. $8(x + 6)$ **61.** $x - 10$ **63.** $(18x - 2)$ ft. **65.** $\frac{x}{6}(7)$ **67.** $(180 - x)°$ **69.** $(173 - 3x)°$ **71.** $2x + 4$
73. $5b^2c^3 + b^3c^3$ **75.** $5x^2 + 9x$ **77.** $-7x^2y$ **79.** answers vary **81.** 2 **83.** -25 **85.** -32

Mental Math, Sec. 2.2
1. $x = 2$ **3.** $n = 12$ **5.** $b = 17$

Exercise Set 2.2
1. yes **3.** no **5.** yes **7.** $x = -13$ **9.** $y = -14$ **11.** $x = -8$ **13.** $x = 11$ **15.** $t = 0$ **17.** $x = -3$
19. $x = 11$ **21.** $w = -30$ **23.** $x = -7$ **25.** $2x + 7 = x + 6, -1$ **27.** $3(x - 10) = 14, 14\frac{2}{3}$ **29.** $x = -2$
31. $y = -10$ **33.** $y = 8.9$ **35.** $b = -0.7$ **37.** $x = -1$ **39.** $t = 12$ **41.** $n = 2$ **43.** $y = 0.2$
45. $x = -12$ **47.** $n = 21$ **49.** $c = \frac{1}{8}$ **51.** $t = -6$ **53.** $y = -25$ **55.** $m = 0$ **57.** $t = 1.83$
59. $3x - 6 = 2x + 8, 14$ **61.** $2(x - 8) = 3(x + 3), -25$ **63.** answers vary **65.** $\frac{2}{7}$ **67.** $\frac{22}{15}$ **69.** $\frac{1}{3}$

Mental Math, Sec. 2.3
1. $a = 9$ **3.** $b = 2$ **5.** $x = -5$

Exercise Set 2.3
1. $x = -4$ **3.** $x = 0$ **5.** $x = 12$ **7.** $x = -12$ **9.** $d = 3$ **11.** $a = -2$ **13.** $k = 0$ **15.** $x = 10$
17. $x = -4$ **19.** $x = -5$ **21.** $y = -6$ **23.** $x = -3$ **25.** $x = 5$ **27.** $x = 2$ **29.** $a = -2$
31. $\frac{1}{2}x = -5, -10$ **33.** $3x = 2x, 0$ **35.** 4 ft. and 8 ft. **37.** $45°$ and $135°$ **39.** $w = -6$ **41.** $z = 4$
43. $h = \frac{3}{4}$ **45.** $a = 0$ **47.** $k = 0$ **49.** $x = \frac{3}{2}$ **51.** $x = 21$ **53.** $x = \frac{11}{2}$ **55.** $x = -\frac{1}{4}$ **57.** $x = -\frac{5}{6}$
59. $n = 1$ **61.** $z = \frac{9}{10}$ **63.** $x = -2$ **65.** $y = -7$ **67.** $z = 1$ **69.** $n = 2$ **71.** $x = -30$ **73.** $\frac{5}{2} = 2\frac{1}{2}$
75. 5 ft. and 16 ft. **77.** -4 **79.** $x = 4$ cm, $2x = 8$ cm **81.** $x = \frac{5}{4}$ **83.** answers vary **85.** $>$ **87.** $=$
89. $=$ **91.** $<$

Calculator Box 2.4
1. $-24 = -24; x = -12$ checks **3.** $19.4 \neq 10.4$; not a solution **5.** $17,061 = 17,061; x = 121$ checks

Exercise Set 2.4
1. $x = 2$ **3.** $x = -5$ **5.** $x = 10$ **7.** $x = 1$ **9.** $n = \frac{9}{2}$ **11.** $x = \frac{9}{4}$ **13.** $x = 0$ **15.** $z = 18$
17. $x = 1$ **19.** all real numbers **21.** no solution **23.** no solution **25.** $2x + \frac{1}{5} = 3x - \frac{4}{5}, 1$
27. $3(x + 5) = 2x - 1, -16$ **29.** -11 and -9 **31.** $x = 4$ **33.** $y = -4$ **35.** $x = 3$ **37.** $y = -2$
39. $c = 4$ **41.** $x = \frac{7}{3}$ **43.** no solution **45.** $z = \frac{9}{5}$ **47.** $y = \frac{4}{19}$ **49.** $a = 1$ **51.** no solution **53.** $x = \frac{7}{2}$
55. $v = -17$ **57.** $t = \frac{19}{6}$ **59.** $t = -\frac{2}{7}$ **61.** $x = 3$ **63.** $x = 13$ **65.** -12 **67.** 15 and 16 **69.** $-\frac{3}{4}$
71. $-1, 0, 1$ **73.** 1 **75.** $x = -\frac{7}{8}$ **77.** $t = 0$ **79.** no solution **81.** answers vary **83.** answers vary
85. $4p + 6$ **87.** $-13y - 1$ **89.** $-2x + 37$ **81.** $10x - 20$

Exercise Set 2.5
1. $h = 3$ **3.** $I = 800$ **5.** $r = 15$ **7.** 140 ft. **9.** 21 meters **11.** $23°$ F **13.** \$360 **15.** 7 hrs.
17. $h = \frac{f}{5g}$ **19.** $W = \frac{V}{LH}$ **21.** $y = 7 - 3x$ **23.** $R = \frac{A - p}{PT}$ **25.** $A = \frac{3V}{h}$ **27.** $b = 5$ **29.** $h = 4$
31. $h = 15$ **33.** $4\frac{4}{5}$ hrs. **35.** 6.72 or 7 packages **37.** $2\frac{1}{2}$ hrs. **39.** 1.34 or 2 gal. **41.** \$2130 **43.** $16''$ pizza

45. 288 cu. in.   **47.** 512 cu. in.   **49.** $a = P - b - c$ **51.** $x = \dfrac{13 + 4y}{5}$ **53.** $h = \dfrac{A}{0.5b}$ **55.** $F = \dfrac{9}{5}C + 32$

57. $y = \dfrac{2x + 3}{7}$ **59.** $y = \dfrac{8}{3} - x$ **61.** $r = 16T - 12$ **63.** answers vary   **65.** $\dfrac{9}{x + 5}$ **67.** $2(10 + 4x)$

69. $\dfrac{1}{2}(x)(5)$ **71.** $\dfrac{2x}{3x}$

Exercise Set 2.6

1. length = 78 ft.; width = 52 ft.   **3.** 2.6 in.   **5.** 3 nickels; 17 dimes   **7.** 4 $10 bills; 8 $20 bills

9. $11,500 @ 8%; $13,500 @ 9%   **11.** $400 @ 9%; $650 @ 10%   **13.** $666\dfrac{2}{3}$ miles   **15.** 160 miles   **17.** 2 gal.

19. $6\dfrac{2}{3}$ lbs.   **21.** 18 ft., 36 ft., 48 ft.   **23.** $30,000 @ 8%; $24,000 @ 10%   **25.** $4500   **27.** 6 nickels

29. 3 adult tickets   **31.** $2\dfrac{2}{9}$ hrs.   **33.** 400 oz.   **35.** $1\dfrac{5}{7}$ hrs.   **37.** $1\dfrac{1}{2}$ hrs.   **39.** answers vary   **41.** -10

43. -9 **45.** -15

Mental Math, Sec. 2.7

1. $x > 2$ **3.** $x \geq 8$

Exercise Set 2.7

1. $x \leq -1$

3. $x > \dfrac{1}{2}$

5. $-1 < x < 3$

7. $0 \leq y < 2$

9. $x < -3$

11. $x \geq -5$

13. $x \geq -2$

15. $x > -3$

17. $x \leq 1$

19. $x > -5$

21. $x \leq -2$

23. $x \leq -8$

25. $x > 4$

27. $-1 < x < 2$

29. $4 \leq x \leq 5$

31. $1 < x < 5$

33. $x > -10$ **35.** 35 cm.

37. $x \geq 20$

39. $x > 16$

41. $x > -3$

43. $x \geq -\dfrac{2}{3}$

45. $x > \dfrac{8}{3}$

47. $x > -13$

49. $x > 0$

51. $x \geq 0$

53. $1 < x < 4$

55. $x > 3$

57. $x \leq 0$

59. $0 < x < \dfrac{14}{3}$

61. 193   **63.** $-3 < x < 3$

65. 10%   **67.**

$x > 1$

69.

$x < \frac{5}{8}$

71.

$x \le 0$

73. answers vary   **75.** 8   **77.** 1   **79.** $\frac{16}{49}$

Chapter 2 Review

1. $6x$ **3.** $4x - 2$ **5.** $3n - 18$ **7.** $-6x + 7$ **9.** $3x - 7$ **11.** $10 - x$ **13.** $x = 4$ **15.** $x = -6$
17. $x = -9$ **19.** 1   **21.** $x = -12$ **23.** $x = -6$ **25.** $x = 7$ **27.** 3   **29.** -4 **31.** $x = -4$
33. $x = -3$ **35.** no solution   **37.** $n = -\frac{8}{9}$ **39.** $y = -1$ **41.** $t = 0$ **43.** $a = -\frac{6}{23}$ **45.** $a = -\frac{2}{5}$
47. $y = 0.7$ **49.** $\frac{5}{7}$ **51.** 5, 7, and 9   **53.** $s = \frac{r + 5}{vt}$ **55.** $y = \frac{2 + 3x}{6}$ **57.** $\pi = \frac{C}{2r}$ **59.** $32\frac{2}{9}°$
61. $35,000 @ 8.5%, $15,000 @ 10.5%   **63.** length = 30 meters   **65.** 48 miles
67.

$x > 0$

69.

$-0.5 \le y \le 1.5$

71.

$x < -4$

73.

$x \le 4$

75.

$-\frac{1}{2} < x < \frac{3}{4}$

77.

$x \le \frac{19}{3}$

79. score must be less than 83

Chapter 2 Test

1. $y - 10$ **2.** $2x - 9$ **3.** $-2x + 10$ **4.** $-15y + 1$ **5.** $x = -5$ **6.** $n = 8$ **7.** $y = \frac{7}{10}$ **8.** $z = 0$

9. $x = 27$ **10.** $y = -\frac{19}{6}$ **11.** $x = 3$ **12.** no solution   **13.** $y = \frac{3}{11}$ **14.** $c = \frac{2}{17}$ **15.** $x = 1$ **16.** $a = \frac{25}{7}$

17. 21   **18.** 7 gal.   **19.** $50,000 @ 8%; $40,000 @ 10%   **20.** 11 $10 bills   **21.** $2\frac{1}{2}$ hrs.   **22.** $h = \frac{V}{\pi r^2}$

23. $t = \frac{W}{6b}$ **24.** $h = \frac{5g - p}{2}$ **25.** $y = \frac{3x - 10}{4}$ **26.**

$x < -2$

27.

$x < 4$

28.

$-1 < x < \frac{7}{3}$

29.

$\frac{7}{4} < x < 4$

30.

$x > \frac{2}{5}$

Chapter 2 Cumulative Review

1. (a) 4; (b) 5; (c) 0; *Sec. 1.2, Ex. 3*   **2.** (a) $\frac{6}{7}$; (b) $\frac{11}{27}$; (c) $\frac{22}{5}$; *Sec. 1.3, Ex. 2*   **3.** (a) $\frac{6}{7}$; (b) $\frac{1}{2}$; (c) 1; (d) $\frac{4}{3}$; *Sec. 1.3, Ex. 5*
4. $\frac{8}{3}$; *Sec. 1.4, Ex. 3*   **5.** (a) 4; (b) $\frac{9}{4}$; (c) $\frac{5}{2}$; (d) 5; *Sec. 1.5, Ex. 1*
6. (a) $\frac{15}{x} = 4$; (b) $12 - 3 = x$; (c) $4x + 17 = 21$; *Sec. 1.5, Ex. 3*   **7.** (a) -10; (b) 17; (c) -21; (d) -12; *Sec. 1.6, Ex. 1*
8. (a) 6; (b) -6; *Sec. 1.6, Ex. 6*   **9.** (a) -12; (b) -3; *Sec. 1.7, Ex. 3*   **10.** (a) -24; (b) -2; (c) 50; *Sec. 1.8, Ex. 1*
11. (a) -6; (b) -24; (c) $\frac{3}{4}$; *Sec. 1.8, Ex. 7*   **12.** (a) $4x$; (b) $11y^2$; (c) $8x^2 - x$; *Sec. 2.1, Ex. 3*
13. (a) $5x + 10$; (b) $-2y - 0.6z + 2$; (c) $-x - y + 2z - 6$; *Sec. 2.1, Ex. 5*   **14.** yes, 2 is a solution; *Sec. 2.2, Ex. 1*
15. $t = -7$; *Sec. 2.2, Ex. 4*   **16.** $y = 140$; *Sec. 2.3, Ex. 3*   **17.** $a = -5$; *Sec. 2.3, Ex. 7*   **18.** $x = 12$; *Sec. 2.4, Ex. 1*
19. $t = \frac{16}{3}$; *Sec. 2.4, Ex. 3*   **20.** $x = 10$; *Sec. 2.4, Ex. 7*   **21.** width = 6 ft.; *Sec. 2.5, Ex. 1*
22. $x = \frac{y - b}{m}$; *Sec. 2.5, Ex. 7*   **23.** *Sec. 2.7, Ex. 7*

$x \le -18$

CHAPTER 3
Exponents and Polynomials

Mental Math, Sec. 3.1
1. base = 3; exponent = 2　　**3.** base = -3; exponent = 6　　**5.** base = 4; exponent = 2

Exercise Set 3.1
1. 49　**3.** -5　**5.** -16　**7.** 16　**9.** x^7　**11.** $15y^5$　**13.** $-24z^{20}$　**15.** p^7q^7　**17.** $\dfrac{m^9}{n^9}$　**19.** $x^{10}y^{15}$

21. $\dfrac{4x^2z^2}{y^{10}}$　**23.** x^2　**25.** 4　**27.** p^6q^5　**29.** $\dfrac{y^3}{2}$　**31.** 1　**33.** -2　**35.** 2　**37.** $5p^2q$　**39.** $81x^2yz^2$

41. $6^5m^4n^3$　**43.** ab^4　**45.** -5　**47.** -64　**49.** 1　**51.** 40　**53.** b^6　**55.** a^9　**57.** $-16x^7$

59. cannot be combined　**61.** $64a^3$　**63.** $36x^2y^2z^6$　**65.** $\dfrac{y^{15}}{8x^{12}}$　**67.** x　**69.** $2x^2y$　**71.** $\dfrac{2a^5y^5}{7}$　**73.** $-\dfrac{3x^3y^4}{2}$

75. $\dfrac{a^{15}}{q^{10}}$　**77.** $\dfrac{y^{18}}{x^3}$　**79.** $\dfrac{y^2}{25x^2}$　**81.** $243x^7$　**83.** $\dfrac{xy}{7}$　**85.** $-\dfrac{25}{q}$　**87.** $20x^5$ sq. ft.　**89.** $27y^{12}$ cubic ft.

91. x^{9a}　**93.** a^{5b}　**95.** x^{5a}　**97.** $x^{5a^2}y^{5ab}z^{5ac}$　**99.** answers vary　**101.** 18　**103.** 6　**105.** 10

Calculator Box 3.2
1. 5.31 EE 03　**3.** 6.6 EE -09　**5.** 1.5×10^{13}　**7.** 6×10^{19}

Mental Math, Sec. 3.2
1. $\dfrac{5}{x^2}$　**3.** y^6　**5.** $4y^3$

Exercise Set 3.2
1. $\dfrac{1}{4^3} = \dfrac{1}{64}$　**3.** $\dfrac{7}{x^3}$　**5.** $4^3 = 64$　**7.** $\dfrac{5}{6}$　**9.** p^3　**11.** $\dfrac{q^4}{p^5}$　**13.** $\dfrac{1}{x^3}$　**15.** z^3　**17.** a^{30}　**19.** $\dfrac{1}{x^{10}y^6}$　**21.** $\dfrac{z^2}{4}$

23. 7.8×10^4　**25.** 1.67×10^{-6}　**27.** 9.3×10^7　**29.** 786,000,000　**31.** 0.0000000008673

33. 6,250,000,000,000,000,000　**35.** 80,000,000,000　**37.** $\dfrac{1}{5^3x^3} = \dfrac{1}{125x^3}$　**39.** $\dfrac{1}{9}$　**41.** $-p^4$　**43.** -2　**45.** r^6

47. $\dfrac{1}{x^{15}y^9}$　**49.** $\dfrac{4}{3}$　**51.** $\dfrac{1}{2^5x^5} = \dfrac{1}{32x^5}$　**53.** $\dfrac{49a^4}{b^6}$　**55.** $a^{24}b^8$　**57.** x^9y^{19}　**59.** $-\dfrac{y^8}{8x^2}$　**61.** 6.35×10^{-3}

63. 1.16×10^6　**65.** 2.0×10^7　**67.** 0.033　**69.** 20,320　**71.** 9,460,000,000,000 k.　**73.** 0.000036

75. 0.0000000000000000028　**77.** 0.0000005　**79.** 200,000　**81.** 15,120,000,000 cubic ft.　**83.** 1.248×10^6

85. a^m　**87.** $27y^{6z}$　**89.** y^{5a}　**91.** $\dfrac{1}{z^{6a+4}}$　**93.** answers vary　**95.** $x = 8$　**97.** $y = -22$　**99.** $x = -4$

Mental Math, Sec. 3.3
1. $-14y$　**3.** $7y^3$　**5.** $7x$

Exercise Set 3.3
1. 1　**3.** 3　**5.** 8　**7.** 1　**9.** (a) 6; (b) 5　**11.** (a) -2, (b) 4　**13.** $23x^2$　**15.** $12x^2 - y$　**17.** $7s$

19. $12x + 12$　**21.** $-3x^2 + 10$　**23.** $-3x^2 + 4$　**25.** $-x + 14$　**27.** $-2x + 9$　**29.** $2x^2 + 7x - 16$

31. $8t^2 - 4$　**33.** $-2z^2 - 16z + 6$　**35.** $2x^3 - 2x^2 + 7x + 2$　**37.** $15a^3 + a^2 + 16$　**39.** $62x^2 + 5$

41. $12x + 2$　**43.** $9xy^2 - x - 27$　**45.** (a) 16; (b) -5　**47.** (a) 1; (b) 57　**49.** (a) 12; (b) -79　**51.** $7x - 13$

53. $-y^2 - 3y - 1$　**55.** $y^2 - 7$　**57.** $2x^2 + 11x$　**59.** $-16x^2 + 8x + 9$　**61.** $8x^2 + 14x + 4$　**63.** $-11x$

65. $27w^2 + 3$　**67.** $-6x^3 - 3x^2 + 3x + 3$　**69.** $2x^2 + 8x - 19$　**71.** $3x - 3$　**73.** $7x^2 - 2x + 2$

75. $4y^2 + 12y + 19$　**77.** $6x^2 - 5x + 21$　**79.** $(x^2 + 7x + 4)$ ft　**81.** $(3y^2 + 4y + 11)$ m

83. (a) $12x^4$; (b) $7x^2$; answers vary　**85.** $<$　**87.** $=$　**89.** $\{-2, 0, 25\}$　**91.** $\{0, 25\}$

Mental Math, Sec. 3.4
1. $10xy$ **3.** x^7 **5.** $18x^3$

Exercise Set 3.4
1. $4a^2 - 8a$ **3.** $7x^3 + 14x^2 - 7x$ **5.** $6x^4 - 3x^3$ **7.** $a^2 + 5a - 14$ **9.** $4y^2 - 16y + 16$
11. $30x^2 - 79xy + 45y^2$ **13.** $4x^4 - 20x^2 + 25$ **15.** $x^3 + 6x^2 + 12x + 8$ **17.** $8y^3 - 36y^2 + 54y - 27$
19. $x^3 - 5x^2 + 13x - 14$ **21.** $x^4 + 5x^3 - 3x^2 - 11x + 20$ **23.** $10a^3 - 27a^2 + 26a - 12$
25. $2x^3 + 10x^2 + 11x - 3$ **27.** $x^4 - 2x^3 - 51x^2 + 4x + 63$ **29.** $2a^2 + 8a$ **31.** $6x^3 - 9x^2 + 12x$
33. $15x^2 + 37xy + 18y^2$ **35.** $x^3 + 7x^2 + 16x + 12$ **37.** $49x^2 + 56x + 16$ **39.** $-6a^4 + 4a^3 - 6a^2$
41. $x^3 + 10x^2 + 33x + 36$ **43.** $a^3 + 3a^2 + 3a + 1$ **45.** $x^2 + 2xy + y^2$ **47.** $x^2 - 13x + 42$ **49.** $3a^3 + 6a$
51. $-4y^3 - 12y^2 + 44y$ **53.** $25x^2 - 1$ **55.** $5x^3 - x^2 + 16x + 16$ **57.** $8x^3 - 60x^2 + 150x - 125$
59. $32x^3 + 48x^2 - 6x - 20$ **61.** $49x^2y^2 - 14xy^2 + y^2$ **63.** $5y^4 - 16y^3 - 4y^2 - 7y - 6$
65. $6x^4 - 8x^3 - 7x^2 + 22x - 12$ **67.** $(4x^2 - 25)$ sq. yds. **69.** $(x^2 + 8x + 16)$ sq. ft.
71. (a) $a^2 - b^2$; (b) $4x^2 - 9y^2$; (c) $16x^2 - 49$ **73.** $2x^2 + 4x - 7$ **75.** $9y^2 + 12y - 1$ **77.** $x^2 - x - 7$

Exercise Set 3.5
1. $x^2 + 7x + 12$ **3.** $x^2 + 5x - 50$ **5.** $5x^2 + 4x - 12$ **7.** $4y^2 - 25y + 6$ **9.** $6x^2 + 13x - 5$
11. $x^2 - 4x + 4$ **13.** $4x^2 - 4x + 1$ **15.** $9a^2 - 30a + 25$ **17.** $25x^2 + 90x + 81$ **19.** $a^2 - 49$
21. $9x^2 - 1$ **23.** $9x^2 - \dfrac{1}{4}$ **25.** $81x^2 - y^2$ **27.** $a^2 + 9a + 20$ **29.** $a^2 + 14a + 49$ **31.** $12a^2 - a - 1$
33. $x^2 - 4$ **35.** $9a^2 + 6a + 1$ **37.** $4x^2 + 3xy - y^2$ **39.** $4a^2 - 12a + 9$ **41.** $25x^2 - 36z^2$
43. $x^2 - 8x + 15$ **45.** $x^2 - \dfrac{1}{9}$ **47.** $a^2 + 8a - 33$ **49.** $x^2 - 4x + 4$ **51.** $6b^2 - b - 35$ **53.** $49p^2 - 64$
55. $\dfrac{1}{9}a^4 - 49$ **57.** $4r^2 - 9s^2$ **59.** $9x^2 - 42xy + 49y^2$ **61.** $16x^2 - 25$ **63.** $x^2 + 8x + 16$ **65.** $a^2 - \dfrac{1}{4}y^2$
67. $\dfrac{1}{25}x^2 - y^2$ **69.** $(4x^2 + 4x + 1)$ sq. ft. **71.** $x^2 + 2xy + y^2 - 9$ **73.** $a^2 - 6a + 9 - b^2$
75. $4x^2 + 4x + 1 - y^2$ **77.** answers vary **79.** $\dfrac{5b^5}{7}$ **81.** $-\dfrac{2a^{10}}{b^5}$ **83.** $\dfrac{2}{3y^2}$

Mental Math, Sec. 3.6
1. a^2 **3.** a^2 **5.** k^3

Exercise Set 3.6
1. $4k^3$ **3.** $3m$ **5.** $-\dfrac{4a^5}{b}$ **7.** $-\dfrac{6x}{7y^2}$ **9.** $5p^2 + 6p$ **11.** $-\dfrac{3}{2x} + 3$ **13.** $-3x^2 \div x - \dfrac{4}{x^3}$
15. $-1 + \dfrac{3}{2x} - \dfrac{7}{4x^4}$ **17.** $5x^3 - 3x + \dfrac{1}{x^2}$ **19.** $x + 1$ **21.** $2x + 3$ **23.** $2x + 1 + \dfrac{7}{x - 4}$ **25.** $4x + 9$
27. $3a^2 - 3a + 1 + \dfrac{2}{3a + 2}$ **29.** $2b^2 + b + 2 - \dfrac{12}{b + 4}$ **31.** $\dfrac{4}{x} + \dfrac{1}{x^2} + \dfrac{9}{5x^3}$ **33.** $5x - 2 + \dfrac{2}{x + 6}$
35. $x^2 - \dfrac{12x}{5} - 1$ **37.** $6x - 1 - \dfrac{1}{x + 3}$ **39.** $4x^2 + 1$ **41.** $2x^2 + 6x - 5 - \dfrac{2}{x - 2}$ **43.** $6x - 1$
45. $-x^3 + 3x^2 - \dfrac{4}{x}$ **47.** $4x + 3 - \dfrac{2}{2x + 1}$ **49.** $2x + 9$ **51.** $2x^2 + 3x - 4$ **53.** $x^2 + 3x + 9$
55. $x^2 - x + 1$ **57.** $-3x + 6 - \dfrac{11}{x + 2}$ **59.** $2b - 1 - \dfrac{6}{2b - 1}$ **61.** $(3x^3 + x - 4)$ ft. **63.** $(7x - 10)$ in.
65. answers vary **67.** $-12a^3 + 16a$ **69.** $4y^3 - 32y^2 - 16y$ **71.** $-36x^2y^2z - 63x^2y^3z - 18xy$
73. $-42s^3r^2 - 63s^2r^3 - 63s^2r^2 - 56sr$

Chapter 3 Review

1. base = 3; exponent = 2 **3.** base = 5; exponent = 4 **5.** 36 **7.** -65 **9.** 1 **11.** 8 **13.** $-10x^5$

15. $\dfrac{b^4}{16}$ **17.** $\dfrac{x^6 y^6}{4}$ **19.** $40a^{19}$ **21.** $\dfrac{3}{64}$ **23.** $\dfrac{1}{x}$ **25.** 5 **27.** 1 **29.** $6a^6 b^9$ **31.** $\dfrac{1}{49}$ **33.** $\dfrac{2}{x^4}$

35. 125 **37.** $1\dfrac{1}{16} = \dfrac{17}{16}$ **39.** $8q^3$ **41.** $\dfrac{s^4}{r^3}$ **43.** $-\dfrac{3x^3}{4r^4}$ **45.** $\dfrac{x^{15}}{8}$ **47.** $\dfrac{a^2 b^2 c^4}{9}$ **49.** $\dfrac{5}{x^3}$ **51.** c^4

53. $\dfrac{1}{x^6 y^{13}}$ **55.** $-\dfrac{15}{16}$ **57.** a^{11m} **59.** $27x^3 y^{6z}$ **61.** 2.7×10^{-4} **63.** 8.08×10^7 **65.** 2.34×10^8

67. 867,000 **69.** 0.00086 **71.** 100,000,000,000,000,000,000 **73.** 0.016 **75.** 7 **77.** 8 **79.** 5 **81.** 6

83. $15a^2 b^2 + 4ab$ **85.** $-6a^2 b - 3b^2 - q^2$ **87.** $8k^2 + 3k + 6$ **89.** $-6m^7 - 3x^4 + 7m^6 - 4m^2$ **91.** $9x^3 y$

93. $6x^2 a^5 y$ **95.** $6x + 30$ **97.** $8a + 28$ **99.** $-7x^3 - 35x$ **101.** $-2x^3 + 18x^2 - 2x$ **103.** $-6a^4 + 8a^2 - 2a$

105. $2x^2 - 12x - 14$ **107.** $4a^2 + 27a - 7$ **109.** $x^4 + 7x^3 + 4x^2 + 23x - 35$ **111.** $x^4 + 4x^3 + 4x^2 - 16$

113. $x^3 + 21x^2 + 147x + 343$ **115.** $x^2 + 14x + 49$ **117.** $9x^2 - 42x + 49$ **119.** $25x^2 - 90x + 81$

121. $49x^2 - 16$ **123.** $4x^2 - 36$ **125.** $\dfrac{4}{3z^2 y}$ **127.** $\dfrac{1}{7} + \dfrac{3}{x} + \dfrac{7}{x^2}$ **129.** $a + 1 + \dfrac{6}{a - 2}$

131. $a^2 + 3a + 8 + \dfrac{22}{a - 2}$ **133.** $2x^3 - x^2 + 2 - \dfrac{1}{2x - 1}$

Chapter 3 Test

1. 32 **2.** 81 **3.** -81 **4.** $\dfrac{1}{64}$ **5.** $\dfrac{y^4}{49x^2}$ **6.** $7x^2 y^2$ **7.** $\dfrac{4}{x^6 y^9}$ **8.** $\dfrac{x^2}{y^{14}}$ **9.** $\dfrac{1}{6xy^8}$ **10.** 5.63×10^5

11. 8.63×10^{-5} **12.** 0.0015 **13.** 62,300 **14.** 0.036 **15.** 5 **16.** $3xyz + 18x^2 y$

17. $16x^3 + 7x^2 - 3x - 13$ **18.** $-3x^3 + 5x^2 + 4x + 5$ **19.** $x^3 + 8x^2 + 3x - 5$ **20.** $3x^3 + 22x^2 + 41x + 14$

21. $2x^5 - 5x^4 + 12x^3 - 8x^2 + 4x + 7$ **22.** $3x^2 + 16x - 35$ **23.** $9x^2 - 49$ **24.** $16x^2 - 16x + 4$

25. $64x^2 + 48x + 9$ **26.** $x^4 - 81b^2$ **27.** $\dfrac{2}{x^2 yz}$ **28.** $\dfrac{x}{2y} + \dfrac{1}{4} - \dfrac{7}{8y}$ **29.** $x + 2$ **30.** $9x^2 - 6x + 4 - \dfrac{16}{3x + 2}$

Chapter 3 Cumulative Review

1. (a) $<$; **(b)** $>$; **(c)** $>$; *Sec. 1.1, Ex. 1* **2.** $\dfrac{2}{39}$; *Sec. 1.3, Ex. 3* **3. (a)** $\dfrac{13}{20}$; **(b)** $\dfrac{12}{11}$; **(c)** $1\dfrac{1}{4}$; *Sec. 1.3, Ex. 7*

4. $\dfrac{14}{3}$; *Sec. 1.4, Ex. 5* **5. (a)** -12; **(b)** -9; *Sec. 1.6, Ex. 3* **6. (a)** -17; **(b)** 11; **(c)** -3; **(d)** 6; *Sec. 1.7, Ex. 1*

7. (a) undefined; **(b)** 0; **(c)** 0; *Sec. 1.8, Ex. 5* **8. (a)** 3; **(b)** -5; **(c)** 0; *Sec. 1.9, Ex. 4*

9. (a) $5x + 7$; **(b)** $-4a - 1$; **(c)** $4y - 3y^2$; **(d)** $7.3x - 6$; *Sec. 2.1, Ex. 4*

10. $x = -3$; *Sec. 2.2, Ex. 5* **11.** -10; *Sec. 2.3, Ex. 5* **12.** no solution; *Sec. 2.4, Ex. 5*

13. 3 yrs; *Sec. 2.5, Ex. 4* **14.** $C = \dfrac{5(F - 32)}{9}$; *Sec. 2.5, Ex. 9*

15. *Sec. 2.7, Ex. 2* $\quad 2 < x \le 4$ **16.** *Sec. 2.7, Ex. 11* $\quad -\dfrac{2}{3} \le x \le \dfrac{2}{3}$

17. (a) y^3; **b)** $\dfrac{q^9}{p^4}$; **(c)** $\dfrac{1}{x^{12}}$; *Sec. 3.2, Ex. 3* **18.** $-4x^2 + 6x + 2$; *Sec. 3.3, Ex. 5* **19.** $4x^2 - 4xy + y^2$; *Sec. 3.4, Ex. 3*

20. (a) $4x^2 - y^2$; **(b)** $36t^2 - 49$; **(c)** $c^2 - \dfrac{1}{16}$; **(d)** $4p^2 - q^2$; *Sec. 3.5, Ex. 6* **21.** $x + 4$; *Sec. 3.6, Ex. 5*

CHAPTER 4
Factoring Polynomials

Mental Math Sec. 4.1
1. $2 \cdot 7$ **3.** $2 \cdot 5$ **5.** 3   **7.** 3

Exercise Set 4.1
1. 4   **3.** 6   **5.** y^2 **7.** xy^2 **9.** 4   **11.** $4y^3$ **13.** $3x^3$ **15.** $9x^2y$ **17.** $3(a + 2)$ **19.** $15(2x - 1)$
21. $6cd(4d^2 - 3c)$ **23.** $-6a^3x(4a - 3)$ **25.** $4x(3x^2 + 4x - 2)$ **27.** $5xy(x^2 - 3x + 2)$ **29.** $(x + 2)(y + 3)$
31. $(y - 3)(x - 4)$ **33.** $(x + y)(2x - 1)$ **35.** $(x + 3)(5 + y)$ **37.** $(y - 4)(2 + x)$ **39.** $(y - 2)(3x + 8)$
41. $(y + 3)(y^2 + 1)$ **43.** $3(x - 2)$ **45.** $-2(4x + 9)$ **47.** $2x(16y - 9x)$ **49.** $4(x - 2y + 1)$
51. $(x + 2)(8 - y)$ **53.** $-8x^8y^5(5y + 2x)$ **55.** $5(x + 2)$ **57.** $-3(x - 4)$ **59.** $6x^3y^2(3y - 2 + x^2)$
61. $-2a^2(a + 3b)$ **63.** $(x - 2)(y^2 + 1)$ **65.** $(y + 3)(5x + 6)$ **67.** $(x - 2y)(4x - 3)$ **69.** $42yz(3x^3 + 5y^3z^2)$
71. $4(y - 3)(y + z)$ **73.** $(3 - x)(5 + y)$ **75.** $3(6x + 5)(2 + y)$ **77.** $2(3x^2 - 1)(2y - 7)$ **79.** answers vary
81. answers vary   **83.** $x = 7$ **85.** $x = -9$ **87.** $x = -18$

Mental Math, Sec. 4.2
1. $(x + 5)$ **3.** $(x - 3)$ **5.** $(x + 2)$

Exercise Set 4.2
1. $(x + 6)(x + 1)$ **3.** $(x + 5)(x + 4)$ **5.** $(x - 5)(x - 3)$ **7.** $(x - 9)(x - 1)$ **9.** not factorable
11. $(x - 6)(x + 3)$ **13.** not factorable   **15.** $(x + 5y)(x + 3y)$ **17.** $(x - y)(x - y)$ **19.** $(x - 4y)(x + y)$
21. $2(z + 8)(z + 2)$ **23.** $2x(x - 5)(x - 4)$ **25.** $7(x + 3y)(x - y)$ **27.** $(x + 12)(x + 3)$ **29.** $(x - 2)(x + 1)$
31. $(r - 12)(r - 4)$ **33.** $(x - 7)(x + 3)$ **35.** $(x + 5y)(x + 2y)$ **37.** not factorable   **39.** $2(t + 8)(t + 4)$
41. $x(x - 6)(x + 4)$ **43.** $(x - 9)(x - 7)$ **45.** $(x + 2y)(x - y)$ **47.** $3(x + 5)(x - 2)$ **49.** $3(x - 18)(x - 2)$
51. $(x - 24)(x + 6)$ **53.** $6x(x + 4)(x + 5)$ **55.** $2t^3(t - 4)(t - 3)$ **57.** $5xy(x - 8y)(x + 3y)$
59. $4(x^2 + x - 3)$ **61.** $(x - 3)(x - 7)$ **63.** $2y(x + 5)(x + 10)$ **65.** $-12y^3(x^2 + 2x + 3)$
67. $(x + 1)(y - 5)(y + 3)$ **69.** answers vary   **71.** $\dfrac{1}{25}$ **73.** $\dfrac{1}{x^{15}}$ **75.** $x^2 - 4$ **77.** $y^2 - 2y + 1$

Mental Math, Sec. 4.3
1. Yes   **3.** No   **5.** Yes

Exercise Set 4.3
1. $(2x + 3)(x + 5)$ **3.** $(2x + 1)(x - 5)$ **5.** $(2y + 3)(y - 2)$ **7.** $(4a - 3)^2$ **9.** $(9r - 8)(4r + 3)$
11. $(5x + 1)(2x + 3)$ **13.** $3(7x + 5)(x - 3)$ **15.** $2(2x - 3)(3x + 1)$ **17.** $x(4x + 3)(x - 3)$ **19.** $(x + 11)^2$
21. $(x - 8)^2$ **23.** $(4y - 5)^2$ **25.** $(xy - 5)^2$ **27.** $(2x + 11)(x - 9)$ **29.** $(2x - 7)(2x + 3)$
31. $(6x - 7)(5x - 3)$ **33.** $(4x - 9)(6x - 1)$ **35.** $(3x - 4y)^2$ **37.** $(x - 7y)^2$ **39.** $(2x + 5)(x + 1)$
41. not factorable   **43.** $(5 - 2y)(2 + y)$ **45.** $(4x + 3y)^2$ **47.** $2y(4x - 7)(x + 6)$ **49.** $(3x - 2)(x + 1)$
51. $(xy + 2)^2$ **53.** $(7y + 3x)^2$ **55.** $3(x^2 - 14x + 21)$ **57.** $(7a - 6)(6a - 1)$ **59.** $(6x - 7)(3x + 2)$
61. $(5p - 7q)^2$ **63.** $(5x + 3)(3x - 5)$ **65.** $(7t + 1)(t - 4)$ **67.** $-3xy^2(4x - 5)(x + 1)$
69. $-2pq(p - 3q)(15p + q)$ **71.** $(y - 1)^2(4x^2 + 10x + 25)$ **73.** answers vary   **75.** $x = \dfrac{2}{3}$ **77.** $x = -1$
79. $x^3 - 8$

Mental Math, Sec. 4.4
1. 1^2 **3.** 9^2 **5.** 3^2 **7.** 1^3 **9.** 2^3

Exercise Set 4.4

1. $(5y - 3)(5y + 3)$ **3.** $(11 - 10x)(11 + 10x)$ **5.** $3(2x - 3)(2x + 3)$ **7.** $(13a - 7b)(13a + 7b)$
9. $(xy - 1)(xy + 1)$ **11.** $(a + 3)(a^2 - 3a + 9)$ **13.** $(2a + 1)(4a^2 - 2a + 1)$ **15.** $5(k + 2)(k^2 - 2k + 4)$
17. $(xy - 4)(x^2y^2 + 4xy + 16)$ **19.** $(x + 5)(x^2 - 5x + 25)$ **21.** $3x(2x - 3y)(4x^2 + 6xy + 9y^2)$
23. $(x - 2)(x + 2)$ **25.** $(9 - p)(9 + p)$ **27.** $(2r - 1)(2r + 1)$ **29.** $(3x - 4)(3x + 4)$ **31.** not factorable
33. $(3 - t)(9 + 3t + t^2)$ **35.** $8(r - 2)(r^2 + 2r + 4)$ **37.** $(t - 7)(t^2 + 7t + 49)$ **39.** $(x - 13y)(x + 13y)$
41. $(xy - z)(xy + z)$ **43.** $(xy + 1)(x^2y^2 - xy + 1)$ **45.** $(s - 4t)(s^2 + 4st + 16t^2)$ **47.** $2(3r - 2)(3r + 2)$
49. $x(3y - 2)(3y + 2)$ **51.** $25y^2(y - 2)(y + 2)$ **53.** $xy(x - 2y)(x + 2y)$ **55.** $4s^3t^3(2s^3 + 25t^3)$
57. $xy^2(27xy - 1)$ **59.** $(x - 2)(x + 2)(x^2 + 4)$ **61.** $(a - 2 - b)(a + 2 + b)$ **63.** $(x - 2)^2(x + 1)(x + 3)$
65. $(x + 6)$ **67.** $(2x + y)$ **69.** $4x^3 + 2x^2 - 1 + \dfrac{3}{x}$ **71.** $2x + 1$ **73.** $3x + 4 - \dfrac{2}{x + 3}$

Exercise Set 4.5

1. $(a + b)^2$ **3.** $(a - 3)(a + 4)$ **5.** $(a + 2)(a - 3)$ **7.** $(x + 1)^2$ **9.** $(x + 1)(x + 3)$ **11.** $(x + 3)(x + 4)$
13. $(x + 4)(x - 1)$ **15.** $(x + 5)(x - 3)$ **17.** $(x - 6)(x + 5)$ **19.** $2(x - 7)(x + 7)$ **21.** $(x + 3)(x + y)$
23. $(x + 8)(x - 2)$ **25.** $4x(x + 7)(x - 2)$ **27.** $2(3x + 4)(2x + 3)$ **29.** $(2a - b)(2a + b)$ **31.** $(5 - 2x)(4 + x)$
33. not factorable   **35.** $(4x - 5)(x + 1)$ **37.** $4(t^2 + 9)$ **39.** $(x + 1)(a + 2)$ **41.** $4a(3a^2 - 6a + 1)$
43. not factorable   **45.** $(5p - 7q)^2$ **47.** $(5 - 2y)(25 + 10y + 4y^2)$ **49.** $(5 - x)(6 + x)$ **51.** $(7 - x)(2 + x)$
53. $3x^2y(x + 6)(x - 4)$ **55.** $5xy^2(x - 7y)(x - y)$ **57.** $3xy(4x^2 + 81)$ **59.** $(x - y - z)(x - y + z)$
61. $(s + 4)(3r - 1)$ **63.** $(4x - 3)(x - 2y)$ **65.** $6(x + 2y)(x + y)$ **67.** $(x + 3)(y - 2)(y + 2)$
69. $(5 + x)(x + y)$ **71.** $(7t - 1)(2t - 1)$ **73.** $(3x + 5)(x - 1)$ **75.** $(x + 12y)(x - 3y)$
77. $(1 - 10ab)(1 + 2ab)$ **79.** $(x - 3)(x + 3)(x - 1)(x + 1)$ **81.** $(x - 4)(x + 4)(x^2 + 2)$ **83.** $(x - 15)(x - 8)$
85. $2x(3x - 2)(x - 4)$ **87.** $(3x - 5y)(9x^2 + 15xy + 25y^2)$ **89.** $(xy + 2z)(x^2y^2 - 2xyz + 4z^2)$
91. $2xy(1 - 6x)(1 + 6x)$ **93.** $(x - 2)(x + 2)(x + 6)$ **95.** $2a^2(3a + 5)$ **97.** $(a^2 + 2)(a + 2)$
99. $(x - 2)(x + 2)(x + 7)$ **101.** answers vary   **103.** -1 **105.** -17 **107.** 5 ft

Mental Math, Sec. 4.6

1. $a = 3$ or $a = 7$ **3.** $x = -8$ or $x = -6$ **5.** $x = -1$ or $x = 3$

Exercise Set 4.6

1. $x = 2$ or $x = -1$ **3.** $x = 0$ or $x = -6$ **5.** $x = -\dfrac{3}{2}$ or $x = \dfrac{5}{4}$ **7.** $x = \dfrac{7}{2}$ or $x = -\dfrac{2}{7}$ **9.** $x = 9$ or $x = 4$

11. $x = -4$ or $x = 2$ **13.** $x = 8$ or $x = -4$ **15.** $x = \dfrac{7}{3}$ or $x = -2$ **17.** $x = \dfrac{8}{3}$ or $x = -9$

19. $x = 0$ or $x = 8$ or $x = 4$ **21.** $x = \dfrac{3}{4}$ **23.** $x = 0$ or $x = \dfrac{1}{2}$ or $x = -\dfrac{1}{2}$ **25.** $x = 0$ or $x = \dfrac{1}{2}$ or $x = -\dfrac{3}{8}$

27. $x = 0$ or $x = -7$ **29.** $x = -5$ or $x = 4$ **31.** $x = -5$ or $x = 6$ **33.** $y = -\dfrac{4}{3}$ or $y = 5$

35. $x = -\dfrac{3}{2}$ or $x = -\dfrac{1}{2}$ or $x = 3$ **37.** $x = -5$ or $x = 3$ **39.** $x = 0$ or $x = 16$ **41.** $y = -\dfrac{9}{2}$ or $y = \dfrac{8}{3}$

43. $x = \dfrac{5}{4}$ or $x = \dfrac{11}{3}$ **45.** $x = -\dfrac{2}{3}$ or $x = \dfrac{1}{6}$ **47.** $x = -\dfrac{5}{3}$ or $x = \dfrac{1}{2}$ **49.** $x = \dfrac{17}{2}$ **51.** $x = 2$ or $x = -\dfrac{4}{5}$

53. $y = -4$ or $y = 3$ **55.** $y = \dfrac{1}{2}$ or $y = -\dfrac{1}{2}$ **57.** $t = -2$ or $t = -11$ **59.** $t = 3$ **61.** $x = -3$

63. $x = \dfrac{3}{4}$ or $x = -\dfrac{4}{3}$ **65.** $t = 0$ or $t = -\dfrac{5}{2}$ or $t = -1$ **67.** -12 or 11   **69.** 9   **71.** $x^2 - 12x + 35 = 0$

73. $x = 0$ or $x = \dfrac{1}{2}$ **75.** $x = 0$ or $x = -15$ **77.** $x = 0$ or $x = 13$ **79.** $\dfrac{47}{45}$ **81.** $\dfrac{17}{60}$ **83.** $\dfrac{15}{8}$ **85.** $\dfrac{7}{10}$

Exercise Set 4.7

1. x and $2x$ **3.** x and $36 - x$ **5.** x = width, $3x - 4$ = length **7.** x = age now, $x - 10$ = age 10 yr ago
9. $25x$ = value of quarters, $10(x + 4)$ = value of dimes **11.** x = second side, $2x - 2$ = first side, $x + 10$ = third side
13. width = 7 in., length = 16 in. **15.** base = 6 m, altitude = 5 m **17.** 5 m **19.** -14 or 13 **21.** 10 and 15
23. 3, 4, 5 or -1, 0, 1 **25.** 18k, 24k, and 30k **27.** 30 miles **29.** width = 11 m, length = 19 m
31. -19 and -21 or 19 and 21 **33.** 3 yd **35.** 10 sides **37.** 27 cm and 36 cm **39.** 12 mm
41. 6 and 8 **43.** width = 7 cm, length = 16 cm **45.** width = 8 in., length = 20 in. **47.** 10%
49. 16 telephones **51.** 8 m **53.** answers vary **55.** $\dfrac{4}{7}$ **57.** $\dfrac{3}{2}$ **59.** $\dfrac{7}{10}$ **61.** $\dfrac{3}{8}$

Chapter 4 Review

1. $2x - 5$ **3.** $4x(5x + 3)$ **5.** $-2x^2y(4x - 3y)$ **7.** $(x + 1)(5x - 1)$ **9.** $(2x - 1)(3x + 5)$ **11.** $(x + 4)(x + 2)$
13. not factorable **15.** $(x + 4)(x - 2)$ **17.** $(x + 5y)(x + 3y)$ **19.** $2(3 - x)(12 + x)$ **21.** $(2x - 1)(x + 6)$
23. $(2x + 3)(2x - 1)$ **25.** $(6x - y)(x - 4y)$ **27.** $(2x + 3y)(x - 13y)$ **29.** $(6x + 5y)(3x - 4y)$
31. $(2x - 3)(2x + 3)$ **33.** not factorable **35.** $(2x + 3)(4x^2 - 6x + 9)$ **37.** $2(3 - xy)(9 + 3xy + x^2y^2)$
39. $(2x - 1)(2x + 1)(4x^2 + 1)$ **41.** $(2x - 3)(x + 4)$ **43.** $(x - 1)(x + 3)$ **45.** $2xy(2x - 3y)$
47. $(5x + 3)(25x^2 - 15x + 9)$ **49.** $(x + 7 - y)(x + 7 + y)$ **51.** $x = -6$ or $x = 2$ **53.** $x = -\dfrac{1}{5}$ or $x = -3$
55. $x = -4$ or $x = 6$ **57.** $x = 2$ or $x = 8$ **59.** $x = -\dfrac{2}{7}$ or $x = \dfrac{3}{8}$ **61.** $x = -\dfrac{2}{5}$ **63.** $t = 3$
65. $t = 0$ or $t = -\dfrac{7}{4}$ or $t = 3$ **67.** $-\dfrac{25}{3}$ and $-\dfrac{47}{3}$ or 7 and 15 **69.** width = 20 cm, length = 25 cm **71.** $-\dfrac{15}{2}$ or 7
73. 32 ft

Chapter 4 Test

1. $3x(3x + 1)(x + 4)$ **2.** prime **3.** prime **4.** $(y - 12)(y + 4)$ **5.** $(3a - 7)(a + b)$ **6.** $(3x - 2)(x - 1)$
7. prime **8.** $(x + 12y)(x + 2y)$ **9.** $x^4(26x^2 - 1)$ **10.** $5x(10x^2 + 2x - 7)$ **11.** $5(6 - x)(6 + x)$
12. $(4x - 1)(16x^2 + 4x + 1)$ **13.** $(6t + 5)(t - 1)$ **14.** $(y - 2)(y + 2)(x - 7)$ **15.** $x(1 - x)(1 + x)(1 + x^2)$
16. $-xy(y^2 + x^2)$ **17.** $x = -7$ or $x = 2$ **18.** $x = -7$ or $x = 1$
19. $x = 0$ or $x = \dfrac{3}{2}$ or $x = -\dfrac{4}{3}$ **20.** $t = 0$ or $t = 3$ or $t = -3$ **21.** $x = 0$ or $x = -4$ **22.** $t = -3$ or $t = 5$
23. $x = -3$ or $x = 8$ **24.** $x = 0$ or $x = \dfrac{5}{2}$ **25.** width = 6 ft, length = 11 ft **26.** 8 and 9 **27.** base = 17 cm
28. -9 and -11 or 9 and 11 **29.** width = 7 in., length = 14 in.

Chapter 4 Cumulative Review

1. (a) $2 \cdot 3 = 6$; **(b)** $8 - 4 \le 4$; **(c)** $\dfrac{10}{2} \ne 6$; *Sec. 1.1, Ex. 4* **2. (a)** $\dfrac{64}{25}$; **(b)** $\dfrac{1}{20}$; **(c)** $\dfrac{5}{4}$; *Sec. 1.3, Ex. 4*

3. (a) 27; **(b)** 2; **(c)** 29; **(d)** 48; **(e)** $\dfrac{1}{4}$; *Sec. 1.4, Ex. 2* **4. (a)** -5; **(b)** 0; **(c)** 6; *Sec. 1.6, Ex. 5*

5. (a) -6; **(b)** 7; **(c)** -5; *Sec. 1.8, Ex. 4* **6.** $-2x - 1$; *Sec. 2.1, Ex. 7* **7.** $x = 17$; *Sec. 2.2, Ex. 2*

8. $x = \dfrac{15}{2}$; *Sec. 2.3, Ex. 4* **9.** $a = 0$; *Sec. 2.4, Ex. 4* **10.** 40 min; *Sec. 2.6, Ex. 4* **11.** *Sec. 2.7, Ex. 5*

12. (a) 8; **(b)** 3; **(c)** 16; **(d)** -16; *Sec. 3.1, Ex. 1* **13. (a)** $\dfrac{1}{9}$; **(b)** $\dfrac{2}{x^3}$; **(c)** 32; *Sec. 3.2, Ex. 1*

$x < -2$

$\leftarrow\!+\!+\!+\!\underset{-5\ -4\ -3\ -2\ -1\ \ 0\ \ 1}{+\!+\!\oplus\!+\!+\!+\!+}\!\rightarrow$

14. (a) 102,000; **(b)** 0.007358; **(c)** 84,000,000; **(d)** 0.00003007; *Sec. 3.2, Ex. 6* **15.** $3t^3 + 2t^2 - 6t + 4$; *Sec. 3.4, Ex. 5*

16. $2y^2 + 11y - 6$; *Sec. 3.5, Ex. 3* **17.** $2xy - 4 + \dfrac{1}{2y}$; *Sec. 3.6, Ex. 3* **18.** $5x^2z\,(5x^2 + 3x + 1)$; *Sec. 4.1, Ex. 6*

19. $(x + 6)(x - 2)$; *Sec. 4.2, Ex. 3* **20.** $(2x - 1)(x + 7)$; *Sec. 4.3, Ex. 3* **21.** $3(2m - n)(2m + n)$; *Sec. 4.5, Ex. 3*
22. $x = 5$ or $x = 4$; *Sec. 4.6, Ex. 2*

CHAPTER 5
Rational Expressions

Mental Math, Sec. 5.1
1. $x = 0$ **3.** $x = 0$ or $x = 1$

Exercise Set 5.1
1. $\dfrac{7}{4}$ **3.** $\dfrac{13}{3}$ **5.** $-\dfrac{11}{2}$ **7.** $\dfrac{7}{4}$ **9.** $-\dfrac{8}{3}$ **11.** $x = -2$ **13.** $x = 4$ **15.** $x = -2$ **17.** none   **19.** $\dfrac{2}{x^4}$

21. $\dfrac{5}{x+1}$ **23.** -5 **25.** $\dfrac{1}{x-9}$ **27.** $5x + 1$ **29.** $\dfrac{x+4}{2x+1}$ **31.** -1 **33.** $-\dfrac{y}{2}$ **35.** $\dfrac{2-x}{x+2}$ **37.** $x + y$

39. $\dfrac{5-y}{2}$ **41.** $-\dfrac{3y^5}{x^4}$ **43.** $\dfrac{x-2}{5}$ **45.** -6 **47.** $\dfrac{x+2}{2}$ **49.** $\dfrac{11x}{6}$ **51.** $\dfrac{1}{x-2}$ **53.** $x + 2$

55. $\dfrac{x+1}{x-1}$ **57.** $\dfrac{m-3}{m+3}$ **59.** $\dfrac{2a-6}{a+3}$ **61.** -1 **63.** $-(x+1)$ or $-x-1$ **65.** $\dfrac{x+5}{x-5}$ **67.** $\dfrac{x+2}{x+4}$

69. $\dfrac{x+3}{x}$ **71.** $x^2 - 2x + 4$ **73.** $\dfrac{x+5}{3}$ **75.** $-x^2 - x - 1$ **77.** answers vary   **79.** $2x^2 + 2x - 1$

81. $-3x - 7$ **83.** $2x + 3$

Mental Math, Sec. 5.2
1. $\dfrac{2x}{3y}$ **3.** $\dfrac{5y^2}{7x^2}$ **5.** $\dfrac{9}{5}$

Exercise Set 5.2
1. $\dfrac{21}{4y}$ **3.** x^4 **5.** $-\dfrac{b^2}{6}$ **7.** $\dfrac{x^2}{10}$ **9.** $\dfrac{1}{3}$ **11.** 1 **13.** $\dfrac{x+5}{x}$ **15.** x^4 **17.** $\dfrac{12}{y^6}$ **19.** $x(x+4)$

21. $\dfrac{3(x+1)}{x^3(x-1)}$ **23.** $m^2 - n^2$ **25.** $-\dfrac{x+2}{x-3}$ **27.** $-\dfrac{x+2}{x-3}$ **29.** $\dfrac{1}{6b^4}$ **31.** $\dfrac{9}{7x^2y^7}$ **33.** $\dfrac{5}{6}$ **35.** $\dfrac{3}{8}$ **37.** $\dfrac{3}{2}$

39. $\dfrac{3x+4y}{2(x+2y)}$ **41.** $-2(x+3)$ **43.** $\dfrac{2(x+2)}{x-2}$ **45.** $\dfrac{(a+5)(a+3)}{(a+2)(a+1)}$ **47.** $-\dfrac{1}{x}$ **49.** $\dfrac{2(x+3)}{x-4}$ **51.** $-(x^2+2)$

53. -1 **55.** $\dfrac{(a+b)^2}{a-b}$ **57.** $\dfrac{3x+5}{x^2+4}$ **59.** $\dfrac{2}{9x^2(x-5)}$ sq ft   **61.** $4x^3(x-3)$ **63.** $\dfrac{x}{2}$

65. $\dfrac{a(2a+b)(3a-2b)}{b^2(a+2b)(a-b)}$ **67.** answers vary   **69.** 1 **71.** $-\dfrac{10}{9}$ **73.** $-\dfrac{1}{5}$

Mental Math, Sec. 5.3
1. 1 **3.** $\dfrac{7x}{9}$ **5.** $\dfrac{1}{9}$ **7.** $\dfrac{7-10y}{5}$

Exercise Set 5.3
1. $\dfrac{a+9}{13}$ **3.** $\dfrac{y+10}{3+y}$ **5.** $\dfrac{3m}{n}$ **7.** $\dfrac{5x+7}{x-3}$ **9.** $\dfrac{1}{2}$ **11.** 4 **13.** $x + 5$ **15.** $x + 4$ **17.** 1 **19.** 33

21. $4x^3$ **23.** $8x(x+2)$ **25.** $6(x+1)^2$ **27.** $x - 8$ or $8 - x$ **29.** $\dfrac{6x}{4x^2}$ **31.** $\dfrac{9ab+2b}{5b(a+2)}$

33. $\dfrac{x(x+1)}{x(x+4)(x+2)(x+1)}$ **35.** $\dfrac{12}{x^3}$ **37.** $-\dfrac{5}{x+4}$ **39.** $\dfrac{x-2}{x+y}$ **41.** 4 **43.** 3 **45.** $\dfrac{1}{a+5}$ **47.** $\dfrac{1}{x-6}$

49. $10x$ **51.** $36x^3y$ **53.** $40x^3(x-1)^2$ **55.** $(2x+1)(2x-1)$ **57.** $(2x-1)(x+4)(x+3)$ **59.** $-\dfrac{5}{x-2}$

61. $\dfrac{7+x}{x-2}$ **63.** $\dfrac{24b^2}{12ab^2}$ **65.** $\dfrac{18}{2(x+3)}$ **67.** $\dfrac{18y-2}{30x^2-60}$ **69.** $\dfrac{x(x-4)}{x(x-4)^2(x+4)}$ **71.** $\dfrac{15x(x-7)}{3x(2x+1)(x-7)(x-5)}$

73. $\dfrac{20}{x-2}$ m   **75.** answers vary   **77.** answers vary   **79.** $x=0$ or $x=-5$ **81.** $x=\dfrac{1}{4}$ or $x=-\dfrac{3}{2}$
83. $x=1$ or $x=5$

Exercise Set 5.4
1. $\dfrac{5}{x}$ **3.** $\dfrac{75a+6b^2}{5b}$ **5.** $\dfrac{6x+5}{2x^2}$ **7.** $\dfrac{21}{2(x+1)}$ **9.** $\dfrac{17x+30}{2(x-2)(x+2)}$ **11.** $\dfrac{35x-6}{4x(x-2)}$ **13.** $\dfrac{5+10y-y^3}{y^2(2y+1)}$
15. $-\dfrac{2}{x-3}$ **17.** $\dfrac{15}{x^2-1}$ **19.** $\dfrac{1}{x-2}$ **21.** $\dfrac{5+2x}{x}$ **23.** $\dfrac{6x-7}{x-2}$ **25.** $-\dfrac{y+4}{y+3}$ **27.** 2   **29.** $3x^3-4$
31. $\dfrac{x+2}{(x+3)^2}$ **33.** $\dfrac{9b-4}{5b(b-1)}$ **35.** $\dfrac{2+m}{m}$ **37.** $\dfrac{10}{1-2x}$ **39.** $\dfrac{15x-1}{(x+1)^2(x-1)}$ **41.** $\dfrac{x^2-3x-2}{(x-1)^2(x+1)}$
43. $\dfrac{a+2}{2(a+3)}$ **45.** $\dfrac{x-10}{2(x-2)}$ **47.** $\dfrac{-2y-3}{(y-1)(y-2)}$ **49.** $\dfrac{-5x+23}{(x-3)(x-2)}$ **51.** $\dfrac{12x-32}{(x+2)(x-2)(x-3)}$
53. $\dfrac{6x+9}{(3x-2)(3x+2)}$ **55.** $\dfrac{2x^2-2x-46}{(x+1)(x-6)(x-5)}$ **57.** 10   **59.** 2   **61.** $\dfrac{25a}{9(a-2)}$ **63.** $\dfrac{x+4}{(x-2)(x-1)}$
65. $\dfrac{2x-16}{(x-4)(x+4)}$ in   **67.** $\dfrac{10y-20}{y(y-5)}$ ft   **69.** $\dfrac{4x^2-15x+6}{(x-2)^2(x+2)(x-3)}$ **71.** $\dfrac{-3x^2+7x+55}{(x+2)(x+7)(x+3)}$
73. $\dfrac{2x^2+2x}{x(x-3)(x^2+3x+9)}$ **75.** $(x-1)(x^2+x+1)$ **77.** $(5z+2)(25z^2-10z+4)$ **79.** $(x+3)(y+2)$

Exercise Set 5.5
1. $\dfrac{2}{3}$ **3.** $\dfrac{2}{3}$ **5.** $\dfrac{1}{2}$ **7.** $\dfrac{4(y-1)}{3(y+2)}$ **9.** $-\dfrac{21}{5}$ **11.** $\dfrac{27}{16}$ **13.** $\dfrac{4}{3}$ **15.** $\dfrac{m-n}{m+n}$ **17.** $\dfrac{2x(x-5)}{7x^2+10}$ **19.** $\dfrac{1}{y-1}$
21. $\dfrac{1}{21}$ **23.** $-\dfrac{4x}{15}$ **25.** $\dfrac{1}{6}$ **27.** $\dfrac{x+y}{x-y}$ **29.** $\dfrac{3}{7}$ **31.** $\dfrac{a}{x+b}$ **33.** $\dfrac{7(y-3)}{8+y}$ **35.** $\dfrac{3x}{x-4}$ **37.** $\dfrac{2x}{x-3}$
39. $-\dfrac{x+8}{x-2}$ **41.** $\dfrac{s^2+r^2}{s^2-r^2}$ **43.** **(a)** 349,445; **(b)** Sunday   **45.** $\dfrac{3x}{x+3}$ **47.** $\dfrac{xy-x}{1+x}$ **49.** $\dfrac{4x+1}{3x+1}$
51. answers vary   **53.** $x=\dfrac{9}{5}$ **55.** $x=\dfrac{3}{2}$ or $x=-\dfrac{3}{2}$ **57.** -1

Mental Math Sec. 5.6
1. $x=10$ **2.** $z=36$

Exercise Set 5.6
1. $x=30$ **3.** $x=0$ **5.** $x=\dfrac{10}{3}$ **7.** $a=5$ **9.** $x=3$ **11.** $a=\dfrac{1}{4}$ **13.** extraneous solution
15. $x=6$ or $x=-4$ **17.** $y=5$ **19.** $x=-\dfrac{10}{9}$ **21.** $x=\dfrac{11}{14}$ **23.** extraneous solution   **25.** $\dfrac{3+2x}{3x}$
27. $\dfrac{x-1}{x(x+1)}$ **29.** $\dfrac{5t-21}{15(t+2)}$ **31.** $R=\dfrac{D}{T}$ **33.** $y=\dfrac{3x+6}{5x}$ **35.** $b=-\dfrac{3a^2+2a-4}{6}$ **37.** $x=-\dfrac{21}{11}$
39. $y=1$ **41.** $x=\dfrac{9}{2}$ **43.** extraneous solution   **45.** $x=-\dfrac{9}{4}$ **47.** $a=-\dfrac{3}{2}$ or $a=4$ **49.** $x=-2$
51. $r=\dfrac{12}{5}$ **53.** $t=-2$ or $t=8$ **55.** $x=\dfrac{1}{5}$ **57.** $a=\dfrac{17}{4}$ **59.** $B=\dfrac{2A}{H}$ **61.** $r=\dfrac{C}{2\pi}$ **63.** $a=\dfrac{bc}{c+b}$
65. $n=\dfrac{m^2-3p}{2}$ **67.** answers vary   **69.** $-2a^2+10$ **71.** $3m^3-3m$ **73.** $(x-3)(x+1)$

Exercise Set 5.7
1. $\dfrac{2}{15}$ **3.** $\dfrac{15}{1}$ **5.** $\dfrac{50}{1}$ **7.** $x=4$ **9.** $x=\dfrac{50}{9}$ **11.** $x=\dfrac{14}{9}$ **13.** $x=5$ **15.** $x=-\dfrac{2}{3}$ **17.** 216 other nuts

19. $106.40   **21.** 9 gal.   **23.** $x = \dfrac{21}{4}$ **25.** $a = 30$ **27.** $x = 7$ **29.** $x = -\dfrac{1}{3}$ **31.** $x = -3$

33. 165 calories   **35.** 20 tablespoons   **37.** 31 miles per gallon   **39.** $19.125 or $19.13   **41.** $1348.58   **43.** 0
45. $3(3x - 4)(x + 1)$

Exercise Set 5.8

1. 2   **3.** -3 **5.** $2\dfrac{2}{9}$ hr   **7.** $1\dfrac{1}{2}$ min   **9.** trip $= \dfrac{5}{4} r$; return $= r + 1$; trip $= 4$ mph; return $= 5$ mph

11. first portion, 10 mph; cooldown, 8 mph   **13.** 6 units   **15.** 5 units   **17.** 2   **19.** $108.00   **21.** 63 mph
23. 21.25 units   **25.** 5   **27.** 217 mph   **29.** 8 mph   **31.** 3 hr   **33.** 20 hr   **35.** $x = 4.4$ ft.; $y = 5.6$ ft.

37. $5\dfrac{1}{4}$ hr   **39.** first pump, 28 min; second pump, 84 min

41. 38 yr old when son was born; son was 42 when Diophantus died   **43.** answers vary   **45.** $z = \dfrac{7}{2}$ or $z = \dfrac{2}{3}$

Chapter 5 Review

1. $x = 2$ or $x = -2$ **3.** $\dfrac{2}{11}$ **5.** $\dfrac{1}{x - 5}$ **7.** $\dfrac{x(x - 2)}{x + 1}$ **9.** $\dfrac{x - 3}{x - 5}$ **11.** $\dfrac{x + 1}{2x + 1}$ **13.** $\dfrac{x + b}{x - c}$

15. $-\dfrac{1}{x^2 + 4x + 16}$ **17.** $-\dfrac{9x^2}{8}$ **19.** $-\dfrac{2x(2x + 5)}{(x - 6)^2}$ **21.** $\dfrac{4x}{3y}$ **23.** $\dfrac{2}{3}$ **25.** $\dfrac{x}{x + 6}$ **27.** $\dfrac{3(x + 2)}{3x + y}$

29. $-\dfrac{2(2x + 3)}{y - 2}$ **31.** $\dfrac{5x + 2}{3x - 1}$ **33.** $\dfrac{2x + 1}{2x^2}$ **35.** $(x - 8)(x + 8)(x + 3)$ **37.** $\dfrac{3x^2 + 4x - 15}{(x + 2)^2(x + 3)}$

39. $\dfrac{-2x + 10}{(x - 3)(x - 1)}$ **41.** $\dfrac{5x - 12}{x^2 - 9}$ **43.** $\dfrac{x - 4}{3x}$ **45.** $\dfrac{x^2 + 2x - 3}{(x + 2)^2}$ **47.** $\dfrac{6}{7}$ **49.** $\dfrac{3y - 1}{2y - 1}$ **51.** $\dfrac{2x^2 + 1}{x + 2}$

53. $\dfrac{b(1 + a)}{a(1 - b)}$ **55.** $\dfrac{x(2x^3 + 1)}{2x^2 + 1}$ **57.** $n = 30$ **59.** $x = 3$ **61.** extraneous (no) solution   **63.** $x = 5$

65. $x = -6$ or $x = 1$ **67.** $y = \dfrac{560 - 8x}{7}$ **69.** $\dfrac{2}{3}$ **71.** $x = 500$ **73.** $c = 50$ **75.** extraneous solution

77. extraneous (no) solution   **79.** $33.75   **81.** 30 mph, 20 mph   **83.** $17\dfrac{1}{2}$ hr   **85.** 3

Chapter 5 Test

1. $x = -1$ or $x = -3$ **2.** $\dfrac{1}{2x - 1}$ **3.** $\dfrac{3}{5}$ **4.** $\dfrac{1}{x - 10}$ **5.** $\dfrac{1}{x + 6}$ **6.** $\dfrac{1}{x^2 - 3x + 9}$ **7.** $\dfrac{2m(m + 2)}{m - 2}$

8. $\dfrac{a + 2}{a + 5}$ **9.** $-\dfrac{1}{x + y}$ **10.** 1   **11.** $\dfrac{(x - 6)(x - 7)}{(x + 7)(x + 2)}$ **12.** 15   **13.** $\dfrac{y - 2}{4}$ **14.** $-\dfrac{1}{2x + 5}$

15. $\dfrac{x^3 - 2x^2 + x}{(x - 3)(x + 3)(x + 2)}$ **16.** $\dfrac{3a - 4}{(a - 3)(a + 2)}$ **17.** $\dfrac{3}{x - 1}$ **18.** $\dfrac{2(x + 5)}{x(y + 5)}$ **19.** $\dfrac{x^2 + 2x + 35}{(x + 9)(x + 2)(x - 5)}$

20. $\dfrac{4y^2 + 13y - 15}{(y + 4)(y + 5)(y + 1)}$ **21.** $y = \dfrac{30}{11}$ **22.** $y = -6$ **23.** extraneous solution   **24.** extraneous solution

25. $\dfrac{b^2 - a^2}{2b^2}$ **26.** $\dfrac{17}{11}$ **27.** 20 cars   **28.** 5 or 1   **29.** 30 mph   **30.** $6\dfrac{2}{3}$ hr

Chapter 5 Cumulative Review

1. $\dfrac{8}{20}$; *Sec. 1.3, Ex. 6*   **2.** (a) 9; (b) 125; (c) 16; (d) 7; (e) $\dfrac{9}{49}$; *Sec. 1.4, Ex. 1*   **3.** (a) -0.06; (b) $-\dfrac{7}{15}$; *Sec. 1.8, Ex. 3*

4. (a) $2x + 6$ (b) $\dfrac{x - 4}{7}$; *Sec. 2.1, Ex. 8*   **5.** $a = 8$; *Sec. 2.2, Ex. 6*   **6.** $x = 2$; *Sec. 2.4, Ex. 2*

7. 15°C; *Sec. 2.5, Ex. 2*   **8.** *Sec. 2.7, Ex. 3* $x \leq -10$

9. (a) 4^7; (b) x^7; (c) y^4; (d) y^{12}; (e) $(-5)^{15}$ or -5^{15}; *Sec. 3.1, Ex. 2*   **10.** (a) x^3; (b) 256; (c) -27; (d) $2x^4y$; *Sec. 3.1, Ex. 6*
11. $3x + 8$; *Sec. 3.3, Ex. 6*   **12.** $x^2 + x - 12$; *Sec. 3.5, Ex. 1*   **13.** $(3m^2n - 1)(a + b)$; *Sec. 4.1, Ex. 8*

14. $(x + 3)(x + 4)$; *Sec. 4.2, Ex. 1*   **15.** $2x^2(2x + 3)(6x + 1)$; *Sec. 4.3, Ex. 5*   **16.** $(2x - 1)(2x + 1)$; *Sec. 4.4, Ex. 1*

17. $(y - 3)(y^2 + 3y + 9)$; *Sec. 4.4, Ex. 6*   **18.** $\dfrac{x + 7}{x - 5}$; *Sec. 5.1, Ex. 5*   **19.** $\dfrac{2}{5}$; *Sec. 5.2, Ex. 2*

20. $\dfrac{9 - 2x^2}{3x(2x + 1)}$; *Sec. 5.4, Ex. 6*   **21.** $\dfrac{39}{20}$; *Sec. 5.5, Ex. 2*   **22.** $x = -3, -2$; *Sec. 5.6, Ex. 2*

23. **(a)** $\dfrac{2}{5}$; **(b)** $\dfrac{3}{4}$; *Sec. 5.7, Ex. 1*

CHAPTER 6
Graphing Linear Equations and Inequalities

Mental Math, Sec. 6.1
1. answers vary; Ex. (5, 5), (7, 3)   **3.** answers vary; Ex. (3, 5) (3,0)

Exercise Set 6.1
1. yes, no, yes   **3.** no, yes, yes   **5.** no, yes, yes   **7.** $(-4, -2)$, $(4, 0)$ **9.** (0, 9), (3, 0)
11. $(11, -7)$; answers vary, Ex. $(2, -7)$
13.   **15.**   **17.**   **19.**

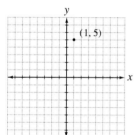   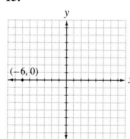   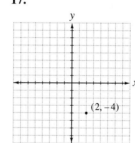   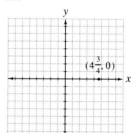

21. yes, no   **23.** no, no   **25.** yes, yes   **27.** yes, yes   **29.** $(3, 1)$; $\left(0, \dfrac{11}{5}\right)$ **31.** $(2, 7)$ **33.** $(0, 0)$

35. $(12, 4)$ **37.** $\left(0, -\dfrac{5}{2}\right)$

39. **(a)** $(0, 2)$ **(b)** $(6, 0)$ **(c)** $(3, 1)$ **41.** **(a)** $(0, -12)$ **(b)** $(5, -2)$ **(c)** $(-3, -18)$ **43.** **(a)** $\left(0, \dfrac{5}{7}\right)$ **(b)** $\left(\dfrac{5}{2}, 0\right)$ **(c)** $(-1, 1)$

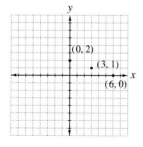   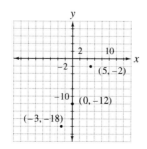   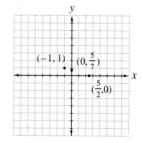

45. (a) (3, 0) **(b)** (3, −0.5) **(c)** $\left(3, \dfrac{1}{4}\right)$ **47. (a)** (0, 0) **(b)** (−5, 1) **(c)** (10, −2)

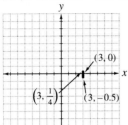

49. A, (0, 0); B, $\left(3\dfrac{1}{2}, 0\right)$; C, (3, 2); D, (−1, 3); E, (−2, −2); F, (0, −1); G, (2, −1)   **51.** 8000   **53.** answers vary

55. answers vary   **57.** answers vary   **59.** $2(x^2 − x − 10)$ **61.** $(x + y)(x + 1)$ **63.** $\dfrac{4x + 8}{3x}$ meters

Exercise Set 6.2

1.

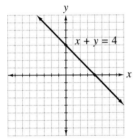

3.

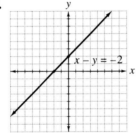

5.

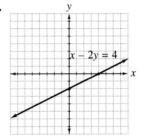

7.

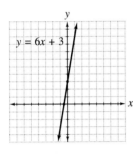

9.

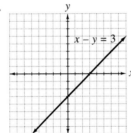

11.

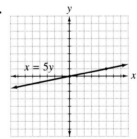

13.

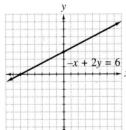

15.

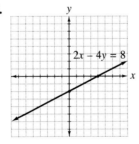

17.

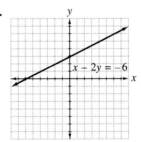

19.

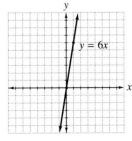

21.

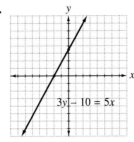

23.

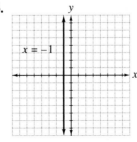

25.
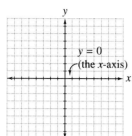
$y = 0$
(the x-axis)

27.
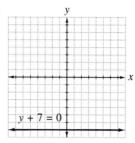
$y + 7 = 0$

29.
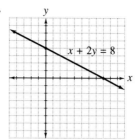
$x + 2y = 8$

31.
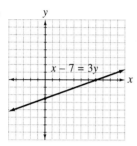
$x - 7 = 3y$

33.
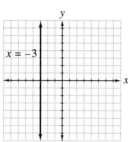
$x = -3$

35.
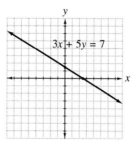
$3x + 5y = 7$

37.
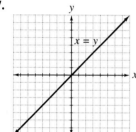
$x = y$

39.
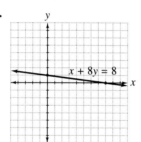
$x + 8y = 8$

41.
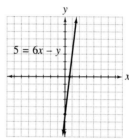
$5 = 6x - y$

43.
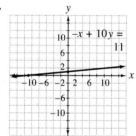
$-x + 10y = 11$

45.
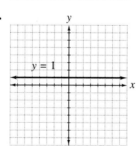
$y = 1$

47.
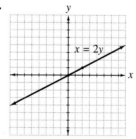
$x = 2y$

49.
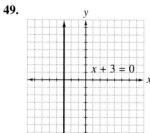
$x + 3 = 0$

51.
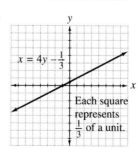
$x = 4y - \frac{1}{3}$
Each square represents $\frac{1}{3}$ of a unit.

53.
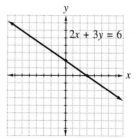
$2x + 3y = 6$

55. answers vary   **57.** answers vary   **59.** $\dfrac{3x}{y}$ **61.** 7   **63.** $\dfrac{x}{4x + 2}$

Mental Math, Sec. 6.3

1. upward   **3.** horizontal

Exercise Set 6.3

1. $\dfrac{8}{7}$ **3.** -1 **5.** $-\dfrac{1}{4}$ **7.** undefined   **9.** $m = 5, b = -2$ **11.** $m = -2, b = 7$ **13.** $m = \dfrac{2}{3}, b = -\dfrac{10}{3}$

15. $m = \dfrac{1}{2}, b = 0$ **17.** neither   **19.** parallel   **21.** perpendicular   **23.** undefined   **25.** 0   **27.** undefined

29. $-\dfrac{2}{3}$ **31.** undefined   **33.** 0   **35.** 2   **37.** -1 **39.** $m = -1, b = 5$ **41.** $m = \dfrac{6}{5}, b = 6$

43. $m = 3, b = 9$ **45.** $m = 0, b = 4$ **47.** $m = 7, b = 0$ **49.** $m = 0, b = -6$

51. undefined slope, no y-intercept   **53.** $-\dfrac{7}{2}$ **55.** $\dfrac{2}{7}$ **57.** $-\dfrac{1}{5}$ **59.** $\dfrac{1}{3}$ **61.** answers vary   **63.** 5

65. $\dfrac{15 + 2x}{5x}$ **67.** $\dfrac{6}{5x}$

Mental Math, Sec. 6.4

1. $m = -4, b = 12$ **3.** $m = 5, b = 0$ **5.** $m = \dfrac{1}{2}, b = 6$ **7.** parallel   **9.** neither

Exercise Set 6.4

1. $y = -x + 1$ **3.** $y = 2x + \dfrac{3}{4}$ **5.** $y = \dfrac{2}{7}x$

7. 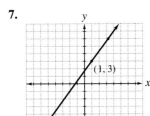   **9.** 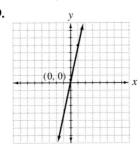   **11.** 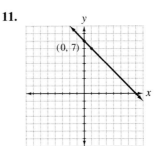   **13.** $6x - y = 10$

15. $8x + y = -13$ **17.** $x - 2y = 17$ **19.** $2x - y = 4$ **21.** $8x - y = -11$ **23.** $4x - 3y = -1$ **25.** $x = 0$
27. $y = 3$ **29.** $x = -7$ **31.** $y = 2$ **33.** $y = 5$ **35.** $x = 6$ **37.** $3x + 6y = 10$ **39.** $x - y = -16$
41. $x + y = 17$ **43.** $y = 7$ **45.** $4x + 7y = -18$ **47.** $x + 8y = 0$ **49.** $-3x + y = 0$ **51.** $x - y = 0$

53. $5x + y = 7$ **55.** $11x + y = -6$ **57.** $x = -\dfrac{3}{4}$ **59.** $y = -3$ **61.** $7x - y = 4$ **63.** $31x - 5y = -5$

65. $2x - 3y = 14$ **67.** $x + y = 8$ **69.** answers vary   **71.** 2, 3   **73.** $\dfrac{-1}{x^3(x + 2)}$ sq in

Mental Math, Sec. 6.5

1. yes   **3.** yes   **5.** yes   **7.** no

Exercise Set 6.5

1. 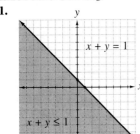   **3.** 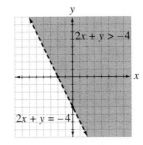   **5.**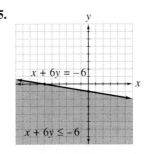

7.
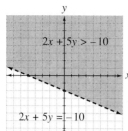

9.
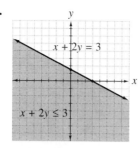

11.
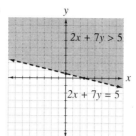

13.
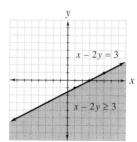

15.
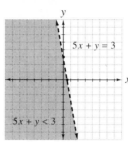

17.
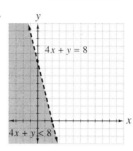

19.
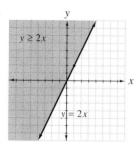

21.
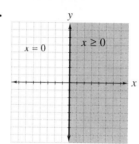

23.
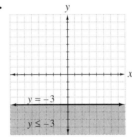

25.
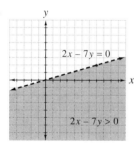

27.
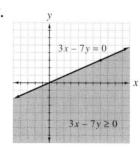

29.
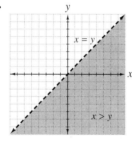

31.
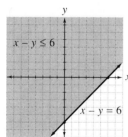

33.
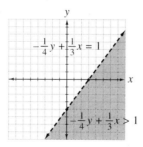

35.
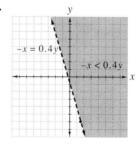

37. $x + y \geq 13$ **39.** answers vary   **41.** 8 cm   **43.** $2x(4x + 1)$ **45.** $(x + 1)(x + y)$

Exercise Set 6.6

1. domain: $\{-7, 0, 2, 10\}$ range: $\{-7, 0, 4, 10\}$ **3.** domain: $\{0, 1, 5\}$ range: $\{-2\}$ **5.** yes **7.** no **9.** yes
11. no **13.** yes **15.** yes **17.** yes **19.** no **21.** no **23.** yes **25.** (a) -9; (b) -5; (c) 1
27. (a) 11; (b) $2\frac{1}{16}$ or $\frac{33}{16}$; (c) 11 **29.** (a) -8; (b) -216; (c) 0 **31.** (a) 7; (b) 7; (c) 0 **33.** all the real numbers
35. all real numbers except -5 **37.** all real numbers **39.** no **41.** yes **43.** no **45.** yes
47. (a) 0; (b) 10; (c) -10 **49.** (a) 245; (b) 5; (c) $3\frac{1}{2}$ or $\frac{7}{2}$ **51.** (a) 0; (b) 6 **53.** (a) 6; (b) 6; (c) 6
55. all real numbers **57.** all real numbers **59.** all real numbers except 2 and -1
61. (a) 11; (b) $2a + 7$; (c) $2a + 11$ **63.** (a) 16; (b) $a^2 + 7$; (c) $a^2 + 6a + 16$
65. (a) answers vary; (b) answers vary; (c) answers vary **67.** $f(x) = x + 7$ **69.** $\dfrac{16x - 8}{(x + 2)(x - 2)}$ in.

71.

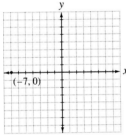

Chapter 6 Review

1.

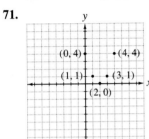

3.

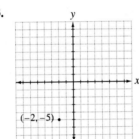

5.

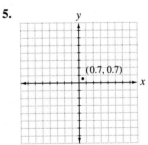

7. no, yes

9. yes, yes **11.** $(7, 44)$ **13.** (a) $(-3, 0)$ (b) $(1, 3)$ (c) $(9, 9)$ **15.** (a) $(0, 0)$ (b) $(10, 5)$ (c) $(-10, -5)$

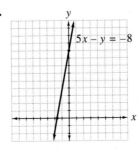

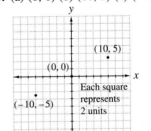

17.

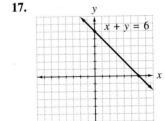

19.

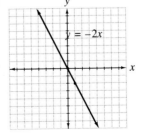

21.

23.

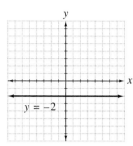

25.

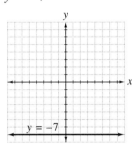

27. $\dfrac{5}{3}$ **29.** -1 **31.** $m = \dfrac{1}{2}, b = -2$

33. $m = \dfrac{4}{5}, b = -\dfrac{9}{5}$

35. $-4x + 5y = 23$

37. $y = 5$

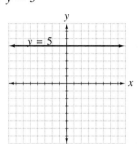

39. $y = -7$

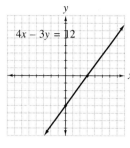

41. perpendicular

43. $4x - y = 8$ **45.** $x - 2y = 7$ **47.** $y = 0$ **49.** $12x - y = 1$ **51.** $x + 2y = -8$ **53.** $y = 8$ **55.** $y = 12$

57.

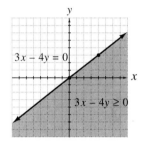

59.

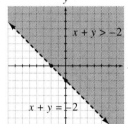

61.

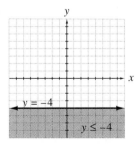

63.

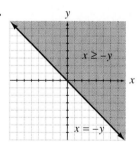

65. yes   **67.** no   **69.** all real numbers except 2   **71.** yes   **73.** (a) -11; (b) 4; (c) -5 **75.** (a) 5; (b) 5; (c) -10

Chapter 6 Test

1. (1, 1)　　**2.** (−4, 17)　　**3.** no　　**4.** no　　**5.** −3　　**6.** −1

7.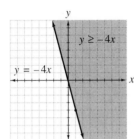

8.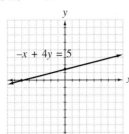

9.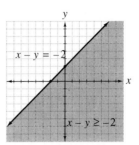

10.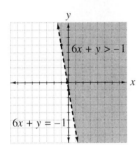

11.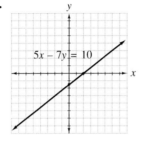

12.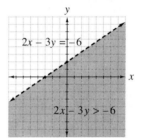

13.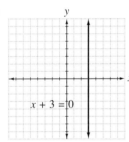

14.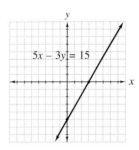

15.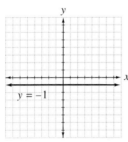

16.

17. yes　　**18.** no　　**19.** no　　**20.** (a) −8; (b) −3.6; (c) −4　　**21.** (a) 0; (b) 0; (c) 60
22. (a) 6; (b) 6; (c) 6　　**23.** $7x + 6y = 0$　　**24.** $8x + y = 11$　　**25.** $x = -5$　　**26.** $x - 8y = -96$　　**27.** $x = -1$
28. $x = 9$　　**29.** (a) yes, domain: all real numbers; (b) no; (c) yes, domain: all real numbers

Chapter 6 Cumulative Review

1. (a) $9 \le 11$; (b) $8 > 1$; (c) $3 \ne 4$; *Sec. 1.1, Ex. 3*　　**2.** (a) −4, (b) 8; (c) −3, *Sec. 1.6, Ex. 2*
3. (a) $6x - 14$; (b) $-4x + 6$; (c) $-11x - 13$; *Sec. 2.1, Ex. 6*　　**4.** $3x + 3$; *Sec 2.1, Ex. 10*　　**5.** $y = -1.6$; *Sec. 2.2, Ex. 3*
6. width = 80 ft; length = 240 ft; *Sec. 2.6, Ex. 1*　　**7.** *Sec. 2.7, Ex. 8*

$$x > \frac{13}{7}$$

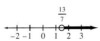

8. (a) x^{10}; **(b)** y^{16}; **(c)** 244,140,625 or $(-5)^{12}$; *Sec. 3.1, Ex. 4*   **9. (a)** $\dfrac{25x^4}{y^6}$, **(b)** x^6; **(c)** $32x^2$; **(d)** a^3b; *Section 3.1, Ex. 8*

10. (a) 3.67×10^8; **(b)** 3.0×10^{-6}; **(c)** 2.052×10^{10}; **(d)** 8.5×10^{-4}, *Sec. 3.2, Ex. 5*   **11.** $4x^3 - x^2 + 7$, *Sec. 3.3, Ex. 6*

12. $9y^2 + 6y + 1$, *Sec. 3.5, Ex. 4*   **13.** $(x + 3)(y + 2)$, *Sec. 4.1, Ex. 9*   **14.** $(x + 6)^2$, *Sec. 4.3, Ex. 7*

15. $(x + 2)(x^2 - 2x + 4)$; *Sec. 4.4, Ex. 5*   **16.** $x = -5, 3$; *Sec. 4.6, Ex. 4*   **17.** $\dfrac{5}{x^2}$; *Sec. 5.1, Ex. 4*

18. 1; *Sec. 5.3, Ex. 2*   **19.** $a = -\dfrac{1}{2}$; *Sec. 5.6, Ex. 5*   **20. (a)** $(0, 12)$; **(b)** $(2, 6)$; **(c)** $(-1, 15)$; *Sec. 6.1, Ex. 2*

21. slope $= -\dfrac{8}{3}$; *Sec. 6.3, Ex. 1*   **22.** *Sec. 6.5, Ex. 1*

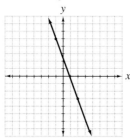   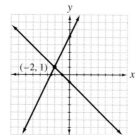

CHAPTER 7
Solving Systems of Linear Equations

Mental Math, Sec. 7.1
1. consistent   **3.** consistent, dependent   **5.** inconsistent   **7.** consistent

Exercise Set 7.1
1. (a) no; **(b)** yes; **(c)** no   **3. (a)** no; **(b)** yes; **(c)** no   **5. (a)** yes; **(b)** yes; **(c)** yes

7. 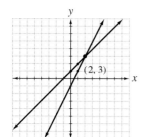   **9.**   **11.**   **13.**

15. intersecting, one solution   **17.** parallel, no solutions   **19.** identical lines, infinite number of solutions

21. 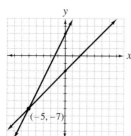   **23.** 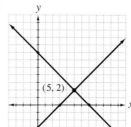   **25.**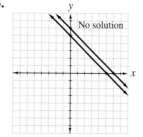

27.

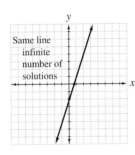

Same line
infinite
number of
solutions

29.

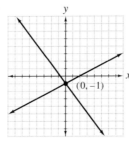

$(0, -1)$

31.

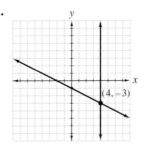

$(4, -3)$

33. intersecting, one solution   **35.** identical lines, infinite number of solutions

37. parallel, no solutions   **39.** answers vary   **41.** $x = -\dfrac{1}{2}$ **43.** $x = 2$ **45.** $x = -2$

Exercise Set 7.2

1. $(2, 1)$ **3.** $(-3, 9)$ **5.** $(4, 2)$ **7.** $(-2, 4)$ **9.** $(-2, -1)$ **11.** no solution

13. infinite number of solutions   **15.** $(10, 5)$ **17.** $(2, 7)$ **19.** $(3, -1)$ **21.** $(3, 5)$ **23.** $\left(\dfrac{2}{3}, -\dfrac{1}{3}\right)$

25. $(-1, -4)$ **27.** $(-6, 2)$ **29.** $(2, 1)$ **31.** no solution   **33.** $\left(-\dfrac{1}{5}, \dfrac{43}{5}\right)$ **35.** answers vary

37. $\dfrac{5 - x}{x + 2}$ **39.** 3   **41.** $x = 10$

Exercise Set 7.3

1. $(1, 2)$ **3.** $(2, -3)$ **5.** $(5, -2)$ **7.** $(-2, -5)$ **9.** $(-7, 5)$ **11.** $\left(\dfrac{12}{11}, -\dfrac{4}{11}\right)$ **13.** no solution

15. $\left(\dfrac{3}{2}, 3\right)$ **17.** $(1, 6)$ **19.** $(6, 0)$ **21.** no solution   **23.** $\left(2, -\dfrac{1}{2}\right)$ **25.** $(6, -2)$

27. infinite number of solutions   **29.** infinite number of solutions   **31.** $(-2, 0)$ **33.** $(2, 5)$ **35.** $(-3, 2)$

37. $(0, 3)$ **39.** $(5, 7)$ **41.** $\left(\dfrac{1}{3}, 1\right)$ **43.** infinite number of solutions   **45.** answers vary

47. $n + (n + 1) + (n + 2) = 66$ **49.** $4(n + 6) = 2n$

Exercise Set 7.4

1. $x + y = 15, x - y = 7$ **3.** $x = $ larger, $y = $ smaller; $x + y = 6500, x = y + 800$ **5.** c   **7.** b   **9.** a

11. 33 and 50   **13.** \$29, adults; \$18, children   **15.** 27 nickels, 53 quarters   **17.** still water, $6\dfrac{1}{2}$ mph; current, $2\dfrac{1}{2}$ mph

19. plane, 455 mph; wind, 65 mph   **21.** $4\dfrac{1}{2}$ oz of 4% solution, $7\dfrac{1}{2}$ oz of 12% solution

23. 113 lb high quality, 87 lb cheaper quality   **25.** 14 and -3 **27.** 23 at original price

29. length $= 15$ ft, width $= 9$ ft   **31.** $4\dfrac{1}{2}$ hr   **33.** \$23 daily fee, \$.21 per mile   **35.**

37. $\dfrac{1}{16}$

$y = 3 - x$

$y < 3 - x$

Exercise Set 7.5

1.

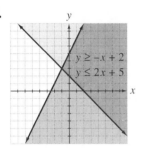

3.

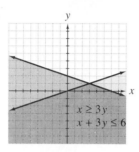

5.

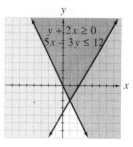

7.

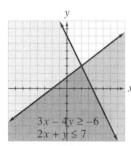

9.

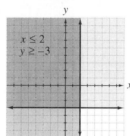

11.

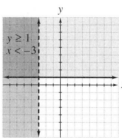

13.

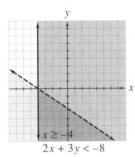

15.

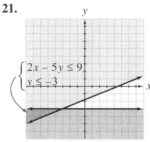

17.

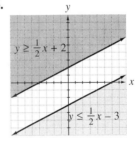

19.

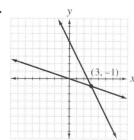

21.

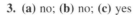

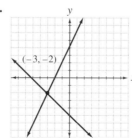

23.

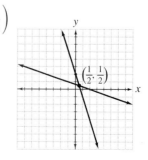

25. $x + y \leq 8, x < 3$ **27.** answers vary   **29.** 9   **31.** $\frac{4}{9}$ **33.** 1   **35.** -5

Chapter 7 Review

1. (a) no; (b) yes; (c) no   **3.** (a) no; (b) no; (c) yes

5.

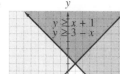

7.

9. $\left(\frac{1}{2}, \frac{1}{2}\right)$

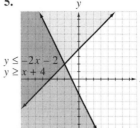

11. consistent   **13.** inconsistent   **15.** intersecting at a single point   **17.** identical   **19.** $(-1, 4)$

21. $(3, -2)$ **23.** no solution   **25.** $(3, 1)$ **27.** $(8, -6)$ **29.** $(-6, 2)$ **31.** $(3, 7)$

33. infinite number of solutions   **35.** $\left(2, -2\frac{1}{2}\right)$ **37.** $(-6, 15)$ **39.** $(-3, 1)$ **41.** -6 and 22

43. ship's rate in still water is 21.1 mph; current is 3.2 mph   **45.** width $= 1.15$ ft, length $= 1.85$ ft

47. one egg, \$.40; one slice bacon \$.65

49.

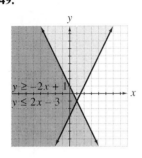

51.

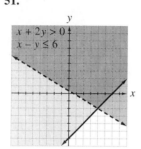

53.

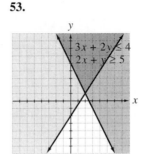

55.

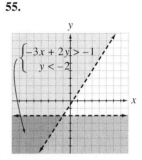

Chapter 7 Test

1. no   **2.** yes   **3.**

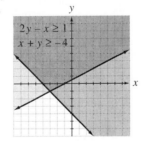

4. $(-4, 1)$ **5.** $\left(\frac{1}{2}, -2\right)$ **6.** $(4, -2)$ **7.** $\left(2, \frac{1}{2}\right)$

8. $(4, -5)$ **9.** $(7, 2)$ **10.** $\left(5, -2\right)$ **11.** 20 \$1.00 bills, 42 \$5.00 bills   **12.** \$1225 at 5%, \$2775 at 9%

13.

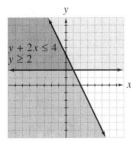

14.

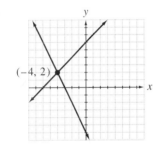

Chapter 7 Cumulative Review

1. (a) $\frac{1}{2}$; (b) 9; *Sec. 1.7, Ex. 4*   **2.** $x = 6$; *Sec. 2.3, Ex. 1*   **3.** shorter piece, 2 ft; longer piece, 8 ft; *Sec. 2.3, Ex. 9*

4. $l = \dfrac{V}{wh}$; *Sec. 2.5, Ex. 6*   **5.** (a) $\dfrac{81}{16}$; (b) $\dfrac{3}{4}$; (c) $\dfrac{1}{16}$; *Sec. 3.2, Ex. 2*   **6.** $x^4 - \dfrac{1}{9}y^4$; *Sec. 3.5, Ex. 7*

7. $(5x + 4y)(5x + y)$; *Sec. 4.3, Ex. 8*   **8.** $(2x + 3)(x - 1)(x + 1)$; *Sec. 4.5, Ex. 2*   **9.** $(x - y)$; *Sec. 5.1, Ex. 9*

10. 1; *Sec. 5.2, Ex. 7*   **11.** $\dfrac{17t + 2}{3t(t + 1)}$; *Sec. 5.4, Ex. 3*   **12.** $x = 8$; *Sec. 5.6, Ex. 4*

13. *Sec. 6.2, Ex. 2*   **14.** *Sec. 6.2, Ex. 6*   **15.** $x = -1$; *Sec. 6.3, Ex. 9*

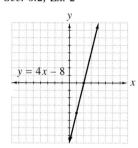

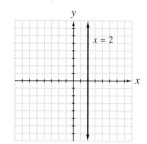

16. *Sec. 6.5, Ex. 4*   **17.** *Sec. 6.5, Ex. 5*   **18.** $\left(6, \dfrac{1}{2}\right)$ *Sec. 7.2, Ex. 2*

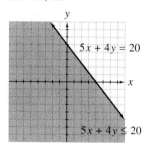

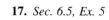

19. no solution; *Sec. 7.3, Ex. 3*   **20.** *Sec. 7.5, Ex. 2*

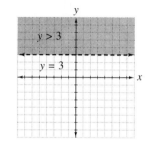

CHAPTER 8
Roots and Radicals

Exercise Set 8.1

1. 4   **3.** 9   **5.** $\dfrac{1}{5}$ **7.** -10 **9.** 4   **11.** -3 **13.** $\dfrac{1}{2}$ **15.** -5 **17.** 2   **19.** 3   **21.** not a real number

23. $\dfrac{1}{3}$ **25.** $\dfrac{3}{5}$ **27.** -7 **29.** z **31.** x^2 **33.** $3x^4$ **35.** xy^3 **37.** 0   **39.** $-\dfrac{1}{2}$ **41.** not a real number

43. -8 **45.** -13 **47.** 1   **49.** $\dfrac{5}{8}$ **51.** 2   **53.** rational, 3   **55.** irrational, 6.083   **57.** rational, 13

59. rational, 2   **61.** x^5 **63.** x^6 **65.** $9x$ **67.** $-12y^7$ **69.** xy **71.** $4x^8$ **73.** 18.0 ft   **75.** 1537 sq ft

77. 1   **79.** 1   **81.** 0   **83.** $5x + y = 14$

Mental Math, Sec. 8.2

1. 3   **3.** x **5.** 0   **7.** $5x^2$

Exercise Set 8.2

1. $2\sqrt{5}$ **3.** $3\sqrt{2}$ **5.** $5\sqrt{2}$ **7.** $\sqrt{33}$ **9.** $\dfrac{2\sqrt{2}}{5}$ **11.** $\dfrac{3\sqrt{3}}{11}$ **13.** $\dfrac{3}{2}$ **15.** $\dfrac{5\sqrt{5}}{3}$ **17.** $x^3\sqrt{x}$ **19.** $\dfrac{2\sqrt{22}}{x^2}$

21. $2\sqrt[3]{3}$ **23.** $\dfrac{\sqrt[3]{5}}{4}$ **25.** $x^5\sqrt[3]{x}$ **27.** $\dfrac{\sqrt[3]{2}}{x^3}$ **29.** $2\sqrt{15}$ **31.** $6\sqrt{5}$ **33.** $2\sqrt{13}$ **35.** $\dfrac{\sqrt{11}}{6}$ **37.** $-\dfrac{\sqrt{3}}{4}$

39. $\dfrac{\sqrt[3]{15}}{4}$ **41.** $2\sqrt[3]{10}$ **43.** $2\sqrt[3]{3}$ **45.** $x^6\sqrt{x}$ **47.** $5x\sqrt{3}$ **49.** $4x^2y\sqrt{6}$ **51.** $\dfrac{2\sqrt{3}}{y}$ **53.** $\dfrac{3\sqrt{x}}{y}$

55. $-2x$ **57.** $2\sqrt[3]{10}$ in **59.** answers vary **61.** $48x^2$ **63.** $3x - 2$ **65.** $-72y^4$

Mental Math, Sec. 8.3

1. $8\sqrt{2}$ **3.** $7\sqrt{x}$ **5.** $3\sqrt{7}$

Exercise Set 8.3

1. $-4\sqrt{3}$ **3.** $9\sqrt{6}$ **5.** $\sqrt{5} + \sqrt{2}$ **7.** $7\sqrt[3]{3}$ **9.** $-5\sqrt[3]{2}$ **11.** $8\sqrt{2}$ **13.** $4\sqrt{6} + \sqrt{5}$ **15.** $5\sqrt[3]{3}$

17. $\sqrt[3]{9}$ **19.** $-5x + 2\sqrt{x}$ **21.** $x\sqrt{x}$ **23.** $5x$ **25.** $\sqrt{5} + \sqrt[3]{5}$ **27.** $-5 + 8\sqrt{2}$ **29.** $8 - 6\sqrt{2}$

31. $14\sqrt{2}$ **33.** $5\sqrt{2} + 12$ **35.** $\dfrac{4\sqrt{5}}{9}$ **37.** $\dfrac{3\sqrt{3}}{8}$ **39.** $9 + 9\sqrt[3]{3}$ **41.** $2\sqrt{5}$ **43.** $-\sqrt{35}$ **45.** $11\sqrt{x}$

47. $\sqrt{x}$ **49.** $5x\sqrt{x}$ **51.** $4x\sqrt{2} + 2x\sqrt[3]{3}$ **53.** $x\sqrt{2}$ **55.** $80\sqrt{6}$ ft **57.** $(8\sqrt{13} + 16\sqrt{5})$ ft

59. $x^2 + 12x + 36$ **61.** $4x^2 - 4x + 1$ **63.** $(4, 2)$

Mental Math, Sec. 8.4

1. $\sqrt{6}$ **3.** $\sqrt{6}$ **5.** $\sqrt{10y}$

Exercise Set 8.4

1. 4 **3.** $5\sqrt{2}$ **5.** $2\sqrt[3]{6}$ **7.** $2\sqrt{5} + 5\sqrt{2}$ **9.** $15 - 12\sqrt{15} - 5\sqrt{2} + 4\sqrt{30}$ **11.** $x - 36$

13. $67 + 16\sqrt{3}$ **15.** 4 **17.** $3\sqrt{2}$ **19.** $5y^2\sqrt{y}$ **21.** $\dfrac{\sqrt{15}}{5}$ **23.** $\dfrac{\sqrt{6}}{6}$ **25.** $\dfrac{\sqrt{10}}{6}$ **27.** $3\sqrt{2} - 3$

29. $2\sqrt{10} + 6$ **31.** $\sqrt{30} + 5 + \sqrt{6} + \sqrt{5}$ **33.** $3 + \sqrt{3}$ **35.** $3 - 2\sqrt{5}$ **37.** $3\sqrt{3} + 1$ **39.** $24\sqrt{5}$

41. 11 **43.** 20 **45.** $36x$ **47.** $\sqrt{30} + \sqrt{42}$ **49.** $4x\sqrt{5} - 60\sqrt{x}$ **51.** $\sqrt{6} - \sqrt{15} + \sqrt{10} - 5$ **53.** -5

55. $x - 9$ **57.** $15 + 6\sqrt{6}$ **59.** $9x - 30\sqrt{x} + 25$ **61.** $5\sqrt{3}$ **63.** $2y\sqrt{6}$ **65.** $2x\sqrt{2}$ **67.** $2xy\sqrt{3y}$

69. $12\sqrt[3]{10}$ **71.** $5\sqrt[3]{3}$ **73.** 3 **75.** $2\sqrt[3]{15}$ **77.** $\dfrac{\sqrt{30}}{15}$ **79.** $\dfrac{\sqrt{15}}{10}$ **81.** $\dfrac{\sqrt{10}}{12}$ **83.** $\dfrac{3\sqrt{2x}}{2}$

85. $-8 - 4\sqrt{5}$ **87.** $-(15 - 5\sqrt{10}) = 5\sqrt{10} - 15$ **89.** $4\sqrt{3} + 4$ **91.** $\dfrac{6\sqrt{5} - 4\sqrt{3}}{11}$

93. $\sqrt{6} + \sqrt{3} + \sqrt{2} + 1$ **95.** $\dfrac{2\sqrt{3} + 2 + 3\sqrt{6} + 3\sqrt{2}}{4}$ **97.** $4 + 2\sqrt{3}$ **99.** $\dfrac{3 + 4\sqrt{2}}{2}$

101. $130\sqrt{3}$ sq meters **103.** $\dfrac{\sqrt[3]{6V\pi^2}}{2\pi}$ **105.** $x + 4$ **107.** $2x - 1$

Exercise Set 8.5

1. $x = 81$ **3.** $x = -1$ **5.** $x = 16$ **7.** $x = 2$ **9.** $x = -3$ **11.** $x = 2$ **13.** no solution **15.** $x = 5$

17. $x = 49$ **19.** no solution **21.** $x = -2$ **23.** $x = \dfrac{3}{2}$ **25.** $x = 4$ **27.** $x = 3$ **29.** $x = 9$ **31.** $x = 12$

33. $x = 3$ or $x = 1$ **35.** $x = -1$ **37.** $x = 2$ **39.** $x = 2$ **41.** no solution **43.** $x = 0$ or $x = -3$

45. 9 **47.** 0.51 km **49.** $3x - 8 = 19, x = 9$ **51.** $2x + 2(2x) = 24,$ length $= 8$ in.

53. $x + y = 20, x - y = 8,$ 6 and 14

Exercise Set 8.6

1. $\sqrt{13}$ **3.** $3\sqrt{3}$ **5.** $c = 15$ **7.** $b = \sqrt{17}$ **9.** 20.6 ft **11.** 5 **13.** $\sqrt{34}$ **15.** $5\sqrt{2}$ **17.** 24 cubic ft

19. 25   **21.** $\sqrt{22}$ **23.** $\sqrt{153}$ **25.** $c = \sqrt{41}$ **27.** $a = 4\sqrt{2}$ **29.** $b = 3\sqrt{10}$ **31.** $\sqrt{29}$ **33.** $\sqrt{73}$

35. $2\sqrt{10}$ **37.** $\dfrac{3\sqrt{5}}{2}$ **39.** 51.2 ft   **41.** 11.7 ft   **43.** $\dfrac{3\sqrt{2}}{2}$ in   **45.** 126 ft   **47.** $r = 360$ ft   **49.** 0.13 km

51. answers vary   **53.** x^6 **55.** x^8 **57.** $\left(\dfrac{1}{2}, \dfrac{5}{2}\right)$

Exercise Set 8.7

1. 2   **3.** 3   **5.** 8   **7.** 4   **9.** $-\dfrac{1}{2}$ **11.** $\dfrac{1}{64}$ **13.** $\dfrac{1}{729}$ **15.** $\dfrac{5}{2}$ **17.** 2   **19.** 2   **21.** $\dfrac{1}{x^{2/3}}$ **23.** x^3

25. 9   **27.** -2 **29.** -3 **31.** $\dfrac{1}{9}$ **33.** $\dfrac{3}{4}$ **35.** 27   **37.** 512   **39.** -4 **41.** 32   **43.** $\dfrac{8}{27}$ **45.** $\dfrac{1}{27}$

47. $\dfrac{1}{2}$ **49.** $\dfrac{1}{5}$ **51.** $\dfrac{1}{125}$ **53.** 9   **55.** $6^{1/3}$ **57.** x^6 **59.** 36   **61.** $\dfrac{1}{3}$ **63.** $\dfrac{x^{2/3}}{y^{3/2}}$ **65.** $\dfrac{x^{16/5}}{y^6}$ **67.** 11,224

69. answers vary   **71.**

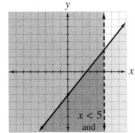

$x < 5$
and
$2x - y \geq 3$

73. $x = -4$ or $x = 2$ **75.** $x = \dfrac{2}{3}$ or $x = -1$

Chapter 8 Review

1. 9   **3.** 3   **5.** $-\dfrac{3}{8}$ **7.** not a real number   **9.** irrational, 8.718   **11.** x^6 **13.** $3x^3y$ **15.** $-2x^2$ **17.** $3\sqrt{6}$

19. $5xy^3\sqrt{6x}$ **21.** $3\sqrt[3]{2}$ **23.** $2y\sqrt[4]{3x^3y^2}$ **25.** $\dfrac{3\sqrt{2}}{5}$ **27.** $\dfrac{3y\sqrt{5}}{2x^2}$ **29.** $\dfrac{\sqrt[4]{9}}{2}$ **31.** $\dfrac{y^2\sqrt[3]{3}}{2x}$ **33.** $\sqrt[3]{2}$

35. $6\sqrt{6} - 2\sqrt[3]{6}$ **37.** $5\sqrt{7} + 2\sqrt[3]{7}$ **39.** $\dfrac{\sqrt{5}}{6}$ **41.** 0   **43.** $30\sqrt{2}$ **45.** $6\sqrt{2} - 18$

47. $3\sqrt{2} - 5\sqrt{3} + 2\sqrt{6} - 10$ **49.** $4\sqrt{2}$ **51.** $\dfrac{x\sqrt{5x}}{2y^4}$ **53.** $\dfrac{\sqrt{30}}{6}$ **55.** $\dfrac{\sqrt{6x}}{2x}$ **57.** $\dfrac{\sqrt{35}}{10y}$ **59.** $\dfrac{\sqrt[3]{21}}{3}$

61. $\dfrac{\sqrt[3]{12x}}{2x}$ **63.** $3\sqrt{5} + 6$ **65.** $4\sqrt{6} - 8$ **67.** $\dfrac{2\sqrt{2} - 1}{7}$ **69.** $-\dfrac{3 + 5\sqrt{3}}{11}$ **71.** $x = 18$ **73.** $x = 25$

75. $x = 12$ **77.** $x = 6$ **79.** $2\sqrt{14}$ **81.** $4\sqrt{34}$ ft   **83.** $\sqrt{130}$ **85.** 2.39 in   **87.** 4   **89.** -2 **91.** -512

93. $\dfrac{8}{27}$ **95.** $\dfrac{1}{5}$ **97.** $a^{5/2}$ **99.** $x^{5/2}$ **101.** 32   **103.** $\dfrac{1}{3^{2/3}}$ **105.** $\dfrac{1}{x^2}$

Chapter 8 Test

1. 4   **2.** 5   **3.** 8   **4.** $\dfrac{3}{4}$ **5.** not a real number   **6.** $\dfrac{1}{9}$ **7.** $3\sqrt{6}$ **8.** $2\sqrt{23}$ **9.** $x^3\sqrt{3}$ **10.** $2x^2y^3\sqrt{2y}$

11. $3x^4\sqrt{x}$ **12.** $2\sqrt[3]{5}$ **13.** $2x^2y^3\sqrt[3]{y}$ **14.** $-8\sqrt{3}$ **15.** $3\sqrt[3]{2} - 2x\sqrt{2}$ **16.** $\dfrac{\sqrt{5}}{4}$ **17.** $\dfrac{x\sqrt[3]{2}}{3}$ **18.** $\dfrac{\sqrt{6}}{3}$

19. $6\sqrt{2x}$ **20.** $\dfrac{\sqrt[3]{15}}{3}$ **21.** $\dfrac{\sqrt[3]{6x}}{2x}$ **22.** $4\sqrt{6} - 8$ **23.** $-1 - \sqrt{3}$ **24.** $x = 9$ **25.** $x = 5$ **26.** $x = 9$

27. $4\sqrt{5}$ in.   **28.** $\sqrt{5}$ **29.** $\dfrac{1}{16}$ **30.** $x^{10/3}$

Chapter 8 Cumulative Review

1. (a) $x + 3$; **(b)** $3x$; **(c)** $2x$; **(d)** $10 - x$; **(e)** $5x + 7$; *Sec. 1.5, Ex. 2*   **2. (a)** -8; **(b)** 2; *Sec. 1.8, Ex. 6*

3. $x = -11$; *Sec. 2.3, Ex. 2*   **4.** infinite solutions; *Sec. 2.4, Ex. 6*   **5.** 45 dimes, 85 quarters; *Sec. 2.6, Ex. 2*

6. $x \geq 1$; *Sec. 2.7, Ex. 9*   **7.** $-6x^7$; *Sec. 3.1, Ex. 3*   **8. (a)** $4x$; **(b)** $13x^2 - 2$; *Sec. 3.3, Ex. 4*

9. $5x^2 - 17x + 14$; *Sec. 3.5, Ex. 2*   **10.** $(4x - 1)(2x - 5)$; *Sec. 4.3, Ex. 2*   **11.** $(2m - 1)^2$; *Sec. 4.3, Ex. 9*

12. $x = -6, -\dfrac{3}{2}, \dfrac{1}{5}$; *Sec. 4.6, Ex. 6*   **13.** $\dfrac{x + 2}{2}$; *Sec. 5.1, Ex. 6*   **14.** $\dfrac{x - 1}{4}$; *Sec. 5.2, Ex. 5*

15. $\dfrac{2m + 1}{m + 1}$; *Sec. 5.4, Ex. 5*   **16. (a)** $x = 4$; **(b)** $\dfrac{9x}{4}$; *Sec. 5.6, Ex. 6*

17. *Sec. 6.2, Ex. 5*

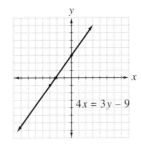

$4x = 3y - 9$

18. $(-2, 0)$; *Sec. 7.2, Ex. 3*   **19. (a)** 2; **(b)** -2; **(c)** -2; **(d)** not a real number; *Sec. 8.1, Ex. 3*

20. (a) $x^2\sqrt{x}$; **(b)** $2y\sqrt{2}$; **(c)** $\dfrac{3\sqrt{5}}{x^3}$; *Sec. 8.2, Ex. 3*

21. (a) $7\sqrt{2}$; **(b)** $9\sqrt{3}$; **(c)** $2 - 3\sqrt{3} - 6\sqrt{2}$; **(d)** $6 - 3\sqrt[3]{2}$; *Sec. 8.3, Ex. 2*   **22.** $\dfrac{4 - \sqrt{2}}{3}$; *Sec. 8.4, Ex. 7*

23. $x = \dfrac{1}{2}$; *Sec. 8.5, Ex. 4*

CHAPTER 9
Solving Quadratic Equations

Exercise Set 9.1

1. $k = \pm 3$ **3.** $m = -5$ or $m = 3$ **5.** $x = \pm \dfrac{9\sqrt{2}}{2}$ **7.** $a = \pm 3$ **9.** $x = \pm 8$ **11.** $p = \pm \dfrac{1}{7}$

13. no real solution   **15.** $x = \pm 5$ **17.** $x = \pm \dfrac{2\sqrt{3}}{3}$ **19.** $x = -2$ or $x = 12$ **21.** $x = -2 \pm \sqrt{7}$

23. $m = 0$ or $m = 1$ **25.** $p = -13$ or $p = 9$ **27.** $y = -4$ or $y = \dfrac{8}{3}$ **29.** no real solution

31. $x = \dfrac{11 \pm 5\sqrt{2}}{2}$ **33.** $q = \pm 10$ **35.** $x = 9$ or $x = 17$ **37.** $z = \pm 2\sqrt{3}$ **39.** $x = -5 \pm \sqrt{10}$

41. no real solution   **43.** $y = \pm \dfrac{\sqrt{22}}{2}$ **45.** $p = -3$ or $p = 8$ **47.** $x = \dfrac{1 \pm \sqrt{7}}{3}$ **49.** $x = \dfrac{7 \pm 4\sqrt{2}}{3}$

51. 15 ft by 15 ft   **53.** $x = 10\sqrt{2}$ cm.   **55.** answers vary   **57.** $\dfrac{12}{7}$ **59.** $\dfrac{3}{5}$ **61.** $(0, 7)$

Mental Math, Sec. 9.2
1. 16   **3.** 100   **5.** 49

Exercise Set 9.2

1. $(x - 2)^2$ **3.** $(k - 6)^2$ **5.** $\left(x - \dfrac{3}{2}\right)^2$ **7.** $\left(m - \dfrac{1}{2}\right)^2$ **9.** $x = 0$ or $x = 6$ **11.** $z = -6$ or $z = -2$

13. $x = -1 \pm \sqrt{6}$ **15.** $x = -\dfrac{1}{2}$ or $x = \dfrac{13}{2}$ **17.** no real solution **19.** $x = \dfrac{3 \pm \sqrt{19}}{2}$

21. $x = -3 \pm \sqrt{34}$ **23.** $z = \dfrac{-5 \pm \sqrt{53}}{2}$ **25.** $x = 1 \pm \sqrt{2}$ **27.** $y = -4$ or $y = -1$

29. $x = -2$ or $x = 4$ **31.** $y = \dfrac{-4 \pm \sqrt{6}}{2}$ **33.** $y = \dfrac{1}{2}$ or $y = 1$ **35.** $y = \dfrac{1 \pm \sqrt{13}}{3}$ **37.** $y = 7$ or $y = -2$

39. $m = -6$ or $m = 3$ **41.** answers vary **43.** $\dfrac{7}{5}$ **45.** $\dfrac{1}{5}$ **47.** $5 - 10\sqrt{3}$ **49.** $\dfrac{3 - 2\sqrt{7}}{4}$

Mental Math, Sec. 9.3

1. $a = 2, b = 5, c = 3$ **3.** $a = 10, b = -13, c = -2$ **5.** $a = 1, b = 0, c = -6$

Exercise Set 9.3

1. -2 or 1 **3.** $\dfrac{-5 \pm \sqrt{17}}{2}$ **5.** $\dfrac{2 \pm \sqrt{2}}{2}$ **7.** $x = 1$ or $x = 2$ **9.** $k = \dfrac{-7 \pm \sqrt{37}}{6}$ **11.** $x = \pm\dfrac{2}{7}$

13. no real solution **15.** $y = -3$ or $y = 10$ **17.** $x = \pm\sqrt{5}$ **19.** $m = -3$ or $m = 4$ **21.** $x = -2 \pm \sqrt{7}$

23. no real solution **25.** $m = 1 \pm \sqrt{2}$ **27.** $p = -\dfrac{3}{4}$ or $p = -\dfrac{1}{2}$ **29.** $a = \dfrac{1}{2}$ or $a = 3$ **31.** $x = \dfrac{5 \pm \sqrt{33}}{2}$

33. $x = -2$ or $x = \dfrac{7}{3}$ **35.** $x = \dfrac{-9 \pm \sqrt{129}}{12}$ **37.** $p = \dfrac{4 \pm \sqrt{2}}{7}$ **39.** $a = 3 \pm \sqrt{7}$ **41.** $x = \dfrac{3 \pm \sqrt{3}}{2}$

43. $x = -1$ or $x = \dfrac{1}{3}$ **45.** $y = -\dfrac{3}{4}$ or $y = \dfrac{1}{5}$ **47.** no real solution **49.** $y = \dfrac{3 \pm \sqrt{13}}{4}$ **51.** $x = \dfrac{7 \pm \sqrt{129}}{20}$

53. no real solution **55.** $z = \dfrac{1 \pm \sqrt{2}}{5}$ **57.** answers vary **59.** $x = \dfrac{3}{5}$ **61.** $x = -\dfrac{11}{30}$ **63.** $z = -4$

Exercise Set 9.4

1. $x = \dfrac{1}{5}$ or $x = 2$ **3.** $x = 1 \pm \sqrt{2}$ **5.** $a = \pm 2\sqrt{5}$ **7.** no real solution **9.** $x = 2$ **11.** $p = 3$

13. $y = \pm 2$ **15.** $x = 0, x = 1$ or $x = 2$ **17.** $z = -5$ or $z = 0$ **19.** $x = \dfrac{3 \pm \sqrt{7}}{5}$ **21.** $m = -1$ or $m = \dfrac{3}{2}$

23. $x = \dfrac{5 \pm \sqrt{105}}{20}$ **25.** $x = \dfrac{7}{4}$ or $x = 5$ **27.** $x = \dfrac{7 \pm 3\sqrt{2}}{5}$ **29.** $z = \dfrac{7 \pm \sqrt{193}}{6}$ **31.** $x = -10$ or $x = 11$

33. $x = -\dfrac{2}{3}$ or $x = 4$ **35.** $x = 0.1$ or $x = 0.5$ **37.** $x = \dfrac{11 \pm \sqrt{41}}{20}$ **39.** $z = \dfrac{4 \pm \sqrt{10}}{2}$ **41.** answers vary

43. $2\sqrt{26}$ **45.** $4\sqrt{5}$ **47.** width = 17 in.; length = 23 in.

Exercise Set 9.5

1. $3i$ **3.** $10i$ **5.** $5i\sqrt{2}$ **7.** $3i\sqrt{7}$ **9.** $-3 + 9i$ **11.** $1 - 3i$ **13.** $8 + 12i$ **15.** $26 - 2i$ **17.** $2 - 3i$

19. $\dfrac{31}{25} + \dfrac{17}{25}i$ **21.** $x = -1 \pm 3i$ **23.** $z = \dfrac{3 \pm 2i\sqrt{3}}{2}$ **25.** $y = -3 \pm 2i$ **27.** $x = \dfrac{-7 \pm i\sqrt{15}}{18}$

29. $m = \dfrac{2 \pm i\sqrt{6}}{2}$ **31.** $15 - 7i$ **33.** $45 + 63i$ **35.** $-1 + 3i$ **37.** $2 - 3i$ **39.** $-7 - 2i$ **41.** 25

43. $\dfrac{2}{5} - \dfrac{9}{5}i$ **45.** $21 + 20i$ **47.** $y = 4 \pm 8i$ **49.** $x = \pm 5i$ **51.** $z = -3 \pm i$ **53.** $a = \dfrac{5 \pm i\sqrt{47}}{4}$

55. $x = -4 \pm i\sqrt{5}$ **57.** $m = \pm 6i$ **59.** $x = -7 \pm i$ **61.** answers vary

63.

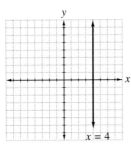

65.

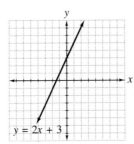

67. $y = \sqrt{39}$ yd

Exercise Set 9.6

1.

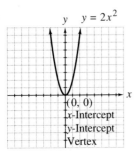

3.

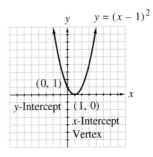

5.

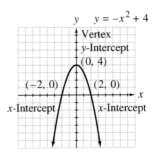

7.

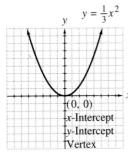

9.

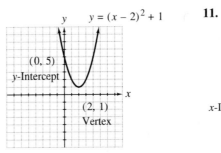

11.

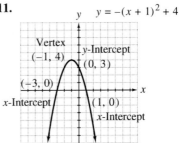

13.

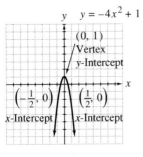

15. f   **17.** a   **19.** h   **21.** b   **23.** d

25.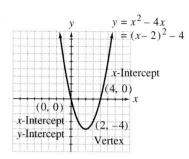

27.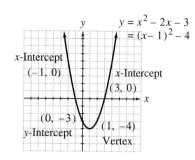

29.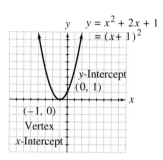

31.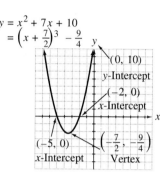

33.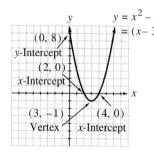

35. $\dfrac{5}{14}$ **37.** $\dfrac{x}{2}$ **39.** $\dfrac{2x^2}{x-1}$ **41.** $-4b$

Chapter 9 Review

1. $x = -\dfrac{3}{5}$ or $x = 4$ **3.** $m = -\dfrac{1}{3}$ or $m = 2$ **5.** $k = \pm 5\sqrt{2}$ **7.** $x = -1$ or $x = 7$ **9.** $x = 4$ or $x = 18$

11. $x = 0$ or $x = \pm 3$ **13.** $p = -3$ or $p = 2$ **15.** $x^2 - 10x + 25 = (x - 5)^2$ **17.** $a^2 + 4a + 4 = (a + 2)^2$

19. $m^2 - 3m + \dfrac{9}{4} = \left(m - \dfrac{3}{2}\right)^2$ **21.** $x = 3 \pm \sqrt{2}$ **23.** $y = -1$ or $y = \dfrac{1}{2}$ **25.** $x = 5 \pm 3\sqrt{2}$

27. $x = -1$ or $x = \dfrac{1}{2}$ **29.** $x = -\dfrac{5}{3}$ **31.** $x = \dfrac{2}{5}$ or $x = \dfrac{1}{3}$ **33.** $x = \dfrac{-1 \pm i\sqrt{39}}{4}$ **35.** $z = \dfrac{-1 \pm \sqrt{21}}{10}$

37. $x = 0$ or $x = \pm\dfrac{1}{2}$ **39.** $x = \dfrac{1}{2}$ or $x = 7$ **41.** $x = \dfrac{1}{3}$ **43.** $x = 3$ **45.** $x = -10$ or $x = 22$

47. $x = \dfrac{3 \pm \sqrt{5}}{20}$ **49.** $x = -5 \pm \sqrt{30}$ **51.** $12i$ **53.** $6i\sqrt{3}$ **55.** $21 - 10i$ **57.** $-8 - 2i$ **59.** $-13i$

61. 25 **63.** $-\dfrac{3}{2} - \dfrac{1}{2}i$ **65.** $\dfrac{2}{5} - \dfrac{9}{5}i$ **67.** $x = \pm 4i$ **69.** $2 \pm 3i$

71. vertex $(0, 0)$; axis of symmetry $x = 0$; opens downward   **73.** vertex $(3, 0)$; axis of symmetry $x = 3$; opens upward

75. vertex $(0, -7)$; axis of symmetry $x = 0$; opens upward   **77.** vertex $(72, 14)$; axis of symmetry $x = 72$; opens downward

79.

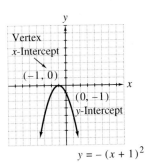

$y = -(x + 1)^2$

81.

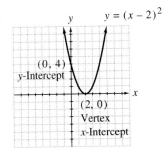

$y = (x - 2)^2$

83.

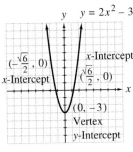

$y = 2x^2 - 3$

85.

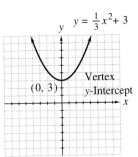

$y = \frac{1}{3}x^2 + 3$

Chapter 9 Test

1. $x = -\frac{3}{2}$ or $x = 7$ **2.** $x = 0$ or $x = -2$ or $x = 1$ **3.** $k = \pm 4$ **4.** $x = \frac{5 \pm 2\sqrt{2}}{3}$ **5.** $x = 10$ or $x = 16$

6. $x = -2$ or $x = \frac{1}{5}$ **7.** $x = -2$ or $x = 5$ **8.** $p = \frac{5 \pm \sqrt{37}}{6}$ **9.** $x = -\frac{4}{3}$ or $x = 1$ **10.** $x = -1$ or $x = \frac{5}{3}$

11. $x = \frac{7 \pm \sqrt{73}}{6}$ **12.** $x = 2 \pm i$ **13.** $x = \frac{1}{3}$ or $x = 2$ **14.** $x = \frac{3 \pm \sqrt{7}}{2}$ **15.** $x = -3$ or $x = 0$ or $x = \frac{1}{2}$

16. $x = 0$ or $x = \pm\frac{1}{3}$ **17.** $5i$ **18.** $10i\sqrt{2}$ **19.** $8 + i$ **20.** $7 - 6i$ **21.** $4i$ **22.** 6 **23.** 13 **24.** $\frac{1 - 7i}{5}$

25.

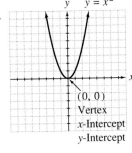

$y = x^2$

26.

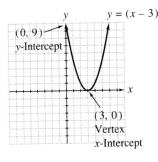

$y = (x - 3)^2$

27.

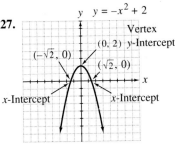

$y = -x^2 + 2$

28.

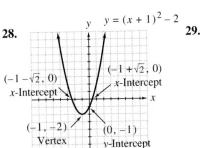

$y = (x + 1)^2 - 2$

29.

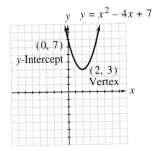

$y = x^2 - 4x + 7$

30.

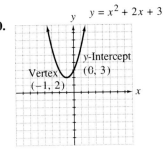

$y = x^2 + 2x + 3$

Chapter 9 Cumulative Review

1. $x = 3$; *Sec. 2.3, Ex. 6*   **2.** $W = \dfrac{P - 2L}{2}$; *Sec. 2.5, Ex.8*   **3.** $1 \le x < 4$; *Sec. 2.7, Ex. 10*

4. **(a)** $s^4 t^4$; **(b)** $\dfrac{m^7}{n^7}$; **(c)** $8a^3$; **(d)** $25x^4 y^6 z^2$; **(e)** $\dfrac{16x^{16}}{81y^{20}}$; *Sec. 3.1, Ex. 5*   **5.** $9x^2 - 6x - 1$; *Sec. 3.3, Ex. 9*

6. **(a)** $10x^4 + 30x$; **(b)** $-15x^4 - 18x^3 + 3x^2$; **(c)** $6n^3 - 10n^2 + 8n$, *Sec. 3.4, Ex. 1*

7. $4x^2 - 4x + 6 - \dfrac{11}{2x + 3}$; *Sec. 3.6, Ex. 7*   **8.** $(5x + y)(2x - 3y)$; *Sec. 4.3, Ex. 4*

9. $9(x - 2)(x + 2)$; *Sec. 4.4, Ex. 3*   **10.** $x = -3, 0, 5$; *Sec. 4.6, Ex. 7*   **11.** $-\dfrac{(x + 2)}{3x + 1}$; *Sec. 5.1, Ex. 8*

12. $\dfrac{2}{x(x + 1)}$; *Sec. 5.2, Ex. 6*   **13.** $\dfrac{3}{x - 2}$, *Sec. 5.4; Ex. 2*   **14.** $\dfrac{3}{z}$; *Sec. 5.5, Ex. 4*

15. $x = \dfrac{5ab + a^2 b}{2b - 3a}$; *Sec. 5.6, Ex. 7*

16. *Sec. 6.2, Ex. 3*   **17.** slope $= 2$; *Sec. 6.3, Ex. 2*

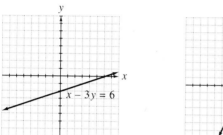
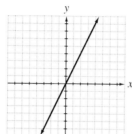

18. $x = -1$; *Sec. 6.4, Ex. 5*   **19.** **(a)** 6; **(b)** 8; **(c)** -5; **(d)** $\dfrac{3}{10}$; **(e)** 0; *Sec. 8.1, Ex. 1*

20. **(a)** $\sqrt{21}$; **(b)** $3\sqrt{5}$; **(c)** $2\sqrt[3]{9}$; **(d)** $20\sqrt{3}$; **(e)** 18; *Sec. 8.4, Ex. 1*   **21.** **(a)** -44; **(b)** $7x + 4\sqrt{7x} + 4$; *Sec. 8.4, Ex. 3*

22. no real solution; *Sec. 8.5, Ex. 3*   **23.** $x = -1$ or $x = 7$; *Sec. 9.1, Ex. 4*   **24.** $x = \dfrac{-1 \pm \sqrt{37}}{6}$; *Sec. 9.3, Ex. 1*

Index

A

Absolute value, 8–9, 47
Acute angle, 427
Addition
  associative property of, 43
  common words expressing, 63
  commutative property of, 42, 204
  of complex numbers, 406
  of decimals, 424
  of fractions, 13–14
  identity element for, 44
  of polynomials, 132–33
  of radicals, 357–60
  of rational expressions, 213–25
  of real numbers, 24–30
  solving linear systems by, 324–29
  symbol for, 3
  See also Distributive property of multipli-
    cation over addition
Addition property of equality,
  59–65, 68–69, 73
Addition property of inequality, 98–99
Additive inverses (opposites), 28, 44, 45, 47
Adjacent angles, 429
Algebraic expression, 21–22
  evaluating, 21
  simplifying, 53–59
Alternate interior angles, 429
Angles, types of
  acute, 427
  adjacent, 429
  alternate interior, 429
  complementary, 427
  corresponding, 429, 430
  exterior, 429
  interior, 429
  obtuse, 427
  right, 427
  straight, 427
  supplementary, 427
  vertical, 428
Approximation, 349
Area of rectangle, formula for, 81, 83
Associative properties, 42–43
Axis of symmetry, 411, 413, 420

B

Base, 16
  common, 115
  of exponential expression, 113, 114
Binomials, 151, 176–80
  defined, 131
  difference of squares, 176–77
  factoring, 176–80
  multiplying, 137, 140–41
  squaring, using FOIL method, 141–42
Boundary line(s), 290
  corner point at intersection of, 336

C

Calculator
  checking possible solutions of equations
    with, 78
  entering negative numbers, 39
  entering number in scientific notation on,
    127
  exponential expressions on, 19
  operations with real numbers, 39
  order of operations on, 19
  square roots using, 350
Cartesian coordinate system,
  260–65, 303
Center
  of circle, 436
  of sphere, 437
Circle, 436
Closure properties, 41
Coefficient, 130, 151
  numerical, 53, 54, 130, 151
Combining like terms, 54–56, 62, 73, 132
Common base, 115
Common denominator
  adding and subtracting rational expres-
    sions with, 213–14
  least, 14, 215–16, 220–22
  lowest, 73
Commutative properties, 41–42
  of addition, 42, 204
  of multiplication, 42

Complement, 427
Complementary angle, 427
Completing the square, 392–97, 420
Complex conjugates, 407
Complex fractions, 225–31, 250
  defined, 225
  parts of, 225
  simplifying, 225–28
Complex numbers, 405–7, 420
  standard form, 406
Complex solutions, solving quadratic
    equations with, 405–10
Compound inequality, 102, 106
Cone, right circular, 437
Congruent triangles, 430–31
Conjugates, 364, 383
  complex, 407
Consecutive integers, 77–78
Consistent system, 312, 313, 315, 340
Constant, 130
  degree of, 131
Coordinate system, Cartesian, 260–65, 303
  x- and y-coordinates, 260, 276
Coplanar lines, 428
Corner point, 336
Corresponding angles, 429, 430
Cross multiplication, 238–39, 250
Cross products, 238, 250
Cube(s), 437
  sum or difference of two, 177–78, 180,
    182
Cube root, 348–49, 383
Cylinder, right circular, 437

D

Decimals
  multiplying signed, 36–37
  operations on, 424–26
Degree
  of polynomial, 151
  of term, 131, 151
Degrees Fahrenheit vs. Celsius, formula for,
  82

Denominator, 10
  adding or subtracting fractions with same, 13
  adding or subtracting rational expressions with common, 213–14
  adding or subtracting rational expressions with unlike, 219–25
  complex numbers in, 407
  least common, 14, 215–16, 220–22
  lowest common (LCD), 73
  of rational expression, 202–3
  rationalizing, 363–64, 383
  zero as, 233
Dependent equations, 312, 322, 340
Descartes, Rene, 260
Descending powers, 131, 151
Difference of squares, factoring, 176–77, 180
Difference of two cubes, 177–78, 180
Distance formula, 81, 84–85
  applications of radical equations involving, 374–75
  word problems using, 92–93, 244–45
Distributive property of multiplication over addition, 43–44, 63
  combining like terms by, 54–56
  to multiply polynomials, 136–37
  to remove grouping symbols, 73, 74, 75, 76
Dividend polynomial, 147
Division
  common words expressing, 63
  of complex numbers, 407
  of decimals, 424–25
  of fractions, 12
  long, 146–48
  multiplication property of inequality and, 100
  of polynomials, 145–50
  power rule for quotients, 116–17, 124, 378
  quotient rule for exponents, 117–18, 122, 124
  of radicals, 362–63
  of rational expressions, 209–10
  of real numbers, 34–41
  sign rules for, 35
  symbol for, 3
  of or by zero, 37
Divisor polynomial, 147, 148
Domain
  of function, 300
  of relation, 297

E

Elimination method, solving linear systems by, 324–29
Equality
  addition property of, 59–65, 68–69, 73
  common words expressing, 63
  multiplication property of, 66–72, 73
  squaring property of, 367
Equal symbol, 1

Equation(s), 106
  calculator to check possible solutions of, 78
  defined, 22
  dependent, 312, 322, 340
  equivalent, 60
  exponential, 16
  of lines, 283–89
  solution (root) of, 60, 106
  solving for variable, 60
  See also Linear equation(s); Quadratic equation(s)
Equations of lines, graphing, 283–89
  point-slope form of, 284–86
Equilateral triangle, 435
Equivalent equations, 60
Equivalent fractions, 13
Even integers, 77
Exponential equation, 16
Exponents, 16, 113–22
  base of exponential expression, 113, 114
  defined, 113
  evaluating exponential expression on calculator, 19
  evaluating number raised to power, 113–14
  negative, 122–24, 378
  power rule for, 115–16, 124, 378
  power rules for products and quotients, 116–17, 124, 378
  product rule for, 114–15, 119, 124, 378
  quotient rule for, 117–18, 122, 124, 378
  rational, 378–82
  scientific notation, using, 125–27
  zero exponent, 118–19, 124, 378
Exterior angles, 429
Extraneous solution, 233, 250, 369
Extremes of proportion, 238, 250

F

Factor(s), 10
  greatest common (GCF), 157–60, 166, 177, 178
Factoring, 10, 11
  by grouping, 160–61, 173–76, 181
  to simplify rational expressions, 203–4
Factoring out, 159–60, 166, 177, 178
Factoring polynomials, 156–99
  binomials, 176–80
  defined, 196
  factoring completely, 166, 180–83
  factoring four–term polynomial by grouping, 160–61
  greatest common factor (GCF), 157–60, 166, 177, 178
  quadratic equations, 183–96
  trinomials, 163–76
FOIL method, 140–42, 362
Formulas, 81–89, 106, 375
  common, 81–82
  defined, 81
  geometric, 89–90
  to solve word problems, 82–84
  solving for one of its variables, 84–86

Fraction(s), 10–15
  adding and subtracting, 13–14
  clearing equation of, 73, 75, 86
  complex, 225–31, 250
  dividing, 12
  equivalent, 13
  multiplying, 12, 36–37
  ratios, 237–41, 245, 250
  signed, 36–37
  simplified or in lowest terms, 11, 203
Fractional exponents, 378–82
Function, 303
  defined, 297
  domain of, 300
  graphing, 297–302
  vertical line test of, 298–99
Function notation, 299–300, 303

G

GCF. *See* Greatest common factor (GCF)
Geometric figures, 435–37
Geometric formulas, 89–90
Geometry, 427
Graphing, 256–309
  Cartesian coordinate system and, 260–65, 303
  equations of lines, 283–89
  finding and plotting ordered pair solutions, 265–67
  functions, 297–302
  intercept method, 268–70
  linear equations, 265–74
  linear inequalities, 97–98, 289–96
  of quadratic equations, 410–19
  slope, 274–82, 303
  solutions of linear systems, 311–18
  solutions of system of linear inequalities, 336–37
  vertical and horizontal lines, 270–71
Greater than or equal to symbol, 2
Greater than symbol, 2
Greatest common factor (GCF), 157–60, 166, 177, 178
  factoring out, 159–60, 166, 177, 178
  finding, 157–59
Grouping, factoring by, 181
  four-term polynomial, 160–61
  trinomials, 173–76
Grouping symbols, 16, 18, 31, 73, 74, 75, 76

H

Half-planes, 290
Horizontal lines, 286
  graph of linear equation as, 271
  slope of, 279–80, 281
Hypotenuse, 192, 193, 372–74, 433

I

Identity, 76
Identity properties for addition and multiplication, 44

Imaginary numbers, 405–6
Imaginary unit *i*, 405, 420
Inconsistent system, 312, 314, 340
Index, 349
Inequality(ies)
  addition property of, 98–99
  compound, 102, 106
  multiplication property of, 99–100
  solution of, 97–98
Inequality(ies), linear, 97–106
  defined, 97
  difference between linear equations and, 97, 99
  graphing, 97–98, 289–96
  solving, 98–101
  system of, 335–39, 340
Inequality symbols, 2
Integers, 47
  consecutive, 77–78
  even, 77
  negative, 6
  odd, 77
  positive, 6
  set of, 6, 7
Intercept method, 268–70
Intercepts, *x*- and *y*-, 268–70, 303
  of parabola, 412–13
  slope-intercept form, 277–78, 283, 286
Interest formula, simple, 81, 84, 91–92
Interior angles, 429
Intersecting lines, 428
Inverses
  additive, 28, 44, 45, 47
  multiplicative, 45, 47
Irrational numbers, 47, 349
  set of, 7
Isosceles trapezoid, 436
Isosceles triangle, 435

L

Least common denominator (LCD), 14
  finding, 215–16, 220–22
Legs of right triangle, 192, 372–74, 433
Less than or equal to symbol, 2
Less than symbol, 2
Like radicals, 357, 383
Like terms, 54, 106
  combining, 54–56, 62, 73, 132
Linear equation(s), 106
  applications of, 89–96
  difference between linear inequalities and, 97, 99
  formulas, 81–89
  in one variable, 60
  point-slope form, 284–86
  slope-intercept form, 277–78, 283, 286
  standard form, 258, 286
  systems of. *See* Linear systems
  in two variables, 257–58, 265–74
Linear equations, graphing, 265–74
  by finding and plotting ordered pair solutions, 265–67
  by intercept method, 268–70
  vertical and horizontal lines, 270–71

Linear equations, solving
  addition property of equality for, 60–63
  combined addition and multiplication properties of equality for, 68–69
  containing rational expressions, 231–35
  general strategy for, 73–76
  multiplication property of equality for, 66–68
Linear inequalities, 97–106
  compound inequalities, 102, 106
  defined, 97
  difference between linear equations and, 97, 99
  graphing, 97–98, 289–96
  solving, 98–101
  systems of, 335–39, 340
  word problems involving, 102–3
Linear systems, 310–45
  addition, solving by, 324–29
  applications of, 329–35
  determining number of solutions of, 315
  graphing, solving by, 311–18
  substitution method of solving, 319–23
Lines
  boundary, 290, 336
  coplanar, 428
  equations of, 283–89
  horizontal, 270–71, 279–80, 281, 286
  intersecting, 428
  parallel, 278–79, 286, 314, 322, 428, 429
  perpendicular, 278–79, 286, 428
  slope of, 274–82, 303
  vertical, 270, 279–80, 281, 286
  See also Linear equation(s)
Long division of polynomials, 146–48
Lowest common denominator (LCD), 73

M

Mathematical statement, 2
Means in proportion, 238, 250
Monomial, 151
  defined, 131
  dividing polynomial by, 145–46
  multiplying, 136
Multiplication
  associative property of, 43
  common words expressing, 63
  commutative property of, 42
  of complex numbers, 406–7
  cross, 238–39, 250
  of decimals, 424
  exponents as repeated, 16
  of fractions, 12
  identity element for, 44
  of polynomials, 136–40
  power rule for products, 116–17, 124, 378
  product rule for exponents, 114–15, 119, 124, 378
  of radicals, 361–62
  of rational expressions, 207–9
  of real numbers, 34–41
  sign rules for, 35
  special products, 140–45

squaring a binomial, 141–42
  of sum and difference of two terms, 142–43
  symbol for, 3
  by zero, 36
  See also Distributive property of multiplication over addition
Multiplication property of equality, 66–72, 73
Multiplication property of inequality, 99–100
Multiplicative inverses (reciprocals), 12, 45, 47

N

Natural numbers, 1, 47
  set of, 5, 7
Negative exponents, 122–24, 378
Negative integers, 6
Negative numbers, 7
  entering in calculator, 39
  square root of, 348
Negative slope, 277, 281
Notation
  function, 299–300, 303
  scientific, 125–27, 151
Not equal symbol, 2
Number line, 1
  absolute value on, 8
  adding real numbers on, 25, 26
  additive inverses on, 28
  order of numbers on, 1–2
  rational numbers on, 6
  real numbers on, 7
  whole numbers on, 6
Numbers
  complex, 405–7, 420
  imaginary, 405–6
  irrational, 7, 47, 349
  natural, 1, 5, 47
  negative, 7, 39
  positive, 7
  prime, 11
  rational, 6, 7, 47
  real. *See* Real numbers
  sets of, 5–10
  signed, 7
  symbols used to compare, 1, 2, 4
  whole, 1, 6, 7, 47
Numerator, 10
Numerical coefficient, 53, 54, 130, 151

O

Obtuse angle, 427
Odd integers, 77
Operation symbols, 3
Opposites (additive inverses), 28, 44, 45, 47
Ordered pair, 257, 303
  finding and plotting ordered pair solutions, 265–67
  finding missing coordinate of, 258–60
  plotting point on rectangular coordinate system, 260–62
  relation as set of, 297

as solution of linear system, 311–12
as solution to equation in two variables, 258
as solution to system of linear inequalities, 335
Order of operation, 17–19
  on calculator, 19
Order property for real numbers, 8
Origin, 260
Original stated problem, checking solutions with, 64

P

Parabola, 410–11, 420
  intercepts of, 412–13
  vertex of, 411–16, 420
Parallel lines, 314, 322, 428
  cut by transversal, 429
  finding equations of, 286
  slope of, 278–79
Parallelogram, 436
Partial product, 138
Perfect squares, 349
Perfect square trinomial, 172–73, 181, 196, 393, 394, 420
Perimeter formulas, 81, 85, 90
Perpendicular lines, 278–79, 286, 428
Plane, 428
  half-plane, 290
Plane figures, 428
Plotting a point, 260
Point-slope form of equation of line, 284–86
Polygon, 430, 435
Polynomial(s), 130–40, 151
  adding, 132–33
  combining like terms in, 132
  defined, 130–31
  degree of, 131, 151
  dividend, 147
  dividing, 145–50
  divisor, 147, 148
  multiplying, 136–40
  prime, 177
  quotient, 147
  remainder, 147, 148
  subtracting, 133–34
Polynomials, factoring, 156–99
  binomials, 176–80
  defined, 196
  factoring completely, 166, 180–83
  four-term, by grouping, 160–61
  greatest common factor (GCF), 157–60
  quadratic equations, 183–96
  trinomials, 163–76
Positive integers, 6
Positive numbers, 7
Positive slope, 277, 281
Power. *See* Exponents
Power rule
  for exponents, 115–16, 124, 378
  for products and quotients, 116–17, 124, 378

Prime numbers, 11
Prime polynomial, 177

Principal square root, 347, 383
Product(s)
  cross, 238, 250
  factored form of polynomial as, 161
  partial, 138
  power rule for, 116–17, 124, 378
  special, 140–45
Product rule
  for exponents, 114–15, 119, 124, 378
  for radicals, 354–55, 361–62
  for square roots, 352–53
Proportion, 238–41, 250
  in similar triangles problem, 245–46
Pythagorean theorem, 192–93, 433
  to solve applications of radical equations, 372–74

Q

Quadrants, 260
Quadratic equation(s), 183–96
  applications of, 189–96
  defined, 183–84, 196
  with degree greater than two, 186–87
  in standard form, 184, 397
  in two variables, 410
Quadratic equations, graphing, 410–19
  of form $y = ax^2 + bx + c$, 411–12
  intercept points of parabola, 412–13
  as parabola, 410–11
  vertex of parabola, 411–16, 420
Quadratic equations, solving, 388–423
  by completing the square, 392–97, 420
  complex solutions, 405–10
  containing rational expressions, 231–35
  by factoring, 184–86, 389, 403
  by quadratic formula, 397–402, 403
  by square root method, 389–92, 398, 403
  summary of methods for, 402–4
Quadratic formula, 397–402, 403
  deriving, 398
Quadrilateral, 435
Quotient
  power rule for, 116–17, 124, 378
  with zero, 37
  See also Rational expressions
Quotient polynomial, 147
Quotient rule
  for exponents, 117–18, 122, 124, 378
  for radicals, 354–55, 362–63
  for square roots, 353–54

R

Radical(s), 346–87
  adding and subtracting, 357–60
  defined, 348
  introduction to, 347–52
  like, 357, 383
  multiplying and dividing, 361–67
  product rule for, 354–55, 361–62
  quotient rule for, 354–55, 362–63
  simplifying, 352–56

Radical equations
  applications of, 372–78
  solving, 367–72
Radical expression, 348
Radical sign, 348, 383
Radicand, 348, 383
Range of relation, 297
Ratio, 237–41, 245, 250
Rational exponents, 378–82
Rational expressions, 200–255
  adding and subtracting, 213–25
  applications of equations containing, 242–49
  complex fractions, 225–31, 250
  defined, 201, 250
  dividing, 209–10
  fundamental principle of, 203
  multiplying, 207–9
  ratio and proportion, 237–41, 245–46, 250
  simplifying, 203–7
  solving equations containing, 231–37
  undefined, identifying, 202–3
  written as equivalent rational expression with given denominator, 216–17
Rationalizing the denominator, 363–64, 383
Rational numbers, 47
  set of, 6, 7
Real numbers, 7, 47
  adding, 24–30
  multiplying and dividing, 34–41
  operations on calculator with, 39
  order property for, 8
  set of, 7
  subtracting, 30–34
Real numbers, properties of, 41–47
  additive and multiplicative identities, 44
  additive and multiplicative inverse properties, 44, 45, 47
  associative property, 42–43
  closure properties, 41
  commutative property, 41–42, 204
  distributive property, 43–44, 63
Reciprocals (multiplicative inverses), 12, 45, 47
Rectangle, 436
  formula for area of, 81, 83
Rectangular coordinate system, 260–65, 303
Rectangular solid, 437
Relation, 297
  domain and range of, 297
Remainder polynomial, 147, 148
Rhombus, 436
Right angle, 427
Right circular cone, 437
Right circular cylinder, 437
Right triangle, 192–93, 433, 435
  Pythagorean theorem and, 192–93, 372–74, 433
Root, 347
  cube, 348–49, 383
  square. *See* Square root
Root (solution) of equation, 60, 106

S

Scalene triangle, 435
Scientific notation, 125–27, 151
  changing to standard form, 126
  defined, 125
  entering number in scientific notation on, 127
  writing number in, 125–27
Sets of numbers, 5–10
  integers, 6, 7
  irrational numbers, 7
  natural numbers, 5, 7
  rational numbers, 6, 7
  real numbers, 7
  whole numbers, 6, 7
Signed numbers, 7
Sign rules for multiplication and division, 35
Similar triangles, 431–32
  word problem involving, 245–46
Simple interest formula, 81, 84
  word problems using, 91–92
Simplified form of radical, 352–56
Simplified (lowest terms) fraction, 11, 203
Slope, 274–82, 303
  defined, 275
  graphing line given its y-intercept and, 283–84
  of horizontal and vertical lines, 279–80, 281
  of line given equation of line, 277–78
  of line given two points on line, 275–77
  negative vs. positive, 277, 281
  of parallel and perpendicular lines, 278–79
  point-slope form of equation of line, 284–86
Slope-intercept form, 277–78, 283, 286
Solution
  of equation, 60, 106
  extraneous, 233, 250, 369
  of inequality, 97–98
  ordered pair as, 258
  of system of linear inequalities, 335
  of system of two equations in two variables, 311–12, 340
Solutions, word problems involving, 93–94
Special products, 140–45
  multiplying binomials using, 140–41
  multiplying sum and difference of two terms, 142–43
  squaring binomials, 141–42
Sphere, 437
Square(s)
  completing the, 392–97, 420
  factoring difference of, 176–77, 180
  perfect, 349
  sum of, 177
  table of, 438
Square (geometric figure), 436
Square root, 347–48
  on calculator, 350
  of negative number, 348
  positive or principal, 347, 383
  product rule for, 352–53
  quotient rule for, 353–54
  rationalizing denominators containing, 363–64, 383

  solving radical equation containing, 367–72
  table of, 438
Square root property, 373
  quadratic equations solved by, 389–92, 398, 403
Squaring a binomial, 141–42
Squaring property of equality, 367
Straight angle, 427
Substitution method, solving linear systems by, 319–23
Subtraction
  addition property of equality and, 61
  addition property of inequality and, 99
  common words expressing, 63
  of complex numbers, 406
  of decimals, 424
  of fractions, 13–14
  of polynomials, 133–34
  of radicals, 357–60
  of rational expressions, 213–25
  of real numbers, 30–34
  symbol for, 3
Sum of squares, 177
Sum of two cubes, 177–78, 180, 182
Supplement, 427
Supplementary angle, 427
Symbols, 1–5
  grouping, 16, 18, 31, 73, 74, 75, 76
  inequality, 2
  operation, 3
  used to compare, 1, 2, 4
Symmetry, axis of, 411, 413, 420
System of linear equations. *See* Linear systems
System of linear inequalities, 335–39, 340

T

Term(s), 53, 106
  combining like, 54–56, 62, 73, 132
  defined, 130
  degree of, 131, 151
  like, 54, 106
Transversal, 429
Trapezoid, 436
Triangle, 430–33, 435
  congruent, 430–31
  defined, 430
  equilateral, 435
  isosceles, 435
  right, 192–93, 372–74, 433, 435
  scalene, 435
  similar, 245–46, 431–32
Trinomials, 151
  defined, 131
  factoring, 163–76
  of form $ax^2 + bx + c$, 168–76
  of form $x^2 + bx + c$, 163–66
  perfect square, 172–73, 181, 196, 393, 394, 420

V

Variable(s), 21–24, 47
  linear equation in one, 60

  linear equations in two, 257–58
  quadratic equation in two, 410
  in radicand of radical expression, 349–50
  solving equation for, 60
  solving formula for one of its, 84–86
Velocity formula, applications of radical equations involving, 375
Vertex of parabola, 411–16, 420
Vertical angles, 428
Vertical lines, 286
  graph of linear equation as, 270
  slope of, 279–80, 281
Vertical line test, 298–99
Volume formula, 82, 85

W

Whole numbers, 1, 47
  set of, 6, 7
Word problems
  direct translation, 242–43
  formulas to solve, 82–84
  geometric formulas to solve, 89–90
  involving distance, 92–93, 244–45
  involving inequalities, 102–3
  involving similar triangles, 245–46
  involving simple interest, 91–92
  involving solutions, 93–94
  involving value of coins, 90–91
  involving work, 243–44
  linear systems in two variables, 329–35
  proportions in, 239–40
  solving, 69–70, 76–78, 89–94
  steps in solving, 89, 243
  translated into quadratic equations, 189–96
  translating, 63–64
Work problems, 243–44

X

X-axis, 260
x-coordinate, 260, 276
x-intercept, 268–70, 303
  of parabola, 412–13

Y

Y-axis, 260
  graph symmetric about, 411
y-coordinate, 260, 276
y-intercept, 268–70, 303
  graphing line given its slope and, 283–84
  of parabola, 412–13
  slope-intercept form, 277–78, 283, 286

Z

Zero, 6
  as denominator, 233
  multiplying by, 36
  quotients with, 37
Zero exponent, 118–19, 124, 378
Zero factor theorem, 184–87, 389

A SCIENTIFIC CALCULATOR

				ON/C
2nd	$\sqrt{x}$ / x^2	10^x / log	e^x / ln	OFF
RAD / DRG	hyp^{-1} / hyp	sin^{-1} / sin	cos^{-1} / cos	tan^{-1} / tan
M− / M+	(	)	1/x	$\sqrt[x]{y}$ / y^x
7	8	9	STO	RCL
4	5	6	×	÷
1	2	3	+	−
0	.	π / EXP	+/−	=